XI.

The order of food is from the most solid to the most light.

XII.

The order of drink is from the mildest to the most heady and the most scented.

XIII.

To say that we should not mix our liquors is a heresy. The tongue becomes saturated, and after the third glass, the finest wine only gives an obtuse sensation.

XIV.

Dessert without cheese, is like a pretty girl with only one eye.

XV.

A cook may be educated, but a "roast cook" must be born such.

XVI.

The most indispensable quality in the cook is punctuality, and such ought to be the duty of the guests.

XVII.

To wait too long for a late guest denotes a lack of consideration to all those who are present.

XVIII.

He who receives guests, and pays no personal care to the repast offered them, is not worthy to have friends.

XIX.

The hostess should always assure herself that the coffee is good, and the host that the *liqueurs* are of the finest quality.

XX.

To invite any one, implies that we charge ourselves with his happiness all the time that he is under our roof.

Biochemistry of Taste and Olfaction

THE NUTRITION FOUNDATION

A Monograph Series

HORACE L. SIPPLE AND KRISTEN W. MCNUTT, EDS.: *Sugars in Nutrition,* 1974

ROBERT E. OLSON, ED.: *Protein-Calorie Malnutrition,* 1975

ANANDA S. PRASAD, ED.: *Trace Elements in Human Health and Disease, Volume I, Zinc and Copper, 1976; Volume II, Essential and Toxic Elements,* 1976

MORLEY R. KARE AND OWEN MALLER, EDS.: *The Chemical Senses and Nutrition,* 1977

JOHN N. HATHCOCK AND JULIUS COON, EDS.: *Nutrition and Drug Interrelations,* 1978

CLIFFORD F. GASTINEAU, WILLIAM J. DARBY, AND THOMAS B. TURNER, EDS.: *Fermented Food Beverages in Nutrition,* 1979

MORLEY R. KARE, MELVIN J. FREGLY, AND RUDY A. BERNARD, EDS.: *Biological and Behavioral Aspects of Salt Intake,* 1980

DONALD S. MCLAREN, ED.: *Nutritional Ophthalmology,* 1980

JENNY T. BOND, L.J. FILER, JR., GILBERT A. LEVEILLE, ANGUS THOMSON, AND WILLIAM B. WEIL, EDS.: *Infant and Child Feeding,* 1981

ROBERT H. CAGAN AND MORLEY R. KARE, EDS.: *Biochemistry of Taste and Olfaction,* 1981

In Preparation

DONALD C. BEITZ AND R. GAURTH HANSON, EDS.: *Animal Products in Human Nutrition*

Biochemistry of Taste and Olfaction

Edited by

ROBERT H. CAGAN

Veterans Administration Medical Center and
Monell Chemical Senses Center and
University of Pennsylvania School of Medicine
Philadelphia, Pennsylvania

MORLEY R. KARE

Monell Chemical Senses Center
and University of Pennsylvania
Philadelphia, Pennsylvania

1981

ACADEMIC PRESS

A Subsidiary of Harcourt Brace Jovanovich, Publishers

New York London Toronto Sydney San Francisco

ACADEMIC PRESS, INC.
111 Fifth Avenue, New York, New York 10003

United Kingdom Edition published by
ACADEMIC PRESS, INC. (LONDON) LTD.
24/28 Oval Road, London NW1 7DX

Library of Congress Cataloging in Publication Data
Main entry under title:

Biochemistry of taste and olfaction.

"Based on the International Symposium on Biochemistry of Taste and Olfaction, held April 21-23, 1980, at the Monell Chemical Senses Center"--Pref.
Includes bibliographies and index.
1. Taste--Congresses. 2. Smell--Congresses. 3. Chemoreceptors--Congresses. 4. Biological chemistry--Congresses. I. Cagan, Robert H. II. Kare, Morley Richard, Date. [DNLM: 1. Smell--Congresses. 2. Taste--Congresses. WV 301 B615 1980]
QP456.B56 612'.86 81-3469
ISBN 0-12-154450-8 AACR2

PRINTED IN THE UNITED STATES OF AMERICA

81 82 83 84 9 8 7 6 5 4 3 2 1

Contents

Part II Taste Receptor Mechanisms

7 Comparative Study of Sweet Taste Specificity

WILLIAM JAKINOVICH, JR.

8 Biochemical Aspects of Sugar Reception in Insects

KAI HANSEN AND HELMUT WIECZOREK

9 A Molecular Approach to Intensity/Time Phenomena in Sugar Sweetness

GORDON G. BIRCH

Part IV Neurotransmitters in Taste and Olfaction

17 Neurotransmitter Biochemistry of the Mammalian Olfactory Bulb

FRANK L. MARGOLIS

18 Neurochemical Studies of the γ-Aminobutyric Acid System in the Olfactory Bulb

MICHAEL R. QUINN AND ROBERT H. CAGAN

19 Neurochemistry of the Olfactory Tubercle

NEIL R. KRIEGER

List of Contributors

Numbers in parentheses indicate the pages on which the authors' contributions begin. Names preceded by an asterisk are discussants only.

L. H. Bannister (7), Department of Biology, Guy's Hospital Medical School, London, SE1 9RT, United Kingdom

J. Bard (85), Memorial Sloan-Kettering Cancer Center, New York, New York 10021

G. K. Beauchamp (85), Monell Chemical Senses Center, Philadelphia, Pennsylvania 19104

Gordon G. Birch (163), National College of Food Technology, University of Reading, Weybridge, Surrey KT 13 ODE, England

E. A. Boyse (85), Memorial Sloan-Kettering Cancer Center, New York, New York 10021

Joseph G. Brand (287), Monell Chemical Senses Center, and Department of Biochemistry, School of Dental Medicine, University of Pennsylvania, Philadelphia, Pennsylvania 19104

Robert H. Cagan (47, 93, 175, 395), Veterans Administration Medical Center, and Monell Chemical Senses Center, and University of Pennsylvania School of Medicine, Philadelphia, Pennsylvania 19104

***Paul Cancalon**, Department of Biological Science, Florida State University, Tallahassee, Florida 32306

John A. DeSimone (213), Department of Physiology, Virginia Commonwealth University, Medical College of Virginia, Richmond, Virginia 23298

George Dodd (333), Department of Chemistry and Molecular Sciences, University of Warwick, Coventry CV4 7AL, Warwickshire, England

***Hans Fisher**, Cook College, Department of Nutrition, Rutgers University, New Brunswick, New Jersey 08903

***Robert C. Gesteland**, Department of Biological Sciences, Northwestern University, Evanston, Illinois 60201

***Thomas V. Getchell**, Department of Anatomy, Wayne State University, Detroit, Michigan 48201

Donald A. Godfrey[1] (443), Department of Pharmacology, and Department of Anatomy and Neurobiology, Washington University School of Medicine, St. Louis, Missouri

Neil I. Goldstein (93), Monell Chemical Senses Center, Philadelphia, Pennsylvania 19104

D. B. Gower (7), Department of Biochemistry and Chemistry, Guy's Hospital Medical School, London, SE1 9RT, United Kingdom

M. R. Hancock (7), Department of Biochemistry and Chemistry, Guy's Hospital Medical School, London, SE1 9RT, United Kingdom

Kai Hansen (139), Institut für Zoologie, Universität Regensburg, D-8400 Regensburg, Federal Republic of Germany

Paul R. Hartig[2] (499), Church Laboratory of Chemical Biology, Division of Chemistry and Chemical Engineering, California Institute of Technology, Pasadena, California 91125

Miles Herkenham (511), Laboratory of Neurophysiology, National Institute of Mental Health, Bethesda, Maryland 20205

James D. Hirsch (311), Department of Biological Research, G. D. Searle and Company, Chicago, Illinois 60680

David E. Hornung (33), Biology Department, St. Lawrence University, Canton, New York 13617

William Jakinovich, Jr. (117), Department of Biological Sciences, Herbert H. Lehman College, City University of New York, Bronx, New York 10468

Manfred L. Karnovsky (1), Department of Biological Chemistry, Harvard Medical School, Boston, Massachusetts 02115

***Robert Koch**, Department of Biochemistry, Mississippi State University, Mississippi State, Mississippi 39762

D. E. Koshland, Jr. (459), Department of Biochemistry, University of California, Berkeley, California 94720

Neil R. Krieger (417), Departments of Pharmacology and Psychiatry, University of Pennsylvania School of Medicine, Philadelphia, Pennsylvania 19104

Stephen J. Kron (287), Faculty of Arts and Sciences, University of Pennsylvania, Philadelphia, Pennsylvania 19104

Kenzo Kurihara (249), Faculty of Pharmaceutical Sciences, Hokkaido University, Sapporo 060, Japan

Howard M. Lenhoff (475), Department of Developmental and Cell Biology, University of California, Irvine, California 92717

Ann M. Maderis (459), Department of Biochemistry, University of California, Berkeley, California 94720

Frank L. Margolis (311, 369), Department of Physiological Chemistry and Pharmacology, Roche Institute of Molecular Biology, Nutley, New Jersey 07110

Franz M. Matschinsky (443), Diabetes Research Center and Department of Biochemistry and Biophysics, University of Pennsylvania School of Medicine, Philadelphia, Pennsylvania 19104

***Bert Menco**, Psychologisch Laboratorium van de Rijksuniversiteit, Varkenmarkt 2, Utrecht, Holland

Michihisa Miyake (249), Faculty of Pharmaceutical Sciences, Hokkaido University, Sapporo 060, Japan

[1] Present address: Department of Physiology, Oral Roberts University, Tulsa, Oklahoma 74171

[2] Present address: Department of Biology, The Johns Hopkins University, Baltimore, Maryland 21218

Hsiao-Ping H. Moore (499), Church Laboratory of Chemical Biology, Division of Chemistry and Chemical Engineering, California Institute of Technology, Pasadena, California 91125

Gregory Mooser (231), University of Southern California, School of Dentistry, Los Angeles, California 90007

Maxwell M. Mozell (33), Physiology Department, Upstate Medical Center, Syracuse, New York 13210

***Randall B. Murphy**, Department of Chemistry, New York University, New York, New York 10003

William T. Norfleet (443), University of Missouri, Columbia, Missouri

Nicholas F. Paoni (459), Department of Biochemistry, University of California, Berkeley, California 94720

Krishna Persaud (333), Department of Chemistry and Molecular Sciences, University of Warwick, Coventry CV4 7AL, Warwickshire, England

Candace B. Pert (511), Section on Biochemistry and Pharmacology, Biological Psychiatry Branch, National Institute of Mental Health, Bethesda, Maryland 20205

***Ernest Polak**, Laboratoire de Neurobiologie Sensorielle, EPHE, CENFARBP No. 6, Fontenay-aux-Roses 92260, France

Steven Price (69), Department of Physiology, Medical College of Virginia, Virginia Commonwealth University, Richmond, Virginia 23298

Michael R. Quinn (395), Monell Chemical Senses Center, Philadelphia, Pennsylvania 19104

Michael A. Raftery (499), Church Laboratory of Chemical Biology, Division of Chemistry and Chemical Engineering, California Institute of Technology, Pasadena, California 91125

Linda D. Rhein (47), Monell Chemical Senses Center, Philadelphia, Pennsylvania 19104

C. David Ross (443), Department of Physiology, Oral Roberts University, Tulsa, Oklahoma 74171

***Susan Schiffman**, Department of Psychiatry, Duke University Medical Center, Durham, North Carolina 27706

David M. Senseman (287), Monell Chemical Senses Center, and Department of Physiology and Pharmacology, School of Dental Medicine, University of Pennsylvania, Philadelphia, Pennsylvania 19104

Lewis Thomas (85), Memorial Sloan-Kettering Cancer Center, New York, New York 10021

Helmut Wieczorek (139), Institut für Zoologie, Universität Regensburg, D-8400 Regensburg, Federal Republic of Germany

***Charles Wysocki**, Monell Chemical Senses Center, Philadelphia, Pennsylvania 19104

M. Yamaguchi[3] (85), Memorial Sloan-Kettering Cancer Center, New York, New York 10021

K. Yamazaki[4] (85), Memorial Sloan-Kettering Cancer Center, New York, New York 10021

Kiyonori Yoshii (249), Faculty of Pharmaceutical Sciences, Hokkaido University, Sapporo 060, Japan

[3] Present address: Monell Chemical Senses Center, Philadelphia, Pennsylvania 19104

[4] Present address: Monell Chemical Senses Center, Philadelphia, Pennsylvania 19104

Foreword

The infrequency of evident scientific interest in and concern for taste and olfaction in recent years stands in striking contrast to the earlier appreciation of the importance of these senses. Shakespeare, in "As You Like It," equated taste with sight and lamented the decline in its acuteness as:

> The sixth age shifts
> Into the lean and slipper'd pantaloon. . .
> Last scene of all,
> That ends this strange eventful history,
> Is second childishness and mere oblivion,
> sans teeth, sans eye, sans taste, sans everything.

One looks in vain in today's textbooks or monographs on physiology, biochemistry, or medicine, with few exceptions, for similar appreciation or meaningful interpretation of the basic aspects of taste and olfaction or for any scientific attention to these senses equivalent to that given to vision and hearing. In fact, the consideration of these chemical senses in texts of physiology and biochemistry has decreased during the last three or four decades.

Even more surprising is the fact that most textbooks on nutrition devote no space to these senses despite the recognition by the lay that taste and smell rank with color in determining whether a food will be considered appealing, edible, highly desirable, or rejected as disgusting. Aberrations of these senses or dysfunctions of them in disease are seldom discussed by physicians.

The 1745 edition of "The Learned M. L. Lemery," as translated by D. Hay, entitled "Treatise of All Sorts of Foods, Both Animal and Vegetable: Also Drinkables: Giving an Account How to Chuse the Best Sort of All Kinds, etc." (printed for T. Osborne in Gray's-Inn, London, 1745) states:

> Each Food hath also a peculiar Taste, whereby 'tis known, and covethed or loathed; this Taste may likewise make Way for us to conjecture, what Principles it contains, the Composition of them and the Effects they are apt to produce. . .

Particularly great importance was attached to these qualities of foods and the role of the senses by the gastronomist, Brillat-Savarin, in his classic book "Physiology of Taste" ("Physiologie du Gout, ou Meditations de Gastronomie," Chez A. Santelet et Cie Libraires, Paris, 1826) from which the end papers in this volume are derived. Brillat-Savarin's aphorisms, examples of which are reproduced in the end papers, illustrate his appreciation of taste and smell in determining the acceptability and enjoyment of foods. The significance of these senses in determining food habits was put in excellent perspective more recently by the too little known treatise of H. D. Renner, "The Origin of Food Habits" (Faber and Faber, Ltd., London, 1944).

Much earlier, about 1500 A.D., the importance of all the senses—taste, smell, touch, sight, and hearing—was beautifully illustrated in the famous tapestries, "The Lady and the Unicorn," in the Musee de Cluny, Paris. These depict "Taste" as "the enchanting lady, with a parakeet on her fist . . . about to sample sweet-meats offered by her servant . . ." and "Smell" as ". . . the lady . . . making a fillet of carnations, those flowers so sweet and spicy to smell." (Margaret B. Freeman, The Unicorn Tapestries, Metropolitan Museum of Art, New York, 1976.) A portion of the "Taste" tapestry in the Cluny series is depicted on the dust jacket of this volume, "Smell" having been an illustration on the dust jacket of the earlier monograph in this series, "The Chemical Senses and Nutrition."

Fortunately, within the last decade, these senses have captured the interests of a group of young biochemists, physiologists, and molecular biologists. This volume exemplifies the imaginative scientific studies characterizing their basic inquiries that advance understanding of fundamental aspects of the phenomena of taste and olfaction which are of much importance in nutrition, medicine, and food science.

It was most appropriate, therefore, that The Nutrition Foundation, the Monell Chemical Senses Center, and the National Institutes of Health jointly supported and sponsored the opportunity for scientists to exchange and review their knowledge pertaining to these phenomena during a symposium at the Monell Chemical Senses Center. The subsequently revised and edited reviews constitute this third monograph on these chemical senses published in The Nutrition Foundation's series.

As President of The Nutrition Foundation, I express our appeciation and gratitude to those who so thoughtfully planned and authored this treatise, thereby, advancing our understanding of long, much-neglected areas of key biological processes that influence our dietary habits, nutritional status, and enjoyment of food, as well as other important social and biological phenomena.

William J. Darby
President
The Nutrition Foundation, Inc.
New York and Washington, D.C.

Preface

This volume comprehensively treats biochemical knowledge of the senses of taste and olfaction. Basic understanding of chemical sensory systems will ultimately come from an interdisciplinary synthesis. Nevertheless, biochemical research on the chemical senses has been so severely neglected that the focused assessment of biochemistry in this monograph is timely and important. Research on these sensory systems had not, until the past several years, reached the biochemical level. Development of suitable animal models, described in this volume, promises to stimulate additional research. But it is not merely to catalogue animal models that this volume was produced. It critically assesses the existing data base in relation to current theories and looks ahead to potentially fruitful research areas.

The authors were asked to emphasize the contributions from their own laboratories and to place them within the context of the research on related aspects from other groups. Thus, some overlap was necessary and even desirable to show different points of view on the interpretation of the same data base.

Biochemical mechanisms at the peripheral receptor level in taste and olfaction are treated. The role of the cell surface is emphasized, although much remains to be learned of the specific chemical nature of receptor molecules. The importance of genetic and immunologic factors is gaining recognition. Transduction events coupled to the initial interaction are discussed, and the possible importance of such factors as cooperative interactions within receptor macromolecules and ion fluxes across the membrane and within the cells are described. Neurotransmitters and other neurochemical aspects of the olfactory system are presented. A selection of examples of chemical sensory systems about which much is now known concludes the volume and serves as a reference point and an aspiration for the field of taste and olfaction.

It is understandable that a field dealing with sensory phenomena should be dominated by viewpoints and experimental approaches of psychology. Indeed, much of the research activity, both past and present, deals with psychology and behavior. These aspects and the questions they pose are important, but understanding the underlying molecular mechanisms of receptor specificity, of transduction, and of synaptic transmission must come from biochemical research. The proportion of biochemical papers in the field is quantitatively miniscule. Yet from the biochemistry, assembled in this treatise for the first time, emerge data and ideas far out of proportion to the number of papers, for herein lies the fundamental workings of these senses. It seems self-evident that biochemical mechanisms underlie the phenomena described and quantified using other approaches.

The field of chemical senses is a microcosm of neurobiology, but it spans a wide range of disciplines—including biochemistry. This volume will help scientists from other disciplines studying taste and olfaction to better understand the systems, and it will serve as a catalyst for the integration of mechanisms into phenomenology. As biochemical understanding of the senses of taste and olfaction continues to grow, this book will serve as a valuable reference source. This work could not have been assembled in this form ten years ago; even five years ago most of the substantive information recorded here simply did not exist.

The book is intended for a wide readership: practicing biochemists, especially those involved with membrane receptors and neurotransmitters; researchers in other disciplines who work within the chemical senses area, including specialists in psychology, neurophysiology, organic chemistry, and nutrition; neuroscientists working with other sensory receptors; those whose activities impinge upon the chemical senses, such as food scientists; and students, not only in biochemistry, but also in physiology, nutrition, and psychology, who are interested in chemical senses but not enough to locate the widely scattered biochemical information.

This treatise, although not a "proceedings," is based on the International Symposium on Biochemistry of Taste and Olfaction held April 21–23, 1980 at the Monell Chemical Senses Center. As organizer of the Symposium, I had the great pleasure of arranging the meeting and welcoming my colleagues to Philadelphia, and with my colleagues here, to host them in this unique biochemical event. Although in one sense it was a culmination of a personal aspiration since I started in this area a decade ago, it is clear that this marked only the beginning of our understanding of these senses.

The following organizations provided financial support for the symposium: The Nutrition Foundation; Fogarty International Center, NIH; National Institute of Neurological and Communicative Disorders and Stroke, NIH; Bureau of Foods, FDA; and Human Nutrition Center, SEA, USDA.

Robert H. Cagan

Some Reflections on Biochemical Approaches to the Phenomena of Taste and Olfaction

MANFRED L. KARNOVSKY

This comprehensive volume deals with many aspects of the biochemistry of taste and olfaction and will, I believe, be just as stimulating as the symposium on which it is based. For obvious reasons it will have a much wider impact. One hopes that the book will fall into the hands, not only of workers in this field, but also into those of a wider group of biochemists who may not yet have had significant exposure to the intriguing topics it covers. Among those readers, I anticipate that there will be biochemically oriented investigators who will be attracted to the pursuit of some of the fascinating problems these chapters suggest.

I am encouraged to express these hopes, even though the complexity of the phenomena of taste and smell, and the fact that relevant research topics are not in the main path of current biochemical investigations, militate somewhat against their realization. However, my sense of encouragement is a result of my own experiences in a "fringe" field, the recounting of which may be of some help in communicating my thoughts.

More than 2 decades ago we began to think about the biochemical basis of a primitive biological phenomenon—phagocytosis. There was little known, and perhaps only two or three biochemists worked in the field. An extraordinarily distinguished visitor to our laboratory commented that our system was far too complicated to yield really meaningful information. We were crestfal-

Biochemistry of Taste and Olfaction

ISBN 0-12-154450-8

len because we had stripped the system down to a suspension of neutrophilic granulocytes, 95% pure, from guinea pigs, in buffered isotonic saline without protein, but with glucose to provide energy. The "trigger" for the cellular phenomenon we wished to study (phagocytosis) was a suspension of inert plastic particles of appropriate size, rather than bacteria or other degradable particles that might have complicated matters. Quantification of phagocytosis was achieved by microscopic observation—admittedly an imprecise measure. It is ironic to recognize that the elimination of factors that we believed might complicate the biochemistry led to defects in the biological approach, as greater sophistication later indicated. The purist is not always completely right—but it is better to start with the purist's approach. Time and even unexpected advances ultimately open up the possibilities of developing controllable systems that are closer to the reality of the native situation. In the case of phagocytosis for example, developments in membrane biology and immunology and the establishment of precise assays of rates of particle ingestion made the collection and interpretation of biochemical data more pointed. One could then ask more meaningful questions regarding the biochemical aspects of the contact of a live bacterium with the phagocyte's surface, the ingestion phase, the bactericidal mechanism(s), the release of lysosomal enzymes, digestion and the concomitants of these events.

When biochemical studies of the phagocytic cell came to maturity, we embarked on a new investigation that was perhaps more hair-raising to the purists in terms of difficulties with adequate experimental models and satisfactory quantification. This was a study of the biochemistry of sleep. Two main questions were posed.

1. Is there a substance that accumulates during wakefulness and later induces sleep? If so, what is it, and how does it work?
2. What biochemical characteristics differentiate the chemistry of the brain during sleep from the situation during wakefulness, and can we define, in chemical terms, the benefits to the brain of periods of sleep?

The first question has great and almost romantic appeal. Given a reasonable assay and the ability to eliminate interfering substances from the preparations of sleep-inducing substances, the problem should be solvable. Assays of sleep are currently more cumbersome than were assays of phagocytosis 2 decades ago. There are acute difficulties in avoiding introduction of the merest traces of interfering contaminants derived from chromatographic columns, reagents, glassware, etc. We require the second, more pedestrian approach mentioned above to give us the satisfactions of small discoveries that nurture hopes of larger ones—a not uncommon situation in the investigation of difficult problems.

These topics have many points of similarity to those included under the umbrella of "taste and smell." All have to do with the molecular events that underlie a physiological response, and in that sense they are classically biochemical. Though ultimately each will include detailed studies of specific proteins, enzymes, effectors, and other matters that are at the core of biochemistry, overall they constitute problems of great complexity. Certainly they all share several disabilities and several advantages. Among the disabilities is the fact that biochemists have, until recently, been less than enthusiastic about research that is not in the mainstream of our still-youthful discipline. Fortunately, however, biochemistry continues in a very dynamic state. As we consider the future, we might note that advances in many new directions, important demonstrations of novel principles, and discoveries that link disparate facts, constantly move the fringes closer to the center. An example in my own field is the way in which massive interest in superoxide, hydroxyl radical, and singlet oxygen—once arcana of inorganic chemists—has swept through organic and biochemical circles to the phagocytizing leukocyte, which cell turns out to be probably the major source of these active oxygen species in mammals. Because leukocytes are easy to obtain and manipulate, and since superoxide is unexpectedly easy to measure, a fringe field has rapidly become heavily populated with investigators, with beneficial results.

However, pioneers have conflicting feelings about an influx of settlers—the pride of belonging to a small and private band is traded for the advantages of the "big time." It is difficult to maintain a fair, judicious, and accepting attitude and simultaneously to insist on adequate quality. Shakespeare said, "things sweet to taste turn in digestion sour," and though this is often true as a small fringe field develops, it need not necessarily be so. I think that the field dealing with the biochemistry of taste and smell will soon become rather popular. Despite the sacrifice of the status of early aristocracy, it will profit from an influx of new investigators. This area of research seems almost to have been waiting for intensive development of many items that are key matters in modern biochemistry. For example, these pages deal with binding sites, receptors, messages, messengers, cyclic nucleotides, histocompatibility, genetics of the immune response, transmitters, membrane biology, ligands, protein conformation, transduction, affinity labeling, ion channels, and gating—all terms from the current biochemical lexicon. The generality of many observations made thus far and reported in these pages is manifest; they include a score of animal species and experimental systems, and a dozen classes of chemical substances. A measureable biological response is available in most of the systems studied, and one is optimistic that increased reliability and precision in that context will lead to major biochemical advances.

One would be remiss not to stress the contribution of the interdisciplinary approach to fringe areas in biochemistry, a matter that is apparent from the list above. The various contributing disciplines enrich each other; enormous stimulation is derived from the combined approaches of and contacts among biochemists, neurobiologists, physiologists, pharmacologists, and behavioral scientists, to name only a small selection. Few will deny the premise that over the past 2 decades biochemistry has become noticeably more biologically oriented. It is hoped that critical attitudes, especially at the chemical level, will not be blunted or diluted by the lure of exciting biological phenomena; however, this does seem to be a manifest danger, and documentable, but sound chemistry is the antidote to the "sourness" to which Shakespeare pointed.

On the last night of the symposium we heard a delightful performance of Mozart's "Hunt" Quartet by four young muscians. I could not help but draw an analogy with the symposium itself. The quartet was talented indeed, and played the notes accurately—but without that ultimate authority and conviction that permits freedom to be truly inventive. During the final session of the symposium, and in the final chapters of this volume, the organizer inserted presentations in fields that relate to taste and smell (as outlined above) and also relate strongly to the current core of biochemistry. Those chapters represent a selection from a sophisticated group of studies in fields where there is a significant "critical mass" of investigators and vital communication with the mother field. This was a masterpiece of planning—to end with the scientific equivalent of the Amadeo or Juilliard or Guarneri Quartets! Those areas of investigation set a standard to which the field of the biochemistry of taste and smell can clearly aspire, with the real hope of success as biochemistry continues to move to envelop what were once fringe areas. The power and scope of that discipline will lead to assimilation and digestion of even the most intricate biological problems in a manner that, to this enthusiast, is almost phagocytic. I am obviously optimistic about progress in the area of this treatise, and such allusive admonitions as may be embedded in my remarks should be taken as the sympathetic suggestions of a kindred spirit who has trodden a not too different path.

PART I

Olfactory Receptor Mechanisms

1

Biochemical Studies on the Boar Pheromones, 5α-Androst-16-en-3-one and 5α-Androst-16-en-3α-ol, and Their Metabolism by Olfactory Tissue

D. B. GOWER, M. R. HANCOCK, AND L. H. BANNISTER

Biochemistry of Taste and Olfaction

ISBN 0-12-154450-8

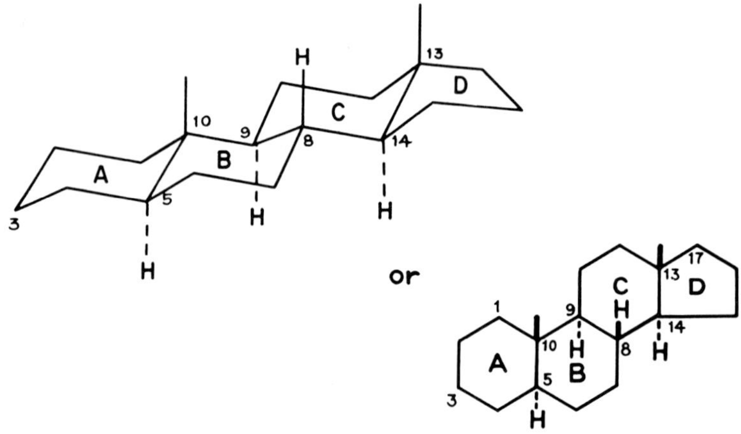

Fig. 1. Three-dimensional structure of 5α-androstane.

I. INTRODUCTION

During the past two decades, the odorous C_{19} steroids, known as androst-16-enes, have attracted a good deal of attention because of their action as sex attractants in pigs and of the possibility of their action as pheromones in human beings. The androst-16-enes are so-named because they are derivatives of the parent hydrocarbon androstane (see Fig. 1) and because they are characterised by unsaturation in ring D between C-16 and C-17. The fact that there is no substituent at C-17 contrasts this group with other C_{19} steroids such as testosterone, which has a 17β-hydroxyl group, or with androst-4-ene-3,17-dione, which has a 17-oxo group. The structures and names of the androst-16-enes, which have been isolated from biological sources, are given in Fig. 2; the structure of testosterone is shown for comparison.

II. ANDROST-16-ENES

A. Physical Characteristics

In their early studies, Prelog and Ruzicka (1944) attempted to isolate testosterone from boar testes. They were unsuccessful in this quest, however, but succeeded in isolating instead two androst-16-enes, namely 5α-androst-16-en-3α-ol and -3β-ol (abbreviated here as an-α and an-β), which occurred in far greater quantities than did testosterone. The Swiss research

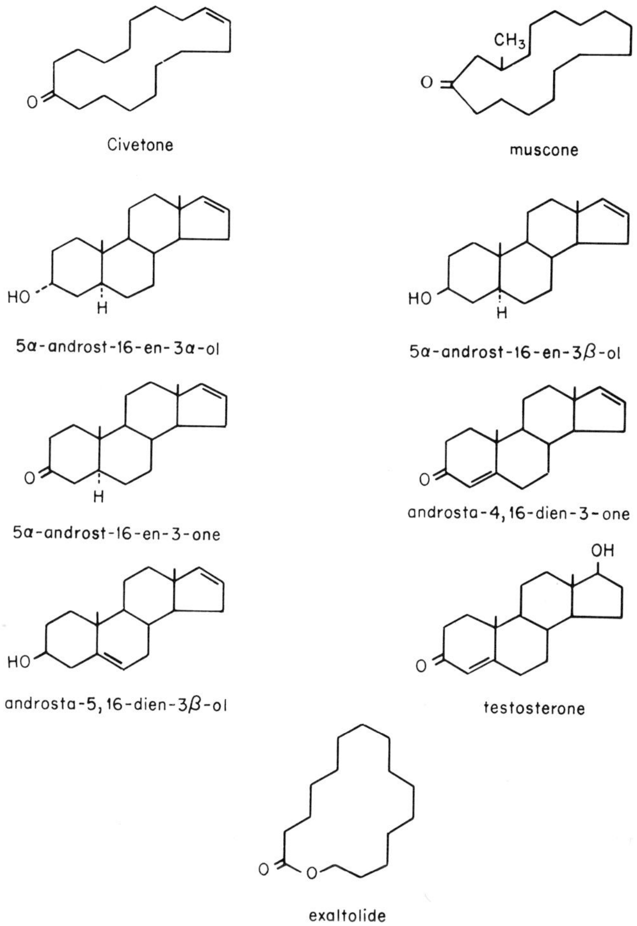

Fig. 2. Structural formulae of androst-16-enes. The formulae of an androgen, testosterone, and some macrocyclic odorous compounds are shown for comparison.

group (Prelog *et al.*, 1945) later commented on the smell of the androst-16-enes. According to them, the alcohols (especially an-α) were musk smelling, whereas the ketones, such as 5α-androst-16-en-3-one (5α-androstenone), were urine smelling. Even earlier than this, Lerche (1936) and Keller (1937) had described the unpleasant smell and flavor of cooked meat taken from an uncastrated or partially castrated boar. Lerche (1936) described the parotid glands as possessing an offensive odor, whereas Keller (1937) considered that the submaxillary glands possessed the unpleasant taste and recommended their removal. Although it is true that both the parotid and submaxillary glands contain and metabolize androst-16-enes (Patterson, 1968b; Katkov *et al.*, 1972; Booth, 1975) (see below), it is now well-documented that the major biosynthetic location is in the testis. Various aspects of the biochemistry and physiology of the androst-16-enes have been reviewed by Gower (1972, 1976, 1981a,b), Booth (1980), and Ewing and Brown (1977).

In addition to their smell, the androst-16-enes are characterized by some degree of volatility. Although this property is of obvious importance from the pheromonal point of view, it has presented problems to researchers because of the analytical losses that may be incurred during evaporation of solutions containing androst-16-enes. In an earlier review, Gower (1972) referred to the problem and recommended that evaporation be carried out on a rotary evaporator under water pump vacuum at temperatures not exceeding 40°C. Quantitative data obtained by Kaufmann *et al.* (1976) showed that, although no loss of 5α-androstenone occurred when the solvent was evaporated at 30°C under a stream of nitrogen (2 min), losses of up to 15% of dry steroid could be encountered if the gas flow was continued for 15 min. Losses during drying *in vacuo* were found to be very small.

The third characteristic feature of the androst-16-enes is their lipophilic nature. They can therefore be extracted readily into nonpolar organic solvents, but they require the use of nonpolar systems for their resolution by column chromatography on alumina (Brooksbank and Haslewood, 1961), on hydroxyalkoxypropyl-Sephadex (Lipidex) (Bicknell and Gower, 1975), or by thin-layer chromatography (Gower, 1964). The lipophilic nature of the androst-16-enes is also of interest with regard to the quality of boar meat. The odorous steroids, 5α-androstenone and an-α, have been shown to occur in the adipose tissue (Patterson, 1968a; Beery and Sink, 1971); when the meat is heated an unpleasant smell is emitted and the meat becomes unpalatable to many individuals.

B. Biosynthesis and Physiological Significance

The biosynthetic pathway for androst-16-enes in boar testis has been studied intensively and is well-documented both *in vitro* and *in vivo* (Fig. 3). The C_{21} steroid pregnenolone is the precursor of androsta-5,16-dien-3β-ol (abbreviated here as andien-β), which is believed to be the first androst-16-ene formed. This reaction occurs rapidly, and the enzyme system "andien-β synthetase," which catalyses it, is tightly bound (Cooke and Gower, 1979) mainly to the smooth endoplasmic reticulum of boar testis (Cooke and Gower, 1977). NADPH and molecular oxygen are required for activity (Katkov and Gower, 1970), and cytochrome *P*-450 may (Mason *et al.*, 1979) or may not (Brophy and Gower, 1973) be a requirement. Once andien-β is formed, it is converted into androsta-4,16-dien-3-one (androstadienone) by means of the 5-ene-3β- hydroxysteroid dehydrogenase-4,5-isomerase and

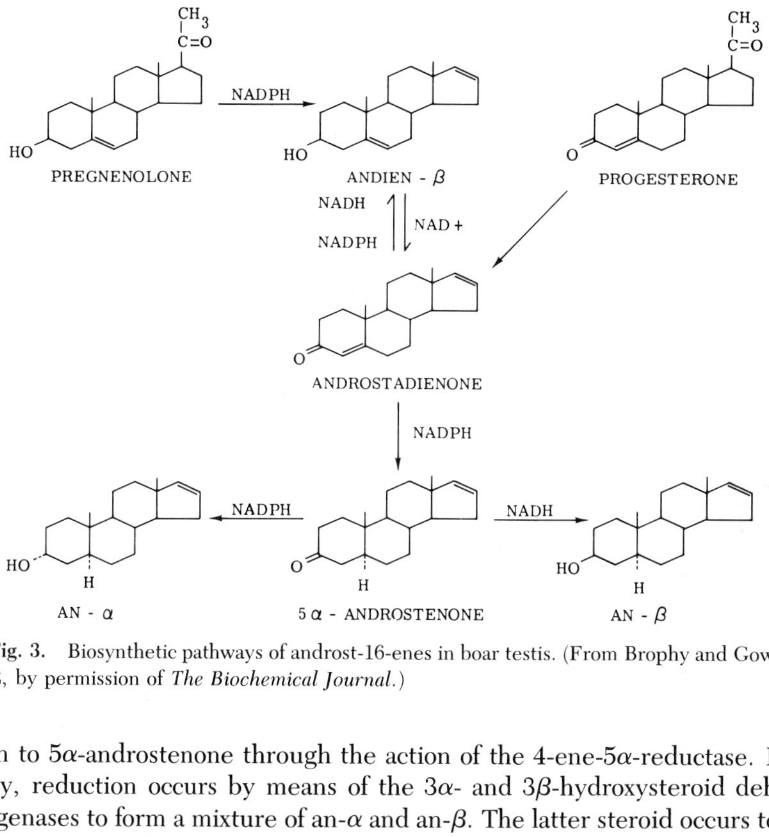

Fig. 3. Biosynthetic pathways of androst-16-enes in boar testis. (From Brophy and Gower, 1972, by permission of *The Biochemical Journal.*)

then to 5α-androstenone through the action of the 4-ene-5α-reductase. Finally, reduction occurs by means of the 3α- and 3β-hydroxysteroid dehydrogenases to form a mixture of an-α and an-β. The latter steroid occurs to a greater extent in boar testis as shown by studies *in vitro* (Brophy and Gower, 1972), *in vivo* (Saat *et al.*, 1972, 1974; Hurden *et al.*, 1979), and by analysis (Booth, 1975). In addition to pregnenolone, the C_{21} steroid, progesterone, can serve to a lesser extent as a precursor of androstadienone (Brophy and Gower, 1972).

Sulfates are also implicated in the biosynthetic scheme; Gasparini *et al.* (1976) have shown that pregnenolone sulfate can be converted into andien-β sulfate. Further, Saat *et al.* (1972, 1974) and Hurden *et al.* (1979) showed that with *in vivo* testicular infusions pregnenolone was converted initially to an-α and an-β sulfates, which were subsequently hydrolysed to the unconjugated alcohols by the action of testicular sulfatases.

It is of particular interest that C_{19} steroids, such as testosterone, 17α-testosterone, and dehydroepiandrosterone (DHA) (for references, see Gower, 1972, 1981a,b), do not serve as precursors for androst-16-enes in boar testis, despite the fact that chemical dehydration of testosterone or 17α-testosterone results in the formation of androstadienone. Pregnenolone and progesterone are, however, metabolized to androgens by entirely independent pathways (Ahmad and Gower, 1968; Loke and Gower, 1972) (Fig. 3).

Recently, Mason *et al.* (1979) proposed an alternative scheme for androst-16-ene biosynthesis in immature porcine testis as follows:

Pregnenolone → 17α-hydroxypregnenolone → 5,16-pregnenolone → andien-β

It has already been shown that ^{3}H-labeled androst-16-enes can be detected in boar spermatic venous plasma within 2 min after [^{3}H]pregnenolone has been infused into the corresponding testicular artery (Saat *et al.*, 1972, 1974; Hurden *et al.*, 1979). The steroids then reach the peripheral circulation; 5α-androstenone occurs to the extent of 1.2–54.1 ng/ml (Andresen, 1974) and an-α to the extent of 11–59 ng/ml (Bicknell and Gower, 1976). Because of their high lipophilic character (see above), both compounds are rapidly taken up by adipose tissue (mean value 2.09 ng/g; Andresen, 1975).

The submaxillary and parotid salivary glands of the boar have also been implicated in androst-16-ene metabolism. In particular, the submaxillary glands contain these steroids (Patterson, 1968b; Katkov *et al.*, 1972; Booth, 1975) and possess all the enzymes necessary for the sequence outlined below (see Fig. 3) (Katkov *et al.*, 1972):

andien-β → androstadienone → 5α-androstenone → an-α plus an-β

Although the submaxillary glands lack the enzyme system for biosynthesis of androst-16-enes from pregnenolone (Katkov *et al.*, 1972), they are nevertheless important in metabolizing existing androst-16-enes obtained from the peripheral blood circulation. As a result, the saliva contains both an-α and 5α-androstenone in a ratio of approximately 10:1 (Patterson and Stinson, cited in Gower, 1972).

Sink (1967) proposed that the androst-16-enes might be involved in porcine reproductive physiology as sex attractants, and the finding of odorous androst-16-enes in boar saliva and submaxillary glands was in keeping with this suggestion. When sexually excited, the boar salivates profusely. The smell of the an-α and 5α-androstenone present reaches the female and elicits the characteristic immobilization response, providing she is in estrus. The male is then able to mount and copulate. It had been known for many years prior to this (for references, see Gower, 1972) that boar odor was only one of the factors necessary to elicit the characteristic mating stance of the sow in estrus when subjected to the usual back-pressure test of Altmann (1941). The work of Melrose *et al.* (1971), however, showed that spraying 5α-androstenone or an-α toward the snouts of recalcitrant females resulted in 50% of the females responding to the Altmann test (1941). It was thus possible to ascertain just when the sow was in estrus—an important consideration

TABLE I Response of Pigs to C_{19} Steroids[a]

Steroid	Concentration (μg/ml)	Animals showing positive response (%)
5α-Androst-16-en-3-one	4.56	48
5α-Androst-16-en-3-one	9.12	58
5α-Androst-16-en-3α-ol	4.3	53
5α-Androst-16-en-3-one, plus 5α-androst-16-en-3α-ol	4.56 4.3	50
Androsta-4,16-dien-3-one	9.12	53
5β-Androst-16-en-3-one	9.12	47
5α-Androst-16-en-3β-ol	9.12	26
5α-Androstan-3-one	9.12	10

[a] Sows in estrus, which were previously negative to the back-pressure test of Altmann (1941), were treated with steroids in aerosol form. The results are expressed as percentages of the animals that then showed the immobilization response. Treatment consisted of two 2-sec sprays directed toward the snout at a distance of 60 cm. Animals showing a positive response were then inseminated artificially. Data compiled from Melrose *et al.* (1971) and Reed *et al.* (1974) by permission of the authors and Messrs. Balliere and Tindall.

if artificial insemination was to be carried out efficiently. Further field experiments by Reed *et al.* (1974) revealed that the response elicited depended on the structure of the steroid used (Table I). For example, 5α-androstan-3-one was only 20% as effective as 5α-androstenone, whereas an-β was 50% as effective as an-α. Surprisingly, 5β-androstenone, which has a completely different three-dimensional structure compared with that of 5α-androstenone, was just as effective.

C. Metabolism of Androst-16-enes in Porcine Nasal Epithelium

Earlier work from this laboratory was concerned with the manner in which the odorous ketones, 5α-androstenone and androstadienone, were metabolized in sensory and nonsensory nasal epithelium of mature and immature sows (Gennings *et al.*, 1974). It was found that the boar pheromone, 5α-androstenone, was reduced to an-α in high yields (approximately 70%) by a cytoplasmic 3α-hydroxysteroid dehydrogenase (3α-OHSDH), which required NADPH as the preferred co-factor. An-β was produced in small quantities (1% or less), whereas androstadienone was converted in low yield into 5α-androstenone. These last two results indicated the presence of low activity 3β-OHSDH and 4-en-5α-reductase. This is in contrast to the testicular metabolism (Fig. 3) where an-β is formed in greater quantities than an-α.

In the work of Gennings *et al.* (1974), no significant differences in 5α-androstenone metabolism could be detected between sensory and nonsensory tissues or between mature and immature animals. It was perhaps significant that the metabolism was directed almost exclusively to the 3α-alcohol [the more active compound pheromonally (Table I)] rather than to an-β.

The mechanism by which an-α or 5α-androstenone produce their response in the sow has not yet been elucidated. It is presumed that the inhaled steroids pass into the mucus, which bathes the nasal tissues, but whether they are bound to it has not yet been ascertained. Our preliminary evidence (D. B. Gower *et al.*, unpublished) indicates that metabolism of 5α-androstenone occurs only in the nasal tissues but not in the mucus. Some evidence, as yet unconfirmed, exists for receptors to 5α-androstenone and an-α occurring in porcine olfactory tissue (Gennings *et al.*, 1977). Similar work using 5α-androstan-3-one with wild and domestic rabbits and with cows has indicated the presence of specific receptors with K_a values of 5×10^7 (Pelosi *et al.*, 1978).

Our group and others have attempted to record electroolfactograms (EOGs) and action potentials from porcine olfactory tissue. The apparatus of Poynder *et al.* (1978) consists of a perfusion chamber that is continually supplied with a Ringer-Locke solution. A piece of olfactory epithelium, on a stainless steel gauze support, was mounted at the surface of the solution and odors were delivered via a multichannel device. When not being stimulated, the tissue was exposed to a humidified stream of gas (O_2:CO_2, 95:5 v/v). The temperature of both tissue and odor delivery system could be maintained at 35°C if desired. EOGs of up to 8 mV were regularly recorded in rat and pig olfactory epithelium with butyl acetate, the value falling to about 4 mV after the tissue had been maintained in this way for 24 hr. The typical rapid EOGs, however, were not detectable (or were only minimally present) when 5α-androstenone and an-α were used. Even when steps were taken to minimize adsorption of these nonpolar steroids to the walls of the delivery tubes and to improve volatility by maintaining the temperature of tissue and delivery system at 35°C, little or no EOG was obtained. In view of the fact that butyl acetate gave a typical response, it seems likely that the reason for the minimal EOG with 5α-androstenone or an-α is their relatively high molecular weights (272 and 273, respectively).

Recently, MacLeod *et al.* (1979) recorded responses from mitral cell-layer neurons of pig olfactory bulb to stimulation of the olfactory epithelium with 5α-androstenone, testosterone, and small molecular weight substances, such as amyl acetate, benzene, and pyridine. Of 37 cells tested with both 5α-androstenone and testosterone, 25 cells showed responses to both steroids. Several cells could clearly discriminate between the two compounds, since four neurons excited by 5α-androstenone were unaffected by testosterone, whereas eight cells were excited by testosterone but showed no response with 5α-androstenone.

D. Androst-16-enes in Humans

The possibility of the existence of pheromones, which may be involved in human social communication, has been suggested for many years (see Comfort, 1974; Gower, 1976, 1981a, and references therein). By analogy with animals that use odor cues to a large extent, it might be expected that humans would produce odorous substances from regions such as the axillae, the genitalia, and the anus. The relationship of body odor to social interactions has been a topic of interest for many years. Kalogerakis (1963) showed that the "oedipal stage" (father–son dominance competition) coincided with a distaste for the odor of the same-sex parent. More recently, McClintock (1971) observed that the menstrual cycles of close female friends or of women living in residence halls become synchronized. Furthermore, infants are now known to identify their mothers using odor cues (Russell, 1976), and the same worker has shown that adults can distinguish between the sexes by smelling garments previously worn by men and women for 24 hr. Male odors were identified as musky, whereas female odors were described as sweet.

During the past two decades, research has shown that men and women biosynthesize odorous androst-16-enes but to a lesser extent than in the pig (for references, see Gower, 1972, 1976, 1981b). It is not surprising, therefore, that these compounds are now thought to be important in human social interactions. Both 5α-androstenone and an-α have been estimated in the peripheral plasma by radioimmunoassay (RIA); Claus and Alsing (1976) found 5α-androstenone at a mean concentration of 3.26 ng/ml in men. Bicknell and Gower (1976) measured an-α and found mean values of 3.08 and 0.66 ng/ml in men and women, respectively. As in the pig, the nonpolar 5α-androstenone accumulates in the adipose tissue of men (mean 103 ng/g) and women (10–30 ng/g) (Claus and Alsing, 1976).

Excretion in human urine of glucuronides of an-α, (Brooksbank and Haslewood, 1961; Cleveland and Savard, 1964), andien-β, and 5β-androst-16-en-3α-ol (Brooksbank and Gower, 1970) has been known for some time. Recently, however, human axillary secretions have been analyzed. Gower (1972) and Claus and Alsing (1976) identified 5α-androstenone, whereas Brooksbank *et al.* (1974) identified an-α. RIA measurement of 5α-androstenone in axillary secretions (Bird and Gower, 1980) have shown values of 12–1134 pmole/24 hr in 14 men and 13–39 pmole/24 hr in 14 women. Only one woman had a high level (550 pmole/24 hr). Freshly secreted human apocrine sweat is known to be odorless and not to contain 5α-androstenone (Labows *et al.*, 1979), so that the source of the odorous steroid is unknown at present.

In view of the finding of 5α-androstenone in axillary secretions from men, it is significant that the olfactory threshold for humans is so low. In a careful and detailed study, Amoore and co-workers (1977) investigated specific anosmias to 5α-androstenone and Exaltolide (ω-pentadecalactone), the urinous and musky "primary odors." Of the subjects, 46% were anosmic to 5α-androstenone, and the olfactory threshold was only 0.18 parts per billion (ppb). The corresponding values for Exaltolide, an-α, and androstadienone (see Fig. 2 for structures) were 1.8, 6.2, and 0.98 ppb, respectively.

These low thresholds would be consistent with the idea that odorous androst-16-enes might be involved in human social interactions, and it is of considerable interest that two recent papers give some experimental support. The work of Cowley *et al.* (1977) provided evidence that in the presence of an-α, women interviewers modified their judgment of men applying for a position of responsibility. The male members of the interviewing panel were, however, unaffected by the odor. A second experiment (Kirk-Smith *et al.*, 1978) showed that the smell of an-α altered the judgement of human subjects who were asked to give their opinions about photographs of normally clothed people, of animals, and of buildings. In the presence of the odor, both men and women judged the photographed women as more attractive, but the judgments as to the attractiveness of the photographed men were modified to only a small extent.

III. DISTRIBUTION OF TISSUES IN THE PORCINE NASAL CAVITY

Since the metabolism of steroids in the nasal cavity depends upon the various tissues in its walls, a brief description of these structures is given (Fig. 4).

As described in various studies (e.g., Negus, 1958; Parsons, 1971), the tissues of the nasal cavity are determined by their position in the cavity. From the surface of the snout, each external naris opens into a narrow vestibule that is lined with nonkeratinizing, stratified squamous epithelium and is moistened by the secretions of various ducts. Posterior to this, the cavity expands dorsoventrally and forms a passage that is encroached upon by the longitudinal ridges of the lateral wall, namely, the nasoturbinate fold (above) and the maxilloturbinate (below). Both of these extend for several centimeters anteroposteriorly and are covered with nonsensory columnar "respiratory" epithelium, consisting of ciliated cells interspersed with

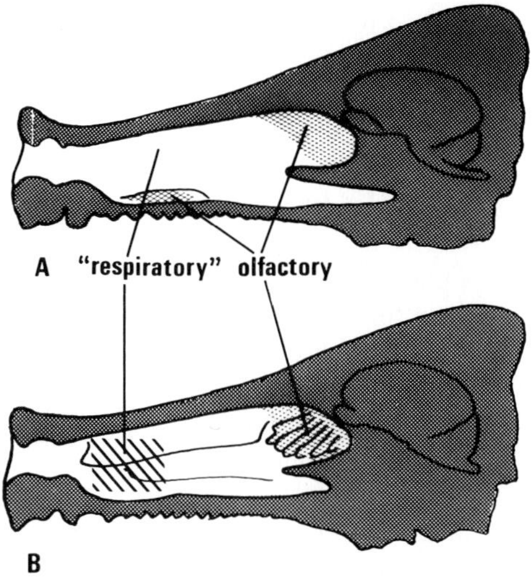

Fig. 4. Diagrams showing the chief features, referred to in the text, of the nasal chamber of the pig. The medial (A) and lateral (B) walls of the nasal cavity are depicted, and the distributions of respiratory and olfactory mucosa are indicated. The diagonally shaded area is that from which the respiratory epithelium was taken.

mucus-secreting goblet cells. The nasopharynx opens ventrally about two-thirds of the way along the nasal cavity, so that a posterior recess extends above the main respiratory pathway at this point. This recess is crowded by the deeply folded ingrowths of the ethmoturbinate bones, which project from the lateral and posterior margins of the cavity. The median septum is a flat plate supported for most of its length by cartilage, but posteriorly by a bony lamina, the mesethmoid, which has a curved anteroventral margin. Posterodorsally, the septum is continuous with the cribriform plate, which separates the nasal cavity from the brain. The anterior parts of the septum are lined with "respiratory" epithelium, but the posterodorsal extremities, arching over to the lateral wall, bear the olfactory (sensory) epithelium, as do the folds of the ethmoturbinates, although the latter also bear patches of respiratory epithelium. The precise boundary between sensory and nonsensory tissues is hard to determine macroscopically. The olfactory area is characterized by pigment of a deep red-brown to gray hue; our histological and electrophysiological observations (W. D. Booth *et al.*, unpublished) show that pigmented areas are often covered with respiratory epithelium for a variable distance within their borders, perhaps due to previous episodes of rhinitic damage to the sensory area. The two types of epithelia can be distinguished, however, in fresh specimens by mounting a piece of mucosa under a cover slip and examining it with phase-contrast or interference-phase microscopy.

The long, regular, and rapidly beating cilia of the respiratory area can be clearly distinguished from the static meshwork of cilia and microvilli that are punctuated by the ends of receptor dendrites and openings of the glands of Bowman, which characterize the sensory area. To ensure a high concentration of olfactory tissue, samples must be taken from the most posterior dorsal recesses of the septum or from the sides of the ethmoturbinate folds. All parts of the nasal cavity, and particularly the anterior lateral and medial walls, are heavily vascularized. Anteriorly, the ventral part of the septum bulges laterally, marking the presence of the tubular vomeronasal organ, which opens at its rostral extremity by a branched duct leading to the buccal and nasal cavities.

The epithelial linings are supported by connective tissue in which are embedded blood vessels, nerves, and numerous glands, the latter extending in some regions several millimeters in thickness. Bojsen-Møller (1967) has analyzed these glandular masses topographically and to some extent histochemically in pigs at various stages of development (Fig. 5). Several major groups can be distinguished on both the medial and lateral walls of the anterior "respiratory" portion, many of them draining anteriorly into, or around, the vestibule. Other glands, such as those of the olfactory area

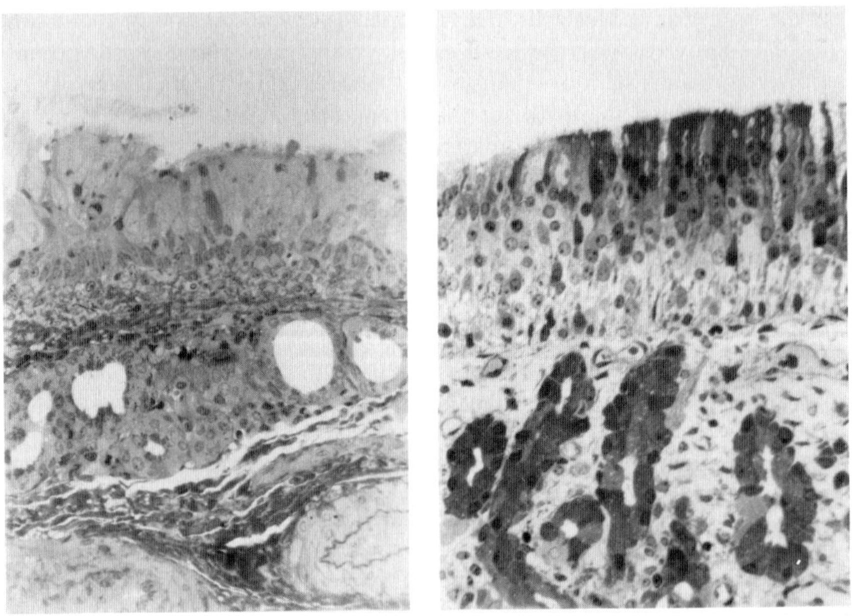

Fig. 5. Light micrographs of vertical sections through the respiratory (left) and septal olfactory (right) areas of the pig, showing the distinctive features of each (see text). Magnification × 350.

(Bowman's glands), secrete locally through narrow ducts onto adjacent surfaces. Bojsen-Møller (1967) found that various groups of glands stained with the periodic acid Schiff method and with alcian blue, indicating the presence of mucosubstances, but that considerable variation occurred in their staining reactions. In view of the wide spectrum of mucosubstances that can be demonstrated histochemically in other species such as the mouse (Cuschieri and Bannister, 1974), it is likely that the metabolism of the glandular tissue is also quite heterogeneous throughout the nasal cavity. Certainly the bulk of the cellular material surrounding the nasal cavity is glandular in nature, and the biochemical activities of the secretory tissue must dominate the metabolism of the nasal tissues in general. However, it should not be overlooked that the respiratory and olfactory epithelia are metabolically active, as reflected in the electron microscopic appearance of their cells (Fig. 6). The fine structure of the nasal epithelia in various mammals (e.g., Graziadei, 1971) shows that large numbers of mitochondria and elements of smooth endoplasmic reticulum occur in both sensory and nonsensory cells of the olfactory epithelium; mitochondria also abound in the respiratory ciliated epithelial cells. Although little histochemical work has been conducted on these tissues (e.g., Cuschieri, 1974) the findings indicate a wide variety of prominent enzymic activities that are yet to be explored biochemically, although it is to be expected that maintenance of the complex nasal cavity environment suitable for respiratory and sensory functions should be a far from simple matter.

The tissues used in the experiments described in the present work were taken from freshly slaughtered female pigs (gilts), which were killed by exsanguination after electrical stunning. The tissue designated "respiratory" was taken from the lateral wall of the anterior half of the nasal chamber; the tissue referred to as "septal" was confined to the heavily pigmented post-

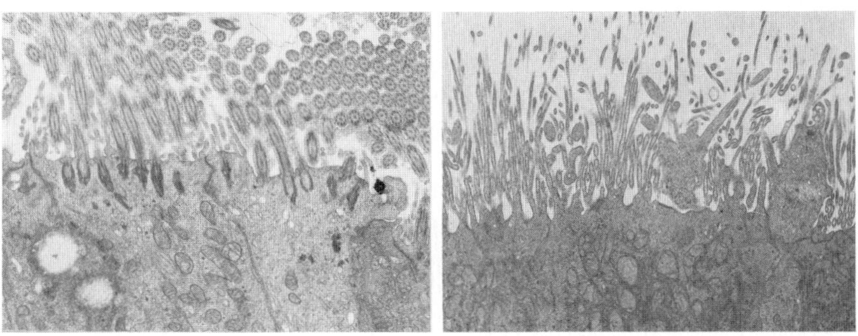

Fig. 6. Electron micrographs of sections through the surface of the ciliated, nonsensory, respiratory epithelium (left), and the olfactory epithelium showing a ciliated sensory ending (right). Magnification × 6600.

erodorsal regions of the septum, typically lined with olfactory sensory epithelium, as was the ethmoturbinate tissue from the posterior recesses of the nasal cavity. As mentioned above, however, the sensory area may also contain some nonsensory tissue of the respiratory type.

Although in our earlier work (Gennings *et al.*, 1977) we showed specific binding of 5α-androstenone and an-α to porcine olfactory tissue, attempts to repeat this work have failed, probably due to the vast capacity of the tissues to take up these steroids nonspecifically. We have now found some of these results to be erroneous, but it is feasible that the reductases present (Gennings *et al.*, 1974) might give rise to binding of some kind.

IV. METABOLISM OF [5α-^{3}H]5α-ANDROSTENONE *IN VITRO* BY PORCINE NASAL EPITHELIUM AND THE EFFECT OF 17β-HYDROXY-5α-ANDROSTAN-3-ONE

[5α-^{3}H]5α-Androstenone (specific radioactivity 14.0 Ci/mmole) was purchased from Isocommerz, Dresden, East Germany and purified before use by column chromatography on alumina (Gower, 1972). The metabolism of this steroid was investigated by incubating it with homogenates of nasal tissues (ethmoturbinate, septum, and respiratory) obtained from the nasal chambers of two female pigs (gilts). Homogenates (10%, w/v) were prepared in 50 m*M* Tris–HCl (pH 7.4) containing 0.25 *M* sucrose. Incubations were continued for 30 min at 37°C in the presence of NADH plus NADPH (each 0.33 m*M*). Reactions were terminated by addition of ethyl acetate (2 ml), and solutions of carrier nonradioactive an-α, an-β, and 5α-androstenone (50 μg of each) were added. After mixing, each tube was centrifuged briefly to separate the ethyl acetate layer, which was removed. Extraction was continued (3 $\times$ 2 ml of ethyl acetate), the organic layer being retained each time. The pooled ethyl acetate fraction was dried (anhydrous Na_2SO_4), filtered, and evaporated *just* to dryness, using a Rotavapor at approximately 40°C with water pump vacuum. Each extract, dissolved in the minimum volume of chloroform, was applied quantitatively to a thin-layer chromatographic plastic sheet (Kieselgel 60, E. Merck, Darmstadt, West Germany) and run twice in benzene-ethyl acetate (9:1, v/v). Authentic androst-16-enes were run at the same time for comparison. After the final drying, the plates were scanned for radioactivity, using a Radiochromatogram Scanner (Panax Ltd., Redhill, Surrey, England). The zones of radioactivity that corresponded in mobility to 5α-androstenone, an-α, and an-β (Fig. 7) were eluted and radioactivity measured in suitable aliquots by scintillation counting. The mass of carrier steroid was determined by gas–liquid chromatography using a Series 104, Model 24, dual flame ionization gas chromatograph (Pye-Unicam, Ltd., Cambridge, England). The silanized glass column (2-m long) was packed with diatomite CLQ (100–120 mesh), coated with OV-1 (2%,

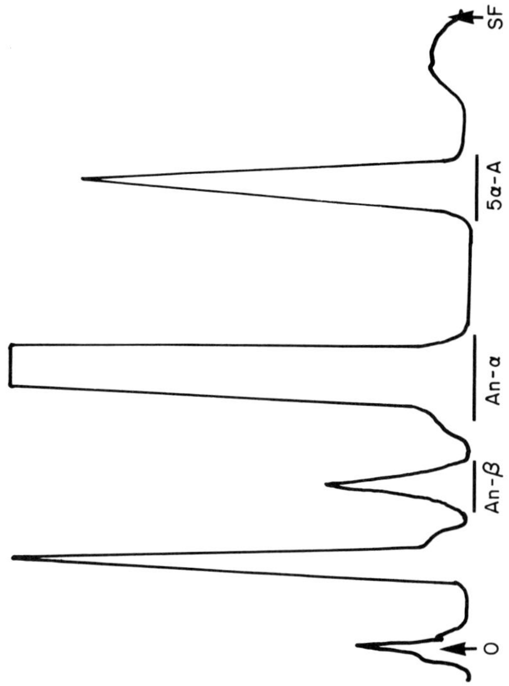

Fig. 7. Radioscan of products of metabolism of [5α-^{3}H] 5α-androst-16-en-3-one separated on a thin-layer chromatographic plastic sheet. The tritiated steroid was previously incubated with soluble portions of sensory nasal epithelium from female pigs (gilts) for 30 min at 37°C in the presence of NADH and NADPH (each 0.33 m*M*). SF indicates solvent front; O, origin; 5α-A, 5α-androstenone; An-α and An-β, 5α-androst-16-en-3α- and 3β-ols; the metabolite running behind An-β is designated "polar," see text.

w/v). Carrier gas (argon) flow rate was 50 ml/min, and the oven temperature was maintained at 200°C. All yields of metabolites were therefore corrected for analytical losses. Further confirmation of the identity of [^{3}H]an-α and [^{3}H]an-β was obtained by subjecting portions of the extracts, obtained from thin-layer chromatography, to combined radio–gas chromatography (Pye-Unicam, Ltd., Cambridge/Panax Ltd., Redhill, Surrey, England) in which the radioactivity was shown to be eluted from the column at the same time as the carrier steroid. For each experiment, a control tube without homogenate was included; it was found previously that if [^{3}H]5α-androstenone is run on a thin-layer plastic sheet streaking occurs, and radioactivity is detectable at all points along the run. For this reason, the radioactivities of an-α, an-β, and 5α-androstenone have been corrected for this small degree of background.

The results (Table II) show that 3α- and 3β-reduction of 5α-androstenone occurs to different extents in different nasal tissues. In septum and respiratory tissues, the predominant product appears to be an-α, whereas in right

TABLE II Metabolism of [5α-^{3}H]5α-Androst-16-en-3-one in Various Nasal Tissues *in Vitro*[a]

Source of tissue	Percentage yield		
	5α-Androstenone	An-α	An-β
Pig 1			
Ethmoturbinate (left)	58	6	31
Ethmoturbinate (right)	42	2	40
Nasal septum	38	51	7
Respiratory	43	46	4
Pig 2			
Ethmoturbinate (left)	36	18	26
Ethmoturbinate (right)	49	18	24
Nasal septum	33	57	5
Respiratory	36	51	6

[a] Portions (1 ml) of homogenates (10%, w/v) of various nasal tissues from two female pigs (gilts) were incubated separately for 30 min at 37°C with [5α-^{3}H]-5α-androstenone (3.5×10^6 dpm) in the presence of NADH plus NADPH (each 0.33 m*M*). After incubation, carrier steroids were added, and the extracted metabolites were separated and quantified. Results are expressed as percentages of the radioactivity initially added.

and left ethmoturbinate tissue, the ratio of an-α to an-β is far more variable. Fig. 7 also shows two additional areas of detectable radioactivity. One polar material, formed in appreciable quantities in some experiments, did not migrate from the origin. The second material, which had a mobility less than that of an-β ("polar"), compares with a metabolite already noted by Gennings *et al.* (1974). The identity of these metabolites has not yet been determined, and it is therefore not possible to postulate their significance.

Having shown that 5α-androstenone is readily reduced in porcine nasal tissues, an experiment was designed to study the possible inhibitory effect of another C_{19} steroid, having an oxo group at C-3 and the 5α-configuration between rings A and B. The steroid chosen was 17β-hydroxy-5α-androstan-3-one (5α-dihydrotestosterone, 5α-DHT). This was incubated with the porcine tissues at 1000 times (25 μ*M*) and 500 times (12.5 μ*M*) the concentration of [5α-^{3}H]5α-androstenone. Yields of an-α, an-β, and unmetabolized substrate were measured and corrected as described above.

In the ethmoturbinate tissue, it was found that the presence of 5α-DHT (25 μ*M*) resulted in 49% (mean value) more 5α-androstenone being unmetabolized compared with control incubations. There was also a dramatic decrease by 71% (mean) in the amount of an-β formed. Changes in the yield of an-α were not consistent, however. In the left ethmoturbinate tissue, the yield increased in the presence of 5α-DHT, whereas the yield from the right ethmoturbinate was diminished by 50%. The reason for this apparent discrepancy is unclear at present. It was also interesting to note that the amount of "polar" metabolite (see Fig. 7) was consistently increased by approximately 280% (no correction for possible analytical losses could be made).

In the septum and respiratory tissues, significantly less 5α-androstenone was metabolized in the presence of 5α-DHT, with a mean decrease of 70% in an-α formation. In this case, the yield of an-β appeared to be unaffected. These results are consistent with those presented in Table II, which show that the formation of the 3α-alcohol predominates in septum and respiratory tissues.

V. SUBCELLULAR LOCATION AND CO-FACTOR DEPENDENCY OF 3α- AND 3β-HYDROXYSTEROID DEHYDROGENASES

For this series of experiments, homogenates of gilt ethmoturbinate epithelium were prepared as described above, and microsomal and soluble fractions were obtained by differential centrifugation as described by Cooke and Gower (1977). The microsomal fraction was washed and recentrifuged to ensure removal of contaminating cytoplasm. Each subcellular fraction, which contained proportionally the same amount of protein as the whole homogenate, was incubated with $[5\alpha\text{-}^3H]5\alpha$-androstenone under conditions described above. The results, summarized in Table III, clearly show that both the 3α-

TABLE III Subcellular Location and Co-factor Dependency of 3α- and 3β-Hydroxysteroid Dehydrogenases in Porcine Ethmoturbinate Tissue[a]

		Percentage yields		
Tissue fraction	Co-factors	5α-Androstenone	An-α	An-β
Homogenate	NADH; NADPH	42.0	10.6	47.4
Homogenate	None	94.9	4.0	1.1
Soluble	NADH; NADPH	39.7	15.7	44.0
Soluble	NADH	60.5	10.3	29.3
Soluble	NADPH	25.1	17.3	57.6
Microsomes	NADH; NADPH	94.3	3.4	2.4
Microsomes	NADH	95.5	3.3	1.2
Microsomes	NADPH	94.9	3.6	1.4

[a] Homogenates (10%, w/v) of ethmoturbinate epithelium from immature female pigs were subjected to differential centrifugation. The microsomal and soluble fractions were incubated for 30 min at 37°C with $[5\alpha\text{-}^3H]$ 5α-androstenone (3.5×10^6 dpm) in the presence of co-factors (0.33 mM) as indicated. See legend to Table II for further details.

and the 3β-OHSDHs occur almost exclusively in the cytosolic fraction, and NADPH is the preferred co-factor. As in the first series of experiments (Table II), an-β formation appears to predominate in ethmoturbinate tissue.

Identical experiments, using septum and respiratory tissues, gave similar results for the subcellular location and co-factor dependency of the 3-OHSDHs.

VI. TIME COURSE OF THE REDUCTION OF [5α-^{3}H]5α-ANDROSTENONE IN PORCINE NASAL TISSUES *IN VITRO*

In further experiments using tissue from an additional group of animals, it was our intention to determine the apparent K_m values for the 3α- and 3β-OHSDHs with 5α-androstenone as the substrate. As a preliminary to this, the time course of the two reductions had to be established. For this purpose, soluble fractions of homogenates (10%, w/v) of epithelia were prepared as described above and incubated separately with [5α-^{3}H]5α-androstenone for periods of up to 60 min. In this series of experiments, both an-α and an-β were formed with septal and respiratory tissues, in contrast with the results obtained earlier. Figure 8 shows the curves obtained for nasal

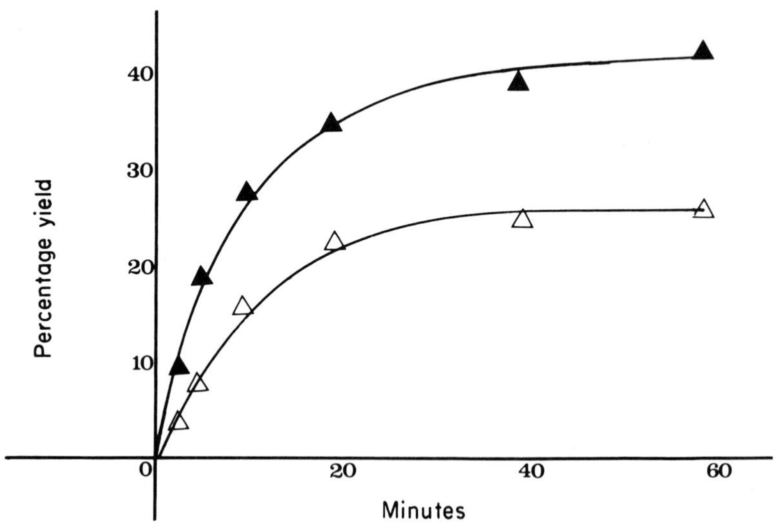

Fig. 8. Time course of reduction of [5α-^{3}H] 5α-androst-16-en-3-one in porcine nasal septum *in vitro*. Soluble fractions of homogenates of nasal septum from female pigs (gilts) were incubated with [5α-^{3}H] 5α-androst-16-en-3-one (0.05 μmole/liter) at 37°C for periods of up to 60 min in the presence of NADH and NADPH (each 0.33 m*M*). ▲—▲, an-α; △—△, an-β.

septum. In the case of an-α, the rate of formation was linear for up to 5 min by which time as much as 20% of product had been obtained. Equilibrium was reached after approximately 40 min of incubation. The rate of production of an-β was slower, the linear portion of the curve extending to about 7 min, when some 12% of product had been obtained.

VII. MEASUREMENT OF APPARENT K_m VALUES FOR 3-HYDROXYSTEROID DEHYDROGENASES IN PORCINE NASAL TISSUES

To measure the apparent K_m values, portions of soluble fraction were incubated with various concentrations (0.05 to 2 μM) of unlabeled 5α-androstenone together with [5α-^{3}H]5α-androstenone for a period of time corresponding to the linear production of metabolites. Lineweaver-Burk plots were drawn, and linear regression analyses were performed on all results. Figures 9 and 10 show the plots obtained for septal tissue, from which the apparent K_m values were 7.2×10^{-7} M and $6.9 \times 10^{-7} M$ for the 3α- and 3β-OHSDHs, respectively. Values of the same order of magnitude were obtained for the other nasal tissues studied.

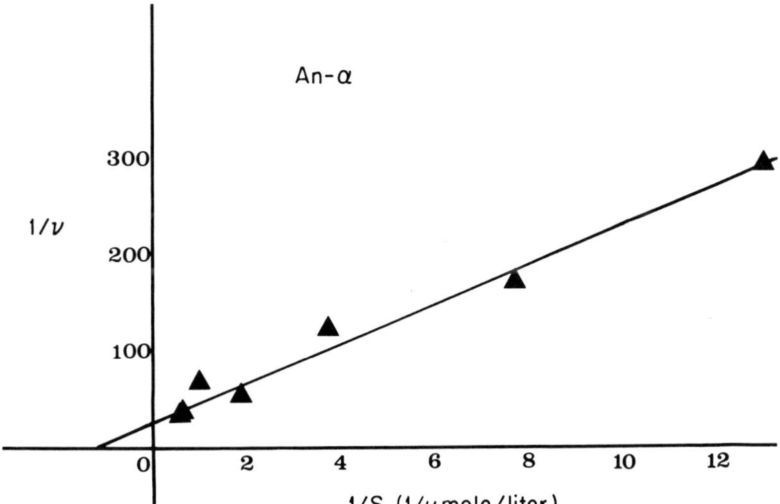

Fig. 9. Estimation of apparent K_m for 3α-hydroxysteroid dehydrogenase in porcine nasal septum. Soluble fractions of homogenates of nasal septum from female pigs (gilts) were incubated with various concentrations (0.05–2 μM) of unlabeled 5α-androst-16-en-3-one together with [5α-^{3}H] 5α-androst-16-en-3-one (3.5×10^6 dpm). The duration of incubation corresponded to the linear production of an-α (Fig. 8).

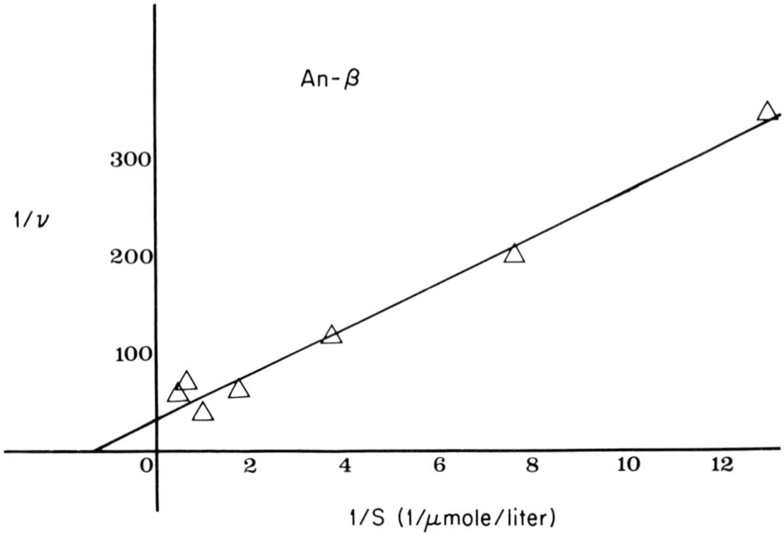

Fig. 10. Estimation of apparent K_m for 3β-hydroxysteroid dehydrogenase in porcine nasal septum. For details see text and legend to Fig. 9. The duration of incubation corresponded to the linear production of an-β (Fig. 8).

VIII. POSSIBLE SIGNIFICANCE OF 3-HYDROXYSTEROID DEHYDROGENASES IN PORCINE NASAL EPITHELIUM

The 3α- and 3β-hydroxysteroid dehydrogenases (or oxidoreductases) are well-known in steroid hormone metabolism. Their usual action, that of reversible conversion of 3-oxosteroids to the 3α- and 3β-alcohols, generally results in loss of activity. This is true, for example, in the case of 17β-hydroxy-5α-androstane-3-one (5α-dihydrotestosterone), when 5α-androstane-3α,17β-diol, and 5α-androstane-3β, 17β-diol are formed by reduction.

The 3α-OHSDHs are found in numerous tissues including liver, adrenals, testes, and ovaries. Recently, accessory sex organs have been studied; Suzuki and Tamaoki (1974) showed that rat seminal vesicles contain the cytosolic enzyme, whereas canine prostate has both cytosolic and a microsomal counterpart (Jacobi *et al.*, 1977). Of particular interest is rat kidney, which contains three distinct 3α-OHSDHs: a cytosolic NADPH-requiring enzyme and two microsomal enzymes, one requiring NADPH and the other NADH (Verhoeven *et al.*, 1976).

Although purification has not yet been attempted, we have shown in the present work that both 3α- and 3β-oxidoreductases occur in porcine nasal tissues and that they can catalyze the reduction of the odorous 5α-androst-16-en-3-one. These data both confirm and extend the earlier studies of Gennings *et al.* (1974). The apparent K_m values for the two enzymes in the various nasal tissues studied were all of the order of 10^{-7} *M*. These compare with those for the 3α-OHSDH of rat seminal vesicles (6.5×10^{-7} *M*, Suzuki and Tamaoki, 1974), of canine prostate (6×10^{-7} *M*, Jacobi *et al.*, 1977), and of the female rat kidney cytosolic enzyme (NADPH-dependent), which has a K_m of around 3×10^{-7} *M*. (Verhoeven *et al.*, 1976). These workers found that the 3α-OHSDHs of female rat kidney were heterogeneous, the microsomal enzyme and its cytosolic counterpart (NAD-requiring) having K_m values of 23.9×10^{-7} and 23.8×10^{-7} *M*, respectively. 5α-Dihydrotestosterone was used as substrate in all cases.

The importance of the two reductases in the olfactory processes of the pig is as yet rather speculative, although both the fact of their presence and of their relatively high efficiency point to a significant role in olfaction. It seems unlikely that conversion of 5α-androstenone to the equally pheromonally active an-α (Table I) would be useful in the presentation of the pheromonal stimulus, since it would presumably not be rapid enough to produce significant amounts of relevant signal. However, clearance of active pheromone from the nasal cavity is a feasible function for the dehydrogenases. 5α-Androstenone is a large molecule with relatively low volatility at blood temperature and is strongly adsorbed to most surfaces. Because of its high solubility in lipids, it would easily diffuse into cells and might remain in the tissues, thus contaminating the nasal cavity so as to interfere with further pheromonal signaling. Conversion to the less pheromonally active an-β (Table I) and of both this and an-α to unidentified polar compounds (Fig. 7) may provide pathways to metabolize active pheromones to behaviorally inactive substances. This phenomenon may be peculiar to the relatively large molecules employed by the pig as pheromones, since normal diffusion processes appear to be sufficient to allow elimination of small molecular weight odors.

The significance of the presence of two metabolic pathways for 5α-androstenone is as yet uncertain. The regional variation in the activity of the two oxidoreductases suggests a difference between olfactory and non-olfactory areas, but the variability from one group of animals to another precludes a definite conclusion. Such variability may be related to pathological changes such as those induced by chronic rhinitis or, more likely, to hormonal status. At the present time, nothing is known about the hormonal control of dehydrogenases in porcine nasal tissues, but the influence of hormonal status on other steroid reductions has been studied. Ghraf *et al.* (1975) have shown that the renal cytosolic 3α-OHSDH is regulated by estrogens. In contrast, the activity of the microsomal counterparts is controlled by androgens. In future studies, it may be profitable to investigate the possible control of dehydrogenases in porcine nasal tissues by steroid hormones and to extend our present knowledge concerning inhibition of the enzymes by other 3-oxosteroids, such as 5α-dihydrotestosterone.

ACKNOWLEDGMENT

We thank Dr. W. D. Booth for help in obtaining tissue and Mrs. D. M. Gower for typing the manuscript. The Agricultural Research Council (grant no. AG 35/33) kindly provided financial support (M.R.H.).

REFERENCES

Ahmad, N., and Gower, D. B. (1968). The biosynthesis of some androst-16-enes from C_{21} and C_{19} steroids in boar testicular and adrenal tissue. *Biochem. J.* **108**, 233–241.

Altmann, M. (1941). Interrelations of the sex cycle and the behaviour of the sow. *J. Comp. Psychol.* **31**, 473–479.

Amoore, J. E., Pelosi, P., and Forrester, L. J. (1977). Specific anosmias to 5α-androst-16-en-3-one and ω-pentadecalactone: the urinous and musky primary odors. *Chem. Senses Flavor* **2**, 401–425.

Andresen, Ø. (1974). Development of a radioimmunoassay for 5α-androst-16- en-3-one in pig peripheral plasma. *Acta Endocrinol. (Copenhagen)* **76**, 377–387.

Andresen, Ø. (1975). A radioimmunoassay for 5α-androst-16-en-3-one in porcine adipose tissue. *Acta Endocrinol. (Copenhagen)* **79**, 619–624.

Beery, K. E., and Sink, J. D. (1971). Isolation and identification of 3α-hydroxy-5α-androst-16-ene and 5α-androst-16-en-3-one from porcine adipose tissue. *J. Endocrinol.* **51**, 223–224.

Bicknell, D. C., and Gower, D. B. (1975). Separation of some 16-androstenes on hydroxyalkoxypropyl-Sephadex (Lipidex™). *J. Chromatogr.* **110**, 210–121.

Bicknell, D. C., and Gower, D. B. (1976). The development and application of a radioimmunoassay for 5α-androst-16-en-3α-ol in plasma. *J. Steroid Biochem.* **7**, 451–455.

Bird, S., and Gower, D. B. (1980). Measurement of 5α-androst-16- en-3-one in human axillary secretions by radioimmunoassay. *J. Endocrinol.* **85**, 8P–9P.

Bojsen-Møller, F. (1967). Topography and development of anterior nasal glands in pigs. *J. Anat.* **101**, 321–331.

Booth, W. D. (1975). Changes with age in the occurrence of C_{19} steroids in the testis and submaxillary gland of the boar. *J. Reprod. Fertil.* **42**, 459–472.

Booth, W. D. (1980). Endocrine and exocrine factors in the reproductive behaviour of the pig. Symp. Zool. Soc. London, No. 45, pp, 289–311. Academic Press, New York.

Brooksbank, B. W. L., and Gower, D. B. (1970). The estimation of 3α-hydroxy-5α-androst-16-ene and other C_{19}-Δ^{16} steroids in urine by gas-liquid chromatography. *Acta Endocrinol. (Copenhagen)* **63**, 79–90.

Brooksbank, B. W. L., and Haslewood, G. A. D. (1961). The estimation of androst-16-en-3α-ol in human urine. Partial synthesis of androstenol and of its β-glucosiduronic acid. *Biochem. J.* **80**, 488–496.

Brooksbank, B. W. L., Brown, R., and Gustafsson, J.-Å. (1974). The detection of 5α-androst-16-en-3α-ol in human male axillary sweat. *Experientia* **30**, 864–865.

Brophy, P. J., and Gower, D. B. (1972). 16-Unsaturated C_{19} 3-oxosteroids as metabolic intermediates in boar testis. *Biochem. J.* **128**, 945–952.

Brophy, P. J., and Gower, D. B. (1973). Further studies on the properties of the enzyme system involved in the conversion of C_{21} steroids into 16-unsaturated C_{19} steroids in boar testis. *Biochem. Soc. Trans.* **1**, 181–183.

Claus, R., and Alsing, W. (1976). Occurrence of 5α-androst-16-en-3-one, a boar pheromone in man and its relationship to testosterone. *J. Endocrinol.* **68**, 483–484.

Cleveland, W. W., and Savard, K. (1964). Studies of excretion of androst-16-en-3α-ol. *J. Clin. Endocrinol. Metab.* **24**, 983–987.

Comfort, A. (1974). The likelihood of human pheromones. In "Pheromones" (M. C. Birch, ed.), pp. 386–396. North-Holland Publ., Amsterdam.

Cooke, G. M., and Gower, D. B. (1977). The submicrosomal distribution in rat and boar testis of some enzymes involved in androgen and 16-androstene biosynthesis. *Biochim. Biophys. Acta* **498**, 265–271.

Cooke, G. M., and Gower, D. B. (1979). Attempted solubilization of boar testis microsomal "Androsta-5,16-dien-3β-ol Synthetase." *Biochem. Soc. Trans.* **7**, 920–922.

Cowley, J. J., Johnson, A. L., and Brooksbank, B. W. L. (1977). The effect of two odorous compounds on performance in an assessment of people test. *Psychoneuroendocrinology* **2**, 159–172.

Cuschieri, A. (1974). Enzyme histochemistry of the olfactory mucosa and vomeronasal organ in the mouse. *J. Anat.* **118**, 477–489.

Cuschieri, A., and Bannister, L. H. (1974). Some histochemical observations on the mucosubstances of the nasal glands of the mouse. *Histochem. J.* **6**, 543–558.

Ewing, L., and Brown, B. L. (1977). Testicular steroidogenesis. *In* "The Testis" (A. D. Johnson and W. R. Gomes, eds.), Vol. 4, pp. 239–287. Academic Press, New York.

Gasparini, F. J., Hochberg, R. B., and Lieberman, S. (1976). Biosynthesis of steroid sulphates by the boar testes. *Biochemistry* **15**, 3969–3975.

Gennings, J. N., Gower, D. B., and Bannister, L. H. (1974). Studies on the metabolism of the odoriferous ketones, 5α-androst-16-en-3-one and 4,16-androstadien-3-one, by the nasal epithelium of the mature and immature sow. *Biochim. Biophys. Acta* **369**, 294–303.

Gennings, J. N., Gower, D. B., and Bannister, L. H. (1977). Studies on the receptors to 5α-androst-16-en-3-one and 5α-androst-16-en-3α-ol in sow nasal mucosa. *Biochim. Biophys. Acta* **496**, 547–556.

Ghraf, R., Lax, E. R., Hoff, H., and Schriefers, H. (1975). The role of the gonads and the hypophysis in the regulation of hydroxysteroid dehydrogenase activities in rat kidney. *Hoppe-Seyler's Z. Physiol. Chem.* **356**, 135–142.

Gower, D. B. (1964). Chromatographic separation of C_{19}-16-dehydrosteroids. *J. Chromatogr.* **14**, 424–431.

Gower, D. B. (1972). 16-Unsaturated C_{19} steroids. A review of their chemistry, biochemistry and possible physiological role. *J. Steroid Biochem.* **3**, 45–103.

Gower, D. B. (1976). Behavioural effects of pheromones. *In* "Modern Trends in Psychosomatic Medicine" (O. W. Hill, ed.), Vol. 3, pp. 42–65. Butterworth, London.

Gower, D. B. (1981a). The biosynthesis and occurrence of 16-androstenes in man. *In* "Hormones in Normal and Abnormal Human Tissues" (K. Fotherby and S. B. Pal, eds.), pp. 1–27. de Gruyter, Berlin.

Gower, D. B. (1981b). Biosynthesis of androgens and other C_{19} steroids. *In* "Biochemistry of Steroid Hormones" (H. L. J. Makin, ed.). Blackwell, Oxford. In press.

Graziadei, P. P. C. (1971). The olfactory mucosa of vertebrates. *In* "Handbook of Sensory Physiology" (L. M. Beidler, ed.), Vol. 4, Part 1, pp. 27–58. Springer-Verlag, Berlin and New York.

Hurden, E. L., Gower, D. B., and Harrison, F. A. (1979). Biosynthesis of 16-androstenes and androgens in boar testis *in vivo*. *J. Endocrinol.* **81**, 161P–162P.

Jacobi, G. H., Moore, R. J., and Wilson, J. D. (1977). Characterization of the 3α-hydroxysteroid dehydrogenase of dog prostate. *J. Steroid Biochem.* **8**, 719–723.

Kalogerakis, M. G. (1963). The role of olfaction in sexual development. *Psychosom. Med.* **25**, 420–432.

Katkov, T., and Gower, D. B. (1970). The biosynthesis of androst-16-enes in boar testis tissue. *Biochem. J.* **117**, 533–538.

Katkov, T., Booth, W. D., and Gower, D. B. (1972). The metabolism of androst-16-enes in boar salivary glands. *Biochim. Biophys. Acta* **270**, 546–556.

Kaufmann, G., Ritter, F., and Schubert, K. (1976). Quantitative determination of the boar taint substance 5α-androst-16-en-3-one in fat. *J. Steroid Biochem.* **7**, 593–597.

Keller, H. (1937). Zur Beseitigung der Geruchs- und Geshmack-sabweichung des Brineneber-gleishes durch Pökelung. *Z. Fleisch- Milchhyg.* **47**, 174–176.

Kirk-Smith, M., Booth, M. A., Carrell, D., and Davies, P. (1978). Human social attitudes affected by androstenol. *Res. Commun. Psychol., Psychiatry Behav.* **3**, 379–384.

Labows, J. N., Preti, G., Hoelzle, E., Leyden, J., and Kligman, A. (1979). Steroid analysis of human apocrine secretion. *Steroids* **34**, 249–258.

Lerche, H. (1936). Geschlechtsgeruch bei Eberkastraten. *Z. Fleisch- Milchhyg.* **46**, 417–420.

Loke, K. H., and Gower, D. B. (1972). The intermediary role of 5-pregnene-3β,20β-diol in the biosynthesis of 16-unsaturated C_{19} steroids in boar testis. *Biochem. J.* **127**, 545–551.

MacLeod, N., Reinhardt, W., and Ellendorf, F. (1979). Olfactory bulb neurons of the pig respond to an identified steroidal pheromone and testosterone. *Brain Res.* **164**, 323–327.

McClintock, M. K. (1971). Menstrual synchrony and suppression *Nature* (*London*) **299**, 244–245.

Mason, J. I., Park, R. J., and Boyd, G. S. (1979). A novel pathway of androst-16-ene biosynthesis in immature pig testis microsomal fractions. *Biochem. Soc. Trans.* **7**, 641–643.

Melrose, D. R., Reed, H. C. B., and Patterson, R. L. S. (1971). Androgen steroids associated with boar odour as an aid to the detection of oestrus in pig artifical insemination. *Br. Vet. J.* **127**, 497–501.

Negus, V. (1958). "Comparative Anatomy and Physiology of the Nose and Paranasal Sinuses." Livingstone, Edinburgh.

Parsons, T. S. (1971). Anatomy of nasal structures from a comparative viewpoint. *In* "Handbook of Sensory Physiology" (L. M. Beidler, ed.), Vol. 4, Part 1, pp. 1–26. Springer-Verlag, Berlin and New York.

Patterson, R. L. S. (1968a). 5α-Androst-16-ene-3-one: Compound responsible for taint in boar fat. *J. Sci. Food Agric.* **19**, 31–38.

Patterson, R. L. S. (1968b). Identification of 3α-hydroxy-5α-androst-16-ene as the musk odour component of boar submaxillary gland and its relationship to the sex odour taint in pork meat. *J. Sci. Food Agric.* **19**, 434–438.

Pelosi, P., Pisanelli, A. M., and Baldaccini, N. E. (1978). Preliminary characterization of the "urinous" olfactory receptor. *Congr. Eur. Chemorecept. Organ., 3rd, Pavia, Italy* Abstr. No. 71.

Poynder, T. M., Bannister, L. H., and Gower, D. B. (1978). Recording of electrical activity from mammalian olfactory epithelium *in vitro:* stimulation with pheromones and other odorants. *Congr. Eur. Chemorecept. Organ., 3rd, Pavia, Italy* Abstr. No. 40.

Prelog, V., and Ruzicka, L. (1944). Über zwei moschusartig riechende Steroide aus Schweinestestes-Extracten. *Helv. Chim. Acta* **27**, 61–66.

Prelog, V., Ruzicka, L., Meister, P., and Wieland, P. (1945). Steroide und Sexual-hormone.

113. Mitteilung. Untersuchungen über den Zusammenhang zwischen Konstitution und Geruch bei Steroiden. *Helv. Chim. Acta* **28**, 618–627.

Reed, H. C. B., Melrose, D. R., and Patterson, R. L. S. (1974). Androgen steroids as an aid to the detection of oestrus in pig artificial insemination. *Br. Vet. J.* **130**, 61–67.

Russell, M. J. (1976). Human olfactory communication. *Nature (London)* **260**, 520–522.

Saat, Y. A., Gower, D. B., Harrison, F. A., and Heap, R. B. (1972). Studies on the biosynthesis *in vivo* and excretion of 16-unsaturated C_{19} steroids in the boar. *Biochem. J.* **129**, 657–663.

Saat, Y. A., Gower, D. B., Harrison, F. A., and Heap, R. B. (1974). Studies on the metabolism of 5α-androst-16-en-3-one in boar testis *in vivo*. *Biochem. J.* **144**, 347–353.

Sink, J. D. (1967). Theoretical aspects of sex odour in swine. *J. Theor. Biol.* **17**, 174–180.

Suzuki, K., and Tamaoki, B. (1974). *In vitro* metabolism of testosterone in seminal vesicles of rats. *J. Steroid Biochem.* **5**, 249–256.

Verhoeven, G., Heyns, W., and De Moor, P. (1976). Interconversion between 17β-hydroxy-5α-androstan-3-one (5α-DHT) and 5α-androstane-3α,17β-diol in rat kidney. *Eur. J. Biochem.* **65**, 565–576.

2

Accessibility of Odorant Molecules to the Receptors

DAVID E. HORNUNG AND MAXWELL M. MOZELL

I. INTRODUCTION

In the higher animals, at least, odorant molecules do not have immediate access to the olfactory receptors. To reach the region of the olfactory receptors, odorant molecules must first be drawn past the extensive sorptive surface presented by an often highly convoluted, multichanneled flow path in the nasal cavity. To then reach the receptors in any given region along the olfactory mucosa itself, the molecules must first pass over the sorptive surfaces of antecedent mucosal regions. Once in the airspace above the olfactory mucosa, the molecules must diffuse through the mucus to gain final access to the receptors themselves. It has been suggested that since these access barriers can affect different odorants differentially, the olfactory system may take advantage of them in the initial stages of odorant discrimination. However, even if these access barriers are not themselves basic mechanisms for

Biochemistry of Taste and Olfaction

ISBN 0-12-154450-8

olfactory discrimination, they would still affect the number of molecules reaching the receptors and could thus influence any primary coding mechanism. In this chapter, some of the causes and effects of differential molecular accessibility to receptors are examined.

II. ODORANT ACCESS

E. D. Adrian (1950a,b), recording electrophysiologically from two locations in the olfactory bulb, observed that different odorous substances excited these two locations differently. Adrian proposed that the excitation patterns observed across the bulb were themselves a reflection of precursory excitation patterns across the olfactory epithelium. He further proposed that these latter patterns were the result of those physicochemical properties of odorants (such as water solubility, lipid solubility, volatility), which can affect the movement of the molecules along the epithelium. Thus, since in many animals the folding of the olfactory receptor sheet provides a large convoluted surface area, molecules of odorants with differing physicochemical properties would be deposited in different space–time patterns across the olfactory mucosa. This spatiotemporal distribution would not only allow for an initial sorting of the molecules of different chemicals across the receptor sheet, but would also affect the accessibility of at least some odorant molecules to the receptors located in the distal regions of the air flow path.

Moncrieff (1955) was the first to supply evidence from the mucosa itself to suggest such a differential sorption. He prepared the head of a freshly killed sheep (Fig. 1) such that an odorized air stream could be introduced into one naris and eluted from the other. He then had human observers smell the emerging air stream. These observers noted that the odorized air, having

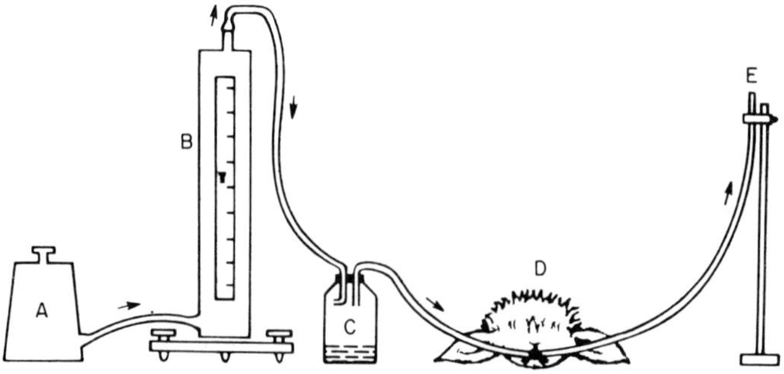

Fig. 1. Diagrammatic representation of the apparatus Moncrieff used to measure the time needed for odorants to pass through the sheep nasal airways. (A) is the air pump, (B) a flowmeter, (C) the odorant, (D) the sheep's head, and (E) the outlet tube from which human observers smelled the emerging air stream. (From Moncrieff, 1967, by permission of CRC Press.)

passed through the sheep's nasal passages, seemed odorless for a time, and this time differed for different odorants. This suggested that different chemicals migrated across the nasal mucosa (if not specifically across the olfactory mucosa) at different rates and thus were differentially sorbed. In addition, Moncrieff (1955) demonstrated that the olfactory mucosa, when cut away from newly killed animals and placed in containers filled with odorized air, could render it odorless. This gave credence to the conclusion (Moncrieff, 1967) that some sorptive process may be the key to olfactory quality discrimination. Moncrieff then suggested that in addition to affecting odorant molecular distributions along the length of the mucosa, these sorptive processes might also affect the accessibility of odorant molecules as they move through the mucus layers toward the olfactory receptors.

Mozell (1966) compared the activity patterns in two separate regions of the frog's olfactory mucosa by simultaneously recording the summated multiunit electrophysiological discharges from the two nerve branches supplying those two regions. The more medial of these branches supplied a region that, being near the external naris, was contacted early by odorant molecules flowing over the olfactory mucosa. The other branch, the more lateral, supplied a region of the mucosa that, by overhanging the internal naris, was positioned farther along the flow path. Mozell proposed that the smaller the ratio (called the LB/MB ratio) of the summated discharge recorded from the lateral nerve branch (LB) to that recorded from the medial nerve branch (MB), the steeper would be the molecular distribution gradient from the external to internal naris. In the initial study (Mozell, 1966) involving four chemicals and in a later one involving 16 chemicals (Mozell, 1970), each chemical produced a characteristic LB/MB ratio that was consistent from frog to frog. Although these LB/MB ratios were affected somewhat by changes in the concentration and flow rate of the stimuli, the effects were small compared with the effect produced by varying the chemical compound. This is important because the activity gradients across the mucosa would be a doubtful "code" for discrimination of odor quality if they were confounded by flow rate and concentration, both of which affect the magnitude of the olfactory nerve discharge (Tucker, 1963). Thus, an odorant-dependent spatial pattern of activity was demonstrated in the peripheral olfaction system.

Mozell (1970) further observed that in general the smaller an odorant's LB/MB ratio, the longer was its retention time when measured with gas chromatography on a Carbowax 20M column. This suggested that a "chromatographic-like" process might underly the separation of odorant molecules across the mucosal sheet. This led Mozell and Jagodowicz (1973) to substitute the *in vivo* frog olfactory sac for the column of a gas chromatograph in an attempt to more directly measure how chemicals migrate across the olfactory mucosa (Fig. 2). These results showed a rather wide range of retention times for 15 different odorants across the mucosal sheet. Thus, in spite of the small size of the frog's olfactory mucosa, molecules of different chemicals do indeed migrate across it at very different rates. Therefore, Mozell and Jagodowicz (1974) predicted that in the time interval of a given sniff, some odorant molecules (those with long retention times) would be unevenly distributed across the mucosal sheet, whereas other odorant molecules (those with short retention times) would be more evenly distributed.

In line with this prediction, Mozell and Jagodowicz (1974) did find a high negative correlation between the earlier measured LB/MB ratios and the newly measured retention times. However, conclusive evidence for the existence of molecular distribution patterns as well as a fuller appreciation of their exact topology required direct determination of the molecular distributions. This requirement was met by using tritium-labeled odorants (butanol, octane, and butyl acetate) to quantitatively map their molecular sorption patterns along the olfactory mucosa (Hornung *et al.*, 1975). Both the number and the mucosal locations of the odorant molecules could thus be quantified.

By use of a flow dilution olfactometer, ^{3}H-labeled odorants were presented through the external naris to the bullfrog's intact olfactory sac. Immediately after stimulation, the animal was frozen in liquid nitrogen. The dorsal surface and eminentia of the olfactory sac were then removed and sawed into 2 mm wide coronal sections. Each section was dissolved in a tissue solubilizer and was counted in a liquid scintillation system. The amount of radioactivity in each section was used to estimate the number of odorant molecules it sorbed. For tritiated butanol, there was a significant decrease in radioactivity from the dorsal surface section containing the external naris to that overhanging the internal naris. For butyl acetate, the gradient was less steep, and for octane, there was a rather even distribution among the several mucosal sections (Fig. 3).

This isotope work showed that a given compound establishes a particular distribution pattern along the intact olfactory sac of the bullfrog. It further showed that within a wide range of stimulus parameters (flow rates, concentrations, and times) the distribution remains fairly constant. This is what Mozell (1970) predicted would be necessary if a "chromatographic-like" differentiation were one of the mechanisms basic to odorant discrimination. However, as mentioned above, even if these molecular distributions are not themselves the basis for olfactory discrimination, they could nevertheless have an important initial role in defining the access of odorant molecules to different receptors. Thus they could have a fundamental impact upon whatever process underlies olfactory discrimination. For example, if selective sensitivities of receptors and/or their regional distributions were the basic

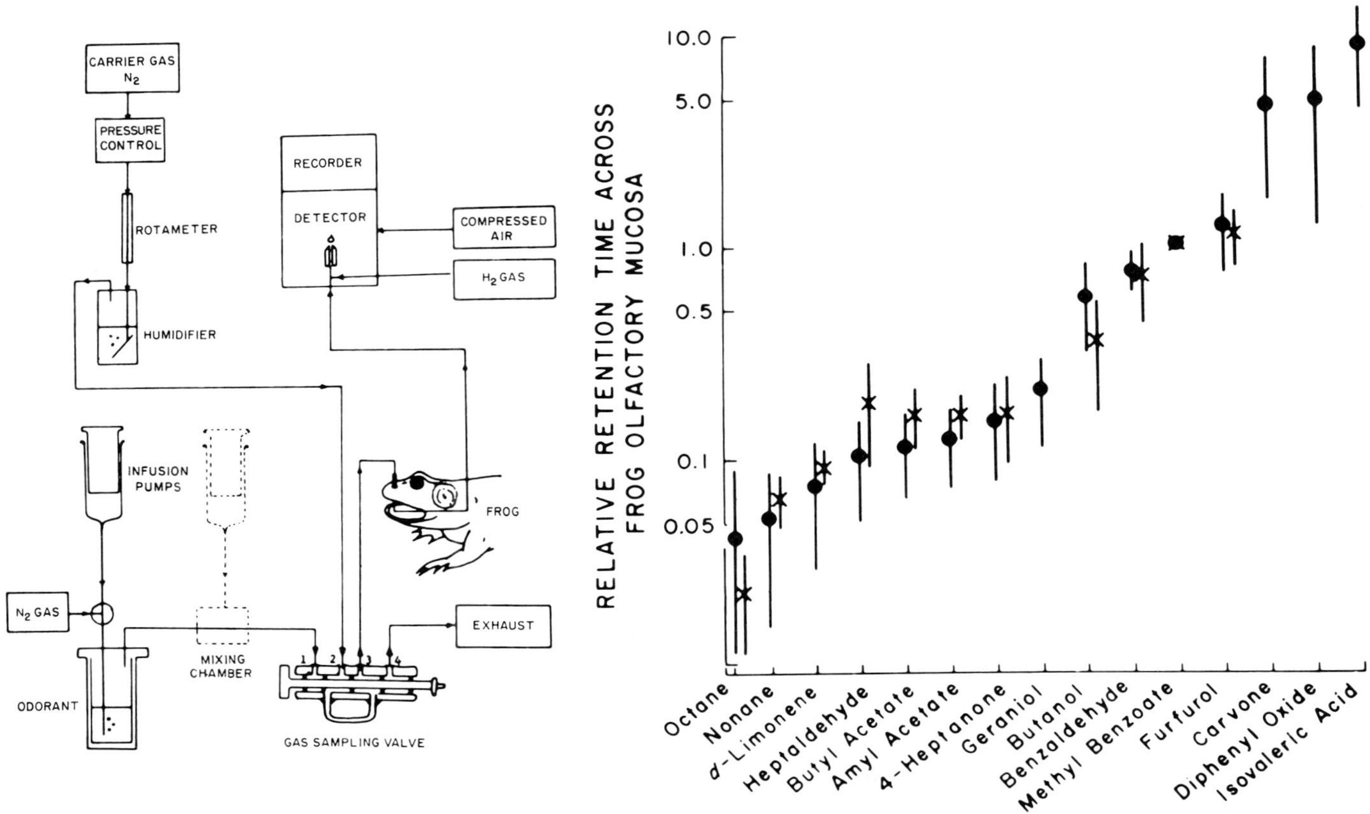

Fig. 2. The drawing to the left shows the diagrammatic representation of the apparatus used by Mozell and Jagodowicz (1973) to measure odorant retention times across the olfactory mucosa of the bullfrog (Mozell and Jagodowicz, 1974, by permission of the New York Academy of Science). The graph on the right side of the figure shows the relative retention times across the olfactory mucosa for the 15 odorants tested. ●, Highest partial pressure for each chemical at room temperature; X, partial pressure ≈ 0.56 mm Hg (Mozell and Jagodowicz, 1973, by permission of the American Association for the Advancement of Science, copyright, 1973.)

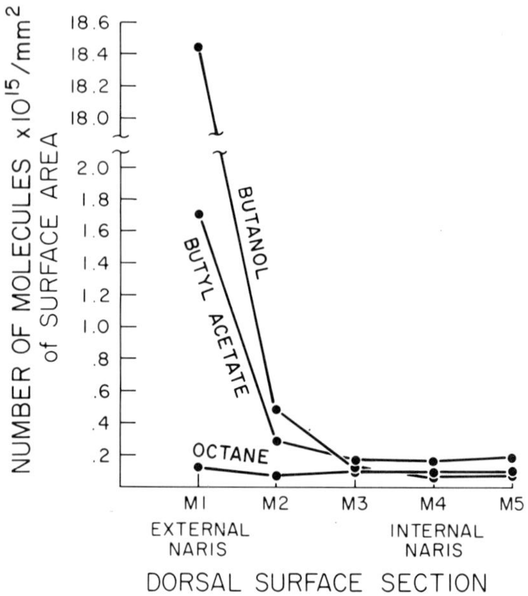

Fig. 3. The relative surface area concentration for each dorsal surface section of the frog mucosa after stimulation with an artifically produced sniff of butanol, octane, or butyl acetate. The stimulus parameters for all odorants were: partial pressure, 6.78 mm Hg; flow rate, 16 cm^2/min; volume 0.42 cm^3.

mechanisms of olfactory discrimination, steep molecular concentration gradients would prevent the molecules of some odorants from reaching otherwise appropriately tuned receptors that are located at more distal points along the mucosal flow path. Therefore, regardless of the mechanisms that ultimately prove to be the basis of both quality and intensity discrimination, this initial distribution of molecules along the olfactory receptor sheet is likely to be of central importance to our full understanding and appreciation of these olfactory mechanisms.

III. MUCOSA/AIR PARTITIONING OF ODORANTS

As described above, the distribution of molecules of different odorants along the olfactory receptor sheet appears to depend upon the partitioning of these molecules between the air phase and the mucosal phase. In chromatographic terms, this is a partitioning between a mobile phase and a stationary phase. The more this partition favors the mucosa the slower (as in a gas chromatograph) the molecules would travel and, within the time interval of a sniff, the steeper would be their concentration gradient from the external naris to the internal naris. The question is how to measure this mucosa/air partitioning, a measurement that may be critical to our understanding of olfactory events. Several standard techniques are available to measure a liquid/air partition coefficient. However, these techniques generally require thorough mixing or bubbling of the air phase with the liquid phase, a process that for the mucosa in its entirety is not applicable. Consequently, since much of the mucosa is aqueous, the water/air partition coefficient has often been used as a first approximation for that of the mucosa/air system.

To better define the mucosal sorptive events, Hornung *et al.* (1979) developed a technique, using ^{3}H-labeled odorants, to measure mucosa/air partition coefficients. For various periods of time, samples of frog mucosa or water were exposed in a closed environment to humidified air saturated with either [^{3}H]butanol, [^{3}H]octane or [^{3}H]butyl acetate. The radioactivity recovered from each of these samples, as determined by liquid scintillation counting, was used as a measure of odorant uptake. For butanol, a water/air partition coefficient determined by the isotope technique was in good agreement with that determined by a more traditional gas-flow technique (Amoore and Buttery, 1978). This would appear to support the validity of the radioisotope procedure for determining mucosal partition coefficients.

After equal exposure times, the water and mucosal samples sorbed equal amounts of butanol. This suggests that for butanol, uptake by olfactory mucosa is not significantly different from uptake by water, and therefore the water/air partition coefficient can be taken to represent the mucosa/air partition coefficient. However, for [^{3}H]octane and [^{3}H]butyl acetate, the mucosal uptake was about 1.5 times that of water, indicating that olfactory mucosa may have an increased ability over water itself to sorb these molecules (Table I). This implies that a multicompartment model with several partitions is required to describe the interaction of odorant molecules with the olfactory mucosa. More importantly, it was found that for the odorants so far tested the mucosal distribution does seem to be related to mucosa/air partitioning. The more the partition favors the mucosa (butanol>butyl acetate>octane), the steeper the gradient from the external to the internal naris (butanol>butyl acetate>octane).

TABLE I Partition Coefficients

Odorant	Water/air ($\times 10^3$)	Olfactory mucosa/air ($\times 10^3$)
n-Butanol	2.80	2.98
n-Butyl acetate	0.37	0.62
Octane	0.0012	0.0021

IV. ODORANT REMOVAL

One might expect that many of the physicochemical properties resulting in the wide range of mucosa/air partition coefficients would also influence the rate at which sorbed odorant molecules are removed from the mucosa. Presumably odorants with low mucosa/air partition coefficients would be readily removed by simply desorbing back into air. On the other hand, the removal of odorants with high mucosa/air partition coefficients might require augmentation by other mechanisms such as flow of mucus away from the olfactory mucosa or uptake by the circulatory system. The radiographic technique developed by Hornung and Mozell (1977b, 1980) was thus modified to quantify the rate at which these three possible mechanisms (desorption into air, mucus flow, and uptake by the circulatory system) are each able to remove odorant molecules from the olfactory sac.

Previous isotope studies (described above) showed that in accordance with its high mucosa/air partition coefficient, [^{3}H]butanol drawn into the olfactory sac by a negative pressure applied to a cannula inserted in the internal naris, established a steep concentration gradient along the olfactory mucosa. To now study the desorption of butanol molecules from the mucosa, the negative pressure was continued in order to draw a stream of nonradioactive room air through the olfactory sac. Butanol molecules that were desorbed into this air stream were collected in a toluene trap located between the internal naris cannula and the negative pressure source. The radioactivity recovered from the trap was readily available for liquid scintillation counting, but to determine the radioactivity remaining in the animal, the mucosa was frozen, cut, and solubilized prior to liquid scintillation counting. A variety of removal times and volumes were tested, but even after 480 cm^3 of room air were delivered during a 30-min period, only 22% of the total recovered butanol molecules were found in the trap. The remaining 78% were found in the mucosa.

Because a cannula was sealed into the internal naris for this air desorption study, the normal flow of mucus from the olfactory sac into the buccal cavity was prevented. Therefore, to quantify the rate at which the flow of mucus into the buccal cavity could itself remove butanol molecules from the olfactory mucosa, some animals were prepared without an internal naris cannula and the odorants were puffed into the sac through the external naris. This stimulus was not followed by a stream of nonradioactive room air. In these animals mucus flow was shown by liquid scintillation counting to remove some butanol molecules; however, after 30 min, 90% of the total recovered counts were found in the olfactory mucosa.

The uptake of butanol molecules from the olfactory mucosa into the circulatory system was determined from the radioactivity found in arterial blood samples. After 30 min, 7% of the initial butanol applied was recovered from the total blood compartment in those animals that had unobstructed mucus flow. Less than 3% of the counts were recovered from the circulatory system in animals with obstructed mucus flow. Thus, mucus flow appears to somehow influence uptake by the circulatory system.

It is noted that the removal by mucus flow and by desorption were studied independently of each other. Therefore, at this time we cannot predict with certainty how these two mechanisms might interact. However, the data would suggest, rather unexpectedly, that even with all the removal mechanisms operating together, a large percentage of the molecules in a sniff of butanol would remain in the mucosa for relatively long periods of time. One must reconcile this with Ottoson's (1956) electrophysiological observation that receptor adaptation to butanol, if it occurs at all, lasts a much shorter time than the 30-min observations of the present experiment. Apparently, although the butanol molecules appear to remain in the mucosa, they do not appear to affect the responsiveness of the receptors. As one possibility, the butanol molecules may be sequestered in a mucosal compartment such that they are unavailable for further receptor interaction.

This is not to imply that all odorants will have the same time course for removal as butanol. Indeed the relative effectiveness of the several removal mechanisms are expected to vary depending on the mucosa/air partition coefficient of the particular odorant involved. This is dramatically illustrated when comparing the removal of [^{3}H]octane (low coefficient) to [^{3}H]butanol (high coefficient). Following a flow of nonradioactive room air, only 0.2% of a [^{3}H]octane application remained in the mucosa for as long as 3 minutes (48cm^3 of room air), whereas 78% of a [^{3}H]butanol application was recovered from the mucosa after 30 min of room air (480 cm^3).

As with butanol, octane removal by the circulatory system was estimated from the radioactivity found in arterial blood samples. About 4% of the sorbed octane stimulus was found in the blood, showing that odorants with relatively low mucosal sorption, like octane, can be removed via the circulatory system. Furthermore, the number of octane molecules in the bloodstream relative to the number sorbed by the mucosa was observed to be in the same proportion as that seen for butanol. Is it possible then, that the circulation removes the molecules of different odorants from the mucosa simply in accordance with their sorbed concentrations?

Investigators interested in long-term effects of exposure to airborne chemicals have, in a sense, been studying the access and egress of possible olfactory stimuli. For example, DiVincenzo and Hamilton (1979) studied uptake by the respiratory system of ^{14}C-labeled butanol in male Beagle dogs. Even after a 6-hr exposure to 50 ppm of *n*-butanol, dogs exhaled in a single expiration relatively little of the total quantity of butanol absorbed during a single inspiration. In addition, of the total amount of *n*-butanol presented to the animal, only 45% was ultimately found in the postexposure expired air even after several hours of collection.

Undoubtedly this tenacious sorption of butanol represents uptake by the mucosa of the entire respiratory tree, including the lungs. The butanol removal data from the frog olfactory mucosa showed a similar tenacious sorption of butanol. Perhaps both sets of data reflect the same basic physicochemical mechanism, i.e., a mucosa/air partition coefficient that favors the mucosa to a degree similar to that for water.

Apparently, certain chemicals once sorbed by either the respiratory or olfactory mucosa might remain somewhere in the mucosa for long periods of time. One of these possible places is the epithelial cells themselves. It is tempting to speculate that the accumulation of these sorbed molecules leads to the continual turnover of olfactory receptor cells as reported by Graziadei (1973).

V. LIMITATION OF THE RADIOISOTOPE PROCEDURE

Following the logic of most tritium studies, Hornung and Mozell (1977a) assumed that the mucosal location of the tritium indexed the location of the original labeled molecule. This assumption might be questioned for butanol when one considers how the butanol was labeled.

The tritiation of *n*-butanol was accomplished (at the New England Nuclear Corporation, Boston, MA) by combining 5μl of 3-buten-1-ol, 1 ml of *n*-butanol (very dry), and 25 mg of platinum black catalyst. Ten curies of tritium were then added, and the reaction was allowed to stir overnight at room temperature. A double distillation was then performed, *in vacuo*, over a molecular sieve. The radiochemical purity of the tritiated *n*-butanol was determined to be in excess of 99%.

This radiochemical synthesis should result in addition of at least one tritium atom at the C3 or C4 position as butenol is converted to butanol. This procedure had the advantage of producing a tritiated butanol having not only a high radiochemical purity, but also having a high specific activity. This latter consideration was important as we needed to present the butanol in the vapor phase, and the number of molecules for detection was therefore relatively small. The disadvantage of this labeling technique is that the hydrogen attached to the oxygen of the butanol may have exchanged with a tritium atom (Malcolme-Lawes, 1979). If such labeling occurred, the tritium would be lost from the butanol as easily as it went on. Thus, tritium could be passed from the butanol to some other molecule. In the case of the olfactory mucosa, this other molecule would most likely be water. Such a series of events would not likely have a major impact on the studies mapping the initial mucosal distribution of butanol, because whether part of the butanol molecule or not, the tritium still indexes where the butanol had been sorbed to the mucosa. If these aspects were to have any impact at all, they would likely do so by affecting those studies that traced the removal of butanol from the mucosa, since the remaining radioactivity may not indicate the remaining butanol. However, although it is not known exactly how much tritium–hydrogen exchange occurred in the synthesis of *n*-butanol, preliminary data suggests that more of the tritium is located on the carbon backbone than on the hydroxyl group.

Precise quantification in the removal studies would seem to require ^{14}C-labeled butanol, an option that has not heretofore been possible. As a result, it is at least possible that the butanol data may be confounded by some tritium exchange. Use of a ^{14}C-labeled compound should resolve the question of what molecule was followed in our butanol studies. It is emphasized that the tritiated octane and butyl acetate studies are less susceptible to the effects of tritium exchange since these molecules contain no hydrogen atoms as labile as that found on the hydroxyl group of butanol.

VI. FURTHER CONSIDERATIONS

The physicochemical properties affecting the mucosa/air partitioning of odorants could play several possible roles in olfaction. For example, as described above, the partitioning of odorants may be a basis for odor discrimination by establishing characteristic odorant distribution patterns across the mucosa. Although considerable evidence supports the existence of spatial patterns (Moulton, 1976), few behavioral studies either support or refute these patterns as a mechanism for discrimination. Bennett (1972) was able to observe the behavioral affects of changing the flow path in one of the nasal passageways by taking advantage of the fact that the rat has an incomplete nasal septum. By inserting artificial nasal septa and unilateral nasal plugs, he altered the flow path of odorants across the olfactory mucosa. This was observed to change the rat's ability to discriminate odorants. Thus, Bennett (1972) suggests that animals use spatial mucosal patterns in making olfactory discriminations.

Another physicochemical determinant of the accessibility of odorant molecules to the receptors has been described by Laffort *et al.* (1974). They suggest that for an odorant molecule to reach the olfactory receptors it is necessary that the stimulus be somewhat soluble in the mucus, but not so "strongly captured" as to prevent receptor interaction. According to Laffort and his co-workers, these two partially opposing properties lead to an effective solubility for odorant molecules. This solubility effect (Laffort *et al.*, 1974) and the chromatographic effect (Mozell, 1970) are similar in at least one regard: They both strongly emphasize mucosal sorptive events as basic to an understanding of the accessibility of odorant molecules to the receptors.

The possibility of a mucosal induced chemical change in the odorant molecules complicates the concept of odorant accessibility (Taylor, 1974). There was a hint of such a change in a finding of Hornung and Mozell (1977c). In the three animals studied, radioactive octane that was passed through the olfactory sac showed a 15-fold increase in water solubility compared with radioactive octane that was not passed through the sac. This could suggest that in the interaction between the odorant molecule and the olfactory mucosa, the odorant was somehow transformed. On the other hand, no such transformation was seen with either [^{3}H]butanol or [^{3}H]butyl acetate. However, the possibility of such a transformation for at least some odorant molecules raises some interesting questions. For instance, can these transformed molecules stimulate a different group of receptors than do the parent molecules? Of perhaps, do the transformed molecules contribute to the detection or indentification of the parent molecule?

VII. CONCLUSION

Getchell and Getchell (1977), in interpreting the slow monophasic voltage response (EOG's), which can be recorded from the surface of the olfactory epithelium of the frog, suggest a multicompartment model for access of odorant molecules to the olfactory receptors. This general theme has also been emphasized by the solubility effect described by Laffort *et al.* (1974) and by the partition coefficient data described by Hornung and Mozell. This multicompartment model for the accessibility of odorant molecules to the receptors from the air phase depends upon the physicochemical principles determining the partitioning between the various compartments. Therefore, a full understanding and appreciation of the olfactory system will not be achieved until these physicochemical principles are fully described for the odor analyzing process.

ACKNOWLEDGMENT

Supported by NIH grant no. NS03904.

REFERENCES

Adrian, E. D. (1950a). Sensory discrimination with some recent evidence from the olfactory organ. *Br. Med. Bull.* 6, 330–333.

Adrian, E. D. (1950b). The electrical activity of the mammalian olfactory bulb. *Electroencephalogr. Clin. Neurophysiol.* **2**, 377–388.

Amoore, J. E., and Buttery, R. G. (1978). Partion coefficients and comparative olfactometry. *Chem. Senses Flavor* **3**, 57–71.

Bennett, M. H. (1972). Effects of asymmetrical mucosal stimulation upon two-odor discrimination in the rat. *Physiol. Behav.* **9**, 301–306.

DiVincenzo, G. D., and Hamilton, M. L. (1979). Fate of *n*-butanol in rats after oral administration and its uptake by dogs after inhalation or skin application. *Toxicol. Appl. Pharmacol.* **48**, 317–325.

Getchell, T. V., and Getchell, M. L. (1977). Early events in vertebrate olfaction. *Chem. Senses Flavor* **2**, 313–326.

Graziadei, P. P. C. (1973). Cell dynamics in the olfactory mucosa. *Tissue Cell* **5**, 113–131.

Hornung, D. E., and Mozell, M. M. (1977a). Factors influencing the differential sorption of odorant molecules across the olfactory mucosa. *J. Gen. Physiol.* **69**, 345–361.

Hornung, D. E., and Mozell, M. M. (1977b). Odorant removal from the frog olfactory mucosa. *Brain Res.* **128**, 158–163.

Hornung, D. E., and Mozell, M. M. (1977c). Preliminary data suggesting alteration of odorant molecules by interaction with receptors. *In* "Olfaction and Taste" (J. LeMagnen and P. MacLeod, eds.), Vol. 6, p. 63. Information Retrieval, London.

Hornung, D. E., and Mozell, M. M. (1980). Tritiated odorants to monitor retention in the olfactory and vomernasal organs. *Brain Res.* **181**, 488–492.

Hornung, D. E., Lansing, R. D., and Mozell, M. M. (1975). Distribution of butanol molecules along bullfrog olfactory mucosa. *Nature (London)* **254**, 617–618.

Hornung, D. E., Serio, J., and Mozell, M. M. (1979). Air/olfactory mucosa partition of odorants. *Neurosci. Abstr.* **5**, 128.

Laffort, P., Patte, F., and Etcheto, M. (1974). Olfactory coding on the basis of physiochemical properties. *Ann. N.Y. Acad. Sci.* **237**, 193–208.

Malcolme-Lawes, D. J. (1979). "Introduction to Radiochemistry." Wiley, New York.

Moncrieff, R. W. (1955). The sorptive properties of the olfactory membrane. *J. Physiol. (London)* **130**, 543–558.

Moncrieff, R. W. (1967). "The Chemical Senses." CRC Press, Cleveland, Ohio.

Moulton, D. G. (1976). Spatial patterning of response to odors in the peripheral olfactory system. *Physiol. Rev.* **56**, 578–593.

Mozell, M. M. (1966). The spatiotemporal analysis of odorants at the level of the olfactory receptor sheet. *J. Gen. Physiol.* **50**, 25–41.

Mozell, M. M. (1970). Evidence for a chromatographic model of olfaction. *J. Gen. Physiol.* **56**, 46–63.

Mozell, M. M. (1976). Processing of olfactory stimuli at peripheral levels. *In* "Chemical Signals in Vertebrates" (D. Müller-Schwarze and M. M. Mozell, eds.), pp. 465–482. Plenum, New York.

Mozell, M. M., and Jagodowicz, M. (1973). Chromatographic separation of odorants by the nose: Retention times measured across *in vivo* olfactory mucosa. *Science* **181**, 1247–1249.

Mozell, M. M., and Jagodowicz, M. (1974). Mechanisms underlying the analysis of odorant quality at the level of the olfactory mucosa, I. Spatiotemporal sorption patterns. *Ann. N.Y. Acad. Sci.* **237**, 76–90.

Ottoson, D. (1956). Analysis of the electrical activity of the olfactory epithelium. *Acta Physiol. Scand., Suppl.* No. 122, 1–83.

Taylor, M. (1974). The origin and function of nasal mucus. *Laryngoscope* **74**, 612–636.

Tucker, D. (1963). Physical variables in the olfactory stimulation process. *J. Gen. Physiol.* **46**, 453–489.

3

Role of Cilia in Olfactory Recognition

LINDA D. RHEIN AND ROBERT H. CAGAN

I. INTRODUCTION

The molecular basis of the initial steps of olfactory sensation has been an enigma to scientists for decades. Many hypotheses of the underlying mechanisms and specificity of the initial interaction have appeared, which are based on electrophysiological recordings of responses to odors or on psychophysical measurements (Beets, 1971; Davies, 1971; Engen, 1971; Ottoson, 1971). These studies have concentrated on the relationship of the structure of the stimulus to its odor quality and intensity in human studies and to its evoked neural response in animals.

Biochemical studies of the mechanisms underlying olfaction have received relatively little attention. It is widely believed that the initial step in olfaction is adsorption of stimulus molecules onto specific receptor sites through a reversible binding interaction (Beets, 1971; Ottoson, 1971). This chapter

Biochemistry of Taste and Olfaction
ISBN 0-12-154450-8

focuses on the role of cilia and on recent biochemical advances in our understanding of the localization of the initial recognition event in olfaction.

Receptor sites for odorants are hypothesized to be located on cell membranes (Lettvin and Gesteland, 1965) at the surface of olfactory receptor neurons. Cilia occur on the distal region of the receptor neurons and project into the overlying mucus layer. It has been postulated for nearly 50 years that the membranes are the loci of odorant receptor sites (Parker, 1922; Hopkins, 1926; Ottoson, 1971). This hypothesis appears reasonable because the cilia are presumably the initial point of odorant contact with the neuron. Earlier experimental attempts to test the hypothesis of cilia involvement in olfactory recognition led to controversial reports. These are described in a section below, along with results of more recent biochemical studies.

II. CILIA IN SENSORY ORGANS

Cilia or modified cilia occur in most sense organs (Barber, 1974). Three exceptions are the eyes of many invertebrates (Eakin, 1972), the taste receptors of vertebrates (Murray, 1973), and the vomeronasal organ (Ciges *et al.*, 1977; Graziadei, 1971). Microvillous projections are found on receptor cell surfaces in each of these organs. It is possible that the microvilli on cells in the taste and vomeronasal organs may carry out similar chemorecognition functions as do cilia on other receptor cells. The recognition function of microvilli has already been demonstrated for visual rhabdomeric receptors (Eakin, 1972; Anderson *et al.*, 1978).

The membrane covering the structural elements at the interface between a receptor cell and its environment is the likely candidate to contain the receptor molecules. This is well-established for visual receptors, which are the best-studied of the sensory receptors from a fundamental viewpoint. The photoreceptive molecules are located in disks that develop by complex invaginations of the membrane of a single cilium (Autrum, 1979; Crescitelli, 1972; Dunn, 1973). These comprise the distal regions of the rod receptor cells. The chemical composition of photoreceptive membranes in the rod outer segments consists of about 50% lipid and 50% protein (Daemen, 1973). One protein, rhodopsin, comprises about 80–85% of the total protein. Upon illumination, the 11-*cis*-retinaldehyde chromophore of rhodopsin absorbs photons, after which it isomerizes to the *trans* isomer. This event precedes a conformational change of the protein, producing several "photointermediates" in succession. Absorption of light by one or more rhodopsin molecules eventually results in a sharp decrease in the Na^+ permeability of the plasma membrane of the rod (Tomita, 1970). The mechanism for this is not understood, but postulates of Ca^{2+} being released and inhibiting Na^+

channels, and of enzymatic degradation of cGMP, resulting in dephosphorylation of Na^+ channels, are under investigation (Hubbell and Bownds, 1979). The decrease in Na^+ permeability leads to a decreased dark current of Na^+ flowing from the inner to the outer segment of the rod cell (Hagins, 1972), which apparently stimulates chemical changes at the synapses between the rod cell and the bipolar cells.

Specific roles of membrane lipids in the visual process are not entirely clear. The overwhelming majority of lipids in rod outer segment membranes are phospholipids (80–90% of the lipid). Their most striking feature is a high content of long-chain highly unsaturated fatty acids (Daemen, 1973). Approximately half of the fatty acids in rod outer segment membranes are polyunsaturated, which is unique among biological membranes. The lipid composition of photoreceptive membranes suggests characteristics of a fluid membrane combined with those of a highly stable organelle. The unsaturation, fluidity, and consequent low viscosity would explain the observed rotational and translational mobility of rhodopsin in the membrane (Cone, 1972; Poo and Cone, 1973). If this freedom of movement is essential for the mechanism of action of rhodopsin, the membrane fluidity, and hence the lipid composition, would be of considerable importance.

III. MORPHOLOGY OF OLFACTORY CILIA

Motile cilia are recognized in ultrastructural cross section by their typical "9 + 2" arrangement of microtubule fibers, which are comprised of nine outer doublet fibers encircling a central pair, all of which are surrounded by a membrane (Fig. 1) (Warner, 1974). Each outer doublet also contains a pair of dynein arms (Gibbons and Grimstone, 1960). These arms contain a Mg^{2+}-stimulated ATPase; the activity of this enzyme is essential for motility of the cilium (Stephens, 1974). Cilia are generally between 5 and 100 μm in length (Warner, 1974).

Cilia on olfactory receptor cells share many common features with other cilia, but some features of olfactory cilia are unique. They are sometimes much longer than other cilia (Andres, 1969; Reese, 1965). In the report of Reese (1965), the longest frog olfactory cilia extended to a length of up to 200 μm. The greater length of olfactory cilia in some species is accompanied by a loss of microtubule fibers. For example, in many animal species, the familiar "9 + 2" complement of fibers occurs in only a short, proximal region, whereas in the long, distal region, the nine outer doublets are reduced to a variable number of singlet fibers (Frisch, 1967; Kerjaschki, 1976; Kratzing, 1978; Loo, 1977; Reese, 1965; Thornhill, 1967). As few as two singlet fibers are present in the distal regions of olfactory cilia in some species (Kerjaschki,

1976; Kratzing, 1978; Loo, 1977). As in other cells with outfoldings of the surface membrane, the surface area is greatly increased, which could be important in olfactory reception.

The presence of dynein arms (Barber, 1974; Gibbons and Grimstone, 1960) throughout the olfactory cilia axoneme has been questioned (Bannister, 1965;Kerjaschki, 1976; Reese, 1965; Thornhill, 1967). Because the dynein arms contain Mg^{2+}–ATPase whose activity is essential for active beating of cilia (Stephens, 1974), their absence from the distal portion would affect their motility, and their complete absence from olfactory cilia would imply that the cilia are immotile. Dynein arms were reported to be absent from the olfactory cilia of minnow and stickleback (Bannister, 1965). On the other hand, lamprey (Thornhill, 1967) and frog (Reese, 1965) cilia contain structures resembling dynein arms, but only in their proximal regions. [Although Reese (1965) did not discuss them, arm-like structures were present in the micrographs.] Kerjaschki (1976) recently reported that dynein arms were absent from the distal segments of mouse olfactory cilia. In contrast with the above reports, dynein arms were repeatedly found associated with the outer doublet fibers in cilia in the sensory regions of the trout olfactory rosette (Figs. 1 and 2).

Active movement of olfactory cilia has been a matter of controversy. Hopkins (1926), Bronshtein (1964), and Vinnikov (1974) had earlier reported motility of olfactory cilia. Blank *et al.* (1978) recently reported that olfactory cilia in the frog are motile and that their motility is synchronized by stimulating the mucosa with an odorant. Døving *et al.* (1977) reported that the method of water transport through the olfactory organ in various fishes, including trout, is via a ciliary mechanism. Our observations of dynein arms in the cilia in the olfactory receptor regions of rainbow trout olfactory rosettes support these findings and the hypothesis that olfactory cilia in the trout are motile.

Some unusual specializations of the membranes of olfactory cilia are observed. "Ovoid dilations" were reported on distal regions of mouse olfactory cilia (Frisch, 1967), "spindle-shaped swellings" on the distal portions of bat and rabbit olfactory cilia (Yamamoto, 1976), "vesicles" along the shafts of distal portions of frog olfactory cilia (Reese, 1965), an "electron opaque halo" around portions of the distal segments of cat olfactory cilia (Andres, 1969), and "tubercles" on the surfaces of cilia from sea trout (Bertmar, 1972, 1973). Our studies (Rhein *et al.*, 1981) revealed that the membrane surrounding cilia in the rainbow trout sensory olfactory epithelium often expands to enclose smooth-walled, clear vesicles or channels (Fig. 2), but their function is not yet understood.

High densities of membrane particles have been reported on the freeze-

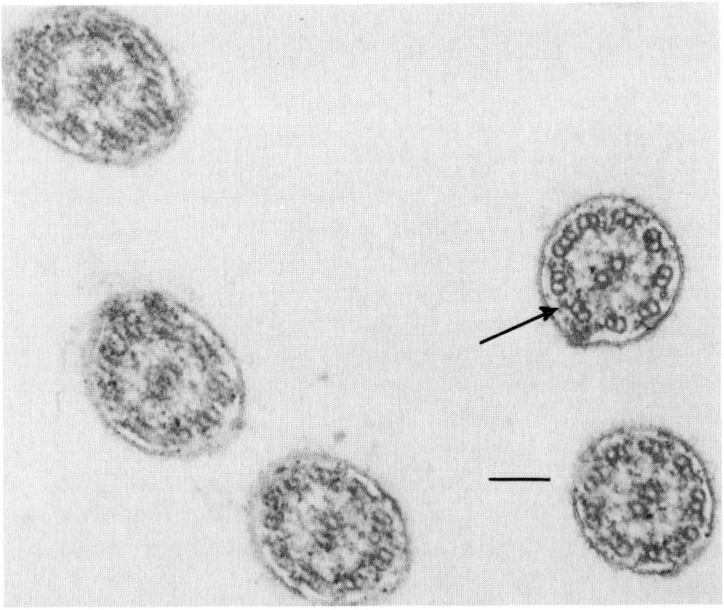

Fig. 1. Cilia *in situ* in the rainbow trout olfactory rosette taken by transmission electron microscopy. The morphological features are typical for cilia showing the "9 + 2" arrangement of microtubule fibers surrounded by a unit membrane. Note the presence of dynein arms (arrow) on the outer doublet fibers. 81,900 ×; bar = 0.1 μm.

fractured surfaces of sensory cilia in bovine (Menco *et al.*, 1976) and mouse (Kerjaschki and Hörandner, 1976) olfactory epithelia. Few particles were found on neighboring (presumably nonolfactory, respiratory) cilia. The particles were speculated to be fragments of proteins associated with the lipid membrane and might be odorant receptors or involved in transducing odorant information (Menco *et al.*, 1976).

In addition to the postulated role of cilia membranes in the initial interaction with odorants, Atema (1973, 1975) proposed a role for cilia microtubules in transducing odorant information. Similar to the postulate for mechanoreceptors (Moran and Varela, 1971), the microtubules in cilia on olfactory receptors were postulated to act as conduction fibers by propagating a wave of conformational change through the protein (tubulin) subunits. Preliminary evidence (Atema *et al.*, 1978) suggests that exposure of lobster olfactory sensillae to colchicine or podophyllotoxin, both of which destroy microtubules, cause a decline in the electrophysiological response. The results encourage further investigation of this hypothesis, studies which could benefit from use of isolated cilia (see below).

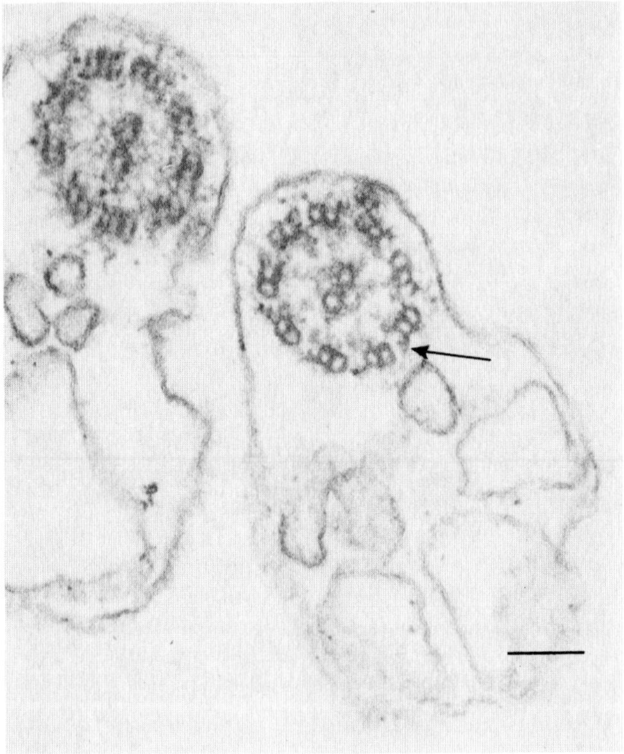

Fig. 2. Transmission electron micrograph of trout cilia *in situ* in the olfactory rosette. The micrograph shows the expanded membrane, which is frequently observed on cilia in the trout olfactory rosette. Arrow depicts a dynein arm. 105,900 ×; bar = 0.1 μm. (Taken in part from Rhein and Cagan, 1980.)

IV. EXPERIMENTAL BASIS FOR ROLE OF CILIA IN OLFACTION

The experimental evidence for direct involvement of olfactory cilia in the initial events of olfaction has been conflicting. In a widely cited preliminary report, Tucker (1967) indicated that removal of cilia from the turtle olfactory epithelium with detergents resulted in a preparation retaining activity towards chemical stimuli. Both electroolfactogram (EOG) and neural responses were used as measures of activity.

Recent experimental results support a direct role of cilia in olfactory sensation. Bronshtein and Minor (1977) showed that brief exposure of the olfactory epithelium of the frog to 0.1–0.15% solutions of Triton X-100 resulted in destruction of cilia and decline of the EOG response to chemical stimuli, e.g., butyl acetate. Electron microscopy revealed that the cilia had been removed. After 2–3 days, the olfactory cilia had regenerated; during the regeneration period, the EOG response was gradually restored. A recent preliminary report (Blank *et al.*, 1978) indicates that application of odorant compounds to frog olfactory epithelium increases ciliary activity and synchronizes their movements. In all of these studies, electrophysiological measures of responsiveness were used.

Earlier experiments by Shibuya (1964) were not conclusive regarding the role of cilia in olfaction. Based on theoretical considerations, Bostock (1974) postulated that cilia are not essential for impulse initiation. He modeled the diffusion characteristics of odorants in relation to the electroolfactogram. Based on epithelial response latencies to chemical stimulation, he concluded that the odor diffuses to the region of the olfactory knob before it interacts with receptor sites. No experimental evidence was presented. A diffusion model is developed by DeSimone (this volume, Chapter 11), which postulates that, if the cilia are involved in the initial interaction, then the proximal region of the cilia should be the active portion. Recent development of an experimental biochemical model (below) to study odorant interactions has allowed a test of the hypothesis that cilia are involved in odorant recognition.

V. ODORANT INTERACTIONS STUDIED BIOCHEMICALLY

Direct verification of the role of cilia in odorant recognition could be accomplished with an isolated, functionally active cilia preparation. Such a preparation would allow detailed examination of the specificity and mechanisms of odorant–cilia interactions. In the past, this approach has been hindered for several reasons. First, a method to assess the functional activity of the isolated cilia with respect to odorant recognition has not been available. A reliable biochemical assay is required to quantify formation of the complex between an odorant compound and an olfactory receptor preparation. Second, a convenient experimental animal with known odorant responses is needed as a biochemical model of vertebrate olfaction. Until recently, neither an animal model nor data to demonstrate the biochemical specificity were available. Third, an isolation procedure is required to selectively remove cilia without destroying the odorant receptor sites. Fourth, criteria to establish the presence of cilia and their degree of enrichment in the isolated preparation are essential. The initial work in our laboratory on this problem dealt with the first two issues and is described immediately below; the third and fourth issues were studied more recently and are described in the following section.

It is well established that salmonid fish have a functional olfactory system and that olfaction serves an important role in aiding the guidance of the final stages of homestream migration (Hasler, 1960; Hasler *et al.*, 1978). Their olfactory system is sensitive to many amino acids as olfactory stimuli, based on electrophysiological studies (Hara, 1973; Sutterlin and Sutterlin, 1971). Morphological studies of the ultrastructure of the olfactory rosettes of rainbow trout (*Salmo gairdneri*) confirm the presence of cilia on the distal regions of cells in the sensory olfactory epithelium presumed to be receptor cells (Rhein *et al.*, 1981). These findings provide a sound morphological and physiological basis for biochemical studies of olfaction in this species. The rainbow trout was initially examined (Zeiger and Cagan, 1975) as a possible biochemical model for vertebrate olfaction. Studies with a sedimentable preparation (Cagan and Zeiger, 1978) enabled examination of odorant-receptor interactions in an attempt to establish biochemically the specificity of binding.

Using the rainbow trout, Cagan and Zeiger (1978) developed a biochemical approach to study odorant–receptor interactions by measuring binding of ^{3}H-labeled odorants to a receptor-enriched preparation. The assay method was based on that used to quantify binding of taste stimuli to a similar type of preparation from catfish taste tissue (Krueger and Cagan, 1976; see also this volume, Chapter 10) and is based on the assay used extensively to measure receptor–ligand interactions in insulin-responsive tissues (Cuatrecasas, 1971a,b). The sedimentable preparation (fraction P2) obtained from a homogenate of the olfactory rosettes is enriched in odorant-binding activity (Table I).

Binding was established as a relevant measure of the initial step in olfactory recognition in trout. The extent of binding of a series of ten odorant amino acids with fraction P2 (Cagan and Zeiger, 1978) was compared with the stimulatory effectiveness of the same ten amino acids measured electrophysiologically from the surface of the trout olfactory bulb (Hara, 1973). The odorant amino acids examined represented stimuli of high, medium, and low electrophysiological effectiveness. The extent of binding correlates favorably with the reported relative stimulatory effectiveness of the odorants (Fig. 3) (correlation coefficient +0.8, $p<0.01$). Similar studies with stereoisomers revealed that L-amino acids, which are somewhat more effective electrophysiological stimuli, bind to a greater extent than do the corresponding D-isomers. Both leucine and alanine were studied.

Binding properties of several amino acid odorants with fraction P2 were examined in more detail by determining the saturation curves (Fig. 4). Scatchard (1949) analyses of binding curves revealed two types of binding sites for most of the amino acids examined (Table I). The values of the apparent dissociation constants for the higher affinity sites are $\sim 10^{-6}$ M, and those for the lower affinity sites are $\sim 10^{-5}$ M. Both L-serine and D-alanine

TABLE I Binding Parameters of Odorant Amino Acids with Fraction P2 from Trout Olfactory Rosettes[a]

	High affinity		Low affinity	
Amino acid	K_D (μM)	B_{max} (pmole/mg)	K_D (μM)	B_{max} (pmole/mg)
L-Threonine	3.7	590	17	1000
L-Serine	4.6	550	—	—
L-Alanine	5.6	440	37	800
L-Histidine	3.9	390	56	1900
L-Lysine	4.2	200	31	600
L-Valine	4.8	140	31	400
β-Alanine	3.9	80	140	800
D-Alanine	—	—	47	800

[a] The values are calculated from Scatchard analyses of the binding data. Taken from Cagan and Zeiger (1978).

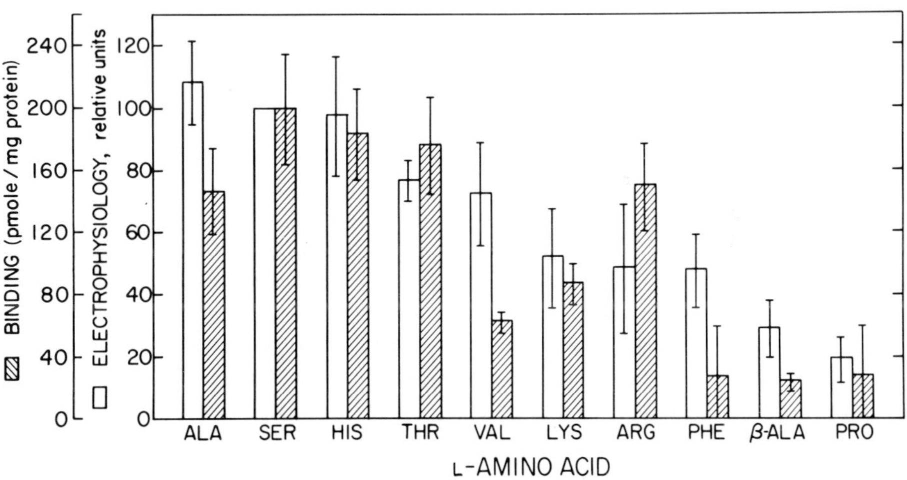

Fig. 3. Comparison of binding and electrophysiological responsiveness of olfactory stimulus amino acids in the rainbow trout. Binding of the ^{3}H-labeled amino acids to fraction P2 from rainbow trout (*Salmo gairdneri*) olfactory tissue was measured with the ligand concentration at 6 μM. The binding data represent four experiments, in each of which binding of every amino acid was measured. The values were corrected for nonspecific binding. The electrophysiological data are from the report by Hara (1973); the relative effectiveness is compared with serine as 100. All data are expressed as mean ± SD. (Figure taken from Cagan and Zeiger, 1978.)

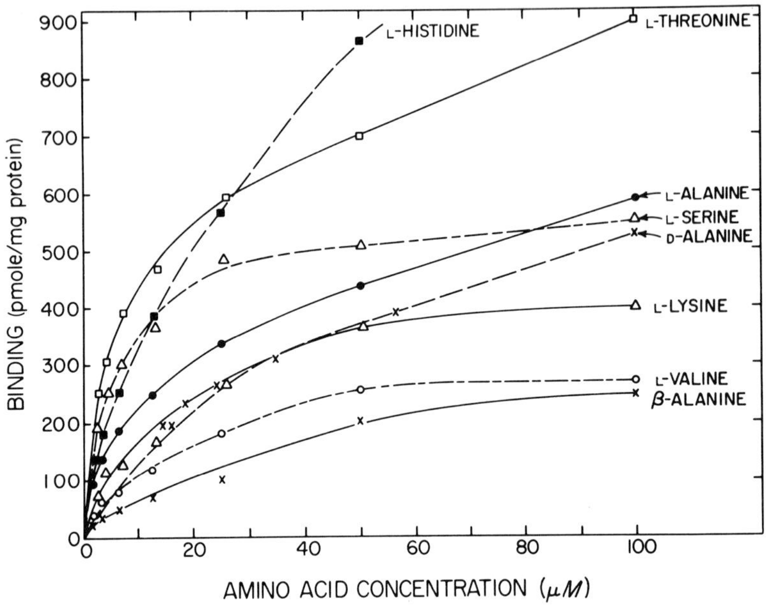

Fig. 4. Binding curves for amino acid olfactory stimuli to fraction P2 from the rainbow trout. The data for each amino acid are the mean values from two separate preparations. The binding value for L-histidine at 0.1 m*M*, 1217 pmole/mg, is omitted from the curve to save space. (Figure taken from Cagan and Zeiger, 1978.)

showed evidence for only a single affinity site with K_D around 10^{-6} *M* for L-serine and 10^{-5} *M* for D-alanine. Results of the Scatchard analyses, in addition to results from a limited series of binding competition studies, led the authors to postulate that a multiplicity of types of odorant binding sites are present in trout olfactory epithelium. They possess different but not necessarily exclusive specificities for various odorants. Among the sites postulated were site TSA, which binds L-threonine, L-serine, and L-alanine, site L, which binds L-lysine, and site A_B, which binds β-alanine. Tentative sites suggested from the results were site V, which binds L-valine; site H, which binds L-histidine; and site A_D which binds D-alanine. These parameters of binding have now been used to establish the functional olfactory activity of the isolated cilia preparation.

VI. ISOLATION AND BIOCHEMICAL CHARACTERIZATION OF OLFACTORY CILIA

Methods to isolate cilia from various cells and tissues have been described. We found (Rhein and Cagan, 1979, 1980) that the method described by

Watson and Hopkins (1962) to isolate *Tetrahymena* cilia and later used by Linck (1973) to obtain mollusk gill cilia, when suitably modified, results in rapid removal of cilia from olfactory rosettes in a reasonable yield. This procedure was therefore selected to obtain trout olfactory cilia (Rhein and Cagan, 1980). A batch of olfactory rosettes (e.g., 100–400) are gently stirred in deciliation medium (10% ethanol, 100 m*M* NaCl, 2 m*M* EDTA, 10 m*M* $CaCl_2$, and 30 m*M* Tris–HCl, pH 8.0) for 20 min on ice. Deciliated rosettes are removed by low-speed centrifugation. The olfactory cilia are then pelleted by centrifugation (10,000 *g*, 10 min) and washed prior to study. The resulting cilia preparation contains about 30 μg of protein per fish (Table II), which represents approximately 0.5% of the whole rosette protein.

No single criterion definitively establishes cilia in an isolated preparation. Therefore several biochemical criteria were selected. In addition, the morphological presence of cilia was demonstrated by electron microscopy, which revealed cilia in the final pellet (Fig. 5). Quantification by this method is unfeasible, and indeed Menco (1977) reported lack of success with attempts to quantify cilia in a preparation derived from a homogenate of olfactory tissue. In our studies, two chemical markers and one enzymatic marker for cilia were used. These biochemical criteria enable more definitive quantification and provide strong evidence for the presence and enrichment of cilia in the preparation.

The markers examined are associated with cilia in other organisms. Axonemal microtubules of cilia contain a well-characterized protein called *tubulin*. Its subunits, α-tubulin and β-tubulin, are separable by polyacrylamide gel electrophoresis in sodium dodecyl sulfate (Kirschner *et al.*, 1975). The isolated olfactory cilia were solubilized in the detergent and electrophoresed (Laemmli, 1970). Although identification by migration dis-

TABLE II Yield of Cilia from Trout Olfactory Rosettes[a]

Sample	Protein per fish (mg)	Protein (mg/gm wet weight)	Recovery (%)
Whole rosettes	6.60	69.5	100
Deciliated rosettes	5.24	65.6	94.4
Supernatant	0.703	8.34	12.0
Cilia	0.0297	0.350	0.5

[a] Homogenates of whole rosettes and of deciliated rosettes were prepared in separate experiments as described (Rhein and Cagan, 1980). Cilia and the resulting supernatant fraction values are from the same preparations used to determine the deciliated rosette values. Each value is the mean of results from two preparations, in each of which three to four replicate determinations were made. The wet weight is that of the whole rosettes from which the sample was obtained.

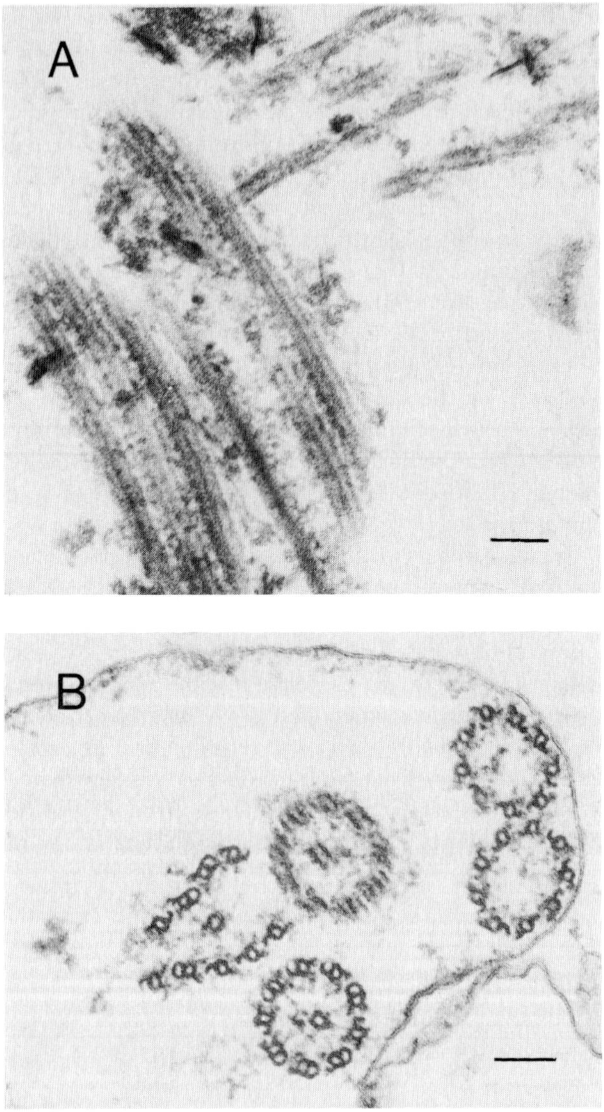

Fig. 5. Transmission electron micrographs of ciliary material in the preparation isolated from trout olfactory rosettes. (A) 77,200 ×; (B) 67,200 ×; bar = 0.1 μm. (Taken in part from Rhein and Cagan, 1980.)

tance upon gel electrophoresis is not conclusive, the results (Rhein and Cagan, 1980) reveal the presence of two major bands in the cilia preparation that co-migrate with authentic α- and β-tubulin. The presence of a significant amount of tubulin suggests substantial enrichment of cilia in the isolated preparation.

Microtubules contain 2 moles of bound guanine nucleotides (present as GTP) per mole of tubulin dimer (Synder and McIntosh, 1976). Associated with the outer doublets of the axonemal microtubule fibers of cilia and flagella are similar amounts of guanine derivatives but no adenine derivatives (Shelanski and Taylor, 1968; Stevens *et al.*, 1967). Substantial enrichment of guanine nucleotides occurs in the isolated cilia preparation; the increase is ten-fold compared with deciliated rosettes (Table III). Because adenine is not preferentially associated with cilia microtubules, it was therefore considered to be a valuable reference compound. The ratio of guanine:adenine is 2.6-fold higher in the cilia than in the whole rosettes and 22-fold higher compared with the deciliated rosettes (Table III). This enrichment of guanine supports the conclusion that the isolated preparation is enriched in cilia.

In addition to the chemical markers, an enzymatic marker was used. Our recent morphological studies (Rhein *et al.*, 1981) showed that dynein arms (Gibbons and Grimstone, 1960) are associated with the outer microtubule doublets of cilia in the sensory regions of trout olfactory rosettes (Figs. 1 and 2). The major constituent of dynein arms of cilia is a Mg^{2+}-dependent ATPase (Stephens, 1974). This enzyme can therefore serve as an appropriate marker for cilia, although it is present also in other cellular structures (Pedersen, 1975; Pullman *et al.*, 1960). The Mg^{2+}–ATPase activity in the cilia is fourfold higher than in the whole rosettes or the deciliated rosettes, even though the isolated cilia represent only 2.3% of the total enzyme activ-

TABLE III Nucleotide Base Content of the Cilia Preparation from Trout Olfactory Rosettes

Sample	Nucleotide base content[a]				Ratio (Guanine:Adenine)
	Guanine (μg/gm wet weight)	Adenine (μg/gm wet weight)	Guanine (μg/mg protein)	Adenine (μg/mg protein)	
Whole rosettes	68.2	104	0.99	1.50	0.66
Deciliated rosettes	8.2	106	0.12	1.55	0.08
Cilia	0.35	0.20	1.31	0.76	1.73

[a] The values expressed relative to the wet weight refer to the original whole rosette sample. Those expressed relative to protein refer to the protein in that fraction. The results are from a single cilia preparation from 385 fish (770 rosettes). Taken from Rhein and Cagan (1980).

ity (Table IV). These results also support the conclusion of enrichment of cilia in the isolated preparation. Although any one criterion alone would not provide conclusive evidence, the three criteria taken together, along with the morphological evidence, strongly support the conclusion that the isolated preparation is substantially enriched in cilia.

Functional activity of the isolated cilia preparation was assessed by quantifying binding of radioactively labeled odorant amino acids. The binding parameters for five amino acids were determined. They were selected based on previous studies (Cagan and Zeiger, 1978) of the specificity of binding to the sedimentable fraction (Fraction P2) from the trout rosettes. As discussed above, the binding parameters and the results of binding competition studies had led to a postulate of the existence of multiple olfactory receptor sites in the trout; these included site TSA, which was postulated to bind L-threonine, L-serine, and L-alanine; site L, which was postulated to bind L-lysine; and site A_D, which was postulated to bind D-alanine. These features of the binding interactions were utilized to determine functional activity of the isolated cilia relevant to odorant recognition.

Binding of ^{3}H-labeled L-threonine, L-serine, L-alanine, L-lysine and of D-[^{14}C]alanine to the isolated cilia was examined. Binding curves were determined (Fig. 6), and analyses were performed according to Scatchard (1949). From these analyses apparent binding affinities (K_D) and maximal binding capacities (B_{max}) were obtained (Table V). The major features of the binding characteristics with the isolated cilia were similar to those obtained with fraction P2 (Cagan and Zeiger, 1978). For example, L-serine showed

TABLE IV Activity of Mg^{2+}-ATPase in the Cilia Preparation from Trout Olfactory Rosettes[a]

Sample	Mg^{2+}-ATPase[b] Total activity (μmole/min)	Specific activity (nmole/min-mg protein)
Whole rosettes	5.16 (100%)	73.9
Deciliated rosettes	4.12 (79.8%)	65.5
Supernatant	0.434 (8.4%)	44.4
Cilia	0.118 (2.3%)	275.0

[a] The values for total activity are expressed per gram of wet weight of the original rosette sample. In parentheses are shown the recoveries expressed as a percentage of the total activity of the whole rosette. The results for cilia are from a single preparation of 300 fish (600 rosettes). Taken from Rhein and Cagan (1980).

[b] In Rhein and Cagan (1980) the correct extinction coefficient was used in the calculations ($6.22 \times 10^3\ M^{-1}\ cm^{-1}$), but a typographical error in the paper showed it as $6.22 \times 10^{-3}\ M^{-1}\ cm^{-1}$.

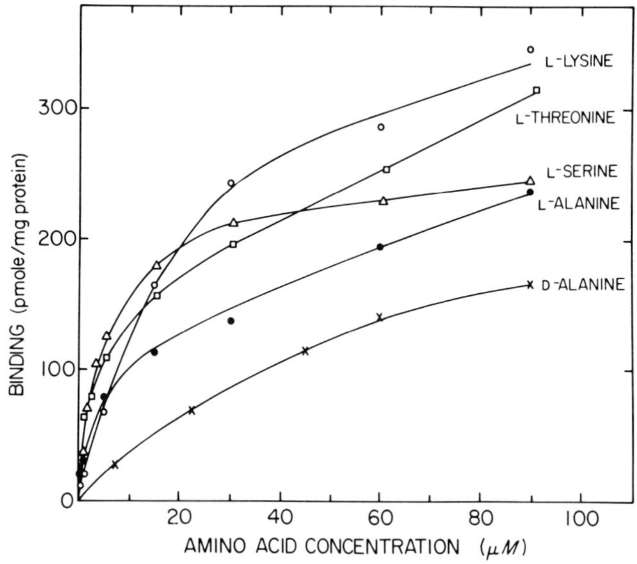

Fig. 6. Binding of odorant amino acids to the cilia preparation from rainbow trout olfactory rosettes. The data for each amino acid are the mean values of five preparations for L-alanine and L-serine, three preparations for L-threonine, two preparations for D-alanine, and one preparation for L-lysine. The binding value of 252 pmole/mg for D-alanine at 180 μM is omitted from the curve in the interests of space. (Taken from Rhein and Cagan, 1980.)

evidence of only a single type of site with $K_D \sim 10^{-6}\ M$; D-alanine also had only a single type of site with $K_D \sim 10^{-5}\ M$. The shape of the Scatchard plots suggested the presence of two affinity sites for each of L-threonine, L-alanine, and L-lysine ($K_D \sim 10^{-6}\ M$ and $K_D \sim 10^{-5}\ M$).

Results of competition studies using the cilia preparation and the four odorants, L-threonine, L-serine, L-alanine, and L-lysine, were very similar to results with fraction P2 (Cagan and Zeiger, 1978). Substantial mutual inhibition of binding is observed among L-threonine, L-serine, and L-alanine. The presence of L-lysine had no effect on binding of L-[^{3}H]threonine, L-]^{3}H]serine, or L-[^{3}H]alanine. These findings support the earlier postulate that odorant receptors for L-threonine, L-serine, and L-alanine are possibly identical or at least are closely associated (postulated site TSA) and that site L (for L-lysine) is independent of TSA.

The values of apparent K_D estimated by Scatchard analyses for the five odorants were similar to those determined for the sedimentable fraction P2 (Cagan and Zeiger, 1978). The binding capacity B_{max} was not enriched compared with fraction P2 (compare Tables I and V), and indeed some denaturation of receptors apparently occurs during cilia isolation due to the 10% ethanol in the deciliation medium. We conclude that a portion of the recep-

TABLE V Binding Parameters of Odorant Amino Acids with the Cilia Preparation from Trout Olfactory Rosettes[a]

Amino acid	High affinity		Low affinity	
	K_D (μM)	B_{max} (pmole/mg)	K_D (μM)	B_{max} (pmole/mg)
L-Threonine (n = 3)	1.6	140	35	480
L-Serine (n = 5)	3.3	220	—	—
L-Alanine (n = 5)	2.4	120	45	360
L-Lysine (n = 1)	6.1	150	65	600
D-Alanine (n = 2)	—	—	60	260

[a] The parameters were calculated from Scatchard plots of each individual preparation comprising the data for the binding curves shown in Fig. 6. The values of n are the numbers of preparations analyzed. Taken from Rhein and Cagan (1980).

tor molecules become denaturated during cilia isolation but that sufficient activity remains to establish that binding activity is associated with the cilia preparation.

The olfactory receptors present in the cilia preparation appear to be the same as those in fraction P2. This is a striking result in view of the diversity of the two isolation procedures. Fraction P2 is a relatively crude cell membrane-containing preparation obtained from homogenates of entire olfactory rosettes. It presumably contains the cilia membranes. To prepare cilia, on the other hand, the tissue is not homogenized. Instead, it is exposed to an ethanol–calcium medium (Linck, 1973), which strips the cilia from the surface.

VII. ISOLATION OF PLASMA MEMBRANES FROM CILIA

It seems likely that olfactory receptor macromolecules are associated with the membranes of the cilia. In current studies, we are attempting to isolate and characterize these membranes. In preliminary experiments (L. D. Rhein, unpublished), membrane fractions were isolated from the cilia preparation using discontinuous sucrose density gradients. One membrane fraction (B) contained high specific odorant binding activity (Table VI); it had a very low buoyant density (1.06/1.08 gm/ml in sucrose gradients) compared with plasma membranes from other sources. For example, plasma membranes from rat liver generally sediment at the interface between densities of 1.16 and 1.18 gm/ml (Emmelot *et al.*, 1974). The density of the olfactory membrane fraction is similar to that of central nervous system myelin (Sheads *et al.*, 1977), which is a very lipid-rich membrane. It is interesting

TABLE VI **Binding of Radioactively Labeled Odorants to Membrane Fractions from Trout Olfactory Cilia**[a]

Fraction	Density (g/ml)	Sucrose molarity (*M*)	Protein (μg/fish)	Specific binding activity (pmole/mg)	
				L-[^{3}H]Alanine	L-[^{3}H]Lysine
A	1.04/1.06	0.32/0.46	0.436	67.0	61.2
B	1.06/1.08	0.46/0.63	0.740	118	100
C	1.08/1.10	0.63/0.81	1.92	61.0	39.7
D	1.10/1.13	0.81/0.99	2.76	36.9	19.7
E	1.13/1.18	0.99/1.38	4.20	17.5	10.7
Pellet	>1.18	>1.38	0.988	11.8	11.4

[a] The density and molar sucrose values are for the two layers of the discontinuous gradient occupying the area immediately above and below the fraction that forms at the interface. Centrifugation was at 25,000 rpm (75,000 *g*) for 30 min at 4°C, using a Beckman L2-65B ultracentrifuge and an SW-41 swinging-bucket rotor. The values are for a single preparation from 250 fish. Radioactive ligands were each at 0.5 μ*M*.

also that the plasma membranes of catfish taste receptors occur at a relatively low buoyant density (≤1.11 gm/ml) in sucrose gradients (Cagan and Boyle, 1978; see also this volume, Chapter 10). The low density of chemosensory membranes could suggest a high lipid-to-protein ratio, or it could be the result of membrane collapse during isolation. Specific binding activities for both odorants L-alanine and L-lysine were highest in this fraction. Recall that no competition is observed between these odorants for binding with either the isolated cilia (Rhein and Cagan, 1980) or the sedimentable fraction P2 (Cagan and Zeiger, 1978) from trout olfactory epithelium, indicating that they bind to different receptor sites. More than a single type of odor receptor molecule thus is present in fraction B.

VIII. FUTURE PROSPECTS

Future research should enable isolation and characterization of odorant receptor macromolecules from the cilia. The receptors will need to be solubilized from the membranes and purified. A key, however, will be the ability to identify the isolated molecules as receptors.

Affinity chromatography has been highly useful for extensively purifying other receptors, and it is beginning to be applied in chemosensory research. An affinity ligand (the odorant) is covalently attached to an insoluble column matrix (usually agarose). The crude solubilized receptor preparation is applied to the column under conditions that allow binding of the receptor to the attached ligand (odorant). Nonreceptor protein will pass through the column, after which adsorbed receptor is displaced by eluting with buffer containing a biospecific ligand. An attempt to purify an anisole-binding protein from dog olfactory mucosa used this technique (Chapter 4, this volume). Affinity chromatographic procedures could be useful for purifying odorant receptors from rainbow trout, such as L-lysine or L-alanine receptors. Chemical and physical characteristics of the odor receptor macromolecules could then be studied.

Further isolation and characterization of the receptor macromolecules and determination of the molecular structure of the active binding site of the receptor can also be studied with the aid of affinity-labeling techniques. These involve covalent attachment of a radioactively labeled ligand to molecular species at the active binding site of the receptor. This will allow identification of the receptor macromolecules during further isolation and purification steps and will permit identification and examination of the active binding site of the receptor. Attempts to apply these techniques to taste receptors are noted in Chapter 10, and future work along these lines could be fruitful. Affinity-labeling reagents used to study interactions of amino acids with active sites on proteins are available (Powers, 1977) and could be useful with the trout odor receptors.

Detailed localization of specific odorant receptor sites along the cilia membrane should be made. It appears likely that a combination of biochemical approaches with electron microscopy could be used to definitively localize the receptors. Covalent attachment of radiolabeled odorants (this volume, Chapter 10) or combination with immunological approaches (this volume, Chapter 6) are exciting possibilities.

Finally, little information is available on the chemical composition of chemosensory membranes. Knowledge of the composition of excitable photoreceptive membranes might provide a basis for interesting comparisons as more information is obtained with olfactory receptor membranes. The sensory membranes that contain the photoreceptors and those that contain the olfactory receptors may have similarities because both originate from cilia. Studies on the chemical composition of olfactory receptor membrane fractions from cilia would provide basic information on the membranes and could lead to a better understanding of the importance also of the lipid portion of the membrane in initial events of olfactory sensation.

ACKNOWLEDGMENT

L. D. Rhein was supported in part by NIH Postdoctoral Fellowship No. 1F32-NS05954 and in part by NSF Grant no. BNS-7924071.

REFERENCES

Anderson, R. E., Benolken, R. M., Kelleher, P. A., Maude, M. B., and Wiegand, R. D. (1978). Chemistry of photoreceptor membrane preparations from squid retinas. *Biochim. Biophys. Acta* **510**, 316–326.

Andres, K. H. (1969). Der olfaktorische saum der katze. *Z. Zellforsch. Mikrosk. Anat.* **96**, 250–274.

Atema, J. (1973). Microtubule theory of sensory transduction. *J. Theor. Biol.* **38**, 181–190.

Atema, J. (1975). Stimulus transmission along microtubules in sensory cells: An hypothesis. *In* "Microtubules and Microtubule Inhibitors" (M. Borgers and M. De Brabander, eds.), pp. 247–257. Am. Elsevier, New York.

Atema, J., Demetria, J., and Oleszko-Szuts, S. (1978). Microtubule-inhibiting drugs alter chemoreceptor function in olfactory sensilla of the lobster. *Biol. Bull. (Woods Hole, Mass.)* **155**, 427 (Abstr.).

Autrum, H. (1979). Introduction. *In* "Handbook of Sensory Physiology" (H. Autrum, ed.), Vol. 7, Part 6A, pp. 1–22. Springer-Verlag, Berlin and New York.

Bannister, L. H. (1965). The fine structure of the olfactory surface of teleostean fishes. *Q. J. Microsc. Sci.* **106**, 333–342.

Barber, V. C. (1974). Cilia in sense organs. *In* "Cilia and Flagella" (M. A. Sleigh, ed.), pp. 403–433. Academic Press, New York.

Beets, M. G. J. (1971). Olfactory response and molecular structure. *In* "Handbook of Sensory Physiology" (L. M. Beidler, ed.). Vol. 4, Part 1, pp. 257–321. Springer-Verlag, Berlin and New York.

Bertmar, G. (1972). Scanning electron microscopy of olfactory rosette in sea trout. *Z. Zellforsch. Mikrosk. Anat.* **128**, 336–346.

Bertmar, G. (1973). Ultrastructure of the olfactory mucosa in the homing baltic sea trout *Salmo trutta trutta*. *Mar. Biol.* **19**, 74–88.

Blank, D. L., Gesteland, R. C., Jansyn, E. M., and Wagner, A. L. (1978). Stimulus-related activity of olfactory receptor cilia. *Fed. Proc., Fed. Am. Soc. Exp. Biol.* **37**, 398. (Abstr.)

Bostock, H. (1974). Diffusion and the frog EOG. *In* "Transduction Mechanisms in Chemoreception" (T. M. Poynder, ed.), pp. 27–36. Information Retrieval, London.

Bronshtein, A. A. (1964). Observation *in vivo* of the motion of cilia of olfactory cells. *Dokl. Akad. Nauk SSSR* **156**, 715–718.

Bronshtein, A. A., and Minor, A. V. (1977). The regeneration of olfactory flagella and restoration of electroolfactogram after treatment of the olfactory mucosa with triton X-100. *Tsitologiya* **19**, 33–39.

Cagan, R. H., and Boyle, A. G. (1978). Plasma membranes from taste receptor tissue: Isolation and binding activity. *Fed. Proc., Fed. Am. Soc. Exp. Biol.* **37**, 1818. (Abstr.)

Cagan, R. H., and Zeiger, W. N. (1978). Biochemical studies of olfaction: [1]. Binding specificity of radioactively labeled stimuli to an isolated olfactory preparation from rainbow trout (*Salmo gairdneri*). *Proc. Natl. Acad. Sci. U.S.A.* **75**, 4679–4683.

Ciges, M., Labella, T., Gayoso, M., and Sanchez, G. (1977). Ultrastructure of the organ of Jacobson and comparative study with olfactory mucosa. *Acta Oto-Laryngol.* **83**, 47–58.

Cone, R. A. (1972). Rotational diffusion of rhodopsin in the visual receptor membrane. *Nature (London)*, New Biol. **236**, 39–43.

Crescitelli, F. (1972). The visual cells and visual pigments of the vertebrate eye. *In* "Handbook of Sensory Physiology" (H. J. A. Dartnall, ed.), Vol. 7, Part 1, pp. 245–363. Springer-Verlag, Berlin and New York.

Cuatrecasas, P. (1971a). Insulin-receptor interactions in adipose tissue cells. Direct measurement and properties. *Proc. Natl. Acad. Sci. U.S.A.* **68**, 1264–1268.

Cuatrecasas, P. (1971b). Perturbation of the insulin receptor of isolated fat cells with proteolytic enzymes. *J. Biol. Chem.* **246**, 6522–6531.

Daemen, F. J. M. (1973). Vertebrate rod outer segment membranes. *Biochim. Biophys. Acta* **300**, 255–288.

Davies, J. T. (1971). Olfactory theories. *In* "Handbook of Sensory Physiology" (L. M. Beidler, ed.), Vol. 4, Part 1, pp. 322–350. Springer-Verlag, Berlin and New York.

Døving, K. B., Dubois-Dauphin, M., Holley, A., and Jourdan, F. (1977). Functional anatomy of the olfactory organ of fish and the ciliary mechanism of water transport. *Acta Zool.* (*Stockholm*) **58**, 245–255.

Dunn, R. F. (1973). The ultrastructure of the vertebrate retina. *In* "The Ultrastructure of Sensory Organs" (I. Friedmann, ed.), pp. 153–265. Am. Elsevier, New York.

Eakin, R. M. (1972). Structure of invertebrate photoreceptors. *In* "Handbook of Sensory Physiology" (H. J. A. Dartnall, ed.), Vol. 7, Part 1, pp. 625–684. Springer-Verlag, Berlin and New York.

Emmelot, P., Bos, C. J., van Hoeven, R. P., and van Blitterswijk, W. J. (1974). Isolation of plasma membranes from rat and mouse livers and hepatomas. *In* "Biomembranes," Part A (S. Fleischer and L. Packer, eds.), Methods in Enzymology, Vol. 31, pp. 75–90. Academic Press, New York.

Engen, T. (1971). Olfactory psychophysics. *In* "Handbook of Sensory Physiology" (L. M. Beidler, ed.), Vol. 4, Part 1, pp. 216–244. Springer-Verlag, Berlin and New York.

Frisch, D. (1967). Ultrastructure of mouse olfactory mucosa. *Am. J. Anat.* **121**, 87–119.

Gibbons, I. R., and Grimstone, A. V. (1960). On flagella structure in certain flagellates. *J. Biophys. Biochem. Cytol.* **7**, 697–715.

Graziadei, P. P. C. (1971). The olfactory mucosa of vertebrates. *In* "Handbook of Sensory Physiology" (L. M. Beidler, ed.), Vol. 4, Part 1, pp. 27–58. Springer-Verlag, Berlin and New York.

Hagins, W. A. (1972). The visual process: Excitatory mechanisms in the primary receptor cells. *Annu. Rev. Biophys. Bioeng.* **1**, 131–158.

Hara, T. J. (1973). Olfactory responses to amino acids in rainbow trout, *Salmo gairdneri*. *Comp. Biochem. Physiol. A* **44**, 407–416.

Hasler, A. D. (1960). Guideposts of migrating fishes. *Science* **132**, 785–792.

Hasler, A. D., Scholz, A. T., and Horrall, R. M. (1978). Olfactory imprinting and homing in salmon. *Am. Sci.* **66**, 347–355.

Hopkins, A. E. (1926). The olfactory receptors in vertebrates. *J. Comp. Neurol.* **41**, 253–289.

Hubbell, W. L., and Bownds, M. D. (1979). Visual transduction in vertebrate photoreceptors. *Ann. Rev. Neurosci.* **2**, 17–34.

Kerjaschki, D. (1976). The central tubuli in distal segments of olfactory cilia lack dynein arms. *Experientia* **32**, 1459–1460.

Kerjaschki, D., and Hörandner, H. (1976). The development of mouse olfactory vesicles and their cell contacts: A freeze-etching study. *J. Ultrast. Res.* **54**, 420–444.

Kirschner, M. W., Suter, M., Weingarten, M., and Littman, D. (1975). The role of rings in the assembly of microtubules *in vitro*. *Ann. N.Y. Acad. Sci.* **253**, 90–106.

Kratzing, J. (1978). The olfactory apparatus of the bandicoot (*Isoodon macrourus*): Fine structure and presence of a septal olfactory organ. *J. Anat.* **125**, 601–613.

Krueger, J. M., and Cagan, R. H. (1976). Biochemical studies of taste sensation. [4]. Binding of L-[^{3}H]alanine to a sedimentable fraction from catfish barbel epithelium. *J. Biol. Chem.* **251**, 88–97.

Laemmli, U. K. (1970). Cleavage of structural proteins during the assembly of the head of bacteriophage T4. *Nature* (*London*) **227**, 680–685.

Lettvin, J. Y., and Gesteland, R. C. (1965). Speculations on smell. *Cold Spring Harbor Symp. Quant. Biol.* **30**, 217–225.

Linck, R. W. (1973). Comparative isolation of cilia and flagella from the lamellibranch mollusc, *Aequipecten irradians. J. Cell Sci.* **12**, 345–367.

Loo, S. K. (1977). Fine structure of the olfactory epithelium in some primates. *J. Anat.* **123**, 135–145.

Menco, B. P. M. (1977). "A Qualitative and Quantitative Investigation of Olfactory and Nasal Respiratory Mucosal Surfaces of Cow and Sheep Based on Various Ultrastructural and Biochemical Methods," Meded. Landbouwhogesch. Wageningen, 77-13. Veenman & Zonen, Wageningen, Netherlands.

Menco, B. P. M., Dodd, G. H., Davey, M., and Bannister, L. H. (1976). Presence of membrane particles in freeze-etched bovine olfactory cilia. *Nature (London)* **263**, 597–599.

Moran, D. T., and Varela, F. G. (1971). Microtubules and sensory transduction. *Proc. Natl. Acad. Sci. U.S.A.* **68**, 757–760.

Murray, R. G. (1973). The ultrastructure of taste buds. *In* "The Ultrastructure of Sensory Organs" (I. Friedmann, ed.), pp. 1–81. Am. Elsevier, New York.

Ottoson, D. (1971). The electro-olfactogram. A review of studies on the receptor potential of the olfactory organ. *In* "Handbook of Sensory Physiology" (L. M. Beidler, ed.), Vol. 4, Part 1, pp. 95–131. Springer-Verlag, Berlin and New York.

Parker, G. H. (1922). "Smell, Taste and Allied Senses in the Vertebrates." Lippincott, Philadelphia, Pennsylvania.

Pedersen, P. L. (1975). Mitochondrial adenosine triphosphatase. *J. Bioenerg.* **6**, 243–275.

Poo, M.-M., and Cone, R. A. (1973). Lateral diffusion of rhodopsin in *Necturus* rods. *Exp. Eye Res.* **17**, 503–510.

Powers, J. C. (1977). Haloketone inhibitors of proteolytic enzymes. *In* "Chemistry and Biochemistry of Amino Acids, Peptides and Proteins" (B. Weinstein, ed.), Vol. 4, pp. 65–178. Dekker, New York.

Pullman, M. E., Penefsky, H. S., Datta, A., and Racker, E. (1960). Partial resolution of the enzymes catalyzing oxidative phosphorylation. I. Purification and properties of soluble, dinitrophenol-stimulated adenosine triphosphatase. *J. Biol. Chem.* **235**, 3322–3329.

Reese, T. S. (1965). Olfactory cilia in the frog. *J. Cell Biol.* **25**, 209–230.

Rhein, L. D., and Cagan, R. H. (1979). Cilia from olfactory epithelium: Isolation, characterization and ^{3}H-labeled odorant binding activity. *Proc. Int. Congr. Biochem., 11th, Toronto* Abstr. p. 576.

Rhein, L. D., and Cagan, R. H. (1980). Biochemical studies of olfaction: [2]. Isolation, characterization and odorant binding activity of cilia from rainbow trout olfactory rosettes. *Proc. Natl. Acad. Sci. U.S.A.* **77**, 4412–4416.

Rhein, L. D., Cagan, R. H., Orkand, P. M., and Dolack, M. K. (1981). Surface specializations of the olfactory epithelium of rainbow trout, *Salmo gairdneri.* In preparation.

Scatchard, G. (1949). The attractions of proteins for small molecules and ions. *Ann. N.Y. Acad. Sci.* **51**, 660–672.

Sheads, L. D., Eby, M. J., Sampugna, J., and Douglass, L. W. (1977). Myelin subfractions isolated from mouse brains. Studies of normal mice during development, quaking mutants, and three brain regions. *J. Neurobiol.* **8**, 67–89.

Shelanski, M. L., and Taylor, E. W. (1968). Properties of the protein subunit of central-pair and outer-doublet microtubules of sea urchin flagella. *J. Cell Biol.* **38**, 304–315.

Shibuya, T. (1964). Dissociation of olfactory neural response and mucosal potential. *Science* **143**, 1338–1340.

Snyder, J. A., and McIntosh, J. R. (1976). Biochemistry and physiology of microtubules. *Annu. Rev. Biochem.* **45**, 699–720.

Stephens, R. E. (1974). Enzymatic and structural proteins of the axoneme. *In* "Cilia and Flagella" (M. A. Sleigh, ed.), pp. 39–76. Academic Press, New York.

Stevens, R. E., Renaud, F. L., and Gibbons, I. R. (1967). Guanine nucleotides associated with the protein of the outer fibers of flagella and cilia. *Science* **156**, 1606–1608.

Sutterlin, A. M., and Sutterlin, N. (1971). Electrical responses of the olfactory epithelium of Atlantic salmon (*Salmo salar*). *J. Fish. Res. Board Can.* **28**, 565–572.

Thornhill, R. A. (1967). The ultrastructure of the olfactory epithelium of the lamprey, *Lampetra fluviatilis. J. Cell Sci.* **2**, 591–602.

Tomita, T. (1970). Electrical activity of vertebrate photoreceptors. *Q. Rev. Biophys.* **3**, 179–222.

Tucker, D. (1967). Olfactory cilia are not required for receptor function. *Fed. Proc., Fed. Am. Soc. Exp. Biol.* **26**, 544 (Abstr.).

Vinnikov, Y. A. (1974). Sensory reception. Cytology, molecular mechanisms and evolution. *In* "Molecular Biology, Biochemistry and Biophysics" (A. Kleinzeller, G. F. Springer, and H. G. Wittmann, eds.), Vol. 17, pp. 24–40, 170–207. Springer-Verlag, Berlin and New York.

Warner, F. D. (1974). The fine struction of the ciliary and flagellar axoneme. *In* "Cilia and Flagella" (M. A. Sleigh, ed.), pp. 11–37. Academic Press, New York.

Watson, M. R., and Hopkins, J. M. (1962). Isolated cilia from *Tetrahymena pyriformis. Exp. Cell Res.* **28**, 280–295.

Yamamoto, M. (1976). An electron microscopic study of the olfactory mucosa in the bat and rabbit. *Arch. Histol. Jpn.* **38**, 359–412.

Zeiger, W. N., and Cagan, R. H. (1975). Biochemistry of olfactory specificity. Binding of labeled stimuli to a preparation from olfactory epithelium. *Am. Chem. Soc., Natl. Meet., 69th, Philadelphia, Pa.* Abstr. No. 8.

4

Receptor Proteins in Vertebrate Olfaction

STEVEN PRICE

I. INTRODUCTION

The function of any sensory receptor is to transduce energy in the environment (the stimulus) into a pattern of nerve impulses. The quality, intensity, and duration of the stimulus are encoded within the pattern. Thus, any sensory receptor must include some physical or chemical structure that can interact with the stimulus and, as a consequence of the interaction, initiate a perturbation within the receptor cell. A receptor cell specialized for detection of chemicals must, then, include some structure with which those chemicals can interact. Olfactory receptor cells are located in the olfactory sensory epithelium, a portion of the lining of the nasal cavity. The bipolar neurons that project from the olfactory mucosa to the brain are the olfactory receptor cells. Thus in olfaction, the processes of recognition, transduction, and encoding occur in a single cell. This is in contrast to taste in which the primary

Biochemistry of Taste and Olfaction

ISBN 0-12-154450-8

receptor cells are not nerves, but cells that secrete some transmitter that stimulates afferent neurons.

During the past 20 years, a general concept of how chemicals stimulate cells has arisen. The initial event is hypothesized to be the interaction of the stimulating chemical with a macromolecule, the "receptor protein," specialized for this function. The receptor protein is widely believed to be on (or in) the cell membrane and to have the property of undergoing a marked change in conformation upon interaction with the stimulus. The conformational change is seen as the initial perturbation leading to the cell's response to the chemical. This general mechanism has been developed most fully as an explanation for the effects of drugs (Ariens *et al.*, 1979) and hormones (Rodbell, 1972). Not surprisingly, it has also been adapted in attempts to explain chemical stimulation of receptor cells for taste and smell (see reviews in Price and DeSimone, 1977; Holley and MacLeod, 1977; see also this volume, Chapter 12).

In this chapter, the evidence bearing on the question of whether the initial interaction of odorants with odor receptor cells occurs at proteinaceous receptors will be examined. The literature in which various preparations that include odor receptor proteins are reported is then reviewed. This review is restricted to reports of studies with verteberate materials. The reader is referred to the reviews of Koshland (1979; see also this volume, Chapter 21) and Hansen (1978; see also this volume, Chapter 8) for introductions to the literature on receptor proteins in bacterial chemoreception and insect olfaction.

II. OLFACTORY RECEPTOR PROTEINS

A. Are Olfactory Receptor Sites Proteinaceous?

1. Theories Involving Lipids as Receptors

It is intellectually attractive to suppose that the receptor protein concept, so widely applicable to stimulations of cells by chemicals, applies to odorant stimulation of olfactory receptor cells. There are alternative hypotheses in which the site of initial interaction is envisioned to be the lipid portion of the cell membrane.

Davies (1971) proposed a mechanism in which odorants, dissolved in the lipid, then left a transient hole in the lipid film when they desorbed. According to this hypothesis, the cell depolarizes as ions flow through the holes. Davies' theoretical treatment was able to account for the varying thresholds of different odorants, but this theory encountered considerable difficulty in attempting to explain the existence of qualitative specificity. Cherry *et al.* (1970) attempted to confirm the predictions of Davies' theory with model lipid membranes, but they were unable to do so. In addition, the resistance changes that odorants induced in their system were far too small to be of significance in olfactory transduction. Koyama and Kurihara (1972) found that odor thresholds are highly correlated with lipid solubility, and they suggested that adsorption to membrane lipid may be the initial step in stimulation of odorant receptor cells, at least in those cells that have a low degree of qualitative specificity (see this volume, Chapter 13). Sperber (1977) has proposed a related mechanism in which odorants are detected by their effects on hypothetical lipid–protein coacervates in membranes.

2. *The Chromatographic Hypothesis of Odor Recognition: The Role of Mucus*

Mozell (1970; Mozell and Jagodowicz, 1973) developed a very interesting hypothesis in which the mucus covering the olfactory epithelium plays a major role in odor recognition (see this volume, Chapter 2). According to this view, different odorants are adsorbed to different degrees in their passage across the surface of the olfactory epithelium. Thus, each will have a characteristic retention time, the nasal passage functioning essentially as a gas chromatograph. According to the hypothesis, retention time contributes to quality recognition. The importance of retention by the mucus in modifying the access of odorants to the olfactory receptor cells is now well-established (Hornung and Mozell, 1977), but the hypothesis that this plays a major role in odor quality recognition is no longer widely held.

Several authors have suggested that odorants may undergo chemical modification in the mucus prior to arrival at the receptor cells (Getchell and Getchell, 1977; Hornung and Mozell, 1980; Vinnikov *et al.*, 1979). If this is true, then the problem of stimulus identification will be complicated even beyond the well-known difficulties presented by the presence of trace amounts of highly stimulatory contaminants in "pure" chemicals. Whether purported chemical reactions undergone by odorants are integral to the recognition process is, of course, completely unknown.

3. *Evidence for Olfactory Receptor Proteins*

There are several lines of evidence to suggest that recognition sites for odorants are in fact on proteins. First, there is the phenomenon of specific anosmia, first described by Guillot (1948). *Specific anosmia* is the situation in which the individual has normal acuity for most odorants, but greatly reduced acuity for a group of compounds that evoke a similar odor quality. It has been studied in detail in humans (Amoore, 1971, 1977) and has been reported to occur in mice (Wysocki *et al.*, 1977; Price, 1977). The existence of specific anosmias implies that there are highly specific receptor sites with which all members of an odor class react when they are used as stimuli at low concentrations. It seems unlikely that such specificity could be achieved without proteins being implicated as receptor sites. Lipids do not appear to offer the structural versatility implied by Amoore's estimate that there are several dozen specific odor classes. The existence of a fairly large number of specific types of odor receptor sites receives further support from the recent work of Baylin and Moulton (1979), who did an electrophysiological study of cross-adaptation to seven odorants by olfactory receptor cells of salamanders. Their results suggest that there are receptor sites that are specific for each of the odorants they used, in addition to receptor sites of low specificity.

Another line of evidence that olfactory receptor sites are proteinaceous is the high degree of correlation of molecular size and shape with odor quality (Amoore *et al.*, 1967; Kier *et al.*, 1977). That is, compounds with similar odor quality tend to have similar molecular dimensions. This implies that there are receptor sites into which they fit, analogous to the fit of substrates into sites on enzymes. It is difficult to imagine sites of specific shapes in the fluid lipid of cell membranes, but the concept of such sites in proteins is well-established. Getchell and Gesteland (1972) reported that the electrophysiological response of frog olfactory mucosa to odorants could be abolished by treating the tissue with *N*-ethylmaleimide, a reagent that reacts with sulfhydryl groups on proteins. When the odorant ethyl *n*-butyrate was present during the exposure to *N*-ethylmaleimide, it specifically protected the epithelium. Olfactory mucosa treated with *N*-ethylmaleimide in the presence of ethyl *n*-butyrate retained nearly full electrophysiological responsiveness to methyl *n*-butyrate and ethyl *n*-butyrate, but it no longer responded to stimulation with other odorants. This suggests that ethyl *n*-butyrate stimulates the frog olfactory receptor by associating with some sulfhydryl group and that derivatizing other sulfhydryl groups prevents other odorants from being olfactory stimuli. These findings were confirmed by Menevse *et al.* (1977), who extended the work and demonstrated that other reagents known to modify proteins are also effective inhibitors of olfactory responses. Among the more interesting of their reagents was mersalyl, which is a sulfhydryl modifier that does not enter cells. Mersalyl's effects mimicked those of *N*-ethylmaleimide, including the protection by ethyl *n*-butyrate, demonstrating that the sulfhydryl groups necessary for odor stimulation are on the mucosal surface of the cell membrane. The evidence supports the hypothesis that olfactory stimuli interact with receptor proteins located on the cell surface, in accord with the general concept of the action of chemical stimuli on receptor proteins in cell membranes.

B. *In Vitro* Preparations Containing Olfactory Receptor Proteins

1. *Identifying Receptor Proteins in Extracts*

A major problem in the development of methods for purifying olfactory receptor proteins is the lack of well-defined, generally accepted criteria by which such proteins can be identified. This problem also exists in studies of taste receptor proteins, and it has received considerable attention (Price and DeSimone, 1977; Cagan and Morris, 1979). With drug and hormone receptors, there is generally a substantial body of data on relative biological potencies of series of structurally similar stimuli. The receptor protein is then identified on the basis of parallelism between the affinity with which stimuli bind to it and their relative potencies (Cuatrecasas, 1974). In the case of vertebrate olfaction, there are few instances in which relative potencies of a series of stimuli are known, so this approach has not found wide applicability. Indeed, in most cases of attempts to extract olfactory receptor proteins, the only criterion used has been the ability of the preparation to bind some odorant with high affinity. Although the ability to form a complex with an odorant is a necessary property of an olfactory receptor protein, possession of this ability is not sufficient evidence to warrant the conclusion that a particular protein functions as a receptor molecule. As discussed later, the absence of criteria for receptor protein identification has been, and continues to be, a major obstacle to progress in this area.

2. *Preparations from Fishes*

Attempts have been made to develop *in vitro* preparations containing olfactory receptor proteins from two species of fish. In both cases, there are electrophysiological data on the relative effectiveness of different stimuli, and comparison of the binding of these stimuli with their potencies has been possible.

Caprio (1977, 1978) has made an electrophysiological study of the olfactory responses of catfish, *Ictalurus punctatus*, to a large number of amino acids. Cancalon (1978) developed a method for isolation of olfactory receptor cells from this species and studied binding of radioactively labeled amino acids to these cells. He observed a highly significant correlation between the amount of a 6×10^{-6} M solution of the amino acid, which was bound by the cells and the activity of that amino acid as an olfactory stimulant at a concentration of 10^{-4} M. No such correlation was observed when isolated cells of respiratory epithelium were used in place of the olfactory receptor cells. When binding of amino acids to catfish olfactory receptor cells was compared with stimulatory effectiveness of the amino acids in trout, a significant correlation was observed. However, the correlation coefficient was much lower than that obtained within the catfish species (Cancalon, 1978). Thus, it appears that the method developed for the isolation of catfish olfactory receptor cells does not result in loss of amino acid receptor protein(s), which could probably be further purified from such preparations.

The other fish species that has been used to study olfactory receptor proteins *in vitro* is the trout, *Salmo gairdneri*. Hara (1973, 1976) has reported the relative effectiveness of a number of amino acids as olfactory stimuli in this species. Cagan and Zeiger (1978) used a procedure, which appears to result in the isolation of a fraction enriched in cell membranes with trout olfactory tissue, and made a detailed study of the binding of amino acids (see this volume, Chapter 3). They found a highly significant correlation between the binding of ten amino acids to this fraction and the electrophysiological response to those amino acids reported by Hara (1973), suggesting that the amino acid receptor proteins are present in their preparation. Cagan and Zeiger (1978) also determined the dependence of binding on amino acid concentration for eight amino acids, and they analyzed the data to determine dissociation constants and numbers of binding sites for each. They found two classes of binding sites. One, the high-affinity sites, had dissociation constants around 4×10^{-6} M for each amino acid, whereas the low-affinity sites had dissociation constants between 1.7×10^{-5} M and 1.4×10^{-4} M. One can calculate the amount of amino acid bound at any concentration from their parameters for dissociation constants and binding capacity, using Eq. 1.

$$B = \frac{B_{\mathrm{m1}}(C)}{K_{\mathrm{d1}} + (C)} + \frac{B_{\mathrm{m2}}(C)}{K_{\mathrm{d2}} + (C)} \tag{1}$$

This equation describes the amount of ligand in the bound state (B) as the total ligand concentration (C) is varied for the case in which there are two binding sites with dissociation constants K_{d1} and K_{d2}, respectively, the concentration of each site being B_{m1} and B_{m2}. Since the electrophysiological responses reported by Hara (1973) were obtained from amino acid solutions at 10^{-4} M, it is interesting to calculate the binding *in vitro* at 10^{-4} M for comparison with the electrophysiological responses. The results of such calculations are shown in Fig. 1. The correlation between the binding to low-affinity sites and electrophysiological responses (Fig. 1, right) is not significant ($r = 0.198$), whereas binding to high-affinity sites (Fig. 1, left) is strongly correlated with the stimulatory potencies of the amino acids ($r = 0.813$). A similar correlation coefficient was obtained by Cagan and Zeiger (1978) in comparing the binding of a series of 10 amino acids, at 10^{-6} M concentrations, with Hara's (1973) results. Since the concentrations of amino acids in Hara's (1973) study are about 20 times the values of the dissociation constants for the high-affinity sites, the results suggest that the sensitivity of

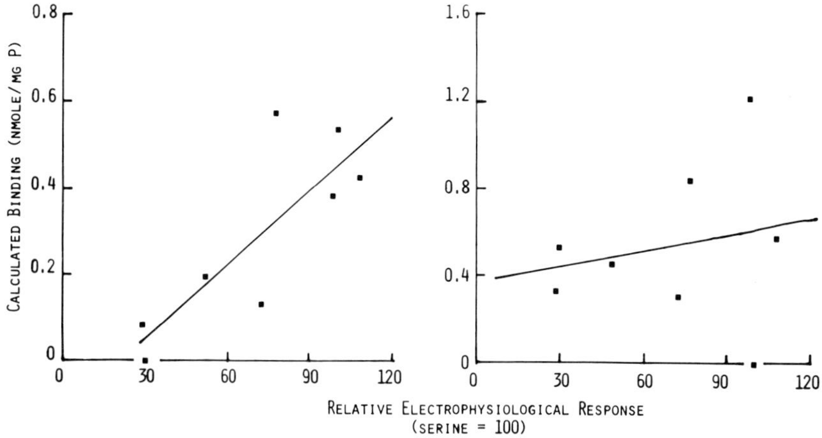

Fig. 1. Binding of amino acids to a cell membrane fraction from trout olfactory tissue is plotted against the electrophysiological response of trout to the same amino acids. The binding at 10^{-4} *M* is calculated from the values of K_D and binding capacity reported by Cagan and Zeiger (1978) for high affinity sites (left) and low affinity sites (right). The electrophysiological responses to 10^{-4} *M* solutions were reported in Hara (1973). The lines are fit by the method of least squares.

trout olfactory receptors for any amino acid simply depend on the number of high-affinity binding sites for that amino acid. This agrees with the model for olfactory acuity proposed by van Drongelen *et al.* (1978) in which thresholds are hypothesized to be determined by spatial summation of small increases in firing frequencies in many receptor cells.

Since the number of binding sites in the preparation differs for different amino acids, it appears likely that not all of the amino acids interact with the same receptor protein. Cagan and Zeiger (1978) explored this issue by studying competition between amino acids for the binding sites in the membrane fraction. Three of the amino acids, L-serine, L-threonine, and L-alanine, compete strongly with each other and are bound by approximately the same number of high affinity sites (5×10^{-10} mole per milligram protein). This suggests that they share a common receptor protein. The other amino acids used either have fewer binding sites or compete much less strongly. Thus, there appears to be at least three different amino acid receptor proteins in trout.

Because Getchell and Gesteland (1972) had presented evidence that sulfhydryl groups were involved in binding of esters by frog olfactory receptors, Cagan and Zeiger (1978) also studied effects of sulfhydryl reagents on amino acid binding by trout olfactory cell membranes. Although mercuric ions inhibited the binding, neither iodoacetate nor *N*-ethylmaleimide were inhibitory at concentrations at which they generally combine with sulfhydryl groups. Thus, it does not appear that amino acid receptor proteins from trout olfactory cell membranes interact with stimuli via a sulfhydryl group.

3. *Preparations from Amphibians*

Gusel'nikov *et al.* (1974) reported a study of spectral changes induced by odorants in isolated cilia from frog olfactory mucosal cells. They obtained clear evidence that proteins in the cilia undergo marked conformational changes when nitrobenzene, *p*-nitrobenzaldehyde, or *p*-dimethylaminobenzaldehyde were present. Humans describe these compounds as having floral odors. The dissociation constant for interaction of *p*-dimethylaminobenzaldehyde with the cilia was determined to be 7×10^{-7} *M*. Since the magnitude of the spectral changes observed suggested that virtually all of the ciliary proteins were involved in conformational changes when the odorants were present (Gusel'nikov *et al.*, 1974), it seems unlikely that information about any specific receptor protein can be inferred from the data. Similar studies using a wider range of odorants with fractions derived from the cilia would be of considerable interest.

Fesenko and his co-workers have published a series of studies of *in vitro* olfactory receptors using amphibian material. In their first report (Fesenko and Pervukhin, 1976), they described the results of incorporating sonicated olfactory mucosal scrapings into artificial lipid membranes. The scrapings caused an increase in the sodium permeability of the membranes when certain odorants, camphor being one of them, were present. They subsequently fractionated the scrapings by column chromatography and gel filtration and obtained an active fraction, which appeared to be nucleoprotein of molecular weight in excess of 10^5 daltons (Fesenko *et al.*, 1977). The fact that it is able to cause an increase in sodium permeability is suggestive of a role for the active component of this fraction in olfactory receptor cell transduction. However, there is no compelling evidence linking this material to initial events in olfaction, except that certain odorants interact with it.

4. *Preparations from Mammals*

Ash (1968) was the first to report an attempt to extract an olfactory receptor protein, observing spectral changes in extracts of rabbit olfactory mucosa when certain odorants were added. The spectral changes were subsequently shown to be due to oxidation of ascorbic acid in the extracts, the oxidation being promoted in some manner by the presence of the odorants (Ash, 1969; Ash and Skogen, 1970). Of a group of 37 odorants, 7 were active in promoting ascorbic acid oxidation in these preparations. That is, the system seemed to have some specificity. No additional evidence linking the odorant-induced oxidation of ascorbic acid to olfactory receptor mechanisms has been presented in the past decade, and this line of research appears to have been abandoned. It is mentioned here because it is the pioneering attempt to prepare extracts containing olfactory receptor proteins.

Gennings *et al.* (1977) reported finding a fraction in extracts of sow olfactory mucosa that bound two pheromones with high affinity (dissociation constants around 10^{-9} *M*), but had low affinity for a structurally related compound that is not a pheromone. The active component in the extracts of olfactory epithelium was absent from extracts of respiratory epithelium. It appeared to be a protein with a molecular weight of 2.3×10^4 daltons, and it probably originated in some membranous portion of the cells, since detergent treatment was required to solubilize it. Further evidence for a role of this protein in olfaction is needed before the conclusion that it is a receptor protein can be accepted (see this volume, Chapter 1), but the findings to date are very promising indeed.

Probably the most extensive biochemical data on a putative olfactory receptor protein are those of Fesenko's group on a camphor-binding protein from rat olfactory mucosa. These workers reported that a constituent of rat olfactory epithelium was able to bind tritiated camphor, using an assay based on equilibrium dialysis. There were two kinds of binding sites in their preparations, one with a high affinity for camphor (dissociation constant around 1.5×10^{-9} *M*) and the other with much lower affinity (dissociation constant around 10^{-7} *M*). The maximum amount of binding of camphor by the low affinity sites was 110 times that by the high affinity sites (Fesenko *et al.*, 1978). The ability of a series of 13 compounds to displace camphor from binding sites in the preparation was studied; the results are shown in Table I. It is clear that the two compounds with camphoraceous odor were more effective in competing with camphor for binding than were any of the other 11 compounds. Fesenko *et al.* (1978) interpret these data as reflecting competition for the high-affinity sites. Their rationale is that the camphor concentration was 5×10^{-10} *M*, a concentration at which they assume that "... the major contribution to binding comes from those structures in the olfactory mucosa that have the dissociation constant of 1.5×10^{-9} *M*" (Fesenko *et al.*, 1978). However, using Eq. (1) to calculate the relative amounts of camphor bound to the two sites from their estimated dissociation constants and the camphor-binding capacity of each, it is found that only 31% of the bound camphor will be on the high affinity site, whereas 69% will be on the low affinity site at a camphor concentration of 5×10^{-10} *M*. That is, the amount bound to the low affinity sites is far from negligible, and data such as that shown in Table I do not suffice to show relative affinities of the compounds for one or the other of the camphor binding sites. Indeed, since camphor bromide and isoborneol, the only two compounds with a camphoraceous odor, are also the only two that are structurally related to camphor, one might expect them to be the most effective of the group at compet-

TABLE I Displacement of [^{3}H]Camphor from an Extract of Rat Olfactory Epithelium by Various Compounds[a]

Compound	Odor	Displacement (%)
Camphor bromide	Camphoraceous	76
Isoborneol	Camphoraceous	68
Methylethyl ketone	Acetone-like	59
Menthone	Minty	55
Anethole	Aromatic, sweet	38
Dibutyl ether	Ethereal	16
Linalool	Floral	15
Menthol	Minty	14
Geraniol	Floral	8
2-Mercaptoethanol	Putrid	5
Ethyl acetate	Floral	None
Citral	Lemon	None
Diisopropyl ether	Ethereal	None

[a] Data and odor descriptions are from Fesenko *et al.* (1978). The concentration of [^{3}H]camphor was 5×10^{-10} *M*, and each compound tested for its ability to displace camphor was at 10^{-8} *M*.

ing with camphor for any site with which camphor interacts. Thus, the evidence that the high-affinity camphor-binding site represents an olfactory receptor protein consists of only two facts: (1) that is was not present in the several tissues other than olfactory epithelium that were examined, and (2) that it has a remarkably high affinity for camphor. These are certainly suggestive, but are not compelling. Indeed, even the "low-affinity" site has a high affinity for camphor.

The camphor binding components could be solubilized with a detergent, Triton X-100, which suggests that they originated on some membranous cellular constituent. At concentrations up to 1%, the detergent did not alter the binding of camphor by the preparations nor did treatment with ribonuclease, phospholipase C, or collagenase. The camphor-binding activity was greatly reduced by treatment with proteolytic enzymes or with low concentrations of modifiers of sulfhydryl groups (Fesenko *et al.*, 1979). All of the above suggests that camphor-binding sites are associated with proteins and that sulfhydryl groups are necessary for binding. The authors subjected the solubilized preparations to isoelectric focusing to determine the isoelectric point for the binding proteins to be around pH 4.8. They also subjected their preparations to electrophoresis and found two camphor-binding fractions, one that binds about three times as much camphor as the other when the camphor concentration was 5×10^{-10} *M* (Fesenko *et al.*, 1979). In view of the foregoing discussion, the component binding the greater amount of camphor must be the low-affinity binding sites; it has a molecular weight of 1.2×10^5 daltons. The other component, presumably the high affinity site, appears to have a somewhat lower molecular weight, around 1×10^5 daltons. Interestingly, borneol was most effective in displacing camphor from the low affinity binding site and was without effect on the high affinity binding of camphor. These conclusions differ from those of the original authors who assumed that the low-affinity binding site could be ignored at the camphor concentration they used.

Another approach to identifying olfactory receptor proteins has recently been introduced in our laboratory. Affinity chromatography on a column to which *O*-methylphenol moieties were attached to a solid support was used to isolate a protein from dog olfactory epithelium. The protein bound tightly to the column but could be displaced from it by *p*-anisic acid, a water soluble *O*-methylpenol. Since anisole is the simplest of the *O*-methylphenols, we have referred to this as "anisole binding protein." It appears to be nearly homogeneous on polyacrylamide gel electrophoresis (Price, 1978). It was absent from adjacent respiratory epithelium. Therefore, it is not some universal cellular constituent, component of extracellular fluid, nor a protein common to all cilia. Since the mucus layer overlying the olfactory epithelium is continuous with that of the respiratory tissue adjacent to it, the protein is unlikely to be from mucus. Use of a detergent, sodium dodecyl sulfate, was required for its extraction, suggesting that it originated on some cellular membrane. Anisole binding protein was injected into a rabbit, and it induced the formation of antibodies (Goldberg *et al.*, 1979). The γ-globulin fraction containing the antibodies was purified. When the olfactory epithelium of a mouse was exposed to a brief wash with 0.1 ml of saline containing less than 1 ng of this material, the electrophysiological response of the tissue to odorants was reduced by 50%. Higher doses abolished the responses completely. γ-Globulin prepared from serum of the same rabbit, drawn before immunization with anisole-binding protein, was used as a control. Even at concentrations as high as 10,000 ng in 0.1 ml, it had no more effect on responses to odorants than did the saline solution in which it was dissolved.

The effect of the antibodies was not specific to responses to *O*-methylphenols. Indeed, the responses to every odorant we have used were inhibited to about the same extent by 0.8 ng of γ-globulin containing the antibodies. Thus, while it has not been proven that anisole-binding protein is an olfactory receptor protein, it is clearly demonstrated to be an element in the process of recognition or transduction. The fact that topically applied antibodies can interact with it suggests that the protein is on the external surface of the membranes of the olfactory tissue cells. In order to explain the fact that the antibodies inhibit the responses to many odorants, it has been hypothesized that olfactory receptor proteins comprise a class of molecules in which a large portion of the structure is common to all. Thus, immunization with one would result in the production of many antibodies. Most of them would be directed against the part of the molecule common to all olfactory receptor proteins, and antibodies would therefore cross-react with them (Goldberg *et al.*, 1979).

Among the older hypotheses concerning the mode of action of chemical stimuli on receptor cells is the suggestion that the stimuli activate or inhibit certain key enzymes, thus altering the metabolic state of the cell (Baradi and Bourne, 1953). This line of thinking was not widely pursued after Beidler (1962) convincingly argued that it could not be the mechanism for stimulation of taste receptor cells by salts. In the case of olfactory receptor cell stimulation, one particular enzyme, the ATPase activated by sodium and potassium, has received attention in this regard. This enzyme includes a subunit that has been reported to be an ionophore (Shamoo and Ryan, 1975). Since changes in the ATPase activity could reflect interactions with this subunit, it is possible to conceptualize a link with the membrane potential and thus with the transduction process. It has been reported that various odorants alter the activity of the ATPase from mammalian olfactory epithelium *in vitro*, but do not affect the activity of the enzyme from other tissues (Koch, 1971; Koch and Desaiah, 1974; Koch and Gilliland, 1977). Interestingly, the antibodies against anisole-binding protein, when present at concentrations around 50 ng/ml, prevent the activation of olfactory epithelial ATPase, which is otherwise caused by 10^{-3} M concentrations of octanone or nonanone. In the absence of the odorants, the antibodies do not affect the ATPase activity, and the control γ-globulin preparation is without effect in the presence or in the absence of the odorants (Koch *et al.*, 1980).

III. RESEARCH NEEDS

It is clear that studies of olfactory receptor proteins in vertebrates are at a very primitive stage. The problem of how to identify a receptor protein in extracts deserves special attention. Correlation of odorant binding with odorant potency has been used in studies of fish amino acid receptors (Cancalon, 1978; Cagan and Zeiger, 1978). Immunological demonstration of a role in olfaction has been applied to a mammalian protein (Goldberg *et al.*, 1979). In every other case, the evidence for receptor roles for proteins rests solely on the protein having high affinity for some odorant and being present in olfactory tissue. The use of multiple criteria (binding paralleling potency as well as immunological demonstration of a role in transduction, for example) will, it is anticipated, become commonplace as the level of sophistication among investigators in the field increases.

As improved confidence in the receptor nature of isolated proteins develops, application of the proteins to unravelling the workings of the olfactory system can be anticipated. For example, the question of where the receptor sites are located has been of interest for many years, but no direct method of visualizing them exists (Ottoson, 1956; Vinnikov, 1974; Bostock, 1974; Getchell, 1977; Getchell *et al.*, 1980). Antibodies against purified receptor proteins could be utilized with modern immunocytochemical techniques to approach this problem in a straightforward way. Another very difficult problem has been to attempt to determine the specificity of olfactory receptor sites from electrophysiological studies of the specificity of neural responses (Holley and MacLeod, 1977; Revial *et al.*, 1978). With purified receptor proteins, this problem could be approached by direct measurement of the affinities of odorants for the sites. Finally, understanding the mechanism by which odorant binding is transduced into changes in receptor-cell potential could be dramatically advanced by studies of the interactions of receptor proteins with membrane phospholipids and membrane enzymes and the effects of odorants upon these interactions.

ACKNOWLEDGMENT

The preparation of this review was supported, in part, by the National Institute of Dental Research under research grant DE-04271.

REFERENCES

Amoore, J. E. (1971). Olfactory genetics and anosmia. *In* "Handbook of Sensory Physiology" (L. M. Beidler, ed.), Vol. 4, Part 1, pp. 245–256. Springer-Verlag, Berlin and New York.

Amoore, J. E. (1977). Specific anosmia and the concept of primary odors. *Chem. Senses Flavor* **2**, 267–281.

Amoore, J. E., Palmieri, G., and Wanke, E. (1967). Molecular shape and odor: pattern analysis by PAPA. *Nature (London)* **216**, 1084–1087.

Ariens, E. J., Beld, A. J., Rodriguez de Miranda, J. F., and Simonis, A. M. (1979). The pharmacon-receptor-effector concept: a basis for understanding the transmission of information in biological systems. *In* "The Receptors" (R. D. O'Brien, ed.), Vol. 1, pp. 33–92. Plenum, New York.

Ash, K. O. (1968). Chemical sensing: an approach to biological molecular mechanisms using difference spectroscopy. *Science* **162**, 452–454.

Ash, K. O. (1969). Ascorbic acid: cofactor in rabbit olfactory preparations. *Science* **165**, 901–902.

Ash, K. O., and Skogen, J. D. (1970). Chemosensing: selectivity, sensitivity and additive effects on a stimulant-induced activity of olfactory preparations. *J. Neurochem.* **17**, 1143–1154.

Baradi, A. F., and Bourne, G. F. (1953). Gustatory and olfactory epithelia. *Int. Rev. Cytol.* **2**, 289–330.

Baylin, F., and Moulton, D. G. (1979). Adaptation and cross-adaptation to odor stimulation of olfactory receptors in the tiger salamander. *J. Gen. Physiol.* **74**, 37–55.

Beidler, L. M. (1962). Taste receptor stimulation. *Prog. Biophys. Biophys. Chem.* **12**, 107–151.

Bostock, H. (1974). Diffusion and the frog EOG. *In* "Transduction Mechanisms in Chemoreception" (T. M. Poynder, ed.), pp. 27–38. Information Retrieval, London.

Cagan, R. H., and Morris, R. W. (1979). Biochemical studies of taste sensation: binding to taste tissue of ^{3}H-labeled monellin, a sweet-tasting protein. *Proc. Natl. Acad. Sci. U.S.A.* **76**, 1692–1696.

Cagan, R. H., and Zeiger, W. N. (1978). Biochemical studies of olfaction: binding specificity of radioactively labeled stimuli to an isolated olfactory preparation from rainbow trout (*Salmon gairdneri*). *Proc. Natl. Acad. Sci. U.S.A.* **75**, 4679–4683.

Cancalon, P. (1978). Isolation and characterization of the olfactory epithelial cells of the catfish. *Chem. Senses Flavor* **3**, 381–396.

Caprio, J. (1977). Electrophysiological distinctions between the taste and smell of amino acids in catfish. *Nature (London)* **266**, 850–851.

Caprio, J. (1978). Olfaction and taste in the channel catfish: an electrophysiological study of the responses to amino acids and derivatives. *J. Comp. Physiol.* **123**, 357–371.

Cherry, R. J., Dodd, G. H., and Chapman, D. (1970). Small molecule lipid-membrane interactions and the puncturing theory of olfaction. *Biochim. Biophys. Acta* **211**, 409–416.

Cuatrecasas, P. (1974). Commentary: insulin receptors, cell membranes and hormone action. *Biochem. Pharmacol.* **23**, 2353–2361.

Davies, J. T. (1971). Olfactory theories. *In* "Handbook of Sensory Physiology" (L. M. Beidler, ed.), Vol. 4, Part 1, pp. 322–350. Springer-Verlag, Berlin and New York.

Fesenko, E. E., and Pervukhin, G. Y. (1976). Frog olfactory transduction modeling on artificial phospholipid membranes. *Stud. Biophys.* **58**, 215–221.

Fesenko, E. E., Novoselov, V. I., Pervukhin, G. Y., and Fesenko, N. K. (1977). Molecular mechanisms of olfactory sensing. II. Studies of fractions from olfactory tissue scrapping capable of sensitizing artifical lipid membranes to actions of odorants. *Biochim. Biophys. Acta* **466**, 347–356.

Fesenko, E. E., Novoselov, V. I., Mjasoedov, N. F., and Sidorov, G. V. (1978). Molecular mechanisms of olfactory reception. III. Binding of camphor by rat and frog olfactory mucosa components. *Stud. Biophys.* **73**, 71–84.

Fesenko, E. E., Novoselov, V. I., and Krapivinskaya, L. D. (1979). Molecular mechanisms of olfactory reception. IV. Some biochemical characteristics of the camphor receptor from rat olfactory epithelium. *Biochim. Biophys. Acta* **587**, 424–433.

Gennings, J. N., Gower, D. B., and Bannister, L. H. (1977). Studies on the receptors to 5α-androst-16-en-3α-ol in sow nasal mucosa. *Biochim. Biophys Acta* **496**, 547–556.

Getchell, M. L., and Gesteland, R. C. (1972). The chemistry of olfactory reception: stimulus-specific protection from sulfhydryl reagent inhibition. *Proc. Natl. Acad. Sci. U.S.A.* **69**, 1494–1498.

Getchell, T. V. (1977). Analysis of intracellular recordings from salamander olfactory epithelium. *Brain Res.* **123**, 275–286.

Getchell, T. V., and Getchell, M. L. (1977). Early events in vertebrate olfaction. *Chem. Senses Flavor* **2**, 313–326.

Getchell, T. V., Heck, G. L., DeSimone, J. A., and Price, S. (1980). The location of olfactory receptor sites: inferences from latency measurements. *Biophys. J.* **29**, 397–412.

Goldberg, S. J., Turpin, J., and Price, S. (1979). Anisole binding protein from olfactory epithelium: evidence for a role in transduction. *Chem. Senses Flavor* **4**, 207–214.

Guillot, M. (1948). Anosmies partielles et odeurs fondamentales. *C. R. Acad. Sci.* **226,** 1307–1309.

Gusel'nikov, V. I., Korolev, A. M., and Frolov, O. Y. (1974). Interaction of odorants and the olfactory hairs of the frog *in vitro. Dokl. Akad. Nauk SSSR* **216,** 213–216.

Hansen, K. (1978). Insect chemoreception. *In* "Receptors and Recognition" (G. L. Hazelbauer, ed.), Ser. B, Vol. 5, pp. 233–292. Chapman & Hall, London.

Hara, T. J. (1973). Olfactory responses to amino acids in rainbow trout, *Salmo gairdneri. Comp. Biochem. Physiol.* A **44,** 407–416.

Hara, T. J. (1976). Structure-activity relationships of amino acids in fish olfaction. *Comp. Biochem. Physiol.* A **54,** 31–36.

Holley, A., and MacLeod, P. (1977). Transduction et codage des informations olfactives chez les vertébrés. *J. Physiol. (Paris)* **73,** 725–848.

Hornung, D. E., and Mozell, M. M. (1977). Factors influencing the differential sorption of odorant molecules across the olfactory mucosa. *J. Gen. Physiol.* **69,** 343–361.

Hornung, D. E., and Mozell, M. M. (1980). Tritiated odorants to monitor retention in the olfactory and vomeronasal organs. *Brain Res.* **181,** 488–492.

Kier, L. B., DiPaolo, T., and Hall, L. H. (1977). Structure-activity studies on odor molecules using molecular connectivity. *J. Theor. Biol.* **67,** 585–596.

Koch, R. B. (1971). Biochemical studies of olfactory tissue: response of ATPase activities to octanol and ascorbic acid. *Chem.-Biol. Interact.* **4,** 195–208.

Koch, R. B., and Desaiah, D. (1974). Preliminary studies on rat olfactory tissue: effects of odorants on Na^+-K^+ ATPase activities. *Life Sci.* **15,** 1005–1015.

Koch, R. B., and Gilliland, T. I. (1977). Responses of Na^+-K^+ ATPase activity from dog olfactory tissue to selected odorants. *Life Sci.* **20,** 1051–1062.

Koch, R. B., Rossi, H., and Price, S. (1980). Effect of antibody against anisole binding protein on odorant perturbation of Na^+-K^+ ATPase activity. *Meet. Assoc. Chemorecept. Sci., 2nd, Sarasota, Fla.*

Koshland, D. E., Jr. (1979). A model regulatory system: bacterial chemotaxis. *Physiol. Rev.* **59,** 811–862.

Koyama, N., and Kurihara, K. (1972). Effect of odorants on lipid monolayers from bovine olfactory epithelium. *Nature (London)* **236,** 402–404.

Menevse, A., Dodd, G., Poynder, T. M., and Squirrel, D. (1977). A chemical-modification approach to the olfactory code. *Biochem. Soc. Trans.* **5,** 191–194.

Mozell, M. M. (1970). Evidence for a chromatographic model of olfaction. *J. Gen. Physiol.* **56,** 46–63.

Mozell, M. M., and Jagodowicz, M. (1973). Chromatographic separation of odorants by the nose: retention times measured across *in vivo* olfactory mucosa. *Science* **191,** 1247–1249.

Ottoson, D. (1956). Analysis of the electrical activity of the olfactory epithelium. *Acta Physiol. Scand., Suppl.* No. 122, 1–83.

Price, S. (1977). Specific anosmia to geraniol in mice. *Neurosci. Lett.* **4,** 49–50.

Price, S. (1978). Anisole binding protein from dog olfactory epithelium. *Chem. Senses Flavor* **3,** 51–55.

Price, S., and DeSimone, J. A. (1977). Models of taste receptor cell stimulation. *Chem. Senses Flavor* **2,** 427–456.

Revial, M. F., Duchamp, A., Holley, A., and MacLeod, P. (1978). Frog olfaction: odour groups, acceptor distribution and receptor categories. *Chem. Senses Flavor* **3,** 23–33.

Rodbell, M. (1972). Cell surface receptor sites. *In* "Current Topics in Biochemistry" (C. B. Anfinsen, R. F. Goldberger, and A. N. Schechter, eds.), pp. 187–218. Academic Press, New York.

Shamoo, A. E., and Ryan, T. E. (1975). Isolation of ionophores from ion-transport systems. *Ann. N.Y. Acad. Sci.* **30**, 83–97.

Sperber, G. O. (1977). Coacervate-like membrane structures and olfactory transduction. *Acta Physiol. Scand.* **99**, 129–139.

van Drongelen, W., Holley, A., and Døving, K. B. (1978). Convergence in the olfactory system: quantitive aspects of odour sensitivity. *J. Theor. Biol.* **71**, 39–48.

Vinnikov, Y. A. (1974). "Sensory Reception. Cytology, Molecular Mechanisms and Evolution." Springer-Verlag, Berlin and New York.

Vinnikov, Y. A., Pyatkina, G. A., Shakhmatova, O. I., and Natochin, Y. V. (1979). Structure and ionic composition of olfactory and respiratory mucus in sturgeon fish and an oxidation hypothesis of olfaction. (In Russ.) *Dokl. Akad. Nauk SSSR* **245**, 750–753.

Wysocki, C. J., Whitney, G., and Tucker, D. (1977). Specific anosmia in the laboratory mouse. *Behav. Genet.* **7**, 171–188.

5

Chemosensation: An Aspect of the Uniqueness of the Individual

K. YAMAZAKI, M. YAMAGUCHI, G. K. BEAUCHAMP,
J. BARD, E. A. BOYSE, AND L. THOMAS

This chapter is concerned with the observation that the major histocompatibility complex of genes (MHC) is the source of chemosensory information that enables mice to identify one another as individuals. In addition to published work, certain studies that are far from complete will be mentioned.

Clearly, any sensory communication system that reflects the genetic composition of individual members of a species and that influences their social and reproductive behavior accordingly must play a vital part in the evolution and biology of that species.

First, a few rudimentary genetical points. The importance of this assembly of genes called the MHC can be judged from the fact that a similar set of genes is believed to operate in all vertebrates. The MHC of the mouse is called *H-2*, and it occupies a segment of chromosome 17 (Klein, 1975). The MHC of man is called HLA, and it occupies a segment of chromosome 6. In viewing the relatedness of mouse and man, it must be understood that, as fellow mammals, their genomes are very similar; they contain virtually the same amounts of DNA (Ohno, 1970), and they may resemble one another even to the preservation of particular homologous genes in particular linkage groups (Lalley *et al.*, 1978). The fact that the karyotypes of man and mouse are not alike visually may be a relatively trivial matter of alternative gross chromosomal arrangements. In fact, certain incipient subspecies of the mouse in nature have karyotypes markedly dissimilar to that of the common laboratory mouse. Old doubts about the wisdom of generalizing from the

Biochemistry of Taste and Olfaction

ISBN 0-12-154450-8

mouse on fundamental biological matters should long have been dispelled.

The MHC comprises many linked genes and can be divided into regions, notably the main regions called *H-2K* (K), *H-2D* (D), and *Tla*. (Although the boundaries of the MHC were previously set at K and D, there are excellent reasons (see Boyse and Cantor, 1978) for including in the MHC the *Tla* region close to D outside these limits, and this convention makes discussion easier.)

When we speak of a mouse's MHC type or H-2 type, we mean the total set of variable alleles of all genes in the MHC region. The set of MHC alleles on a given chromosome 17 is called a *haplotype*, and clearly a vast number of haplotypes is possible. The haplotype is denoted by a superscript letter or letters, such as $H\text{-}2^b$ (abbreviated *b*) or $H\text{-}2^k$ (abbreviated *k*).

The MHC is best known from studies on tissue transplantation, because incompatibility of MHC types causes extreme rejection of grafts. The fate of kidney transplants in man depends mainly on MHC compatibility. Throughout the MHC region, there are also genes that determine the degree of response to particular antigens, and other genes that are selectively expressed in lymphocytes. Thus, the MHC is concerned with the many aspects of how immune cells—lymphocytes equipped with specific receptors for antigen—handle chemical information from the environment. We wonder whether there is any mechanistic and evolutionary connection with the reception and handling of chemosensory information by the nervous system (Boyse and Cantor, 1978; Boyse, 1979), because this also involves the MHC, as we shall see.

Genes in the MHC also have certain effects of no evident immunological significance. For example, the physiology of steroid hormones (testosterone and cortisone) is affected by genetic variation in the MHC region (for review, see Ivanyi, 1978).

The use of inbred congenic mice has been essential in work on MHC-associated chemosensory communication. An H-2 congenic strain is produced by crossing two selected inbred strains of different *H-2* types and by backcrossing many times to one of these strains, serologically selecting for the donor *H-2* type in each generation (a simple outline is given in Boyse, 1977). Consequently, the final congenic strain is genetically identical to the base strain except for a segment of chromosome 17 bearing the *H-2* haplotype introduced from the donor strain. Any difference between the base strain and its congenic partner strain, if proved to be genetic, must be due to genes in the *H-2* region, because this is the only genetic difference between the inbred base strain and its congenic partner.

One further point requires emphasis. Being inbred, both strains are homozygous for all their genes; each chromosome pair is identical. If the inbred and congenic strains are crossed, a genetically uniform F_1 hybrid population is produced, all being heterozygous, let us say *bk*, for *H-2*.

If these identical F_1 mice are crossed among themselves, an F_2 population is produced, comprising mice of three categories, distinguishable by *H-2* typing. These are *bb* homozygotes, genetically identical to one grandparent strain, and *kk* homozygotes, genetically identical to the other (congenic) grandparent strain. The remaining 50% of the F_2 population are *bk* heterozygotes, and these are discarded. If any trait that distinguishes inbred and congenic strains is reproduced with these F_2 segregants, then this trait must be determined by *H-2*. In genetic terms, the trait in question has segregated with *H-2*. Fortuitous nongenetic differences that may have arisen in the separate inbred and congenic colonies can be discounted.

Again, both sets of F_2 homozygotes are progeny of the same genetically uniform heterozygous F_1 breeding pairs. Therefore the total environment of both *bb* and *kk* F_2 segregant mice, from conception to weaning, has been rendered uniform; they were born of identical F_1 parents.

The original observation, made at Memorial Sloan-Kettering Cancer Center, was that mice of a particular inbred strain seemed socially more reactive to congenic mice of a different *H-2* type than to members of their own strain. To substantiate this impression, it was decided to set up a test system in which inbred males were presented simultaneously with two females, both in estrus, the two females being congenic for *H-2* and therefore differing genetically only in their *H-2* types. The trios were observed until the male had mated with one of the females, as indicated by a vaginal plug.

It transpired that there is indeed an *H-2*-associated mating bias, commonly in favor of the female whose *H-2* type differs from that of the male. For example, taking the *H-2* types *b* and *k* (mentioned above) for purposes of illustration, B6 males, which are $H\text{-}2^b$, mated preferentially with congenic B6-Hw^k females, rather than with the alternative B6 females, and B6-$H\text{-}2^k$ males mated preferentially with B6 females (Yamazaki *et al.*, 1976). MHC-associated preference was demonstrated also among F_2 segregants (Yamazaki *et al.*, 1978), bred as described above.

The next study was intended to determine whether this *H-2*-associated communication system involved chemical sensation. For this purpose, we used a Y maze in which air is drawn through two odor boxes, containing *H-2* congenic mice. The air is then conducted to the left and right arms of the maze, which are thereby scented differentially by mice whose only genetic difference is *H-2*. The incentive to run the maze was a drop of water, the test subject mice having been deprived of water for 23 hours beforehand. Neither male nor female showed any spontaneous *H-2* preference in this Y maze system, but both could be trained to do so by reinforcement, as mentioned. We shall return to this point later. The trained mice exhibited equally good discrimination with *H-2*-typed F_2 segregant mice as the odor sources. This study, which was conducted with four trained B6 mice, a male and a female reinforced for $H\text{-}2^b$ and a male and female reinforced for $H\text{-}2^k$, has been published (Yamazaki *et al.*, 1979).

In a further study, the same four trained mice were used in Y maze trials with urines from *H-2* typed *b* and *k* F_2 mice. The results, which are being prepared for publication, indicate that urine is a rich source of *H-2* differentiating odors. Various biological materials other than urine, including homogenates of various tissues, have not so far proved effective.

Thus there remains no doubt that the MHC defines differential odors that individually distinguish mice and that urine is a prime odor source. It is not known whether and to what extent other genes may contribute to chemosensory identification.

Evidence that MHC-controlled odors are multiple comes from various experiments in which the participating mice differed from one another not throughout the entire MHC region but only at one particular subregion, such as *K, D*, or *Tla*. In each case discrimination has been observed (Andrews and Boyse, 1978; Yamaguchi *et al.*, 1978; and unpublished data), though much further work is required to substantiate this fully for each subregion. Thus the chemosensory identity conferred by MHC types may consist of characteristic profiles for a set of odors, the MHC identity of each individual mouse being established by either quantitative or qualititative genetically determined differences in the components of the odor set (Boyse *et al.*, 1980).

The use of F_2 segregants involves two separate issues that should not be confused. One concerns the genetic determination of the odor profile: F_2 segregant mice, or their urines, are as effective as odor sources as are the inbred and congenic mice, or their urines. This means that no mechanism other than genetic polymorphism of MHC genes need be considered in accounting for the information used in this communication system. The mode of odor specification by MHC, direct or indirect, is unknown. Biochemical analysis is needed to shed light on this.

The response of F_2 segregant mice to this chemical information, in the mating preference assay, is an entirely different matter. This involves behavior, which the production of the information itself does not. In the studies on MHC-associated mating preference, there was a hint that the preferences of F_2 segregants may in some cases differ in degree from those of their genetically identical inbred and congenic counterparts (Yamazaki *et al.*, 1978). This may suggest imprinting, because although B6 mice and F_2 $H\text{-}2^b$ homozygotes, for example, are genetically identical, their prenatal and postnatal environments are not. Thus inbred B6 embryos and weanlings experience only $H\text{-}2^b$, whereas genetically identical F_2 $H\text{-}2^b$ segregants, born of heterozygous ($H\text{-}2^b/H\text{-}2^k$) F_1 parents, experience both $H\text{-}2^b$ and $H\text{-}2^k$ from conception onward. In this context there are reports that infant female mice can be imprinted on their sire, leading in later life to avoidance of males of the same strain as the sire (Yanai and McClearn, 1973). One would wish to see such studies extended to inbred, congenic, and F_2 systems in which genetic variation would be confined to the MHC complex.

We saw that MHC-associated mating preference is spontaneous, whereas MHC discrimination in the Y maze must be learned. This may suggest that chemoperception of MHC types is interpreted by the mouse in the context of other information, doubtless perception of estrus under the circumstances of mating. Indeed, it seems altogether probable that a highly evolved mammal like the mouse should possess advanced faculties for the integration and processing of combined sensory information.

The system in which we chose to view MHC-associated communication from this perspective concerns the phenomenon known as pregnancy block, or the Bruce Effect, after its discoverer, Hilda Bruce (1959). If a female mouse is separated from her mate after fertilization, the risk that pregnancy will be blocked—that the embryos will not implant—is increased if she is exposed to a male of a strain different from her first mate (for reviews, see Whitten, 1966; Godowicz, 1970) or to the urine of such a male (Dominic, 1966a). The mechanism presumably is neuroendocrine, because blocking can be prevented by prolactin (Dominic, 1966b), a hormone that prepares the uterine epithelium for implantation of the cleaving embryos on the fourth or fifth day after conception. But what is the sensory signal?

There is a pheromone (or pheromones) in male urine that have various effects on the reproductive physiology of females (Whitten, 1956; Vandenbergh, 1969). But if male pheromone alone caused blocking of pregnancy, why should it make a difference whether the intruding male, or urine, is of the same or a different strain? A likely reason, we think, among various possible explanations, is that the female perceives not only the male pheromone but additional chemical information concerning the identity of the male, including his MHC type. We have the following evidence that this is so, but these data are most preliminary, no more than an indication that we are on the right track.

All females in this pilot study were of the inbred strain BALB, whose $H\text{-}2$ type is *d*. The stud males were either B6 ($H\text{-}2^b$), or congenic B6-$H\text{-}2^k$. The second (test) male to which the fertilized females were exposed was either the same stud male, a male of the same strain as the stud male, or a congenic male (B6 if the stud male was B6-$H\text{-}2^k$, and B6-$H\text{-}2^k$ if the stud male was B6).

Day 0 was the day of mating, when a vaginal plug was observed. On day 1 the female was isolated in a new cage. On days 2–4, she occupied one side of a divided cage. On the other side of a perforated screen was a test male, which was either the stud male again, another male of the stud strain, or an *H-2* congenic male. On days 5–7, she was returned to the cage she occupied on day 1 and was examined for return of estrus until day 7, when the uterus was removed for inspection. Initiation of the estrous cycle during these 7 days was taken to indicate a blocked pregnancy; where there was no evidence of estrus, pregnancy was considered to be preserved.*

We found that the incidence of blocking from exposure to males of an unfamiliar *H-2* type was higher than from re-exposure to the stud male or from exposure to a male of the stud's *H-2* type. Thus we shall use this system to test our proposition that the sensing of a male of novel *H-2* type triggers in the female a neuroendocrine response that frustrates implantation and blocks pregnancy, perhaps providing a model of genetically based behavior serving to promote outbreeding and heterozygosity in a species whose ecology tends to undesirable homozygosity from inbreeding (Klein, 1975).

ACKNOWLEDGMENT

This work was supported in part by grant CA-23676 from the National Institutes of health and by grant RF-77042 from the Rockefeller Foundation. EAB is American Cancer Society Research Professor of Cell Surface Immunogenetics.

REFERENCES

Andrews, P. W., and Boyse, E. A. (1978). Mapping of an H-2-linked gene that influences mating preference in mice. *Immunogenetics* 6, 265–268.

Boyse, E. A. (1977). The increasing value of congenic mice in biomedical research. *Lab. Anim. Sci.* 27, 771–781.

Boyse, E. A. (1979). Conspectus. Problems of development: Immunology as master and servant. *Curr. Top. Dev. Biol.* 13, xv–xix.

Boyse, E. A., and Cantor, H. (1978). Immunogenetic aspects of biologic communication: A hypothesis of evolution by program duplication. *In* "The Molecular Basis of Cell–Cell Interactions" (R. A. Lerner and D. Bergsma, eds.), Birth Defects Original Article Series, Vol. 14, pp. 249–283. Alan R. Liss, New York.

Boyse, E. A., Yamazaki, K., Yamaguchi, M., and Thomas, L. (1980). Sensory communication among mice according to their MHC types. *In* "The Immune System: Functions and Therapy of Dysfunction" (G. Doria and A. Eshkol, eds.), pp. 45–53. Academic Press, New York.

*This form of the blocking assay includes blocking of pseudopregnancy as well as of true pregnancy (Dominic, 1966c).

Bruce, H. M. (1959). An exteroceptive block to pregnancy in the mouse. *Nature (London)* **184**, 105.

Dominic, C. J. (1966a). Observations on the reproductive pheromones of mice. I. Source. *J. Reprod. Fertil.* **11**, 407–414.

Dominic, C. J. (1966b). Observations on the reproductive pheromones of mice. II. Neuroendocrine mechanisms involved in the olfactory block to pregnancy. *J. Reprod. Fertil.* **11**, 415–421.

Dominic, C. J. (1966c). Block to pseudopregnancy in mice caused by exposure to male urine. *Experientia* **22**, 534–535.

Godowicz, B. (1970). Influence of the genotype of males on pregnancy block in inbred mice. *J. Reprod. Fertil.* **23**, 237–241.

Ivanyi, P. (1978). Some aspects of the H-2 system, the major histocompatibility system in the mouse. *Proc. R. Soc. London, Ser. B* **202**, 117–158.

Klein, J. (1975). "Biology of the Mouse Histocompatibility-2 Complex." Springer-Verlag, Berlin and New York.

Lalley, P. A., Minna, J. D., and Francke, U. (1978). Conservation of autosomal gene synteny groups in mouse and man. *Nature (London)* **274**, 160–162.

Ohno, S. (1970). "Evolution by Gene Duplication." Springer Publ., New York.

Vandenbergh, J. G. (1969). Male odor accelerates female sexual maturation in mice. *Endocrinology* **84**, 658–660.

Whitten, W. K. (1956). Modification of the oestrous cycle of the mouse by external stimuli associated with the male. *J. Endocrinol.* **13**, 399–404.

Whitten, W. K. (1966). Pheromones and mammalian reproduction. *Adv. Reprod. Physiol.* **1**, 155–177.

Yamaguchi, M., Yamazaki, K., and Boyse, E. A. (1978). Mating preference tests with the recombinant congenic strain BALB.HTG. *Immunogenetics* **6**, 261–264.

Yamazaki, K., Boyse, E. A., Mike, V., Thaler, H. T., Mathieson, B. J., Abbott, J., Boyse, J., Zayas, Z. A., and Thomas, L. (1976). Control of mating preferences in mice by genes in the Major Histocompatibility Complex. *J. Exp. Med.* **144**, 1324–1335.

Yamazaki, K., Yamaguchi, M., Andrews, P. W., Peake, B., and Boyse, E. A. (1978). Mating preferences of F_2 segregants of crosses between MHC-congenic mouse strains. *Immunogenetics* **6**, 253–259.

Yamazaki, K., Yamaguchi, M., Baranoski, L., Bard, J., Boyse, E. A., and Thomas, L. (1979). Recognition among mice: Evidence from the use of a Y maze differentially scented by congenic mice of different major histocompatibility types. *J. Exp. Med.* **150**, 755–760.

Yanai, J., and McClearn, G. E. (1973). Assortative mating in mice. III. Genetic determination of female mating preference. *Behav. Genet.* **3**, 75–84.

6

The Major Histocompatibility Complex and Olfactory Receptors

NEIL I. GOLDSTEIN AND ROBERT H. CAGAN

I. INTRODUCTION

The immune response comprises a complex series of reactions initiated upon recognition of a foreign substance or antigen as "non-self." Recognition of "self" or "non-self"—of what is innate and what is foreign—is mediated by a locus of linked genes designated the major histocompatibility complex (MHC) (Benacerraf and Katz, 1975; Gill *et al.*, 1978; Klein, 1979). A simplified diagram of the murine MHC (called the histocompatibility-2 or *H-2* locus) is shown in Fig. 1.

The MHC is involved in a number of important immunologic functions, including transplantation graft rejection, cell–cell interactions, immune responses (*Ir*) to complex antigens (e.g., synthetic polypeptides) and to certain infectious agents (viruses and bacteria), susceptibility to tumors, and control of complement activity (Banacerraf and Katz, 1975; Gill *et al.*, 1978). In addition, the MHC is associated with traits not currently classified as im-

Biochemistry of Taste and Olfaction
ISBN 0-12-154450-8

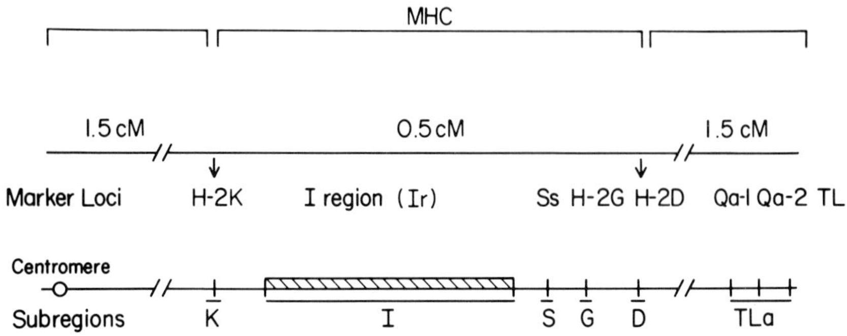

Fig. 1. Schematic diagram of the murine major histocompatibility complex (MHC) designated histocompatibility-2 or *H-2*. The diagram shows the partial genetic fine structures. The boundaries of the *H-2* complex are defined by the arrows in the middle drawing. A description of the regions and marker loci (alleles) is found in the text. The term cM is an abbreviation for centiMorgan and is defined as the map distance between two loci showing 1% crossing over between them.

munological. These include levels of plasma testosterone and weights of steroid-sensitive organs (Iványi *et al.*, 1972a,b), developmental effects (Goldman *et al.*, 1977), binding of viruses and bacteria to MHC surface antigens (Helenius *et al.*, 1978), levels of cAMP in cells and binding of hormones to cell membranes (Meruelo and Edidin, 1975; Lafuse and Edidin, 1978), cell-to-cell adhesion (Bartlett and Edidin, 1978), and mating preferences in inbred strains of laboratory mice (Yamazaki *et al.*, 1976; see also this volume, Chapter 5). It has been suggested that the MHC exerts its influence on the above immunological and nonimmunological traits through its gene products, especially by means of serologically distinct cell surface antigens. These macromolecules may function at the cell surface to modify or moderate the interactions of specific cell receptors with their ligands (Meruelo and Edidin, 1980).

The murine MHC, or *H-2* (Fig. 1), is localized to chromosome 17 and is composed of a series of regions and subregions. The *K* and *D* regions define the boundaries of the MHC and contain the genes *H-2K* and *H-2D*, which code for specific, serologically defined surface glycoproteins (Cunningham *et al.*, 1976; Silver *et al.*, 1976; Gill *et al.*, 1978; Klein, 1979). Each gene is highly polymorphic, having multiple alleles coding for both general (public or shared) and specific (private) antigenic specificities. These specificities are expressed on specific cell surface glycoproteins.

The *I* region (immune response or *Ir*) is located between the *K* and *D* regions and is itself divided into several subregions. The *Ir* region controls a number of reactions involving cellular recognition and interaction, including the graft-versus-host reaction, the mixed lymphocyte reaction, cell-mediated lymphocytotoxicity, and the interaction among T cells, B cells, and macrophages (Benacerraf and Katz, 1975). The gene products of the *Ir* region are serologically distinct glycoproteins designated as the *Ia* or immune-associated antigens.

Other regions within the *H-2* complex are less clearly understood. These include the *S* region, which codes for the C4 component of complement, and the *G* region, which has not been fully defined (Gill *et al.*, 1978). In addition, a number of important loci are associated with the MHC but are not within the complex. On the centromeric side is the *T* locus, which controls several developmental processes and differentiation antigens, and the *Tf* locus, which is a recessive mutation characterized by repeated loss and regrowth of hair in mice. To the right of the *H-2* complex is the *TLa* locus, which controls the expression of a thymic antigen (Boyse *et al.*, 1964) and appears to be involved in murine mating behavior (Andrews and Boyse, 1978).

How does MHC regulate immune recognition of self and non-self? A number of laboratories have devised theoretical models for *H-2* modulation of the immune response (Zinkernagel *et al.*, 1978; Langman, 1978; Talmage, 1979). The models are based on the observation that killing by cytotoxic T cells is restricted to target cells containing both self (*H-2* haplotypes) and non-self (antigenic) determinants (Zinkernagel and Doherty, 1975). Several researchers have described a "dual receptor" theory, which involves both positive and negative selection of T cells in the thymus gland (Langman, 1978; Talmage, 1979). They speculate that these T cells are "imprinted" with specific receptors against both antigen and the MHC while in the thymus, and they are activated only in the presence of complementary determinants on a stimulating (target) cell.

II. SELF, NON-SELF, AND OLFACTION

The idea that a relationship exists between the immunological concept of "self-versus-non-self" and the sense of olfaction was first proposed by Lewis Thomas in his prize winning book, "The Lives of a Cell" (Thomas, 1974). In it he states:

> Olfactory receptors for communication between different creatures are crucial for the establishment of symbiotic relations. . . . The system is evidently an ancient one, long antedating the immunologic sensing of familiar or foreign forms of life by the antibodies on which we now depend so heavily for our separateness. It has recently been learned that the genes for the marking of self by cellular antigens and those for making immunologic responses by antibody formation are closely linked. It is possible that the invention of antibodies evolved from the earlier sensing mechanisms needed for symbiosis, perhaps designed, in part, to keep the latter from getting out of hand.

Thomas described various phylogenetic parallels between immunology and olfaction: (1) the high sensitivities of both sensing systems to the specific chemical signals that act as stimuli (antigens or odorants); (2) the involvement of receptor-like macromolecules in the initial recognition event; and (3) the memory response.

The first experimental evidence for Thomas' thesis that a relationship exists between immune recognition and olfaction was presented by Yamazaki and co-workers (1976). Their work is described in the preceding chapter (Chapter 5), but a brief description of some of the salient points is given here. These researchers observed that mating preferences in inbred strains of mice are influenced by the major histocompatibility complex, *H-2*. They defined "strain preference" as the tendency of males of one *H-2* haplotype to prefer females of the same or another (congenic) *H-2* haplotype. For example, $H\text{-}2^k$ male and $H\text{-}2^b$ female matings occurred in about 70% of the trials. *Consistency of choice* was defined as the specific mating tendency of an individual male during a series of trials. The authors proposed a theoretical model involving two linked genes in the *H-2* region, one for the signal and one for the receptor. In addition, they postulated that the cues reside in the female urine and that the entire process proceeds via the olfactory system (see this volume, Chapter 5).

The evolutionary significance inferred from these studies concerns the selective advantages of outbreeding in nature. For example, heterozygosity of genes in the vicinity of the *H-2* region, such as the *Ir* genes, would increase the range of an individual's immune response and thereby contribute to their survival probability in the face of potentially hazardous materials.

In subsequent studies, Yamazaki *et al.* (1978) confirmed the localization in the MHC of the genes responsible for male *H-2* associated mating preferences. They also postulated that female mice are involved in the mating choice through mediation by genes in the *Qa-TLa* region (Andrews and Boyse, 1978) (Fig. 1), and they also postulated the presence of two genetic loci; signal genes (*Ris*), which code for the cues, and recognition genes (*Rir*), which probably code for receptor macromolecules. In males, these genetic loci are located within the MHC, and in females they are in the *Qa-TLa* region (Andrews and Boyse, 1978; Yamaguchi *et al.*, 1978; Yamazaki *et al.*, 1978).

III. MONOCLONAL ANTIBODY PRODUCTION

We postulate that *Rir* (recognition genes) code for sex-specific surface receptors. These may be isolatable using immunological and biochemical techniques. The experimental paradigm we are using to isolate these MHC-associated olfactory receptors is shown in Fig. 2. This approach takes advantage of recent advances in immunologic technology, specifically the ability to produce monoclonal antibody by the hybridoma technique of Köhler and Milstein (1975, 1976).

Isolation of many macromolecules have been facilitated by the use of antibodies as experimental reagents because an antibody is specific for its immunizing antigen. Classical immunologic techniques are usually used to obtain antibody; this involves sequential injection of the antigen into a suitable animal vector (e.g., rabbit, goat, sheep) and collection of blood at intervals. Antibody-containing serum is separated from the red blood cells by centrifugation and the immunoglobulins are precipitated using ammonium sulfate (Garvey *et al.*, 1977). Although large quantities of material can be obtained using these methods, the final product is not homogeneous but contains an unknown number of heterologous antibodies with varying affinities for the antigen. Furthermore, the purity of an antiserum is inversely proportional to the complexity of the immunizing antigen and the composition of antisera varies among lots. A source of homogeneous antibody would be extremely valuable.

Recently, Köhler and Milstein (1975, 1976) developed a method to obtain large quantities of homogeneous or "monoclonal" antibody, even when the original antigen is highly complex. This methodology involves formation of

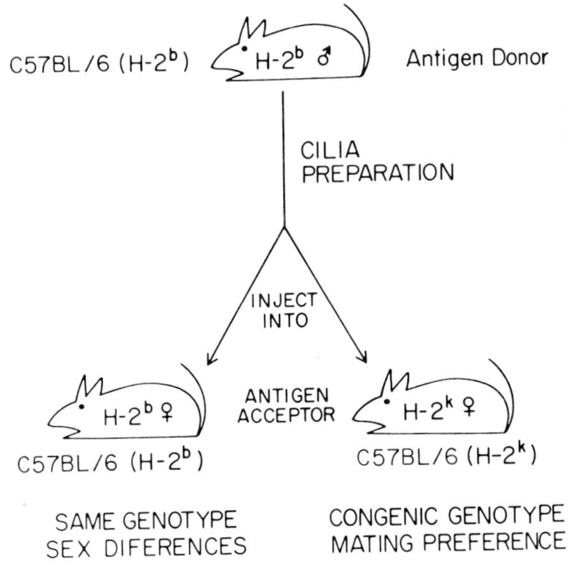

Fig. 2. The experimental protocol for isolating MHC-associated olfactory receptors in mice. See text for details.

"lymphocyte hybridomas," which result from the somatic cell fusion between splenic B lymphocytes and cultured myeloma cells. A myeloma is a primary malignant tumor of bone marrow; the myeloma cells used for the fusion are of murine plasmacytoma origin. When the hybridomas are subsequently cloned, cell lines can be derived which synthesize monoclonal antibody. Thereby, the multiple components of a heterogeneous population of hybrid cells, producing a heterogeneous population of antibody molecules, can ultimately be resolved into specific antibody-producing cell lines that can be maintained in tissue culture.

In developing the hybridoma technology, Köhler and Milstein (1975, 1976) took advantage of classic immunological techniques and advances in the field of somatic cell hybridization. The procedure is shown in Fig. 3. Briefly, an animal's immune system is activated by injecting an antigen. This initiates the primary humoral response in which synthesis of antigen-specific antibody follows a latent period of several days. In time, the antibody titer decreases due to clearance of the antigen from the body by the antibody. The levels of antibody, however, do not drop to the preimmune state, so that upon reimmunization a secondary or anamestic (memory) response is generated. The antibody titer increases rapidly and eventually surpasses that of the primary response. At this point (3–5 days following the secondary injection), the animal's spleen, a major source of B lymphocytes, is excised and the splenocytes (B lymphocytes) are collected. These lymphocytes are fused with a cultured myeloma (plasmacytoma) cell line using polyethylene glycol as the fusogen (Pontecorvo, 1975; Kennett *et al.*, 1978). The resulting fusion products are called *hybridomas*.

Following hybridization, the cells are serially diluted into microtiter plates, and the hybridomas are selected with HAT (hypoxanthine:aminopterin:thymidine) medium (Littlefield, 1964). This selection procedure works because the myeloma cells lack the enzyme hypoxanthine phosphoribosyltransferase (HPRT), which is involved in a salvage pathway of DNA synthesis. Aminopterin blocks the regular synthetic pathway so that those cells that lack HPRT will die in aminopterin-containing medium; in addition, unfused lymphocytes are viable for only 72–96 hr when grown *in vitro* without stimulation by mitogens. Only those cells that result from the fusion of a myeloma cell and a lymphocyte will survive in HAT medium because of complementation of the defective HPRT gene in the myeloma cell by the lymphocyte genome (Littlefield, 1964). A fortuitous factor in this technique is that the greatest percentage of B lymphocytes that undergo fusion to form hybridomas are those that have been activated, i.e., are antibody producing (Köhler *et al.*, 1978).

Hybridomas are visualized as macroscopic clumps 7–21 days following fusion. At this point, it becomes necessary to assay the culture medium for

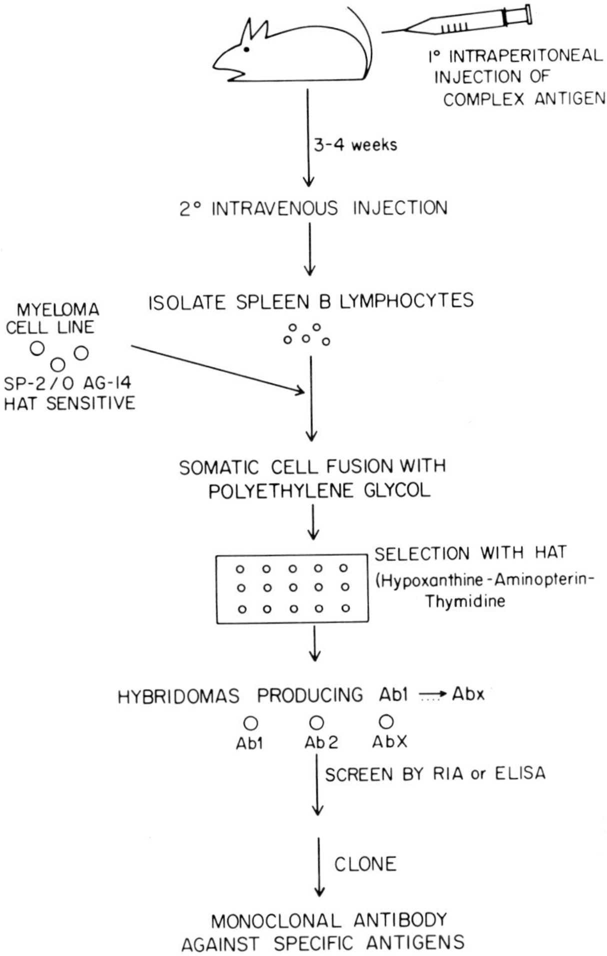

Fig. 3. The experimental procedure to produce lymphocyte hybridomas. (Based on Köhler and Milstein, 1975, 1976.) See text for details.

the presence of the desired antibody. RIA (radioimmunoassay) (Manson *et al.*, 1978) or ELISA (enzyme-linked immunosorbent assay) procedures (Voller *et al.*, 1978; A. H. Smith, personal communication) have been used in our studies to select hybridomas that are positive for the immunizing antigen. Positive colonies are cloned by limiting dilution over a feeder layer of splenocytes following which clonal cell lines that produce monoclonal antibody can be derived.

Large quantities of antibody can, in principle, be obtained by injecting these cell lines, which retain the malignant characteristics of the parental myeloma cell, into syngeneic animals in which the cells grow as ascites tumors. The ascites fluid can be periodically tapped with a syringe and the antibody isolated from the fluid by ammonium sulfate precipitation. The purified monoclonal antibody might then be used to isolate the original immunizing antigen using appropriate biochemical techniques, such as immunoprecipitation with Staph A protein (Kessler, 1975) or immunoaffinity chromatography on Sepharose columns (Cuatrecasas, 1970).

IV. EXPERIMENTAL STUDY OF RELATIONSHIPS BETWEEN OLFACTORY RECEPTORS AND THE MHC

In our experimental paradigm (Fig. 2), a complex antigen was used. A preparation that apparently contains olfactory cilia from C57BL/6 $H\text{-}2^b$ male mice was used to immunize female mice of both $H\text{-}2^b$ and $H\text{-}2^k$ (congenic) haplotypes. Hybridoma cell lines were then isolated that produced monoclonal antibody against components in cilia.

A preparation was isolated from male mice using the methodology of Rhein and Cagan (1980; see also this volume, Chapter 3) which is a modification of the technique used to isolate mollusk gill cilia (Linck, 1973). We presume the mouse preparation to be enriched in cilia, although it has not yet been characterized. In using this preparation as the antigen, it is assumed that male cilia contain receptors that are absent from homologous cilia from syngeneic or congenic females. Male cilia are postulated to contain macromolecules that are antigenically foreign to females. These antigens would therefore elicit an immune response upon injection into female mice with the concomitant synthesis of antibody. Injecting $H\text{-}2^b$ male cilia into an $H\text{-}2^b$ female would reveal sex differences; injecting the same cilia preparation into an $H\text{-}2^k$ female would reveal congenic differences (Fig. 2).

Before its injection into the female recipients, the cilia preparation was uniformly mixed with complete Freund's adjuvant (1:1, v/v). Approximately 100–200 μgm of ciliary protein were injected per animal via the intraperitoneal route. After 3 weeks, each animal was given a secondary intravenous injection of the cilia preparation (100 μgm of protein) in phosphate-buffered saline solution. Three days following the intravenous injection, the animals were killed by cervical dislocation and the spleens were carefully removed from each. Lymphocytes were collected by perfusing each spleen with tissue culture medium. Red blood cells in the perfusate were lysed with 0.17 *M* ammonium chloride (Kennett *et al.*, 1978), and the lymphocytes were then washed with a saline solution. Cultured myeloma cells (SP2/0 Ag-14, HPRT negative; obtained from Dr. Roger Kennett) (Shulman *et al.*, 1978) were added to the lymphocytes at a ratio of 1:10, and

TABLE I Isolation of Hybridomas

Experiment number	Antigen	Recipient	Number of hybridomas[a]	Positive at first screen[b]	Positive at second screen[b]
1	Male Olfactory Cilia from C57BL/6 $H\text{-}2^b$	Female $H\text{-}2^b$	10/576	5 (RIA)	1 (RIA)-F
		Female $H\text{-}2^k$	42/576	14 (RIA)	2 (RIA)-C 2 (RIA)-F
2	Male Olfactory Cilia from C57BL/6 $H\text{-}2^b$	Female $H\text{-}2^b$	21/480	3 (ELISA)-C 12 (ELISA)-F	— —

[a] Number of wells with colonies/total number plated.
[b] RIA, radioimmunoassay; ELISA, enzyme-linked immunosorbent assay; F, frozen in liquid nitrogen; C, cloned.

TABLE II Cloning of Hybridomas[a]

Hybridoma colony number	From fusion	Number of clones/number of wells plated	Number of positive clones/number of clones	Number of clones subcultured/number positive clones	Positive reassay
MCSK-1	$H\text{-}2^b$ Male cilia → $H\text{-}2^k$ female	15/96	6/15	6/6	2
MCSK-2	$H\text{-}2^b$ Male cilia → $H\text{-}2^k$ female	27/96	5/27	5/5	2
MCSB-1	$H\text{-}2^b$ Male cilia → $H\text{-}2^b$ female	34/96	11/34	6/11	3
MCSB-2	$H\text{-}2^b$ Male cilia → $H\text{-}2^b$ female	24/96	14/24	6/14	2
MCSB-3	$H\text{-}2^b$ Male cilia → $H\text{-}2^b$ female	58/96	41/58	9/41	3

[a] Colonies MCSK-1 and MCSK-2 resulted from experiment No. 1 in Table I. Colonies MCSB-1 to -3 resulted from experiment No. 2 in Table I. All assays were carried out using an ELISA procedure.

fusion was performed using 30% polyethylene glycol as the fusogen (Kennett *et al.*, 1978). Following the fusion step the cells were plated into six 96-well microtiter plates, and hybridomas were selected with HAT medium.

In our first series of experiments, hybridoma colonies grew in 10 wells of the 576 seeded for the fusion involving splenocytes from $H\text{-}2^b$ females injected with an $H\text{-}2^b$ male cilia preparation and in 42 wells for $H\text{-}2^k$ females (Table I). Colonies were assayed using a modified RIA procedure (Manson *et al.*, 1978) in which $H\text{-}2^b$ male cilia were bound to filter disks activated with cyanogen bromide. Among the female $H\text{-}2^b$ hybridomas, five colonies were positive for male cilia and 14 were positive with the $H\text{-}2^k$ hybridomas. The positive colonies were subcultured into 20-well microtiter plates, and when the hybridomas reached confluency, a second RIA was done. As shown in Table I, there was a decrease in the number of positive colonies; this is probably due to the poor viability of antibody secreting hybridomas before they are cloned (Kennett *et al.*, 1978; N. I. Goldstein, unpublished observations). Based on the results of the RIA, two $H\text{-}2^k$ colonies were chosen for cloning by serial dilution. The remainder were frozen in liquid nitrogen for future cloning.

In addition to the experiment described above, a second fusion was done using splenocytes from $H\text{-}2^b$ females. This resulted in 21 hybridoma colonies (Table I), of which 15 were found to be positive for $H\text{-}2^b$ male olfactory cilia using an ELISA procedure. Three colonies from this group were cloned based on the results of the ELISA. The remainder were stored in liquid nitrogen for future work.

The results of the cloning experiments are shown in Table II. Cells were serially diluted to an approximate number of 10 cells/ml and 0.1 ml of the suspension was plated over a feeder layer of $2\text{–}3 \times 10^6$ lymphocytes in tissue culture medium (Kennett *et al.*, 1978). Many of the clones that grew were found to be positive for $H\text{-}2^b$ male olfactory cilia using an ELISA assay. A manageable number of clones (32 clones) were subcultured following which weakly positive colonies were discarded based on a second series of ELISA assays. The resulting clones are shown in the right hand column of Table II. We hope to use these hybridomas to isolate large amounts of monoclonal antibody that is directed against specific antigens on male olfactory cilia.

V. FUTURE PROSPECTS

Further experimentation into the biology and biochemistry of the olfactory receptors involved in MHC associated mating behavior offer exciting prospects. Preliminary experiments with monoclonal antibody isolated from hybridoma clones that are positive for male olfactory cilia are presently being explored as immunoaffinity reagents to attempt to isolate specific macromolecules. These can then be studied biochemically. In addition, these antibodies can be used as competing ligands in biochemical assays to study the nature of the chemical signals involved in murine mating behavior.

A number of related studies are also underway. Olfactory cilia from female *H-2*^{b} mice have been injected into male *H-2*^{b} and *H-2*^{k} animals to study the receptors coded for by the *Qa-TLa* region (Andrews and Boyse, 1978). We are exploring monoclonal antibodies to study receptors from the vomeronasal organ of mice. Finally, work is in progress to isolate cell lines of olfactory receptor cells in tissue culture which may then be used to study the mating receptors at the molecular and genetic level.

ACKNOWLEDGMENT

The cultured myeloma cells (SP2/0 AG-14) were provided by Dr. Roger H. Kennett of the University of Pennsylvania Cell Center, supported under Grant No. GM-20138 from the National Institute of General Medical Sciences, NIH.

REFERENCES

Andrews, P. W., and Boyse, E. A. (1978). Mapping of an *H-2*-linked gene that influences mating preferences in mice. *Immunogenetics* **6**, 265–268.

Bartlett, P. F., and Edidin, M. (1978). Effect of the *H-2* gene complex rates of fibroblast intercellular adhesion. *J. Cell Biol.* **77**, 377–388.

Benacerraf, B., and Katz, D. H. (1975). The histocompatibility-linked immune response genes. *Adv. Cancer Res.* **21**, 121–173.

Boyse, E. A., Old, L. J., and Luell, S. (1964). Genetic determination of the *TL* (thymus-leukemia) antigen in the mouse. *Nature (London)* **201**, 779.

Cuatrecasas, P. (1970). Protein purification by affinity chromatography. Derivatizations of agarose and polyacrylamide beads. *J. Biol. Chem.* **245**, 3059–3065.

Cunningham, B. A., Henning, R., Milner, R. J., Reske, K., Ziffer, J. A., and Edelman, G. M. (1976). Structure of murine histocompatibility antigens. *Cold Spring Harbor Symp. Quant. Biol.* **41**, 351–362.

Garvey, J. S., Cremer, N. E., and Sussdorf, D. H. (1977). "Methods in Immunology." Benjamin, New York.

Gill, T. J., III, Cramer, D. V., and Kunz, H. W. (1978). The major histocompatibility complex-comparison in the mouse, man and the rat. *Am. J. Pathol.* **90**, 735–777.

Goldman, A. S., Katsumata, M., Yaffe, S. J., and Gasser, D. L. (1977). Palatal cytosol cortisol-binding protein associated with cleft palate susceptibility and *H-2* genotype. *Nature (London)* **265**, 643–645.

Helenius, A., Morein, B., Fries, E., Simons, K., Robinson, P., Schirrmacher, V., Terhorst, C., and Strominger, J. L. (1978). Human (*HLA-A* and *HLA-B*) and murine (*H-2K* and *H-2D*) histocompatibility antigens are cell surface receptors for Semliki Forest virus. *Proc. Natl. Acad. Sci. U.S.A.* **75**, 3846–3850.

Iványi, P., Gregorová, S., and Micková, M. (1972a). Genetic differences in thymus, lymph node, testes and vesicular gland weights among inbred mouse strains. Association with the major histocompatibility (*H-2*) system. *Folia Biol.* (*Prague*) **18**, 81–97.

Iványi, P., Hámpl, R., Stárka, L., and Micková, M. (1972b). Genetic association between *H-2* gene and testosterone metabolism in mice. *Nature* (*London*), *New Biol.* **238**, 280–281.

Kennett, R. H., Denis, K. A., Tung, A. S., and Klinman, N. R. (1978). Hybrid plasmacytoma production: Fusions with adult spleen cells, monoclonal spleen fragments, neonatal spleen cells and human spleen cells. *Curr. Top. Microbiol. Immunol.* **81**, 77–91.

Kessler, S. W. (1975). Rapid isolation of antigens from cells with a staphylococcal protein A-antibody adsorbent: Parameters of the interaction of antibody-antigen complexes with protein A. *J. Immunol.* **115**, 1617–1624.

Klein, J. (1979). The major histocompatibility complex of the mouse. *Science* **203**, 516–521.

Köhler, G., and Milstein, C. (1975). Continuous cultures of fused cells secreting antibody of predefined specificity. *Nature* (*London*) **256**, 495–497.

Köhler, G., and Milstein, C. (1976). Derivation of specific antibody-producing tissue culture and tumor lines by cell fusion. *Eur. J. Immunol.* **6**, 511–519.

Köhler, G., Hengartner, H., and Shulman, M. J. (1978). Immunoglobulin production by lymphocyte hybridomas. *Eur. J. Immunol.* **8**, 82–88.

Lafuse, W., and Edidin, M. (1978). *H-2* Haplotype affects the binding of glucagon to mouse liver plasma membranes. *Fed. Proc., Fed. Am. Soc. Exp. Biol.* **37**, 1600.

Langman, R. E. (1978). Cell-mediated immunity and the major histocompatibility complex. *Rev. Physiol. Biochem. Pharmacol.* **81**, 1–37.

Linck, R. W. (1973). Comparative isolation of cilia and flagella from the lamellibranch mollusc, *Aequipecten irradians*. *J. Cell Sci.* **12**, 345–367.

Littlefield, J. W. (1964). Selection of hybrids from matings of fibroblasts *in vitro* and their presumed recombinants. *Science* **145**, 709–710.

Manson, L. A., Verastegui-Cerdan, E., and Sporer, R. (1978). A quantitative disc radioimmunoassay for antibodies directed against membrane-associated antigens. *Curr. Top. Microbiol. Immunol.* **81**, 232–234.

Meruelo, D., and Edidin, M. (1975). Association of mouse liver adenosine 3′:5′-cyclic monophosphate (cyclic AMP) levels with *Histocompatibility-2* genotype. *Proc. Natl. Acad. Sci. U.S.A.* **72**, 2644–2648.

Meruelo, D., and Edidin, M. (1980). The biological function of the major histocompatibility complex: Hypotheses. *Contemp. Top. Immunobiol.* **9**, 231–253.

Pontecorvo, G. (1975). Production of mammalian somatic cell hybrids by means of polyethylene glycol treatment. *Somat. Cell Genet.* **1**, 397–400.

Rhein, L. D., and Cagan, R. H. (1980). Biochemical studies of olfaction: Isolation, characterization and odorant binding activity of cilia from rainbow trout olfactory rosettes. *Proc. Natl. Acad. Sci. U.S.A.* **77**, 4412–4416.

Shulman, M., Wilde, C. D., and Köhler, G. (1978). A better cell line for making hybridomas secreting specific antibodies. *Nature* (*London*) **276**, 269–270.

Silver, J., Cecka, J. M., McMillan, M., and Hood, L. (1976). Chemical characterization of products of the *H-2* complex. *Cold Spring Harbor Symp. Quant. Biol.* **41**, 369–377.

Talmage, D. W. (1979). Recognition and memory in the cells of the immune system. *Am. Sci.* **67**, 173–177.

Thomas, L. (1974). "The Lives of a Cell," pp. 40–41. Viking Press, New York.

Voller, A., Bartlett, A., and Bidwell, D. E. (1978). Enzyme immunoassays with special reference to ELISA techniques. *J. Clin. Pathol.* **31**, 507–520.

Yamaguchi, M., Yamazaki, K., and Boyse, E. A. (1978). Mating preference tests with the recombinant congenic strain BALB.HTG. *Immunogenetics* **6**, 261–264.

Yamazaki, K., Boyse, E. A., Miké, V., Thaler, H. T., Mathieson, B. J., Abbott, J., Boyse, J., Zayas, Z. A., and Thomas, L. (1976). Control of mating preferences in mice by genes in the major histocompatibility complex. *J. Exp. Med.* **144**, 1324–1335.

Yamazaki, K., Yamaguchi, M., Andrews, P. W., Peake, B., and Boyse, E. A. (1978). Mating preferences of F_2 segregants of crosses between MHC-congenic mouse strains. *Immunogenetics* **6**, 253–259.

Zinkernagel, R. M., and Doherty, P. C. (1975). *H-2* Compatibility requirement for T-cell-mediated lysis of target cells infected with lymphocytic choriomeningitis virus. Different cytotoxic T-cell specificities are associated with structures coded for in *H-2K* or *H-2D*. *J. Exp. Med.* **141**, 1427–1436.

Zinkernagel, R. M., Callahan, G. N., Althage, A., Cooper, S., Klein, P. A., and Klein, J. (1978). On the thymus in the differentiation of "*H-2* self-recognition" by T-cells: Evidence for dual recognition? *J. Exp. Med.* **147**, 882–896.

PART I

Discussion

Gesteland: A general question is the degree to which we are observing the olfactory message-generating process or responses that are not tissue specific. Apparent dissociation constants and Michaelis–Menten constants are used often and interpreted by some as being receptor specific, while others state them to indicate no receptor specificity. We don't have apparent constants computed for such processes as diffusion, partitioning, and nonequilibrium processes. How do the binding constants in trout and pigs compare? Dr. Gower concludes that the steroids probably are not binding to specific receptors.

Gower: We compared the apparent K_m values with the binding constants, which were 10^{-7}–10^{-8} M. They are lower than with the amino acids.

Karnovsky: You earlier suggested that the hydrogenases themselves are not the receptor molecules. And yet you now state that they *are* the receptors.

Gower: You're quite right. We have recently completed further studies and obtained K_m values of 10^{-7} M for the enzymes.

Schiffman: The binding results with the amino acids are interesting because the groupings are consistent with psychophysical *taste* data. The binding data indicate that L-threonine, L-serine, and L-alanine are grouped together, and L-histidine and L-lysine are separate, while D-alanine showed little binding. Qualitatively, threonine, serine, and alanine are all quite sweet. Histidine and lysine taste similar to one another, and D-alanine tastes very different from the others. Thus, in terms of qualitative taste psychophysics, the olfactory binding looks surprisingly similar.

Rhein: Lysine does not inhibit binding of threonine, serine, or alanine. There definitely have to be two different receptor recognition sites for these odorants.

Gesteland: Do cilia other than olfactory cilia show specific uptake mechanisms for these amino acids? Other tissues seem to have active transport accumulation systems for amino acids.

Rhein: I only briefly examined this question. Amino acid transport systems in vertebrates are usually Na^+-dependent. We found no effect of 10 mM Na^+ on the measured binding. Neither did 10 mM K^+, used as a control, affect the binding. We have no evidence for active transport.

Margolis: Presumably trout, like many other animals, have ciliated respiratory epithelia. Do the cilia from respiratory epithelia show the same biochemical characteristics and the same binding patterns?

Rhein: We have not studied other types of cilia. In the trout, unfortunately there are no cilia on the gills.

Cagan: We considered the initial problem to be obtaining cilia in an active form with respect to odorant binding. Now that this is achieved, it would be interesting to compare them with other epithelial sources to determine if the binding might reflect more widespread mechanisms.

Cancalon: Do the olfactory cells in the trout have cilia? Do some have only microvilli?

Rhein: Our ultrastructural studies of the intact rosettes show the cells in the olfactory regions of the trout to be ciliated. Some microvilli are present on some other cell types but they are sparse.

Cancalon: In a catfish I've characterized two types of receptor cells present, one type with only cilia and one type with microvilli.

Rhein: A microvillous cell is described in fish but its function is not known. We see very few of these in the trout.

Cagan: How do you know that cells without cilia are olfactory receptor cells? What are the criteria?

Cancalon: Taking pictures of the entire cell from the microvilli to the axon.

Cagan: How do you know that the cells don't have both microvilli and cilia, and that you have sectioned through a portion without cilia? Have you done serial sections?

Cancalon: I've isolated both types of cells. Also, by scanning electron microscopy of the surface of the olfactory lamellae, cells with only cilia or microvilli can be seen. My question regards removing only olfactory cilia and not cilia from respiratory cells.

Rhein: For the initial question we posed, it is insignificant whether or not respiratory cilia are present in the preparation. The important point is that the biochemical criteria show the preparation is enriched in ciliary material. Only the olfactory cilia would presumably bind the odorants, but this cannot be selectively demonstrated with the techniques used.

Cagan: There are two questions. One is the functional role of cilia in detecting odorants, and the other is whether or not there are cilia in the preparation that are not functional in odorant recognition. At the current state of the art, it is not possible to remove only one type of cilium rather than another, or to identify them. We were pleased to obtain cilia with odorant binding activity!

Rhein: When the cilia are removed, they are probably fragmented. It is not possible to distinguish among various types of cilia.

Price: The question of odorant binding proteins on cilia other than olfactory receptor cells assumes that what makes it an olfactory receptor cell is the presence of these sites. This is not necessarily true. One can conceive of a common ciliary constituent being an amino acid binding site. If the cell does not have a nerve projecting to the olfactory bulb it will not be an olfactory receptor. If you prepared respiratory cilia and discovered them to bind amino acids as do olfactory cilia, then that would still not be a good reason to conclude that they are not receptor sites.

Matschinsky: How do binding constants compare with comparable constants for internal transport systems, such as in liver, muscle, or heart? I would expect them to be 1 or 2 orders of magnitude different because in the internal milieu these amino acids are around 0.1–0.5 mM. It would be interesting to know if they are comparable or if these are specializations in the cilia.

DeSimone: I believe that there are only two mammalian tissues that are Na^+-dependent for active transport of amino acids—the kidney and the gut, while other tissues take up amino acids without apparently requiring Na^+.

Cagan: But some microorganisms require Na^+ for transport.

DeSimone: For mammalian tissues, the Na^+ requirement is not universal.

Menco: Na^+ depletion results in insertion of rod-shaped intramembranous particles, observed using freeze-fracture, in the luminal cells of the midportion of chick cloaca. Olfactory supporting cells also show such rod-shaped particles in their luminal regions, but they are not present in all instances. Hence it cannot be excluded that supporting cells play a role in regulating the ionic environment surrounding olfactory receptor endings.

Mooser: How do the binding activities compare for the cilia relative to trout Fraction P2?

Rhein: The K_m values are the same, but the B_{max} values were higher with the P2 preparation. We found that when Fraction P2 was exposed to deciliation medium, there was a marked decrease in binding activity (B_{max}). We conclude that the receptors are being denatured to some extent.

Mooser: Do you lose a considerable portion of the total amount of binding sites during isolation?

Cagan: It's difficult to measure the total number of binding sites in the whole rosettes because the homogenate is so messy to assay. There is considerable variability. It seems, however, that the ethanol does denature the receptors. That's a problem.

Rhein: Even with the denaturation, we obtain considerable binding activity which allows verification of some of the binding features.

Mooser: But little was lost in terms of affinity?

Rhein: The affinity remains the same.

Price: With the dorsal epithelium of the frog, what is the receptor distribution? Are most of the olfactory receptors in the dorsal region?

Hornung: We always use the dorsal epithelium around the external naris. We do not know how the receptor density varies across the dorsal mucosal sheet.

Gesteland: The tissue presumably contains a certain fraction by weight of lipid, possibly 1%. Wouldn't you expect the octanol partitioning to eventually show an accumulation in the lipid?

Hornung: Upon examining uptake with time using pieces exposed for 16–18 hours, the curve does reach a plateau. The plateau is above that for water, and I suspect the difference is reflected in the uptake by the lipid compartment. Why more octane is not found in the mucosa due to its lipid solubility remains yet to be explained.

Karnovsky: Why is there so much concern about the exchangeable tritium of the butanol? It could be removed by making a derivative and purifying to remove the hydroxyl tritium. Is there concern about the α-carbon tritiums and the possibility that the butanol is actually metabolized by dehydrogenation? With respect to a more unlikely metabolic reaction the octane might have been hydroxylated. What could the octane be converted to?

Hornung: We determined that for butanol most of the label is on the carbon backbone rather than the hydroxyl group by taking the butanol to butyl acetate. However, it is possible that 5–10% of the label may be on the —OH group. The fate of this hydroxyl label has been our concern for octane. It is at least possible that following interaction with the mucus, this odorant may be changed chemically. Other explanations are that something from the mucosa may either dissolve in the water and thereby increase water's ability to dissolve octane or dissolve in the benzene and decrease its ability to dissolve octane.

Karnovsky: Sounds unlikely.

Hornung: We collected those data 2 years ago and still don't have a satisfactory explanation. We're open to suggestions!

Brand: In determining the partition coefficients using olfactory tissue in the teflon apparatus, did you ever reverse the tissue so that it was upside down to measure partitioning into the cut side?

Hornung: No. We did examine partitioning with various other liquids. For example, a watery mucus collected from the dog respiratory epithelium had the same uptake as water.

Brand: Have you studied other tissues from that area of the frog?

Hornung: We have not.

Brand: What are the differences or similarities between the receptors of the sow olfactory epithelium and the respiratory epithelium with respect to the presence of the reductase enzymes and their activity? Might this affect the relative distribution of the steroids?

Gower: We don't know whether these reductases have any significant effect in the olfactory process. That's the main problem. Why we find different portions of 3α- and 3β-alcohols in the different tissues at the moment escapes us, apart from the suggestion that there might be a refuse disposal activity.

Hirsch: When Scatchard first determined the utility of the graphical plot, he dealt with a system in which essentially 90% or greater of the total binding was specific. In dealing with impure membrane preparations, such as those described here and those that are often used in the central nervous system, sometimes a Scatchard plot is not applicable. For example, when the specific binding represents 50% or less of the total binding it may be somewhat dangerous to use a Scatchard plot, while other plots such as the double-reciprocal plot may be more appropriate.

Rhein: In calculating the values of K_D and B_{max} we first correct for nonspecific binding. Total binding is measured at each desired ligand concentration; for nonspecific binding, excess unlabeled ligand is added. This is now a common procedure. Nonspecific binding generally represents 10% or less of the total binding. For the studies described we selected amino acids with low nonspecific binding.

Hirsch: Is it possible to denervate the olfactory rosette of the trout by cutting the olfactory nerves? Once the neurons have degenerated, the percentage of cilia in the preparation that were associated with the olfactory cells could be determined.

Rhein: It would be an interesting study. It was done with the catfish taste system.

Cagan: As a technical matter it is difficult to do with the trout. Typically a modest size preparation requires rosettes from 50 fish heads (100 rosettes). For example, the preparation illustrating the 22-fold enrichment of guanine:adenine compared with the deciliated rosettes used 385 fish! The scale of operation necessary for carrying out the biochemistry makes it difficult. It would be a nice experiment, but our holding facilities for trout are not adequate.

Getchell: Olfactory nerve section in the frog is straightforward and could be used to determine whether the spatial gradient of absorption changes across the surface of the epithelium, as it may after approximately 10 days. Electrophysiological responses show that presumably all olfactory receptor cells have degenerated. Would you predict a change in the sorptive properties of the regional pattern?

Hornung: I do not think that the regional pattern would change much.

Getchell: Why not? One can construct alternative arguments.

Hornung: I don't believe that the distribution pattern is established as a result of a receptor site interaction. I think it is simply based on the air/mucosa partition coefficient and that the receptor sites contribute little to the mucosal distribution pattern.

Getchell: Regarding the question of transformation of the odor by NEM, what does "transformed" mean? Chemically bound?

Hornung: We mean only that there is more tritium in the water phase.

Getchell: It's very perplexing. Also, in conceptualizing the problem of access of molecules to the receptor cells, it may be advantageous to think of it as two problems. Considering the olfactory epithelium as a flat sheet, the first question is the distribution of absorption of molecules across that epithelial surface in the Y plane. Second is the partitioning and distribution of molecules in the X plane, toward the presumed receptor sites. These two conceptual portions are not necessarily mutually exclusive.

Hornung: I agree.

Mozell: I also agree. You may recall, we have been suggesting this for a number of years. Most people speak of the Y plane, but we have stressed the X plane also. In this regard, and going back to an earlier question, I think that the receptor cell as a whole could play a prominent role in the sorption process. Our data do show that for some chemicals there is a greater sorption by the mucosa than by water, and part of this increase could be sorption by receptor cell surfaces. Indeed, in the limiting situation, one can think of this surface (which includes the cilia) as analogous to the stationary phase in chromatography, having in this function no neural impact but allowing odorants to be separated along the mucosa. The neural function of the receptor cells would then signal the effect of this sorting out.

Karnovsky: Regarding the preparation of cilia, the difficulty in getting a really clean preparation of cilia has in the past worked against approaching the very sophisticated and delicate problem of what their real role is in olfaction. The enhancement of tubulin was quite small, the enhancement of guanine nucleotides was not quite clear, but the enhancement of the ATPase was really clear, showing a three- to fourfold increase. To take account of the good with the bad—the ethanol treatment in the original deciliation might actually enhance ATPase by facilitating entry of substrate ATP to the ATPase. Isn't it possible to do the experiment the other way around? You could prepare an analog of one of the amino acids and photo-attach it or use a thiocyano group attached to an amino acid to obtain covalent linkage. With ^{14}C-labeled amino acid derivatives it wouldn't be that hard.

Cagan: We'd be delighted to have a suitable photoaffinity label. Currently we don't. One problem is where to derivatize a small molecule, such as alanine, without destroying its odorant activity. Regarding the comments on enhancement of the markers, tubulin is enriched but we have not quantified it. The increase in the guanine:adenine ratio for the cilia is threefold higher than for the whole rosettes and 22-fold higher than for the deciliated rosettes.

Rhein: Also, guanine may actually dissociate from the cilia during deciliation, as has been reported.

Cagan: It is difficult to identify criteria to establish both the presence of cilia, which we have done, and also their quantitative degree of enrichment, which is more difficult. This may be a reason why there had not been research previously with isolated olfactory cilia. We would welcome additional criteria or approaches for this. The possibility of covalent labeling is one that we are discussing; we have attempted it with the catfish taste receptor system. We usually work out new methods with the taste tissue because the cilia material is so preciously small in amount.

Polak: You mentioned that anisic acid displaces the bound protein fraction. How specific is this acid; can you substitute it by other acids, such as acetic?

Price: We haven't used many other acids. The carboxylic acid derivative of benzaldehyde, *p*-carboxybenzaldehyde, does not displace it. Nor does the protein bind to a column with *p*-carboxybenzaldehyde coupled to it. We haven't used acetic acid, but I doubt that it would work.

Karnovsky: In the polyacrylamide gel electrophoresis in SDS of the dog olfactory protein, there may be only polypeptide units or subunits of the important proteins. In the gels there is clearly a major component, but other proteins are also there. Might they act in concert with the major band leading to the lack of specificity you observe?

Price: There are other weakly staining regions. A problem with immunological preparations is that a small amount of protein contaminant, perhaps 2%, may be very antigenic.

Cancalon: Have you separated the material obtained from affinity chromatography using two-dimensional gels?

Price: We have not used two-dimensional gels. However, if the reducing agent is omitted the pattern doesn't change much, and without the SDS there is still a single band.

Mozell: Could the specificity be hidden by the EOG recording technique you used? Perhaps you would have seen better discrimination with a more finely focused method of recording.

Price: We initially recorded from specific places in the olfactory mucosa, but were concerned about access of antibody. If it did not affect responses to benzaldehyde, for example, how would you know that it got near the benzaldehyde receptors? We decided to record mass responses for that reason.

Jakinovich: You made antibodies to dog epithelium and you did the physiology experiments in the mouse. Is that relevant? Why not use the dog as the test animal?

Price: I hope it's relevant. We don't have facilities for recording EOGs from the dog.

Dodd: A general point in many biochemical studies with smell is the appropriate control tissue. You use respiratory tissue, but it is important in such a study to use other types of neuron cells.

Price: We used only the respiratory epithelium. We selected it because it has cilia and mucus on its surface. One could use brain or other tissues, but respiratory epithelium seems as close to olfactory tissue as possible.

Goldstein: In using SDS in the extraction procedure, isn't it possible that the detergent would denature the binding sites? SDS binds avidly and cannot be removed except electrophoretically. Therefore the binding of the protein to the column may be a nonspecific adsorption.

Price: We know that the protein binds *O*-methylphenols because it adsorbs to the affinity column. If it's denatured, which undoubtedly it is in SDS, it at least has not lost that property.

Cagan: What are the binding properties *in vitro* of the isolated dog epithelial protein?

Price: We have not examined binding to odorants *in vitro*.

Cagan: What then constitute criteria for identification of an isolated receptor? Also, with the Na^+,K^+-ATPase you used $10^{-3}M$ odorant, which is perhaps 10^7 or 10^8 times higher than the receptors require for a response to those compounds.

Koch: Maximum stimulation of Na^+,K^+-ATPase activity was observed, in our early studies, to occur at $10^{-3}M$ odorant. This concentration is used routinely. I realize the vast discrepancy in odorant concentration needed to stimulate an intact organism and a disrupted and dispersed enzyme system in solution. The enzyme response for a particular odorant varies with preparations from different portions of the olfactory epithelium. Odor sensing could be an ensemble of these different responses. Sometimes differential responses are seen as low as $10^{-8}M$. The Na^+,K^+-ATPase response to a single odorant is variable, e.g., $10^{-3}M(+30\%)$, $10^{-5}M(-20\%)$, $10^{-7}M(+15\%)$, and $10^{-8}M(+45\%)$.

Dodd: The relatively hydrophobic ligands, such as nonanone and decanone, at the concentrations used give rise to membrane stabilization effects. These are nonspecific, involving adsorption of a hydrophobic ligand into the phospholipid region, which could account for the difference in ATPase activity.

Koch: Nonanone and decanone show no stimulating effects on Na^+,K^+-ATPase in brain, kidney, and heart. We studied the effect of phospholipid replacement in olfactory nerve ending particles using density gradient centrifugation and observed that odorant response could be changed by changing the phospholipid content. Disruption of secondary bonding forces in the membrane by odorants could result in a change of Na^+,K^+-ATPase activity that would be specific for each different odorant. I do not propose that perturbation of enzyme activity per se is the cause of nerve signal perturbation, but that it may be associated with a Na^+ ionophore effect which could be responsible for depolarization. The latter is pure speculation.

Gesteland: Could the species-specific or strain-specific signals discussed by Dr. Yamazaki consist of slight differences in ratios of amounts of a large number of small molecules, rather than a single or few pheromone substances? A significant component of the signal could be the amount of lipophilic substance present, which would disorder the lipid membrane and have nothing to do with protein receptors. If so, will genetic experiments lead to the mechanism?

Dodd: I would expect perhaps small differences in phospholipid composition in the membranes of the population of neurons which could be the basis of a coding mechanism.

Goldstein: The answer must await the description of the chemical signals themselves. The receptors are not well defined as are insulin or acetylcholine receptors. One approach is to identify the signals, which Dr. Yamazaki is doing with mouse urine. At that point we can hopefully determine what direction the problem will take and whether there are specific cues and specific receptors that are responsible for the type of mating behavior.

Gesteland: Is there any demonstration in vertebrates that there is a specific cue substance?

Gower: The sow exhibits a specific sexual response to 5α-androstenone and the closely related 3α-alcohol, both of which are found in boar saliva. We have investigated the possibilities of a specific receptor molecule in the sow olfactory epithelium but have not, as yet, achieved any reliable results.

PART II

Taste Receptor Mechanisms

7

Comparative Study of Sweet Taste Specificity

WILLIAM JAKINOVICH, JR.

I. INTRODUCTION

In this chapter, the specificity of mammalian and insect taste responses to sweeteners is compared electrophysiologically and behaviorally. One section deals with sugars; another deals with nonsugar sweeteners.

Biochemistry of Taste and Olfaction

ISBN 0-12-154450-8

In mammals, chemical stimuli are believed to be adsorbed to taste cell membranes, producing a depolarization, which in turn leads to stimulation of a taste nerve ending across a synapse (Beidler and Gross, 1971). An alternate view is that the site of chemical interaction is at the nerve endings within the taste bud (Robbins, 1970).

In insects, chemicals are believed to be adsorbed to (Morita, 1972), or oxidized by (Norris, 1969), the membranes of taste receptor cell dendrites. These interactions lead to cellular depolarization and generation of action potentials, which are conducted to the insect's central nervous system.

II. SPECIFICITY OF SUGAR TASTE RESPONSE

In most species, it is uniformly the sugars that are the tasted and preferred sweeteners. A comparison of their taste responses indicates that sugar receptor sites have evolved nearly identical receptor specificity.

A. Disaccharides

In most animal species, sucrose is the best taste stimulus of the disaccharides. In humans, it is the sweetest and has the lowest taste threshold of the disaccharides (Fabian and Blum, 1943). In electrophysiological experiments with other animals, sucrose stimulates taste receptors at lower concentrations than do other disaccharides, for example, the squirrel monkey (Snell, 1965), the fleshfly, *Boettcherisca peregrina* (Morita and Shiraishi, 1968), the blowfly, *Phormia regina* (Omand and Dethier, 1969), and the Mongolian gerbil (Jakinovich, 1976).

The linkage between the monosaccharides of a disaccharide appears to be important because the disaccharides containing α-glucopyranosides (i.e., maltose and maltitol) are more effective taste stimuli than those containing β-glucopyranosides (i.e., cellobiose and cellobiitol) (Dethier, 1955; Schoonhoven, 1969; Moskowitz, 1974; Jakinovich, 1976; Lee, 1977; Wieczorek, 1978).

On the other hand, lactose, the β-galactopyranoside, is sweeter to man than melibiose, the α-galactopyranoside (Moskowitz, 1974). Similarly, in the Mongolian gerbil, lactose and lactitol, the β-galactopyranosides, were found electrophysiologically to have a stronger taste stimuli than melibiose and melibiitol, the α-galactopyranosides (Jakinovich, 1976).

B. Monosaccharides

1. *Reducing Monosaccharides*

These sugars, when dissolved in water, undergo mutarotation, which reaches equilibrium in a few hours. The solution contains a mixture of isomers, the concentrations of which are dependent upon time of dissolution, type of sugar, and temperature. Among the reducing monosaccharides, fructose is the sweetest and has a lower taste threshold (0.02 *M*) compared with glucose (0.045 *M*) (Fabian and Blum, 1943). A comparison of electrophysiological thresholds in many animals shows the same pattern as the human, namely that fructose has a lower taste threshold than other monosaccharides (e.g., Jakinovich and Goldstein, 1976).

2. *Methyl Glycosides*

The methyl glycosides of simple sugars are ideal for studying the specificity of taste. Unlike the reducing sugars, they do not mutarotate to form a mixture of isomers but adopt well-defined conformations and configurations (Eliel *et al.*, 1965).

In comparing the taste responses in many species to methyl D-glycopyranosides behaviorally as well as electrophysiologically, it appears that methyl α-D-glucopyranoside is a stronger taste stimulant than its anomer, methyl β-D-glucopyranoside or its epimers methyl α-D-mannopyranoside, methyl α-D-allopyranoside, and methyl α-D-galactopyranoside (Table I). This indicates that the axial methoxy group at C-1 and equatorial hydroxyl groups at C-2, C-3, and C-4 of the sugar pyran ring may be required for the maximal binding and stimulation of the receptor cell.

3. L-*Sugars*

The requirement for the three equatorial hydroxyl groups may explain why certain L-sugars stimulate the sugar receptor of the fly (Jakinovich *et al.*, 1971; Hanamori *et al.*, 1974). In these instances, models of each sugar can be superimposed over a model of methyl α-D-glucopyranoside in such a manner that the required hydroxyl groups are equivalent (Fig. 1). However, this requirement does not explain why the fly's sugar receptor cells are not stimulated by methyl β-L-glucopyranoside (Jakinovich *et al.*, 1971).

4. *Reducing Anomers*

In behavioral and electrophysiological experiments using reducing sugars where the state of mutarotation is known, α-anomers of D-pyranose sugars in most animals are more effective taste stimuli than β-anomers (Shallenberger and Acree, 1971; Hanamori *et al.*, 1974; Jakinovich and Goldstein, 1976).

TABLE I Summary of Taste Responses to Methyl D-Glycopyranosides[a]

Animal	Position				Method	Reference
	C-1	C-2	C-3	C-4		
Bee, honey	α	Glu	—	Glu	b	von Frisch (1935)
Caterpillars						
Mammestra brassica	α	—	—	—	e	Wieczorek (1976)
Bombyx mori	α	Glu	—	—	e,b	Ishikawa (1967)
Cockroach, *Periplaneta americana*	α	—	—	—	b	Wieczorek (1978)
Flies						
Boetcherisca perigrina	α	Glu	—	—	e	Hanamori *et al.* (1972)
Phormia regina	α	Glu/man[b]	—	—	b	Dethier (1955)
Phormia terraenovae	α	Glu	—	—	b	Pflumm (1972)
Sarcophaga bullatta	α	Glu	Glu	Glu	e	Jakinovich *et al.* (1971)
Gerbils						
Meriones unguiculatus	α	Glu	—	Glu	e	Jakinovich and Oakley (1975)
Meriones shawi	α	—	—	—	e	Jakinovich and Goldstein (1976)
Meriones libycus	α	—	—	—	e	
Psammomys obesus	α	—	—	—	e	
Hamster	α	—	—	Glu	e	Noma *et al.* (1974)
Human	α	Glu	—	—	b	Birch (1976)

[a] Comparisons are between methyl α-D-glucopyranoside (α) and the respective anomer or epimers (man, allo, gal). The most effective configuration is shown. —, compound not tested; e, electrophysiological; b, behavioral; man, methyl α-D-mannopyranoside; allo, methyl α-D-allopyranoside; gal, methyl α-D-galactopyranoside.

[b] It could not be determined from the published results which glucoside was most stimulating.

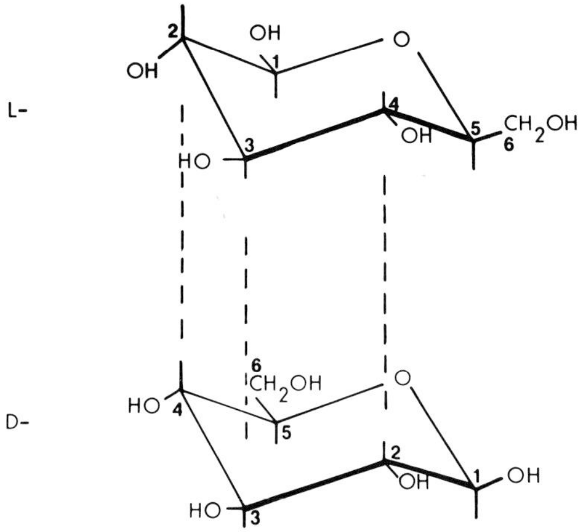

Fig. 1. Superposition of L-glucose over D-glucose to accentuate the structural similarities.

However, it has been shown that an exception, β-D-galactopyranose, is a better electrophysiological stimulus than α-D-galactopyranose in the fly (Hanamori *et al.*, 1974) and the golden hamster (Noma *et al.*, 1974). A possible explanation is shown when a model of the β-sugar is superimposed over a model of the methyl α-D-glucopyranoside; the sugar's one axial and three equatorial hydroxyl groups are in the appropriate orientations for maximal stimulation (Hanamori *et al.*, 1974).

5. *Deoxy Sugars*

Replacing a single hydroxyl group with a hydrogen atom at various positions on the D-glucopyranose molecule (deoxy sugars) usually results in a less intense taste stimulus. Such a replacement at C-1 results in a reduction of the electrophysiological response by 10% in flies (Hanamori *et al.*, 1974; Jakinovich *et al.*, 1971) and 5% in the gerbil (Jakinovich and Goldstein, 1976). Compared with D-glucose, the 2-deoxy derivative is 70% less effective in the flies and 30% less effective in the gerbil (Fig. 2). Replacement at C-3 results in a completely ineffective stimulus in the fleshfly. The 6-deoxy derivative is only 50% as effective as D-glucose in the fleshfly. Behavioral responses by the fly to the deoxy sugars have been shown to be somewhat greater than recorded responses (Evans, 1963).

Sweetness intensity of the monodeoxy sugars has not been determined in humans, but the dideoxy derivatives have been shown to be not sweet (Birch, 1976), indicating that certain hydroxyl groups of the sugar are required for sweet taste stimulation.

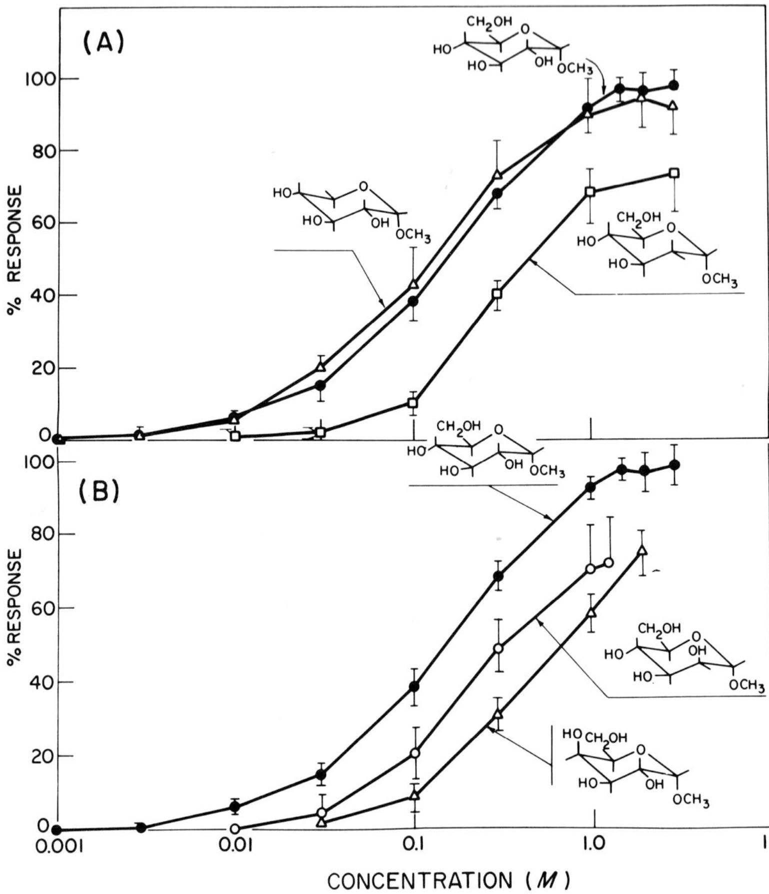

Fig. 2. (A) Comparison of integrated chorda tympani nerve responses in gerbils to methyl α-D-glucopyranoside (●), methyl α-D-xylopyranoside (△), and methyl 2-deoxy-α-D-*arabino*-hexopyranoside (□) solutions flowed over the tongue. Bars represent 95% confidence intervals. (B) Taste responses of gerbils to methyl α-D-glycopyranoside (●), methyl α-D-mannopyranoside (○), and methyl α-D-galactopyranoside (△). Responses relative to sucrose response of 100%. (Jakinovich and Goldstein, 1976.)

6. *O-Methyl Derivatives*

Replacement of a hydroxyl proton in D-glucose with a methyl group results in a less intense taste stimulus except at C-1, where replacement of the proton has an enhancing effect. In the fly, the C-2 and C-6 derivatives are about 50% less effective stimuli than D-glucose, the C-3 is about 80–100% less effective, and the C-4 does not stimulate at all (Evans, 1963; Jakinovich *et al.*, 1971). The impulse discharge of a hamster chorda tympani nerve fiber is about 23% less in response to 3-*O*-methyl glucose than to α-D-glucose (Noma *et al.*, 1974). The taste intensities of the mono-*O*-methyl derivatives are not known in man, but the dimethyl derivatives are tasteless (Birch, 1976), indicating again the need for free hydroxyl groups in sugar taste stimulation.

7. *Fructose*

In contrast to the glucopyranosides, the molecular requirements for maximal stimulation of taste receptors by fructose differ among species. In the human, β-D-fructopyranose is the sweetest fructose isomer (Birch, 1976; Shallenberger and Acree, 1971). β-D-Fructofuranose is the fructose isomer responsible for fly taste receptor stimulation (Hanamori *et al.*, 1974). (These conclusions were derived from experiments utilizing mutarotating sugar solutions and, therefore, are not absolutely conclusive). In the gerbil, methyl β-D-fructofuranoside produces the strongest electrophysiological response of the fructosides (Jakinovich and Goldstein, 1976). The methyl fructosides do not stimulate the fly's taste receptor (Jakinovich *et al.*, 1971), nor are they preferred by bees (von Frisch, 1935).

C. Polyols

Even though there is no correspondence between sweetness and the structure of cyclic and linear polyols in man (Birch, 1976; Moskowitz, 1974), a correspondence is seen in other animals. For example, electrophysiological experiments with the hamster (Hardiman, 1964) and gerbil (Jakinovich and Oakley, 1976) suggest that there is a relationship between polyol chain length and stimulatory effectiveness. In the gerbil, the effectiveness of linear polyols increases as the length of the chain increases to five carbons (Fig. 3) and then levels off. No difference was observed that could be attributed to molecular structure. By comparison, *myo*-inositol, the cyclic polyol, is a better stimulus than any of the linear polyols, probably because it resembles methyl α-D-glucopyranoside, one of the most effective stimulants.

The insect's taste receptors are not usually stimulated by linear polyols, but they are stimulated by cyclic polyols. In fact many caterpillar larvae have a *myo*-inositol taste receptor cell (Schoonhoven, 1974), which adult flies do not. *Myo*-inositol stimulates the fly's sugar receptor cell (Hodgson, 1957) in a manner parallel with the D-pyranosides (Figs. 4 and 5) (Jakinovich *et al.*, 1971). This can be explained by comparing models of the compounds. The most effective polyol, *myo*-inositol, has five equatorial and one axial hydroxyl group, which can be superimposed over the hydroxyl groups of methyl α-D-glucopyranoside, the most effective monosaccharide.

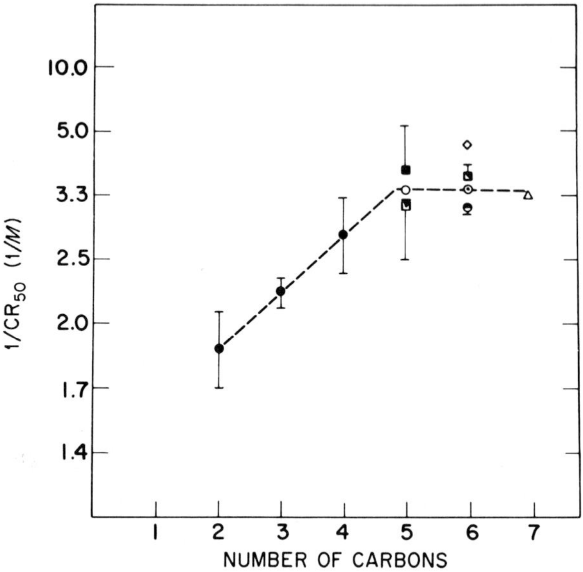

Fig. 3. Relationship between the number of carbons in sugar alcohols and the reciprocal concentration that elicited a 50% response in the gerbil (CR_{50}). Bars indicate 95% confidence intervals. Ethylene glycol (●); glycerol (●); erythritol (●); D-ribitol (■); L-arabinitol (▼); D-arabinitol (○); D-xylitol (□); D-sorbitol (⊙); D-galactitol* (◓); D-mannitol* (◩); *myo*-inositol* (◇); perseitol*(△); sucrose (▽). Asterisk indicates sugars whose insolubility prevented direct determination of maximum response. The CR_{50} for these compounds was estimated from K_d. (From Jakinovich, 1979.)

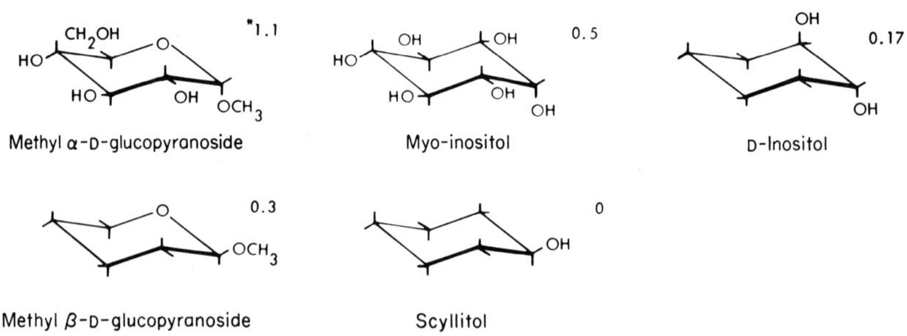

Fig. 4. Comparison of molecular configuration and taste effectiveness of methyl glucosides and equivalent cyclitols in the fly's sugar taste receptor. Effectiveness is represented as the maximum electrophysiological response evoked by a compound compared with the maximum response evoked by D-glucose. Identical substituent groups are removed from the sugar and cyclitol to show differences (Jakinovich *et al.*, 1971). Asterisk indicates response relative to D-glucose.

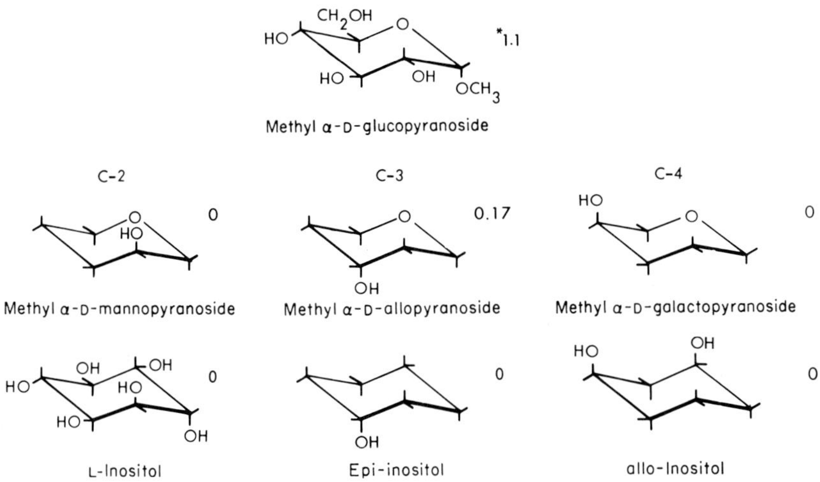

Fig. 5. Comparison of molecular configurations and taste effectiveness of methyl glycosides and equivalent cyclitols in the fly's sugar taste receptor. The substituent groups that are identical in methyl α-D-glucopyranoside and *myo*-inositol are removed in order to show differences. For the meaning of effectiveness see legend to Fig. 4. (From Jakinovich *et al.*, 1971.) Asterisk indicates response relative to D-glucose.

D. Evolution of Sugar Taste Specificity

Comparing the taste responses of different species to many sugars, it would seem that certain aspects of sugar receptor site specificity have evolved in a parallel manner. This is especially true with taste responses to glycopyranosides. If propagation of a species depends, in part, upon consumption of nutrient-laden foods whose palatability depends upon the presence of certain sugars, then survival depends upon how well an animal can taste those sugars. Two likely candidates that would influence the evolution of sugar receptor specificity are sucrose (a glucopyranoside) and glucose (glucopyranose) because they are two of the most common soluble sugars found in plants and seeds (Arnold, 1968; Pazur, 1970). The desert rodents, which are extremely sensitive to sucrose (Jakinovich and Oakley, 1975), are an excellent example of taste adaptation to satisfy physiological requirements. Schmidt-Nielsen (1964) has shown that the kangaroo rat's survival can depend upon choosing high carbohydrate seeds; high protein seeds produce negative water balance.

Presently I am unable to account for the species differences in taste response to fructose. Perhaps this compound represents taste specialization.

III. NONSUGAR SWEETENERS

In contrast to sugars, most of the non-sugar sweeteners are neither tasted uniformly by all species nor even preferred by many species. Insects do not prefer them nor are their sugar receptors usually stimulated by these compounds, except for amino acids (Schoonhoven, 1974).

A. Saccharin

1. Sodium Saccharin (Na-Saccharin)

This sweetener is the most extensively studied of the nonsugar sweeteners (Table II). Taste aversion studies indicate that Na-saccharin tastes sweet to the hamster (Nowlis and Frank, 1977) and sweet-bitter to the rat (Morrison and Jessup, 1977; Nowlis and Frank, 1977). These behavioral studies are consistent with both hamster and rat electrophysiological experiments where Na-saccharin, depending upon concentration, stimulated taste neurons sensitive to sucrose, sodium chloride, hydrochloric acid, and quinine (Ogawa *et al.*, 1969). Neurons sensitive to sucrose and sodium chloride were stimulated by low concentrations of Na-saccharin. Neurons sensitive to hydrochloric acid, quinine, and sodium chloride were stimulated by high concentrations of Na-saccharin. These single-neuron studies would explain the taste preference behavioral responses to Na-saccharin in different species. Most animals prefer Na-saccharin to water at low concentrations, and avoid it at high concentrations. A few animals, cats and the squirrel monkey, avoid, rather than prefer, Na-saccharin.

2. Saccharinic Acid (H-Saccharin)

Based on Deutsch and Hansch's (1966) proposal that relative sweetness increases with hydrophobic bonding, one would expect H-saccharin to be sweeter than sodium saccharin. This view is supported by electrophysiological experiments in hamsters which show H-saccharin to be a more effective stimulus than Na-saccharin (Hardiman, 1964). In humans, however, a comparison of taste thresholds indicates Na-saccharin is sweeter than H-saccharin. The threshold to H-saccharin is 68 μM (Blakeslee and Salmon, 1935) and to Na-saccharin is $23 \mu M$ (Pfaffmann, 1959).

B. Sodium Cyclamate

Taste-aversion experiments indicate that sodium cyclamate tastes like sodium chloride (salty) to the hamster and like quinine (bitter) to the rat (Nowlis and Frank, 1977). In preference experiments, this behavior is consistent with the rejection of sodium cyclamate by rats (Murray *et al.*, 1953), pigs (Baldwin, 1976), and cats (Beauchamp *et al.*, 1977). A single exception is the black mouse (*Mus domestica*) where 1% sodium cyclamate is preferred to water, but higher concentrations are rejected (Smith and Ross, 1960). This preference may be due to the mouse's sodium preference.

C. Dulcin

Taste responses to dulcin do not follow any consistent pattern. For example, this compound is reportedly preferred (Fisher *et al.*, 1965; Dua-Sharma and Smutz, 1977) or rejected (Glaser, 1972) by the squirrel monkey. Neither the rat nor the hamster respond behaviorally to dulcin (Carpenter, 1956; Fisher *et al.*, 1965). This observation is in accordance with rat single-taste neuron studies in which dulcin did not stimulate units sensitive to sucrose (Ogawa *et al.*, 1969). In contrast, dulcin stimulates sucrose-sensitive neurons in the macaque monkey (Sato *et al.*, 1975). Unfortunately, the macaque monkey's behavior toward dulcin has not been reported.

D. Stevioside

The macaque monkey's chorda tympani nerve responds well when this compound is applied to the tongue (Sato *et al.*, 1977). Behaviorally, stevioside is palatable at the single concentration (0.5 m*M*) tested in this monkey.

E. Chloroform

Saturated solutions of chloroform applied to rat tongue produces an electrophysiological response in the chorda tympani nerve (Yamamoto and Kawamura, 1974). However, it may or may not taste sweet to the rat because it was observed to produce responses in sucrose-sensitive neurons as well as neurons sensitive to other taste qualities.

F. Lead Acetate

This compound is rejected by rats in preference experiments (Mason and Safford, 1965). Lead acetate probably tastes sour to the dog because it stimulates taste neurons that are sensitive to citric acid but not to sucrose (Andersson *et al.*, 1950). On the other hand, the macaque monkey's perception of lead acetate may resemble man's (sweet and sour) because it stimulates taste neurons sensitive to sucrose and those sensitive to hydrochloric acid (Sato *et al.*, 1975).

TABLE II Taste Responses of Different Mammals to Sodium Saccharin[a]

Animal	Electrophysiological threshold (*M*)	Behavioral threshold (*M*)	Behavior	Reference
Cats				
Felis sp.	+	0.01	Avoidance	Beauchamp *et al.* (1977)
Panthera sp.	—	—	Avoidance	Carpenter (1956); Nagaki *et al.* (1964)
Others				
Dog	+	—	Preference Avoidance	Andersson *et al.* (1950); Jacobs and Sharma (1969); Kare (1971)
Opposum	—	0.0004	p → a	Pressman and Doolittle (1966)
Rabbit	—	0.02	p → a	Carpenter (1956); Ganchrow (1979)
Primates				
Human	+	0.000023	Sweet	Pfaffmann (1959); Zotterman (1971)
Macaque monkey	0.0002	—	—	Ogawa *et al.* (1972)
Rhesus monkey	+	0.0004	p → a	Gordon *et al.* (1959); Weiskrantz (1966)
Squirrel monkey	—	0.0005	Avoidance p → a	Dua-Sharma and Smutz (1977); Fisher *et al.* (1965)

Rodents				
Hamster	0.0001	0.0005	p → a	Carpenter (1956); Hardiman (1964)
Gerbil	+	—	Indifference	Oakley *et al.* (1979); Pierson *et al.* (1973)
Guinea pig				
Cavia porcellus	—	0.0004	p → a	Jacobs (1978)
Cavia aperea	—	0.0005	p → a	
Mouse	—	0.0000085	Preference	Hoshishima *et al.* (1962); Pelz *et al.* (1973)
		0.002	Indifference	
Rat	0.0003	0.0003	p → a	Fisher *et al.* (1965); Ogawa *et al.* (1969)
Ruminants				
Sheep	+	0.01	p → a	Bell and Kitchell (1966); Goatcher and Church (1970)
Pig	—	0.005	p → a	Baldwin (1976); Kare *et al.* (1965)
Cow	+	—	Indifference preference	Bell and Kitchell (1966)
Goat	+	—	—	Bell and Kitchell (1966)

[a] —, No data; +, sodium saccharin was observed to stimulate the chorda tympani taste nerve. The threshold was not measured; p → a indicates preference for sodium saccharin at low concentrations and avoidance at higher concentrations.

G. Amino Acids

The taste receptors of mammals and insects respond differently to amino acids. For example, of the amino acids that taste sweet to man most are D isomers, whereas the L-amino acids are tasteless, sweet, bitter, or bittersweet (Shallenberger and Acree, 1971; Solms *et al.*, 1965). This is surprising in an evolutionary sense because the essential L-amino acids, including valine, leucine, isoleucine, methionine, phenylalanine, and tryptophan (Rose, 1949), are unpalatable. In direct contrast, none of the amino acids sweet to man, D or L, stimulate the fly's sugar receptor (Shiraishi and Kuwabara, 1970). Instead, the ones that do stimulate the fly's receptor are L-amino acids: valine, leucine, isoleucine, methionine, phenylalanine, and tryptophan, all of which may be essential in the fly (Goldrich, 1973).

A few investigators have studied the taste responses of rats to amino acids (Halpern *et al.*, 1962; Tateda and Hidaka, 1966; Tapper and Halpern, 1968). Behavioral taste-aversion experiments showed that the rat responds to DL-alanine and glycine in the same way that it does to sodium saccharin. In preference experiments, the rat rejects DL-valine and DL-methionine and shows the typical preference-avoidance response to high and low concentrations of glycine and DL-alanine. In electrophysiological experiments, glycine and alanine (in the D, L, and DL forms), as well as DL-valine, produce taste responses in the rat's chorda tympani nerve, whereas DL-methionine and DL-tryptophan do not.

H. Aspartame

Aspartame (L-aspartyl-L-phenylalanylmethyl ester) produces an electrophysiological response in the macaque monkey's chorda tympani nerve (Sato *et al.*, 1977) but not in the hamster's (Nowlis and Frank, 1977). Behaviorally, the compound is palatable to the monkey (Sato *et al.*, 1977) but it is bitter to the hamster and rat, as shown in a taste aversion experiment (Nowlis and Frank, 1977).

I. Proteins

There appears to be a clear divergence in mammalian taste response to the sweet proteins, monellin, thaumatin, and miraculin. Only certain primates respond behaviorally or electrophysiologically to monellin and thaumatin (Table III) (Glaser *et al.*, 1978). Thaumatin does not produce electrophysiological responses in certain nonprimate mammals such as the pig (Glaser *et al.*, 1978), dog, hamster, rabbit (Hellekant, 1976), guinea pig, and rat, but thaumatin is preferred by the rat at 0.5 μM (Brouwer *et al.*, 1973).

TABLE III Responses of Primates to the Proteins Monellin and Thaumatin[a]

Family	Genus	Number animals tested	Monellin[b]	Thaumatin[b]	Method[c]
Tupaiidae	*Tupaia glis*	4	−	−	b
Lemuridae	*Lemur catta*	8	−	−	b
	L. variegatus	2	−+	−	b
	L. mongoz	4	+	−	b
	Cheirogaleus medius	2	−	−	b
	Microcebus murinus	1	−	−	b
Lorisidae	*Loris tardigradus nycticeboides*	2	−	−	b
	Nycticebus coucang	1	−	−	b
Galagidae	*Galago senegalensis*	1	−	−	b
Tarsiidae	*Tarsius syrichta carbonarius*	2	−	−	b
Callitrichidae	*Callithrix j. jacchus*	13	−	−	b
	Cebuella pygmaea	1	−+	−	b
	Saguinus midas tamarin	23	−	−	eb
	S. fuscicollis nigrifrons	3	−+	−	b
	S. oe. oedipus	5	−	−	b
Cebidae	*Aotus trivirgatus*	9	−	−	b
	Saimiri sciureus	3	−+	−	b
	Ateles geoffroyi	5	−	−	b
	Lagothrix lagotricha	3	−	−	b
Cercopithecidae	*Macaca fascicularis*	2	+	+	eb
	M. mulatta	14	+	+	eb
	M. nigra	3	+	+	b
	Theropithecus gelada	1	+	+	b
	Cercopithecus aethiops	12	+	+	eb
	C. diana	3	+	+	b
	C. ascanius	3	+	+	b
	Erythrocebus patas	2	+	+	b
	Presbytis obscurus	2	+	+	b
Hylobatidae	*Symphalangus syndactylus*	4	+	+	b
Pongidae	*Pongo p. pygmaeus*	3	+	+	b
	P. p. abelii	4	+	+	b
	Pan troglodytes	4	+	+	b
	P. paniscus	4	+	+	b
	Gorilla g. gorilla	2	+	+	b
Hominidae	*Homo sapiens*		+	+	b

[a] Reprinted by permission. From Glaser *et al.* (1978).
[b] −, No response; −+, slight preference; +, preference.
[c] e, Electrophysiological; b, behavioral.

Likewise, monellin does not produce electrophysiological responses in the guinea pig or rat (Brouwer *et al.*, 1973). Finally, miraculin tastes sweet to man in the presence of an acid and produces electrophysiological responses in the chorda tympani nerve in both man and monkey but not in the rat (Diamant *et al.*, 1972).

IV. RECEPTOR SITE MODELS

Receptor site models have been proposed to account for the specificity of the sweet taste in humans (Shallenberger and Acree, 1967; Horowitz and Gentili, 1971; Hodge and Inglett, 1974) and the sugar responses of the gerbil (Jakinovich and Oakley, 1976) and the fly (Shimada *et al.*, 1974; see also this volume, Chapter 8).

A. AH-B Site

Examination of the molecular structures of a large number of sweeteners by Shallenberger and Acree (1967) has led to the observation that sweeteners have a common feature—the AH,B system. The AH,B system consists of two electronegative atoms A and B separated by a distance of 2.5 to 4 Å. The H is a hydrogen atom attached to the electronegative atom A by a covalent bond. The A and B are usually either oxygen or nitrogen, but can be carbon, chlorine, or a center of unsaturation. According to the hypothesis, there must be a complimentary AH,B System in the receptor site (an AH-B site) such as a protein peptide bond or a glutamine or asparagine amino group.

In addition to the AH,B System, a lipophilic binding group, titled X (Kier, 1972) or γ (Shallenberger and Lindley, 1977), has been proposed to explain the different sweetness intensities of various compounds. To taste sweet, a compound does not require the X group but must have at least an AH,B system (e.g., ethylene glycol and chloroform are sweet).

Many compounds that possess AH,B systems are not sweet, such as L-amino acids; three-dimensional properties of the sweetener receptor site have been proposed by Shallenberger *et al.* (1969) to account for these variations of taste.

B. Multiple Receptor Sites

The AH,B system is simple but it does not account for chemoreceptor responses to sweeteners in all organisms. The sugars are tasted and preferred by many organisms, but other sweeteners are not (Kare, 1971). One explanation for this difference is that sweetness reception involves a multiple receptor site system.

A comparison of behavioral and biochemical chemosensory studies indicates that multiple sites are the rule. For example, the bacterium *E. coli* uses nine different receptor sites to detect sugars in its medium (Adler *et al.*, 1973). These sites are sensitive to *N*-acetylglucosamine, fructose, galactose, glucose, maltose, mannitol, ribose, sorbitol, and trehalose. Some of the receptors are very specific, whereas others interact with many different sugars. This bacterium does not detect saccharin or cyclamate.

The fleshfly's labellar sugar-receptor cell has a glucose, fructose, and a carboxylate anion site (Shimada *et al.*, 1974). The glucose site binds sugars that resemble α-D-glucopyranose. The fructose site binds sugars that resemble β-D-fructofuranose as well as L-phenylalanine and L-tryptophan (Shimada, 1975). The carboxylate anion site binds L-valine and L-leucine (Shimada and Isono, 1978).

Evidence for the presence of different sweetener sites in mammals is seen when comparing single chorda tympani nerve responses. Some nerve fibers respond to certain sweeteners but not to others, suggesting the presence or absence of receptor sites on the cells that the neuron innervates (e.g., Andersson *et al.*, 1950; Pfaffmann, 1969). Similarly, the whole nerve responds to certain sweeteners in some animal species but not others, indicating the presence of certain types of receptor sites on the animal's receptor cells (Hellekant *et al.*, 1974). Finally, alloxan inhibits the rat's electrophysiological taste responses to sugars but not to sodium saccharin, glycine, or sodium cyclamate, indicating the presence of at least two classes of receptor sites (Zawalich, 1973).

V. RESEARCH NEEDS

Research into sensory reception of sweeteners by mammals is in a primitive state compared with the work on insects. The reason, of course, is the difficulty of working with the mammalian taste system, as pointed out in several chapters of this volume, and the multitude of sweeteners to which most mammals may respond. The first research goal should be to identify the compounds that stimulate the receptors and taste sweet to the experimental animals. To date, most conclusions about taste quality have been drawn from taste preference experiments. Since taste preference may not be related to sweetness, a more definitive behavioral technique, such as conditioned taste aversion (Nowlis and Frank, 1977), would be very useful in the characterization of the sweeteners.

A second goal would be to determine how many sites are present on the receptor cells and the specificity of each. This goal can be attained by using competitive and noncompetitive inhibitors (e.g., Zawalich, 1973), mixtures (Jakinovich and Goldstein, 1976), neural adaptation (Smith and Frank, 1972), and single neuron analysis (Pfaffmann, 1969).

Under optimal conditions with simultaneous availability of both physiological and behavioral data, the biochemist can attempt to isolate the elusive receptor site macromolecules.

ACKNOWLEDGMENT

This paper was supported in part by grants from the City University of New York PSC-BHE Research Award Program and the National Institutes of Health (Grant 1R01NS16022-01 CMS).

REFERENCES

Adler, J., Hazelbauer, G. L., and Dahl, M. M. (1973). Chemotaxis toward sugar in *Escherichia coli*. *J. Bacteriol.* **115**, 824–847.

Andersson, B., Landgren, S., Olsson, L., and Zotterman, Y. (1950). The sweet taste fibres of the dog. *Acta Physiol. Scand.* **21**, 105–119.

Arnold, W. N. (1968). The selection of sucrose as the translocate of higher plants. *J. Theor. Biol.* **21**, 13–20.

Baldwin, B. A. (1976). Quantitative studies on taste preferences in pigs. *Proc. Nutr. Soc.* **35**, 69–73.

Beauchamp, G. K., Maller, O., and Rogers, J. G., Jr. (1977). Flavor preferences in cats (*Felis catus* and *Panthera sp.*). *J. Comp. Physiol. Psychol.* **91**, 1118–1127.

Beidler, L. M., and Gross, G. W. (1971). The nature of taste receptor sites. *Contrib. Sens. Physiol.* **5**, 97–127.

Bell, F. R., and Kitchell, R. L. (1966). Taste reception in the goat, sheep and calf. *J. Physiol. (London)* **183**, 145–151.

Birch, G. G. (1976). Structural relationships of sugars to taste. *Crit. Rev. Food Sci. Nutr.* Vol. 8, pp. 57–95.

Blakeslee, A. F., and Salmon, T. N.(1935). Genetics of sensory thresholds: individual taste reactions for different substances. *Proc. Natl. Acad. Sci. U.S.A.* **21**, 84–90.

Brouwer, J. N., Hellekant, G., Kasahara, Y., Van der Wel, H., and Zotterman, Y. (1973). Electrophysiological study of the gustatory effects of sweet proteins monellin and thaumatin in monkey, guinea pig and rat. *Acta Physiol. Scand.* **89**, 550–557.

Carpenter, J. A. (1956). Species differences in taste preferences. *J. Comp. Physiol. Psychol.* **49**, 139–144.

Dethier, V. G. (1955). The physiology and histology of the contact chemoreceptors of the blowfly. *Q. Rev. Biol.* **30**, 348–371.

Deutsch, E. W., and Hansch, C. (1966). Dependence of relative sweetness on hydrophobic bonding. *Nature (London)* **211**, 75.

Diamant, H., Hellekant, G., and Zotterman, Y. (1972). The effect of miraculin on the taste buds of man, monkey and rat. *In* "Olfaction and Taste" (D. Schneider, ed.), Vol. 4, pp. 241–244. Wiss. Verlagsges., Stuttgart.

Dua-Sharma, S., and Smutz, E. R. (1977). Taste acceptance in squirrel monkeys (*Saimiri sciureus*). *Chem. Senses Flavor* **2**, 341–352.

Eliel, E. L., Allinger, N. L., Angyal, S. J., and Morrison, G. A. (1965). "Conformational Analysis." Wiley (Interscience), New York.

Evans, D. R. (1963). Chemical structure and stimulation by carbohydrates. *In* "Olfaction and Taste" (Y. Zotterman, ed.), Vol. 1, pp. 165–176. Pergamon, Oxford.

Fabian, F. W., and Blum, H. B. (1943). Relative taste potency of some basic food constituents and their competitive and compensatory action. *Food Res.* **8**, 179–193.

Fisher, G. L., Pfaffmann, C., and Brown, E. (1965). Dulcin and saccharin taste in squirrel monkeys, rats and men. *Science* **150**, 506–507.

Ganchrow, J. R. (1979). Taste preferences in rabbits for acids, sucrose, saccharin and quinine. *Physiol. Behav.* **22**, 457–460.

Glaser, D. (1972). Die reactionen bei einigen primaten auf zwei künstliche süsstoffe und H_2O dest. *Folia Primatol.* **18**, 433–443.

Glaser, D., Hellekant, G., Brouwer, J. N., and Van der Wel, H. (1978). The taste response in primates to the proteins thaumatin and monellin and their phylogenetic implications. *Folia Primatol.* **29**, 56–63.

Goatcher, W. D., and Church, D. C. (1970). Taste responses in ruminates. I. Reactions of sheep to sugars, saccharin, ethanol and salts. *J. Anim. Sci.* **30**, 777–783.

Goldrich, N. R. (1973). Behavioral response of *Phormia regina* (Meigen) to labellar stimulation with amino acids. *J. Gen. Physiol.* **61**, 74–88.

Gordon, G., Kitchell, R., Strom, L., and Zotterman, Y. (1959). The response pattern of taste fibres in the chorda tympani of the monkey. *Acta Physiol. Scand.* **46**, 119–132.

Halpern, B. P., Bernard, R. A., and Kare, M. R. (1962). Amino acids as gustatory stimuli in the rat. *J. Gen. Physiol.* **45**, 681–701.

Hanamori, T., Shiraishi, A., Kijima, H., and Morita, H. (1972). Stimulation of labellar sugar receptor of the fleshfly by glucosides. *Z. Vgl. Physiol.* **76**, 115–124.

Hanamori, T., Shiraishi, A., Kijima, H., and Morita, H. (1974). Structure of monosaccharides effective in stimulation of the sugar receptor of the fly. *Chem. Senses Flavor* **1**, 147–166.

Hardiman, C. W. (1964). Rat and hamster chemoreceptor responses to a large number of compounds and the formation of a generalized chemosensory equation. Ph.D. Thesis, Florida State Univ., Univ. Microfilms, Ann Arbor, Michigan.

Hellekant, G. (1976). On the gustatory effects of monellin and thaumatin in dog, hamster, pig and rabbit. *Chem. Senses Flavor* **2**, 97–105.

Hellekant, G., Hagstrom, E. C., Kasahara, Y., and Zotterman, Y. (1974). On the gustatory effects of miraculin and gymnemic acid in the monkey. *Chem. Senses Flavor* **1**, 137–145.

Hodge, J. E., and Inglett, G. E. (1974). Structural aspects of glycosidic sweeteners containing ($1'\rightarrow2$)-linked disaccharides. *In* "Symposium: Sweeteners" (G. E. Inglett, ed.), pp. 216–234. Avi, Westport, Connecticut.

Hodgson, E. S. (1957). Electrophysiological studies of arthropod chemoreception. II. Responses of labellar chemoreceptors of the blowfly to stimulation by carbohydrates. *J. Insect Physiol.* **1**, 240–247.

Horowitz, R. M., and Gentili, B. (1971). Dihydrochalcone sweeteners. *In* "Sweetness and Sweeteners" (G. G. Birch, L. F. Green, and C. B. Coulson, eds.), pp. 69–80. Applied Science Publ., London.

Hoshishima, K., Yokoyama, S., and Seto, K.(1962). Taste sensitivity in various strains of mice. *Am. J. Physiol.* **202**, 1200–1204.

Ishikawa, S. (1967). Maxillary chemoreceptors in the silkworm. *In* "Olfaction and Taste" (T. Hayashi, ed.), Vol. 2, pp. 761–777. Pergamon, Oxford.

Jacobs, H. L., and Sharma, K. N. (1969). Taste versus calories: sensory and metabolic signals in the control food intake. *Ann. N.Y. Acad. Sci.* **157**, 1084–1125.

Jacobs, W. W. (1978). Taste response in wild and domestic guinea pigs. *Physiol. Behav.* **20**, 579–588.

Jakinovich, W., Jr. (1976). Stimulation of the gerbil's gustatory receptors by disaccharides. *Brain Res.* **110**, 481–490.

Jakinovich, W., Jr. (1979). The specificity of sugar taste responses in the gerbil. *In* "Carbohydrate–Protein Interaction" (I. J. Goldstein, ed.), ACS Symposium Series, No. 88, pp. 123–130. Am. Chem. Soc., Washington, D.C.

Jakinovich, W., Jr., and Goldstein, I. J. (1976). Stimulation of the gerbil's gustatory receptors by monosaccharides. *Brain Res.* **110**, 491–504.

Jakinovich, W., Jr., and Oakley, B. (1975). Comparative gustatory response in four species of gerbilline rodents. *J. Comp. Physiol.* **99**, 89–101.

Jakinovich, W., Jr., and Oakley, B. (1976). Stimulation of the gerbil's gustatory receptors by polyols. *Brain Res.* **110**, 505–513.

Jakinovich, W., Jr., Goldstein, I. J., von Baumgarten, R. J., and Agranoff, B. W. (1971). Sugar receptor specificity in the fleshfly (*Sarcophaga bullata*). *Brain Res.* **35**, 369–378.

Kare, M. (1971). Comparative study of taste. *In* "Handbook of Sensory Physiology" (L. M. Beidler, ed.), Vol. 4, Part 2, pp. 278–292. Springer-Verlag, Berlin and New York.

Kare, M. R., Pond, W. C., and Campbell, J. (1965). Observations on the taste reactions in pigs. *Anim. Behav.* **12**, 265–269.

Kier, L. B. (1972). A molecular theory of sweet taste. *J. Pharm. Sci.* **61**, 1394–1397.

Lee, C. (1977). Structural functions of taste in the sugar series: Taste properties of sugar, alcohols and related compounds. *Food Chem.* **2**, 95–105.

Mason, D. J., and Safford, H. R. (1965). Palatability of sugar of lead. *J. Comp. Physiol. Psychol.* **59**, 94–97.

Morita, H. (1972). Primary processes of insect chemoreception. *Adv. Biophys.* **3**, 161–198.

Morita, H., and Shiraishi, A. (1968). Stimulation of the labellar sugar receptor of the fleshfly by mono- and disaccharides. *J. Gen. Physiol.* **52**, 559–583.

Morrison, G. R., and Jessup, A. (1977). A dual taste for saccharin in the rat. II. Taste change following alloxan injection. *Chem. Senses Flavor* **2**, 395–400.

Moskowitz, H. R. (1974). The psychology of sweetness. *In* "Sugars in Nutrition" (H. L. Sipple and K. W. McNott, eds.), pp. 37–64. Academic Press, New York.

Murray, E. J., Wells, H., Martin, K., and Miller, N. E. (1953). Sodium sucaryl: a substance which tastes sweet to human subjects but is avoided by rats. *J. Comp. Physiol. Psychol.* **46**, 134–137.

Nagaki, J., Yamashita, S., and Sato, M. (1964). Neural response of cat to taste stimuli of varying temperatures. *Jpn. J. Physiol.* **14**, 67–89.

Noma, A., Sato, M., and Tsuzuki, Y. (1974). Taste effectiveness of anomers of sugars and glycosides as revealed from hamster taste responses. *Comp. Biochem. Physiol. A* **48**, 249–262.

Norris, D. M. (1969). Transduction mechanism in olfaction and gustation. *Nature (London)* **222**, 1263–1264.

Nowlis, G. H., and Frank, M. (1977). Qualities in hamster taste: neural and behavioral responses. *In* "Olfaction and Taste" (J. LeMagnen and P. MacLeod, eds.), Vol. 6, pp. 241–248. Information Retrieval, London.

Oakley, B., Jones, L. B., and Kaliszewski, J. M. (1979). Taste response of the gerbil IXth nerve. *Chem. Senses Flavor* **4**, 79–87.

Ogawa, H., Sato, M., and Yamashita, S. (1969). Gustatory impulse discharges in response to saccharin in rats and hamsters. *J. Physiol.* (*London*) **204**, 311–329.

Ogawa, H., Yamashita, S., Noma, A., and Sato, M. (1972). Taste responses in the macaque monkey chorda tympani. *Physiol. Behav.* **9**, 325–331.

Omand, E., and Dethier, V. G. (1969). An electrophysiological analysis of the action of carbohydrates on the sugar receptor of the blowfly. *Proc. Natl. Acad. Sci. U.S.A.* **62**, 136–143.

Pazur, J. H. (1970). Oligosaccharides. *In* "The Carbohydrates: Chemistry and Biochemistry" (W. Pigman and D. Horton, eds.), pp. 69–137. Academic Press, New York.

Pelz, W. E., Whitney, G., and Smith, J. C. (1973). Genetic Influence on saccharin preference of mice. *Physiol. Behav.* **10**, 263–265.

Pfaffmann, C. (1959). The sense of taste. *In* "The Handbook of Physiology," (J. Field, ed.) Vol. 1, Sect. 1, pp. 507–533. Am. Physiol. Soc., Washington, D.C.

Pfaffmann, C. (1969). Taste preference and reinforcement. *In* "Reinforcement and Behavior" (J. T. Tapp, ed.), pp. 215–241. Academic Press, New York.

Pflumm, W. W. (1972). Molecular structure and stimulating effectiveness of oligosaccharides and glycosides. *In* "Olfaction and Taste" (D. Schneider, ed.), Vol. 4, pp. 364–370. Wiss. Verlagsges., Stuttgart.

Pierson, S. C., Schaeffer, R. W., and King, G. D. (1973). The consumption of saccharin and glucose solutions by Mongollian gerbils. *Bull. Psychon. Soc.* **2**, 389–391.

Pressman, T. G., and Doolittle, J. H. (1966). Taste preferences in the Virginia oppossum. *Psychol. Rep.* **18**, 875–878.

Robbins, N. (1970). Are taste bud nerve endings the site of gustatory transduction? *In* "Oral Sensation and Perception" (J. F. Bosma, ed.), pp. 71–79. Thomas, Springfield, Illinois.

Rose, W. C. (1949). Amino acid requirements of man. *Fed. Proc., Fed. Am. Soc. Exp. Biol.* **8**, 546–557.

Sato, M., Ogawa, H., and Yamashita, S. (1975). Response properties of macaque monkey chorda tympani fibers. *J. Gen. Physiol.* **66**, 781–810.

Sato, M., Yasutake, H., and Ito, H. (1977). Taste discrimination in the monkey. *In* "Olfaction and Taste" (J. LeMagnen and P. MacLeod, eds.), Vol. 4, pp. 233–248. Information Retrieval, London.

Schmidt-Nielsen, K. (1964). "Desert Animals." Oxford Univ. Press (Clarendon), London and New York.

Schoonhoven, L. M. (1969). Gustation and foodplant selection in some lepidopterous larvae. *Entomol. Exp. Appl.* **12**, 555–564.

Schoonhoven, L. M. (1974). Comparative aspects of taste receptor specificity. *In* "Transduction Mechanism in Chemoreception" (T. M. Poynder, ed.), pp. 189–200. Information Retrieval, London.

Shallenberger, R. S., and Acree, T. E. (1967). Molecular theory of sweet taste. *Nature* (*London*) **216**, 480–482.

Shallenberger, R. S., and Acree, T. E. (1971). Chemical structure of compounds and their sweet and bitter taste. *In* "Handbook of Sensory Physiology" (L. M. Beidler, ed.), Vol. 4, Part 2, pp. 222–277. Springer-Verlag, Berlin and New York.

Shallenberger, R. S., and Lindley, M. G. (1977). A lipophilic-lipophobic attribute and component in the stereochemistry of sweetness. *Food Chem.* **2**, 145–153.

Shallenberger, R. S., Acree, T. E., and Lee, C. Y. (1969). Sweet taste of D- and L-sugars and amino-acids and the steric nature of their chemoreceptor site. *Nature* (*London*) **221**, 555–556.

Shimada, I. (1975). Two receptor sites and their relation to amino acid stimulation in the labellar sugar receptor of the fleshfly. *J. Insect Physiol.* **21**, 1675–1680.

Shimada, I., and Isono, K. (1978). The specific receptor site for aliphatic carboxylate anion in labellar sugar receptor of the fleshfly. *J. Insect Physiol.* **24**, 807–811.

Shimada, I., Shiraishi, A., Kijima, H., and Morita, H. (1974). Separation of two receptor sites in a single labellar sugar receptor of the fleshfly by treatment with p-chloromercuribenzoate. *J. Insect Physiol.* **20**, 605–621.

Shiraishi, A., and Kuwabara, M. (1970). The effects of amino acids on the labellar hair chemosensory cells of the fly. *J. Gen. Physiol.* **56**, 768–782.

Smith, D. V., and Frank, M. (1972). Cross adaptation between salts in the chorda tympani nerve of the rat. *Physiol. Behav.* **8**, 213–220.

Smith, M. P., and Ross, S. (1960). Acceptance of sodium sucaryl by C57 black mice. *J. Genet. Psychol.* **96**, 101–104.

Snell, T. C. (1965). The response of the squirrel monkey chorda tympani to a variety of taste stimuli. MS. Thesis, Brown Univ., Providence, Rhode Island.

Solms, J., Vuataz, L. and Egli, R. H. (1965). The taste of L- and D-amino acids. *Experientia* **21**, 692–694.

Tapper, D. N., and Halpern, B. P. (1968). Taste stimuli, a behavioral classification. *Science* **161**, 708–710.

Tateda, H., and Hidaka, I. (1966). Taste responses to sweet substances in rat. *Mem. Fac. Sci., Kyushu Univ., Ser. E* **4**, 137–149.

von Frisch, K. (1935). Über den Geschmackssin der Biene. *Z. Vgl. Physiol.* **21**, 1–156.

Weiskrantz, L. (1966). Effects of medial temporal lesion on taste preference in the monkey. *Nature (London)* **187**, 879–880.

Wieczorek, H. (1976). The glycoside receptor of the larvae of *Mamestra rassicae L.* (Lepidoptera, Noctuidae). *J. Comp. Physiol.* **106**, 153–176.

Wieczorek, H. (1978). Biochemical and behavioral studies of sugar reception in the cockroach. *J. Comp. Physiol.* **124**, 353–356.

Yamamoto, T., and Kawamura, Y. (1974). Chloroform responses of the chorda tympani nerve in the rat. *Physiol. Behav.* **13**, 245–250.

Zawalich, W. S. (1973). Depression of gustatory sweet response by alloxan. *Comp. Biochem. Physiol. A* **44**, 903–909.

Zotterman, Y. (1971). The recording of the electrical response from human taste nerves. *In* "Handbook of Sensory Physiology" (L. M. Beidler, ed.), Vol. 4, Part 2, pp. 102–115. Springer-Verlag, Berlin and New York.

8

Biochemical Aspects of Sugar Reception in Insects

KAI HANSEN AND HELMUT WIECZOREK

I. INTRODUCTION

The sugar-receptor cell of the fly's taste hairs is the physiologically best studied system among invertebrate contact chemoreceptors. This is due to the fact that the receptor cells are accessible to several basically different methods of investigation. Sixty years ago, long before the responsible sense organs had been identified, Minnich (1921) detected that in hungry insects the stimulation by sugars is coupled with a behavioral feeding response.

Biochemistry of Taste and Olfaction

ISBN 0-12-154450-8

From this feeding response, von Frisch (1935) and later Dethier and co-workers (Dethier, 1955a, 1976) determined the pattern of specificity and threshold concentrations for numerous carbohydrates. After the introduction of electrophysiological methods in the 1950s, it became clear that one receptor cell, called the sugar receptor, is mainly responsible for the fly's sugar specificity. The elaboration of the electrophysiological technique by Morita and co-workers resulted in the possibility of determining the exact concentration dependence of spike responses, and this led to the development of kinetic models for the sugar receptor (Morita, 1972a). On the basis of electrophysiological progress, Shimada *et al.* (1972) initiated pharmacological studies on the sugar receptor. This led to the detection of different receptor sites. Another approach is represented by experiments to test the hypothesis that glucosidases might act as receptor proteins in sugar reception (Hansen, 1969, 1978).

The intent of this chapter is to integrate results from recent biochemical, biophysical, and physiological approaches into a holistic picture of the function of the sugar receptor.

II. GENERAL FEATURES OF TASTE HAIRS

A. Fine Structure of Taste Hairs

As is common in insect sensilla, taste hairs contain several axially oriented receptor cells below their cuticular parts (Fig. 1). These are surrounded concentrically by three sheath cells (for details, see Hansen, 1978; Altner and Prillinger, 1980). The two outer sheath cells, the tormogen (TOC) and the trichogen cell (TRC) form cuticular parts of the hair, the hair shaft (HS), and the hair socket (SR) during ontogeny. Thereafter both cells develop apical invaginations, which form the lymph cavity II (LCII). The plasma membranes show a complex arrangement of folded membranes (MF) toward the lymph cavity. Lymph cavity II is continuous with canal II (CII) of the hair shaft. It is separated from another lymph cavity (LCI), which surrounds the dendrites with the tubelike dendrite sheath (DS). This sheath is formed by the innermost sheath cell, called the thecogen cell (T). All cells are apically connected by intercellular junctions (J). The distal parts of the receptor cells, the dendritic outer segments (DOS, referred to in this chapter as dendrites), are modified cilia. Their only cytoplasmic structures are microtubules (MT). These dendrites, which have a diameter of only 0.5 μm, pass through the whole length of the hair shaft (400 μm maximally) to the pore at its tip. This terminal pore (TP) is a feature common to all contact chemosensory sensilla. It enables the stimulating molecules to reach the dendritic membranes from

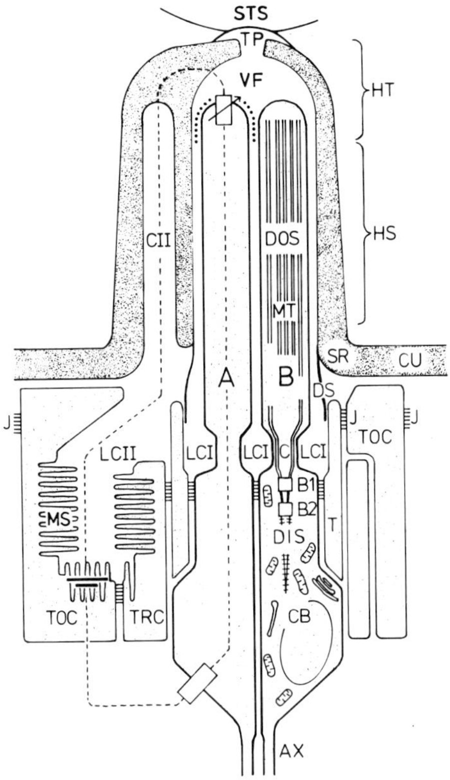

Fig. 1. Schematic organization of a taste hair. The right receptor cell (B) shows details of the fine structure; the left one (A) is a simplified functional diagram. Most proportions are distorted for reasons of clarity. The dotted line at the dendrite's tip represents the area of the primary process. The broken line indicates the flow of the receptor current, which is controlled by the conductance change of the membrane of the dendritic tip, shown here as a variable resistance (modified after Thurm, 1974). For further explanations see text. Abbreviations: AX, axon; B1/B2, ciliary basal bodies in tandem arrangement; C, ciliary structures at the transition region between distal and proximal dendritic segment; CB, receptor cell body; CU, cuticle of the taste hair bearing segment; CII, canal II within the hair shaft; DIS, dendritic inner segment; DOS, dendritic outer segment; DS, dendritic sheath; HS, hair shaft (real length 100–400 μm); HT, hair tip (the distance between pore and dendritic membranes is less than 1 μm); J, intercellular junctions; LCI/II, lymph cavities I/II; MS, system of folded plasma membranes; MT, microtubules; SR, hair socket region, not shown in detail; STS, applied stimulus solution; T, thecogen cell (formerly dendritic sheath forming cell); TOC, tormogen cell; TP, terminal pore; TRC, trichogen cell; VF, viscous fluid of canal I representing the extracellular milieu of the dendrites.

outside. Proximally, the dendritic outer segment (DIS) has a short ciliary region (C) with a $9 \times 2 + 0$ arrangement of the ciliary microtubules and two basal bodies (B1, B2). The inner segment is an elongated protrusion of the receptor cell body (CB) or perikaryon. More proximally, the receptor cells develop axons (AX) which run to the CNS.

B. Physiology of Taste Hairs

Stimulating molecules applied in solution (STS) to the hair tip must first pass the so-called "viscous fluid" (VF), which represents the outer ionic milieu of the dendrites. It can be assumed that the interaction between the stimulating molecules and the dendritic membranes results in a change in the conductance of the membrane (receptor cell A in Fig. 1). This change induces a receptor current to flow, which in turn controls the spike generator in the proximal cell membranes of the receptor cell (Morita, 1972b). The specificity of a receptor cell depends only on the binding properties of the dendritic membrane. In contrast, the conductance change, and the subsequent electrical events are rather uniform in all insect sensilla, regardless of modality.

During the last decade, evidence has accumulated that indicates that the main voltage source for the receptor current may be an electrogenic potassium pump outside the receptor cell. This pump is located in the folded membrane system (MS) of the tormogen and trichogen cell (Thurm, 1974). It produces a voltage of about 100 mV. Because of the resistive properties of the taste hair (high resistance across the epithelium caused by intercellular junctions, low resistance through canal II of the hair shaft), this voltage also lies to a great extent over the dendritic membrane at the hair tip. The diffusion potential at the dendritic membrane seems to be rather small; the fluid of the lymph cavity and probably even the dendritic outlet milieu at the hair tip of taste hairs might also contain a rather high potassium concentration equal to the intracellular concentration, as is the case for olfactory sensilla (Kaissling and Thorson, 1980). Therefore, this potential should play a minor role as a driving force for the receptor current.

III. SPECIFICITY OF THE SUGAR RECEPTOR

Sugar receptors exist in sensilla on the mouth parts, legs, or antennae of representatives of most insect orders (Frings and Frings, 1949). As sugars are always of plant origin, the widespread occurrence of sugar receptors reflects the fact that sugars are not only nutrients, but also signal components that trigger the feeding response by way of the sensory system. On the other hand, the widespread occurrence of sugar receptors is an indication that in an early stage of their evolution insects were plant feeders. Most of them still are today. Modern ferns, whose forerunners were important food plants in the Carboniferous, contain high concentrations of sucrose, glucose, fructose, and galactose (Berti and Bottari, 1968). The same sugars, except galactose, occur freely and in appreciable concentrations in the leaves and nectar of angiosperm plants. Therefore, during the phylogenetic development of the sugar receptors and even today, sucrose, fructose, and glucose were primarily responsible for exerting a selective pressure on the formation of sugar specificity in insect taste receptors.

In the following, we confine ourselves to the most intensely studied sugar receptor—that of the fly. According to an earlier concept, the four receptor cells of the taste hairs of the fly were classified into sugar, water, cation, and anion cells (for literature, see Dethier, 1976). This implied that each taste receptor cell reacts specifically with only one chemical class of substances. However, in the last few years, a broad specificity pattern of the sugar receptor became evident; it includes a selected number of monosaccharides, oligosaccharides, glycosides, open-chain sugar alcohols, cyclitols, amino acids, and fatty acids (Table I). Therefore, the well-established term "sugar receptor" is a misleading oversimplification. The sugar receptor of the fly cannot be stimulated by a variety of secondary plant substances (Dethier, 1980), by alkali halides, or by several artificial sweeteners such as saccharin, dulcin, cyclamate, and the sweet-tasting (for man) protein thaumatin (Schoonhoven, 1974). As the number of substances tested is limited, our knowledge might still be incomplete.

A semiquantitative evaluation of the specificity pattern of the sugar receptor, in other words of the "sweetness" of the stimuli, is given by the determination of behavioral thresholds. This has been performed with more than 200 compounds (for a selection see Table I, column 2).

A more appropriate measure of receptor specificity and sensitivity is the electrophysiological single-cell recording of concentration-response curves. These are obtained by plotting the spike frequency as a function of the logarithm of the stimulus concentration. Each response to a certain sugar is characterized by the following three independent parameters: (a) by the K_b value representing the concentration evoking half maximal response, (b) by the height of the maximal response, R_{max}, obtained at high concentration, and (c) by the slope of the concentration-response curve (Hill coefficient). However, until now all three parameters have been obtained only for about 10 sugars, while maximal responses alone are known for about 25; the greatest dynamic range is covered by the K_b values: mannose, 700 mM (Morita and Shiraishi, 1968), p-nitrophenyl-α-glucoside, 2 mM (Hanamori *et al.*, 1972), and L-phenylalanine, 0.6 mM (Shiraishi and Kuwabara, 1970).

TABLE I Specificity Pattern of the Fly Sugar Receptor

	1[a]		2[b]	3[c]	4[d]
A	Arabinose		++	P(F)	i
	Xylose		+	P	
	Fucose		++	F	
	L-Arabinose		+	P	
	L-Fucose		++	P	
B	Glucose		++	P	i
	L-Glucose		++	P	
	Mannose		+	P	
	Galactose		+	F(P)	
	Fructose		++	F	ni
	L-Sorbose		++	P	i
C	Sucrose	Glcα-1-2βFruf	+++	P	++
	Turanose	Glcα-1-3 Frup	+++	P	++
	Palatinose	Glcα-1-6 Fru	+++	P	+
	Maltose	Glcα-1-4 Glc	+++	P	++
	Trehalose	Glcα-1-1 Glc	++	P	(++)
	Cellobiose	Glcβ-1-4 Glc	+	P	−
	Gentiobiose	Glcβ-1-6 Glc	+		−
	Melibiose	Galα-1-6 Glc	+		−
D	Raffinose		++	P	
	Melezitose		++	P	+
E	Methyl α-Glcp		++		+
	Methyl β-Glcp		+		−
	Nitrophenyl-α-Glcp		++++		++
	Nitrophenyl-β-Glcp		+		−
F	Sorbitol		+		
	Mannitol		+		
G	Myo-inositol		+		
H	Phenylalanine		++++	F	
	Tryptophan		+++	F	
	Leucine		+++	C	
	Isoleucine			C	
	Methionine		+++	C	
	Valine		++++	C	
I	Butyrate		+++	C	
	Valerate		+++	C	
	Isovalerate		+++	C	
	Caproate		+++	C	

[a] The compounds tested belong to the following classes of chemical substances: A, pentoses; B, hexoses; C, dissaccharides; D, trisaccharides; E, glycosides; F, linear hexitols; G, cyclitols; H, amino acids; I, fatty acids. All carbohydrates belong to the D series and amino acids to the L series, if not stated otherwise.

[b] Behavioral thresholds from tarsal taste hairs of *Phormia* (for literature, see Hansen, 1978) except the substances of classes H and I. The latter values are extrapolated from behavioral tests or electrophysiological studies on labellar hairs of Boettcherisca or

The values of the maximal responses differ only by seven-fold (see Hansen, 1978, Table 8.2), whereas the Hill coefficients lie between 1.0 and 2.6 (Morita *et al.*, 1977).

It must be emphasized that in addition to the sugar receptor of the fly several other such receptors exist in insects, which differ from that of the fly in their specificity as well as in their sensitivity. However, these comparative data are not dealt with here (see Chapter 7). On the other hand, the evidence is increasing that in the same species of fly, only sugar cells of one labellar hair type always exhibit the same constant and reproducible specificity pattern and/or concentration response curves. Those of other labellar hair types may differ to a greater or lesser extent (e.g., Morita, 1972a; Wieczorek, 1981).

Quantitative knowledge of the specificity pattern is the precondition for all further investigation of the primary processes of the sugar receptor. Some aspects are dealt with in the following sections.

IV. TRANSDUCTION

A. General Considerations

In the sugar receptor, only the extreme tip of the dendritic outer segment, which is about 3 μm^2 and is thus less than 2% of the total length, is involved in transduction (Hansen, 1978). For a theoretical concept of the processes of transduction in the sugar receptor, we should consider three principles that are derived mainly from knowledge of transduction in the subsynaptic membrane of the cholinergic synapse (Barrantes, 1979). The procedure seems to be justifiable, as the postsynaptic part of a synapse behaves like a transmitter-specific chemoreceptor.

Principle 1: The chemosensitive membrane is occupied by a large number of sites that interact specifically with stimulus molecules forming weak reversible bonds. The concentration dependence of the membrane conductivity change, which controls spike frequency and reflects the fraction of sites occupied by stimulus molecules (Colquhoun, 1973).

Phormia (Shiraishi and Kuwabara, 1970; Goldrich, 1973; Shimada, 1978). The "threshold" represents that concentration of a test sugar to which 50% of a population of water-satiated, hungry flies respond. The symbols correspond to the following threshold concentration +, 200–2000 m*M*; ++, 20–200 m*M*; +++, 2–20 m*M*; ++++, <2 m*M*.

[c] Stimuli of the pyranose site are denoted P, those of the furanose site F, and those of the aliphatic carboxylate site C.

[d] Oligosaccharides split (+, ++) or not split (−) by glucosidases of crude extracts of tarsi; monosaccharides inhibiting (i) or with no effect (ni) on glucosidase activity.

Principle 2: The sites making direct contact with the stimulus molecules represent defined areas of receptor proteins that are integral constituents of the membrane. As a consequence of the isolation of several receptor proteins in other systems, their existence is now generally accepted (acetylcholine receptor protein of the cholinergic synapse, see this volume, Chapter 23; rhodopsin of visual receptor cells, Knowles and Dartnall, 1977; O'Brien, 1978). In addition, there are theoretical reasons to postulate proteins as receptor molecules in the primary processes: Protein-stimulus interaction is the only model that can explain the variety of known sensitivities and specificities. Specificity is obtained by a set of weak bonds, e.g., hydrogen bonds. The arrangement of the bonds, together with flexibility of the protein and of the stimulating molecule, determines whether a complex is to be formed at all, and, if so, whether it is of high or low affinity (Greaves, 1976).

Principle 3: The increase of membrane conductance during stimulation is caused by the opening of ion channels. These channels are structural parts of proteins that extend across the entire thickness of the membrane (Montal, 1979). The channels are closed in the resting state in which no stimulus molecules are bound to the receptor proteins. In the presence of stimulus molecules, the formation of a complex triggers a temporary opening of a channel, resulting in a short elementary current pulse. The pulse amplitude depends on the conductance of a single channel and on the voltage across the membrane. For the acetylcholine receptor of the frog muscle membrane, the following values were obtained: single channel conductance about 20 pS (picosiemens $\triangleq 20 \times 10^{-12}\Omega^{-1}$), or 2 pA (picoamperes) at 100 mV membrane voltage, duration about 10 msec (Neher and Stevens, 1977). The total receptor current through the membrane is the sum of the elementary current pulses.

B. Semiquantitative Calculations

1. *Number of Receptor Protein Molecules*

From the electrical analog of the taste hair, a receptor current of the order of 60 pA can be estimated (H. Wieczorek and K. Hansen, unpublished). This current generates a spike frequency of 200 spikes/sec. If we assume that the channels of the sugar receptor resemble those of the acetylcholine receptor, then only 30 channels would be simultaneously open. If channels and receptor proteins are supposed to exist in a 1:1 ratio, neglecting intervals when channels are closed, then a minimum of 30 receptor protein molecules are involved in transduction. With the proposed minimal receptive area of 3 μm^2, a density of only 10 receptor protein molecules per μm^2 results.

In the next step, we will attempt to approximate the amount of receptor protein in taste hair-rich tissue for four different situations: In cases 1 and 3, only the membrane of the dendritic tip, a few microns in length, is considered to bear receptor protein molecules; in cases 2 and 4, the whole membrane of the dendritic outer segment, 100–300 μm in length, is occupied by receptor protein. In cases 1 and 2, the above mentioned density of 10 molecules/μm^2 will be assumed. In cases 3 and 4, the same calculation is done with a one-thousand times higher density of 10^4 molecules/μm^2, a value which is discussed for cholinergic subsynaptic membranes (Barrantes, 1979).

Case 1. The pair of labellar lobes of one fly bears 250 taste hairs (Wilczek, 1967), each taste hair containing 30 receptor molecules. This makes a total of about 7.5×10^3 receptor molecules or, assuming a molecular weight of 10^5, 10^{-15}gm receptor protein.

Case 2. From the data of Wilczek (1967), it can be further calculated that the 250 hairs have a total length of 40 mm. Then the total surface of the dendritic sugar receptor membrane of a pair of lobes is $6 \times 10^4 \mu m^2$, if the mean diameter of the dendrites is set to 0.5μm. These $6 \times 10^4 \mu m^2$ are occupied by 10 receptor molecules/μm^2. The total number in one labellum is then 6×10^5 receptor molecules, corresponding to 8×10^{-14}gm receptor protein. In cases 3 and 4, the resulting values are one-thousand times higher than those obtained in cases 1 and 2, respectively.

It is practicable to prepare the labellar lobes of 1000 flies for one biochemical experiment. That means that maximally 10^{-10}gm membrane-bound receptor protein are to be expected in case 2 (10^{-7}gm in case 4), which is an unusually small amount. However, this situation is by no means different from that of receptor proteins in taste buds of mammals: Assuming a total number of 20,000 taste buds per bovine tongue (Davies *et al.*, 1979) and assuming further that there are 40 receptor cells per taste bud as well as a maximal receptive area of 10 μm^2 per receptor cell (calculated from Murray, 1971), we arrive at about 10^8 (10^{11} in case 4) receptor molecules per tongue, if, as in the fly, 10 (10^4 in case 4, respectively) receptor molecules per μm^2 are assumed. According to Cagan (1971), it is practicable to use one tongue in an experiment. That means that 2×10^{-11}gm (2×10^{-8}gm in case 4) receptor protein are to be expected, if we assume a molecular weight of 10^5, as in the fly. These values, calculated for cases 2 and 4, respectively, indicate that even in the future, taste receptor proteins may not be available in the quantities necessary to apply modern biochemistry to elucidate the amino acid composition, sequence, and protein structure.

2. *Gain in Transduction*

The energy content of the hydrogen bonds between the sugar and the receptor protein may be of the order of 5 kcal/mole of formed complexes or 3

$\times\ 10^{-20}$W·sec/single complex. Through a single open channel, a cation-mediated current of 2 pA flows for 10 ms at a voltage drop of probably 60 mV. This is equivalent to an energy of 10^{-15}W·sec. Comparing both energies the gain of the system is 3×10^4.

V. THE PHARMACOLOGICAL APPROACH

In this section, the effects of substances are discussed, which themselves are not stimulating but which modulate—inhibit or enhance—the activity of the stimulated sugar receptor. It should be emphasized that in the case of the taste hairs, the drugs are applied by placing a micropipette filled with a solution of the drug over the hair tip. As do the stimulating molecules, those of the drug enter by diffusion through the viscous fluid. The drugs interact exactly with those areas of the dendritic membranes where transduction occurs. During short (<5 min) periods or treatment, no interference occurs with other membranes, which lie—in the long labellar taste hairs of flies—about 300 μm away at the hairbase below the cuticle. We can assume that spreading over such distances occurs only by diffusion and requires at least several minutes. Therefore, the dendrites of taste hairs seem to be especially well-suited for pharmacological studies of transduction phenomena.

A. Identification of Different Sites

In competition experiments, mannose inhibits only the response to fructose but not to glucose, as was first shown by Dethier *et al.* (1956) in behavioral tests. These results were incompatible with a one-site model of sugar reception and marked the beginning for all discussions about a multiple-site model (Evans, 1963). Later, Morita and Shiraishi (1968) monitored the same mannose effect electrophysiologically. The indirect evidence for the existence of more than one site was definitely confirmed by treating the dendrites with the sulfhydryl group blocker *p*-mercuribenzoate (*p*-MB or PCMB), which influences the responses to glucose and fructose differently (Shimada *et al.*, 1972). The sequence of procedures is demonstrated in Fig. 2. First, the response of the receptor to glucose (point 1) and fructose (point 2) is tested in the absence of *p*-MB. Then the hair tip is treated for 3 min with 0.5 m*M* *p*-MB (TR), optimized by varying time and concentration. *p*-MB itself evokes no discharge of the receptor even after prolonged exposure. After a washing step (W), the response to glucose is inhibited (points 3,5, Fig. 2), whereas that to fructose remains unaffected

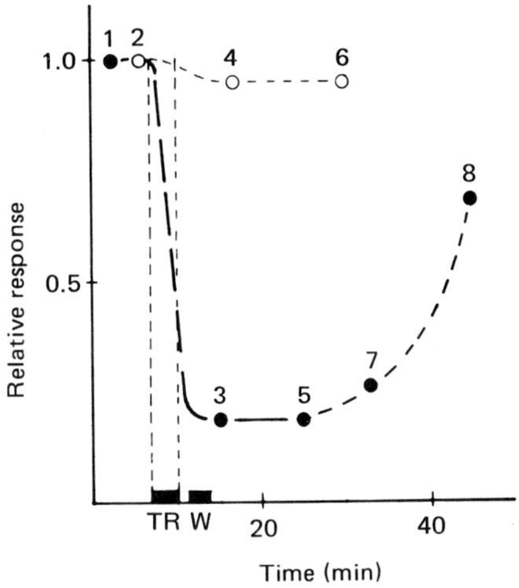

Fig. 2. Sequence of procedures for testing the effect of *p*-mercuribenzoate (*p*-MB) on the response of the sugar receptor to glucose (closed circles) and fructose (open circles); TR, Treatment period with *p*-MB; W, washing with balanced salt medium. For further explanation see Section V,A. (After Shimada *et al.*, 1974.)

(points 4,6, Fig. 2). The different behavior of the sugar receptor reveals the existence of at least two sites, a *p*-MB sensitive "pyranose site" and an insensitive "furanose site" (see below). This *p*-MB sensitivity of the pyranose site has been confirmed by the following additional control experiments. The depression is reversible; it disappears after 30–100 min (points 7,8, Fig. 2). This time interval might reflect the slow dissociation of the protein–SH–*p*-MB complex. The depressed response is reversed much more rapidly, and the response is completely restored by a short treatment with such thiol-protecting reagents as β-mercaptoethanol or cysteine. This confirms SH-group specificity. By varying the sucrose concentration before and after treatment with *p*-MB, it was shown that the maximal response was lowered rather than the K_b increased. *p*-MB does not bind to the glucose binding site itself, since high concentrations of glucose (3 *M*) applied simultaneously with *p*-MB exhibit only a rather small protecting effect (Shimada *et al.*, 1972, 1974).

1. The Pyranose Site

Shimada *et al.* (1974) tested the *p*-MB sensitivity of the sugar cell responses to numerous sugars. As shown in column 3 of Table I, the following monosaccharides react with the *p*-MB sensitive pyranose site: glucose, arabinose, xylose, mannose, L-fucose, L-arabinose, L-sorbose, and L-glucose. Fructose, fucose, and galactose interact with the furanose site.

Binding to the pyranose site requires equatorial hydroxyl groups at the C-2, C-3, and C-4 positions of a monosaccharide, as was formulated by Jakinovich *et al.* (1971; see also this volume, Chapter 7). The more general rule (Hanamori *et al.*, 1974) that three adjacent equatorial hydroxyl groups are necessary for binding to the pyranose site is not appropriate, since fucose and galactose meet this condition, although they bind predominantly to the furanose site. Responses to several oligosaccharides, the glycosidic sugar of which is glucose, are also depressed by *p*-MB, regardless of whether they contain a fructopyranose, a fructofuranose, or a glucose as the nonglycosidic part (Ninomiya and Shimada, 1976). For further discussion, see Section VI.

2. *The Furanose Site*

Pharmacologically, the furanose site is not only characterized by its insensitivity to treatment with *p*-MB, but also by its sensitivity to 2,4,6-trinitrobenzenesulfonic acid (TNBS) and *N*-bromosuccinimide (NBS). The lysine-specific TNBS suppresses to a greater degree the response to fructose than to glucose. The tryptophan-specific NBS inhibits responses both to glucose and to fructose, but the response to fructose recovers faster than that to glucose (Shimada *et al.*, 1974).

The designation "furanose" site was chosen because only the furanoid conformation of fructose has a stimulating effect. This was demonstrated by comparing the concentration response curve for freshly dissolved fructose, which has a furanose content of 5% and a pyranose content of 95%, with that for equilibrium fructose, with a furanose content of 32%. The corresponding K_b values are 200 mM for the solution of low furanose content but 28 mM for that of high furanose content (Hanamori *et al.*, 1974).

However, the structural demands of the furanose site are not yet well-understood. Pharmacological treatment with *p*-MB indicated that galactose binds predominantly to the furanose site and weakly to the pyranose site (Shimada *et al.*, 1974). But experiments with mutarotating galactose revealed that β-galactopyranose seems to be the adequate stimulus (Hanamori *et al.*, 1974). Furthermore, the fructose derivatives methyl α-fructofuranoside and methyl β-fructofuranoside are without effect as stimulants (Jakinovich *et al.*, 1971). Furthermore, the furanose site is inhibited by mannose and glucose, which in solution contain no furanose (Angyal, 1969), and by β-fructopyranose (Morita and Shiraishi, 1968; ;Morita *et al.*, 1977; Hanamori *et al.*, 1974).

An independent argument for an autonomous furanose site is its existence in the water receptor. If the response to water is suppressed by a salt solution, the cell responds like a sugar receptor upon stimulation by sugars. It is highly specific for fructose, fucose, and galactose, whereas most sugars reacting with the pyranose site have little or no effect (Wieczorek and Köppl, 1978). It has been demonstrated for the water receptor that the furanose component of fructose is the stimulating molecular conformation (Wieczorek, 1980). In the bug *Dysdercus*, no furanose site seems to exist. Neither fructose nor fucose stimulate, and galactose shows only a slight effect (Bresch, cited in Hansen, 1978).

3. *Reception of Amino Acids and the Aliphatic Carboxylate Site*

The sugar receptor is stimulated by six amino acids (Phe, Trp, Val, Leu, Ile, Met) (Table I). The response to these is not inhibited by *p*-MB. However, this does not mean that all of them bind to the furanose site. A partial differentiation was obtained by a short treatment of the taste hair with Pronase E. Thereupon, the response to Val, Leu, Ile, and Met is depressed, whereas that to fructose, Phe, and Trp is not affected. Furthermore, after treatment with TNBS, the response to Phe is inhibited in the same way as that to fructose. Therefore Shimada and Isono (1978) differentiate between the amino acids Phe and Trp, which stimulate the furanose site and those which bind to another site, called the "aliphatic caboxylate site." The latter is also stimulated by certain fatty acids such as butyrate, valerate, isovalerate, and caproate (Shimada, 1978).

B. Influence of Cyclic Nucleotides

Cyclic nucleotides and their corresponding enzymes are discussed as being involved in processes of photoreception (Pober and Bitensky, 1979). In vertebrate rod outer segments (which are of ciliary origin as are the dendritic outer segments of the insect sugar cell), light absorption by rhodopsin is coupled with the activation of a cGMP-specific phosphodiesterase (PDE), leading to a rapid decrease of cGMP concentration. The role of cGMP is not at all understood. Although several hypotheses exist, it is not clear in which part of a possible causal chain of events cGMP is important, whether in transduction itself or in processes of adaptation. Intracellular recordings from toad rods revealed that short-term applications of 0.5 m*M* dibutyryl cGMP or of the PDE inhibitor isobutylmethylxanthine at 5 m*M* lead to a decrease of the resting potential and to an increase of the response amplitude. Long-term exposure to the drugs results in a further decrease of the resting potential and in a loss of responsiveness to light (Lipton *et al.*, 1977).

With the intent to examine the involvement of cyclic nucleotides in the dendritic outer segments of the fly's taste hairs, Daley and Vande Berg

(1976) monitored electrophysiologically the effects on the sugar receptor of cAMP, cGMP, and the PDE inhibitor aminophylline. They used the largest labellar hairs of *Phormia regina*, and reported the following results:

1. 0.5 m*M* cAMP as well as its dibutyryl derivative inhibit the sugar receptor, when the drugs are applied together with 0.2 *M* sucrose; the resulting response is about 70% of the effect of 0.2 *M* sucrose alone.
2. Under the same conditions 1 m*M* dibutyryl cGMP induces a response of 135%.
3. The receptor response to a mixture of 2 m*M* aminophylline and 0.2 *M* sucrose is decreased to about 65% of the original value.

Corresponding experiments in our laboratory (Table II) confirmed the effect of aminophylline. In mixtures with glucose concentrations, which themselves induce responses of the sugar receptor in the range of the K_b (Wieczorek, 1980), 10 m*M* aminophylline reduces the response to about 20% of that to glucose alone. With respect to the modulating effect of the cyclic nucleotides our results differ from those of Daley and Vande Berg (1976). Only dibutyryl cGMP (but not cGMP, cAMP, and dibutyryl cAMP) has an effect: The response to a mixture of 5 m*M* dibutyryl cGMP and glucose is about 75% of that to glucose alone.

The discrepancies between the results of Daley and Vande Berg (1976) and ours are not explainable. Such basic mechanisms should be the same in every case and should not depend on the species used, hair type, or sugar. However, this work is in its very early stages, so it remains speculative as to whether the external application of cyclic nucleotides and PDE inhibitors prevents the lowering of cyclic nucleotide level inside the receptor cell as discussed above for the vertebrate photoreceptors. In any case, it should be noted that specific influences exist (only dibutyryl cGMP and aminophylline are modulating the sugar receptor response) and that the receptor response is affected already about 100 msec after the application of the drugs.

C. Involvement of Microtubules in Transduction

The dendritic outer segments contain no other organelles except large numbers of microtubules. This peculiarity led to the hypothesis that, in addition to a function as cytoskeleton, microtubules are also involved in transduction (Moran and Varela, 1971; Atema, 1973). Matsumoto and Farley (1978) examined this hypothesis for the sugar and salt receptors of flies. They tested electrophysiologically the dependence of salt and sugar receptor responses upon treatment with vinblastine and colchicine. The periods of treatment were exceptionally long (up to 4 hr). The receptors of the treated

TABLE II Effect of Cyclic Nucleotides and Aminophylline on the Glucose Response of the Fly Sugar Receptor[a]

Compound	Relative response	n	p
5 mM cyclic AMP 300 mM glucose	1.11 ± 0.05	12	>0.05, ns
5 mM dibutyryl cyclic AMP 400 mM glucose	1.04 ± 0.11	10	>0.05, ns
5 mM cyclic GMP 300 mM glucose	1.00 ± 0.08	14	>0.05, ns
5 mM dibutyryl cyclic GMP 300 mM glucose	0.76 ± 0.06	9	<0.01, s
10 mM aminophylline 300 mM glucose	0.20 ± 0.05	8	<0.01, s
10 mM aminophylline 400 mM glucose	0.18 ± 0.07	7	<0.01, s

[a] Influence of mixtures of glucose and either cyclic nucleotides (Na salts) or aminophylline on the response of the sugar receptor in the large labellar hairs of *Protophormia terraenovae* (hair nomenclature according to Wilczek, 1967). Mean values ± SEM of n-tested receptors are shown. p, calculated using Student's t-test; ns, not significant; s, significant. Each experimental value originates from a series of three successive tests: glucose → glucose + drug → glucose. After checking that both the enveloping responses to pure glucose are approximately equal, the response to the mixture of glucose and the drug was related to the mean response to the enveloping glucose stimuli. Unit of response is the number of impulses during a period of 300 msec beginning 50 msec after the onset of stimulus. The analysis of responses in the interval that was chosen by Daley and Vande Berg (1976, 1–2.5 sec after the onset of stimulus) leads to identical results. Tip recording was applied; the electrolyte in the solution was 5 mM NaCl. Further information on methodological details are given in Wieczorek and Köppl (1978).

hairs exhibited a reduced spike frequency compared with hairs treated for the same length of time with insect Ringer solution. However, the effects on individual hairs were not always the same. Moreover, electron microscopic analysis of treated hairs revealed not only disruption of the microtubules, but also destruction of the dendritic membrane. As the authors point out, the reason for the reduced responsiveness of the receptors still remains unclear. We feel that the long periods of drug treatment also influenced the sheath cells; therefore, the transepithelial potential, which may play some role in the generation of receptor currents, is lowered.

D. Inhibition by Hydrophobic Effectors

Dethier and Chadwick (1947) observed that sugar solutions applied to the tarsal taste hairs do not elicit the proboscis reflex if the solutions contain low concentrations of higher aliphatic alcohols. They then systematically determined the rejection thresholds of anosmic flies for mixtures of 0.1 *M* sucrose with about 80 aliphatic alcohols, ketones, and aldehydes (Chadwick and Dethier, 1949; Dethier and Chadwick, 1950). The effectiveness increased with increasing chain length of the alkyl substituents. The best inhibitors, heptanol and octanol, were rejected at concentrations below 1 m*M*.

Only when Steinhardt *et al.* (1966) investigated this inhibition problem electrophysiologically did it become clear that the response of the sugar receptor is specifically inhibited, and that no stimulation of any receptor occurred. For a more detailed study, octylamine was chosen because of its effectiveness at very low concentrations. At $10^{-5} M$ and pH 6.5, where it acts as octylammonium cation, it inhibited the response of the sugar receptor to 80 m*M* sucrose by about 50%. Kinetic data of the inhibition revealed a competitive type; however, the nature of the interaction still remains unsolved.

Quinine, a typical stimulant for bitter taste in man, is rejected by bees if offered at 0.5 m*M* in 1-*M* sucrose solution (von Frisch, 1935). On the other hand, administered alone it stimulates the deterrent receptor of the *Pieris* caterpillar at 0.1 m*M* (Ma, 1972). In the sugar receptor of the fly, Morita *et al.* (1977) tested the influence of quinine on the concentration response curves for sucrose, glucose, and fructose. Here, quinine acted at the exceptionally low concentration of 20 μM, by shifting the K_b values for sucrose and glucose 4 times toward higher concentrations; the Hill coefficient thereby increased twofold, whereas the maximal responses remained unaffected. For a fructose stimulus, the K_b shift was still greater, but the influence upon the Hill coefficient was lower. Quinine does not interact only with sugar specific sites, because inhibitory effects were observed in the salt receptor, too. In a manner similar to quinine, but at concentrations three orders of magnitude higher, *p*-nitrophenyl-α-mannoside and *p*-nitrophenyl-β-glucoside influenced the response of the sugar receptor.

VI. GLUCOSIDASES AS POSSIBLE RECEPTOR PROTEINS OF THE PYRANOSE SITE

A special property of the pyranose site is that the disaccharides sucrose, turanose, maltose, and palatinose are the best stimulating sugars (Table I). They exhibit threshold values about 10 times lower than those of other disaccharides, such as cellobiose, melibiose, and gentiobiose. All members of the first group of disaccharides possess glucose as their glycosidic moiety and contain an α-linkage, whereas the members of the second group are either β-glucosides or galactosides. The importance of the α-linkage is also seen when comparing the behavioral thresholds of the glycosides, such as α- and β-methylglucoside or nitrophenyl-α- and -β-glucoside. The same rules of threshold graduations hold for the bee and the mosquito (von Frisch, 1935; Salama, 1966). The observation that glucosidic disaccharides with an α-linkage exhibit lower thresholds than glucose itself is correlated with the fact that the glycosidic oxygen acts as a hydrogen acceptor and the corresponding hydroxyl group of the glucose as a donor.

This structure activity relationship induced Hassett *et al.*. (1950) to remark that the disaccharide specificity of the sugar receptor might parallel the substrate specificity of an intestinal α-glucosidase. No one would have looked for the existence of glucosidases in tarsal taste hairs if Dethier (1955b) had not serendipitously detected that the legs of flies contain a sucrose-splitting enzyme, accessible from the outside. It was later shown that this enzyme is a typical α-glucosidase (EC 3.2.1.20) of broad specificity, which splits all the highly stimulating disaccharides, whereas the less effective ones are not split. The same correlation is observed comparing nitrophenyl-α-glucoside (threshold 1 m*M*) and nitrophenyl-β-glucoside (threshold 200–500 m*M*; this value is extrapolated due to its low solubility) (see Table I, column 4; see also Hansen, 1969). Furthermore, several monosaccharides that react with the pyranose site, such as glucose, sorbose, xylose, and arabinose, inhibit the glucosidase activity and are therefore bound specifically to the enzyme. On the other hand, fructose is not inhibitory. The activity of the enzyme is about 5 times higher in crude extracts of taste hair-rich tarsi (units per dry weight) than in taste hair-poor tibiae. An improved test of this experiment is illustrated in Fig. 3. By extraction with the detergent Triton X-100, the activities of soluble as well as of membrane-bound glucosidases have been determined. Furthermore, neither glyceraldehyde-3-phosphate dehydrogenase (GAPDH: EC 1.1.1.8) nor malate dehydrogenase (MDH: EC 1.1.1.37) show such a difference in their activity between tarsi and tibiae. These enzymes serve as indices of the tarsal cytoplasm volume (GAPDH) and of tarsal mitochondria (MDH), respectively. It can therefore be exluded that the higher tarsal glucosidase activity reflects only a higher content of tissue inside the chitinous skeleton.

This correlation of glucosidase distribution with the taste hair number, as well as the similarities noted between the glucosidase and the pyranose site, led to the working hypothesis that the pyranose site is identical with an α-glucosidase ("glucosidase hypothesis," Hansen, 1969). That means that the specificity of binding should be the same in both systems.. There is no

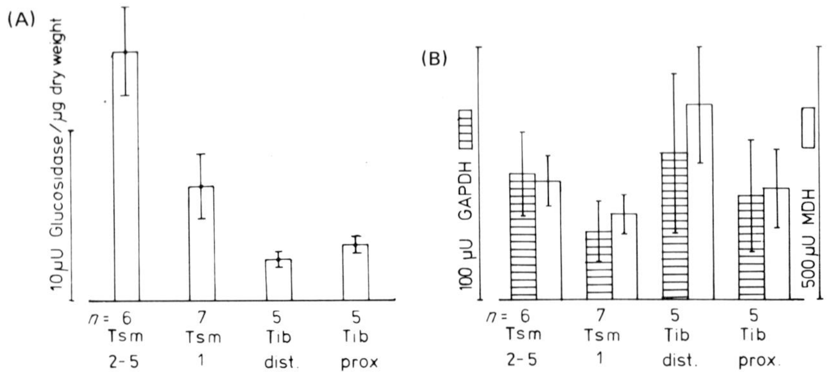

Fig. 3. Comparison of the activities of (A) α-glucosidase and of (B) glyceraldehyde-3-phosphate dehydrogenase (GAPDH) and malate dehydrogenase (MDH) in different segments of the first leg of *Protophormia terraenovae*. All activities are given in μU per μgm dry weight with standard deviations and number of independent experiments. Tsm 2–5, 2nd, 3rd, 4th, and 5th tarsomeres were used; TSM 1, 1st tarsomere; Tib dist, tibia, distal half; Tib prox, tibia proximal half.

reason to assume that the hydrolytic cleavage is part of the causal chain of the primary process.

The examination of the glucosidase hypothesis initiated a series of more detailed investigations of glucosidase from taste hair-rich segments. These experiments had two focal points; (1) one set of experiments was performed under *in vivo* conditions, and (2) another set was performed with such conventional biochemical methods as chromatographic separations and assays with isolated membranes.

1. *In Vivo Experiments*

An appreciable percentage of the glucosidase activity is accessible from the outside of freshly amputated tarsi or labella if they are floated in groups on the surface of an incubation medium. The activity is not extracted during incubation, but remains fixed to the labella and legs (Koizumi *et al.*, 1973). There is some evidence that the enzymes are accessible only through the terminal taste hair pores, because upon application of hydrostatic pressure to the hemolymph space of tarsi, microscopically small droplets appear only at the tip of the taste hairs, but not elsewhere (Hansen unpubl.). Kijima *et al.*, (1973) refined the experiments of Koizumi *et al.* (1973) and slipped micropipettes containing the incubation medium over a pair of the largest labellar hairs. Using an ultramicromethod for the determination of glucose, glucosidase activities of about 1 pmole/hr and hair were measured with sucrose as the substrate. The authors state that the activity that cannot be extracted is bound to the dendritic membranes. In addition, this "tip enzyme" exhibits some interesting properties with respect to its possible function as a receptor protein. In the same manner as the electrophysiologically obtained receptor response, it is far less inhibited by tris and nojirimycin (5-amino-5-deoxyglucose) than other glucosidases. As is the receptor, it also is completely independent of pH over the surprisingly broad range between pH 3-8 (Koizumi *et al.*, 1974; Morita *et al.*, 1977). Therefore the tip enzyme experiments were considered for a time to be direct proof of the glucosidase hypothesis. However, it is questionable if the reaction is really limited to the hair tip during the whole incubation time of 1 hr. Hanamori (1976) has shown that [^{14}C]glucose applied to the hair tip appears rapidly (within less than 20 min) in the inner parts of the labellum, presumably by diffusion. In the enzyme assay, the substrate might therefore reach deeper regions of the taste hair and might react there with the highly active soluble glucosidase of low K_m (probably the P-III enzyme), which was observed by Kijima *et al.* (1973) after transsecting the hair shafts at their midpoints. Diffusion is generally accompanied by dilution of the externally applied substrate or inhibitor solution. This was evident in the short taste hairs of the maxillary palp of the cockroach (Wieczorek, 1978). The dilution would easily explain why the "tip enzyme" is less affected by inhibitors and why it is rather pH-independent.

In summary, at present it is not clear whether the tip enzyme exists as a separate enzyme entity, which has aberrant properties due to its specialized functional employment, or whether the glucosidases of other loci of the sensillum are accessible from the outside of the hair.

2. *Chromatographic Isolation of Several Glucosidases*

The fractionation of glucosidases from taste hair-rich tarsi and labella by DEAE ion exchange chromatography has revealed at least ten glucosidases to date. These ten enzymes cannot be separated completely by single-stage chromatography; they are eluted from the ion-exchange column as four main peaks that have been further separated by gel chromatography (Amakawa *et al.*, 1972; Kühner and Hansen, 1975; Kijima *et al.*, 1977; K. Hansen and H. Bührer, unpublished). These main peaks are denoted differently by Amakawa *et al.* (1972), who use the prefix P and by Hansen (1974, 1978): P-I of Amakawa and co-workers = enzyme I of Hansen, P-II = III + IV, P-III = V. The P-II enzyme contains three components, which differ in their molecular weight (P-II M, P-II D, P-II T; Kijima *et al.*, 1977). Our peak III contains three enzymes (III_1, III_2, III_3; see Fig. 8.11 in Hansen, 1978), peak IV contains three to four enzymes, whereas peak V (V_2 in Hansen, 1974) is a single enzyme. Several other enzymes (e.g., enzyme II) of lower activity are not mentioned here. Glucosidase activities of the labellar hemolymph space are removed by washing the labella isotonically prior to homogenization. All the glucosidases of our classification (except for two components of peak IV) differ considerably from each other by their specificity, by their K_m, or by their pH optima.

3. *Which Glucosidase Can Be Considered a Receptor Protein?*

The following generalizations are possible concerning the distribution of the enzymes in taste hair-rich and -poor segments (Amakawa *et al.*, 1972; Hansen, 1974):

1. The activities of peaks III and IV (= P-II) are very low in the taste hair-less femora and the taste hair-poor tibia, but they represent 50% of the whole activity in the tarsi and still more in the labella. Therefore, the distribution is correlated with taste hair density.
2. Antennae, which contain 1000–2000 small olfactory sensilla but no taste hairs, exhibit a glucosidase activity one-third that of the tarsomeres. One-fourth of this activity belongs to the enzyme III. Therefore, only enzymes of peak IV are specific components of the tarsal and labellar taste hairs. Peak III, which is highly active in tarsi and labella, also displays low activity in other chemosensory sensilla (K. Hansen, unpublished).
3. P-I and P-III of *Phormia regina* and the enzymes I and V of Protophormia, respectively, occur not only in taste hair-rich segments, but also in tibiae, femora, and antennae, which have few or no taste hairs. On the other hand, the P-III enzyme can be eluted from transsectioned taste hairs (Kijima *et al.*, 1973). Therefore, the enzyme P-III may exist in the lymph cavities of sensilla generally. These correlations show that the taste hairs contain certain glucosidases in high activities, but only the peak IV group is taste hair-specific.

Morita (1972a) has chosen another approach. The K_b values of the largest labellar hairs for sucrose and *p*-nitrophenyl-α-glucoside are 60 and 3 mM, respectively. For theoretical reasons, the K_b value is always smaller than the respective dissociation constant of the sugar receptor–protein complex. If the pyranose site is identical with the active center of the glucosidase, the dissociation constant should be equal to the Michaelis constant (K_m). Therefore only those glucosidases might act as receptor proteins, the K_m values of which lie above the just mentioned K_b values. This is valid only for P-II (our enzymes of peaks III and IV). On the other hand, enzymes of P-I and P-III can be excluded because of their low K_m values.

The latest and perhaps most promising approach is the investigation of membrane-bound glucosidases (Amakawa *et al.*, 1975; Kijima *et al.*, 1977). Kijima and co-workers reported that within the P-II-enzyme group there are three membrane-bound enzymes that have been separated by gel chromatography after solubilization by detergents. All three enzymes exhibit a similar and rather broad specificity for sucrose, turanose, nitrophenyl-α- and phenyl-α-glucoside and show the necessary high Michaelis constants. One of these enzymes (P-II-D) was obtained from preparations of taste hair-rich labellar integument representing an especially high percentage of the total activity. This is therefore discussed as being the possible receptor protein.

However, as stated by Morita *et al.* (1977), there is at present no explanation for the discrepancy that the enzyme P-II D is strongly inhibited by low concentrations of tris and by nojirimycin, whereas the sugar receptor is influenced only at much higher concentrations. A possible solution might be to demonstrate membrane-bound glucosidases in isolated hair shafts that contain only dendritic membranes.

In summary, despite several suggestive parallels between the glucosidases and the pyranose site, the investigations have revealed many complex features of taste hairs, which make both the verification and rejection of the glucosidase hypothesis difficult.

ACKNOWLEDGMENT

Supported by the Deutsche Forschungsgemeinschaft SFB 4/C1.

REFERENCES

Altner, H., and Prillinger, L. (1980). Ultrastructure of invertebrate chemo-, thermo- and hygroreceptors and its functional significance. *Int. Rev. Cytol.* **67**, 69–139.

Amakawa, T., Kawabata, K., Kijima, H., and Morita, H. (1972). Isozymes of α-glucosidase in the proboscis and legs of flies. *J. Insect Physiol.* **18**, 541–553.

Amakawa, T., Kijima, H., and Morita, H. (1975). Insoluble α-glucosidase: possible pyranose site of the sugar receptor of the labellar hairs of the blowfly, Phormia regina. *J. Insect Physiol.* **21**, 1419–1425.

Angyal, S. J. (1969). Zusammensetzung und Konformation von Zuckern in Lösung. *Angew. Chem.* **81**, 172–182.

Atema, J. (1973). Microtubule theory of sensory transduction. *J. Theor. Biol.* **38**, 181–190.

Barrantes, F. J. (1979). Endogenous chemical receptors: some physical aspects. *Annu. Rev. Biophys. Bioeng.* **8**, 287–321.

Berti, G., and Bottari, F. (1968). Constitutents of ferns. *Prog. Phytochem.* **1**, 589–685.

Cagan, R. H. (1971). Biochemical studies of taste sensation. I. Binding of ^{14}C-labeled sugars to bovine taste papillae. *Biochim. Biophys. Acta* **252**, 199–206.

Chadwick, L. E., and Dethier, V. G. (1949). Stimulation of tarsal receptors of the blowfly by aliphatic aldehydes and ketones. *J. Gen. Physiol.* **32**, 445–452.

Colquhoun, D. (1973). The relation between classical and cooperative models for drug action. *In* "Drug Receptors" (H. P. Rang, ed.), pp. 149–182. Macmillan, New York.

Daley, D. L., and Vande Berg, J. S. (1976). Apparent opposing effects of cyclic AMP and dibutyryl-cyclic GMP on the neuronal firing of the blowfly chemoreceptors. *Biochim. Biophys. Acta* **437**, 211–220.

Davies, R. O., Kare, M. R., and Cagan, R. H. (1979). Distribution of taste buds on fungiform and circumvallate papillae of bovine tongue. *Anat. Rec.* **195**, 443–446.

Dethier, V. G. (1955a). The physiology and histology of the contact chemoreceptors of the blowfly. *Q. Rev. Biol.* **30**, 348–371.

Dethier, V. G. (1955b). Mode of action of sugar-baited fly traps. *J. Econ. Entomol.* **48**, 235–239.

Dethier, V. G. (1976). "The Hungry Fly." Harvard Univ. Press, Cambridge, Massachusetts.

Dethier, V. G. (1980). Evolution of receptor sensitivity to secondary plant substances with special reference to deterrents. *Am. Nat.* **115**, 45–66.

Dethier, V. G., and Chadwick, L. E. (1947). Rejection thresholds of the blowfly for a series of aliphatic alcohols. *J. Gen. Physiol.* **30**, 247–253.

Dethier, V. G., and Chadwick, L. E. (1950). An analysis of the relationship between solubility and stimulating effect in tarsal chemoreception. *J. Gen. Physiol.* **33**, 589–599.

Dethier, V. G., Evans, D. R., and Rhoades, M. V. (1956). Some factors controlling the ingestion of carbohydrates by the blowfly. *Biol. Bull. (Woods Hole, Mass.)* **111**, 204–222.

Evans, D. R. (1963). Chemical structure and stimulation by carbohydrates. *In* "Olfaction and Taste" (I. Y. Zotterman, ed.), Vol. 1, pp. 165–176. Pergamon, Oxford.

Frings, H., and Frings, M. (1949). The loci of contact chemoreceptors in insects. *Am. Midl. Nat.* **41**, 602–658.

Goldrich, N. R. (1973). Behavioral responses of Phormia regina (Meigen) to labellar stimulation with amino acids. *J. Gen. Physiol.* **61**, 74–88.

Greaves, M. F. (1976). Cell surface receptors, a biological perspective. *In* "Receptors and Recognition" (P. Cuatrecasas and M. F. Greaves, eds.), Ser. A, Vol. 1, pp. 1–32. Chapman & Hall, London.

Hanamori, T. (1976). Penetration of sugars from the chemosensory hair tip in the fleshfly. *Chem. Senses Flavour* **2**, 229–239.

Hanamori, T., Shiraishi, A., Kijima, H., and Morita, H. (1972). Stimulation of labellar sugar receptor of the fleshfly by glycosides. *Z. Vgl. Physiol.* **76**, 115–124.

Hanamori, T., Shiraishi, A., Kijima, H., and Morita, H. (1974). Structure of effective monosaccharides in stimulation of the sugar receptor of the fly. *Chem. Senses Flavour* **1**, 147–166.

Hansen, K. (1969). The mechanism of insect sugar reception, a biochemical investigation. *In* "Olfaction and Taste" (C. Pfaffmann, ed.), Vol. 3, pp. 382–391. Rockefeller Univ. Press, New York.

Hansen, K. (1974). α-Glucosidases as sugar receptor proteins in flies. *Colloq. Ges. Biol. Chem.* **25**, 207–232.

Hansen, K. (1978). Insect chemoreception. *In* "Receptors and Recognition" (G. L. Hazelbauer, ed.), Ser. B, Vol. 5, pp. 231–292. Chapman & Hall, London.

Hassett, C. C., Dethier, V. G., and Gans, J. (1950). A comparison of nutritive values and taste thresholds of carbohydrates for the blowfly. *Biol. Bull. (Woods Hole, Mass.)* **99**, 446–453.

Jakinovich, W., Jr., Goldstein, I. v. J., Baumgarten, R. J., and Agranoff, B. W. (1971). Sugar receptor specificity of the fleshfly Sarcophaga bullata. *Brain Res.* **35**, 369–378.

Kaissling, K. E., and Thorson, J. (1980). Insect olfactory sensilla: structural, chemical and electrical aspects of the functional organisation. *In* "Receptors for Neurotransmitters, Hormones, and Pheromones in Insects" (D. B. Satelle *et al.*, eds.), pp. 261–282. Elsevier, Amsterdam.

Kijima, H., Koizumi, O., and Morita, H. (1973). α-Glucosidase at the tip of the contact chemosensory seta of the blowfly, Phormia regina. *J. Insect Physiol.* **19**, 1351–1362.

Kijima, H., Amakawa, T., Nakashima, M., and Morita, H. (1977). Properties of membrane-bound α-glucosidases: possible sugar receptor protein of the blowfly, Phormia regina. *J. Insect Physiol.* **23**, 469–479.

Knowles, A., and Dartnall, H. J. A. (1977). The photobiology of vision. *In* "The Eye" (H. Davson, ed.), 2nd ed. Vol. 2B, pp. 1–689. Academic Press, New York.

Koizumi, O., Kijima, H., Kawabata, K., and Morita, H. (1973). α-Glucosidase activity on the outside of labella and legs of the fly. *Comp. Biochem. Physiol. B* **44,** 347–356.

Koizumi, O., Kijima, H., and Morita, H. (1974). Characterization of a glucosidase at the tips of the chemosensory setae of the fly, Phormia regina. *J. Insect Physiol.* **20,** 925–934.

Kühner, J., and Hansen, K. (1975). Chromatographic isolation of α-glucosidases from the taste hair rich tarsi of the fly Protophormia terraenovae. Their possible role as sugar receptor proteins. *J. Comp. Physiol.* **99,** 257–270.

Lipton, S. A., Rasmussen, H., and Dowling, J. E. (1977). Electrical and adaptive properties of rod photoreceptors in Bufo marinus II. Effects of cyclic nucleotides and prostaglandins. *J. Gen. Physiol.* **70,** 771–791.

Ma, W. C. (1972). Dynamics of feeding responses in Pieris brassicae Linn. as a function of chemosensory input: a behavioural, ultrastructural and electrophysiological study. *Meded. Landbouwhogesch. Wageningen* **72,** 1–162.

Matsumoto, D. E., and Farley, R. D. (1978). Alterations of ultrastructure and physiology of chemoreceptor dendrites in blowfly taste hairs treated with vinblastine and colchicine, *J. Insect Physiol.* **24,** 765–776.

Minnich, D. E. (1921). An experimental study of the tarsal chemoreceptors of two nymphalid butterflies. *J. Exp. Zool.* **33,** 173–203.

Montal, M. (1979). Rhodopsin in model membranes. *Biochim. Biophys. Acta* **559,** 231–257.

Moran, D. T., and Varela, F. G. (1971). Microtubules and sensory transduction. *Proc. Natl. Acad. Sci. U.S.A.* **68,** 757–760.

Morita, H. (1972a). Properties of the sugar receptor site of the blowfly. *In* "Olfaction and Taste" (D. Schneider, ed.), Vol. 4, pp. 357–363. Wiss. Verlagsges., Stuttgart.

Morita, H. (1972b). Primary processes of insect chemoreception. *Adv. Biophys.* **3,** 161–198.

Morita, H., and Shiraishi, A. (1968). Stimulation of the labellar sugar receptor of the fleshfly by mono- and disaccharides. *J. Gen. Physiol.* **52,** 559–583.

Morita, H., Enomoto, K. I., Nakashima, M., Shimada, I., and Kijima, H. (1977). The receptor sites for sugars in chemoreception of the fleshfly and blowfly. *In* "Olfaction and Taste" (J. Le Magnen and P. MacLeod, eds.), Vol. 6, pp. 39–46. Information Retrieval, London.

Murray, R. G. (1971). Ultrastructure of taste receptors. *In* "Handbook of Sensory Physiology" (L. M. Beidler, ed.), Vol. 4, Part 2, pp. 31–50. Springer-Verlag, Berlin and New York.

Neher, E., and Stevens, C. F. (1977). Conductance fluctuations and ionic pores in membranes. *Annu. Rev. Biophys. Bioeng.* **6,** 345–381.

Ninomiya, M., and Shimada, I. (1976). Stereospecificity for oligosaccharides of two receptor sites in a labellar sugar receptor of the fleshfly. *J. Insect Physiol.* **22,** 483–487.

O'Brien, P. J. (1978). Rhodopsin, a light sensitive membrane glycoprotein. *In* "Receptors and Recognition" (P. Cuatrecasas and M. F. Greaves, eds.), Ser. A, Vol. 6, pp. 107–150. Chapman & Hall, London.

Pober, J. S., and Bitensky, M. W. (1979). Light regulated enzymes of vertebrate retinal rods. *Adv. Cyclic Nucleotide Res.* **11,** 265–301.

Salama, H. S. (1966). The function of mosquito taste receptors. *J. Insect Physiol.* **12,** 1051–1060.

Schoonhoven, L. M. (1974). Comparative aspects of taste receptor specificity. *In* "Transduction Mechanisms in Chemoreception" (T. M. Poynder, ed.), pp. 189–201. Information Retrieval, London.

Shimada, I. (1978). The stimulating effect of fatty acids and amino acid derivatives on the labellar sugar receptor of the fleshfly. *J. Gen. Physiol.* **71,** 19–36.

Shimada, I., and Isono, K. (1978). The specific receptor site for aliphatic carboxylate anion in the labellar sugar receptor of the fleshfly. *J. Insect Physiol.* **24,** 807–811.

Shimada, I., Shiraishi, A., Kijima, H., and Morita, H. (1972). Effects of sulphydryl reagents on the labellar sugar receptor of the fleshfly. *J. Insect Physiol.* **18**, 1845–1855.

Shimada, I., Shiraishi, A., Kijima, H., and Morita, H. (1974). Separation of two receptor sites in a single labellar sugar receptor of the fleshfly by treatment with p-chloromercuribenzoate. *J. Insect Physiol.* **20**, 605–621.

Shiraishi, A., and Kuwabara, M. (1970). The effects of amino acids on the labellar hair chemosensory cells of the fly. *J. Gen. Physiol.* **56**, 768–782.

Steinhardt, R. A., Morita, H., and Hodgson, E. S. (1966). Mode of action of straight chain hydrocarbons on primary chemoreceptors of the blowfly, Phormia regina. *J. Cell. Physiol.* **67**, 53–62.

Thurm, U. (1974). Basics of the generation of receptor potentials in epidermal mechanoreceptors of insects. *In* "Symposium Mechanoreception" (J. Schwartzkopff, ed.) *Abh. Rhein.-Westf. Akad. Wiss.* **53**, pp. 355–385.

von Frisch, K. (1935). Über den Geschmackssinn der Biene. *Z. Vgl. Physiol.* **21**, 1–156.

Wieczorek, H. (1978). Biochemical and behavioral studies of sugar reception in the cockroach. *J. Comp. Physiol.* **124**, 353–356.

Wieczorek, H. (1980). Sugar reception by an insect water receptor. *J. Comp. Physiol.* **138**, 167–172.

Wieczorek, H. (1981). Sugar receptors in the labellar taste hairs of the fly. Symp. Neurobiol. Invert., Tihany 1980 (in press).

Wieczorek, H., and Köppl, R. (1978). Effect of sugars on the labellar water receptor of the fly. *J. Comp. Physiol.* **126**, 131–136.

Wilczek, M. (1967). The distribution and neuroanatomy of the labellar sense organ of the blowfly Phormia regina Meigen. *J. Morphol.* **122**, 175–201.

9

A Molecular Approach to Intensity/Time Phenomena in Sugar Sweetness

GORDON G. BIRCH

I. INTRODUCTION

Intensity/time relationships in sugar sweetness are of practical importance in reference to the new intense sweeteners, but they are also of critical theoretical importance in elucidating the function of receptor sites. Intensity/time curves may reflect total flux of stimuli at the receptor and hence magnitude of ionophor activity. It is possible that multidisciplinary progress in the understanding of chemoreception has been limited by a failure to account satisfactorily for the time course of receptor events.

The rate theory of pharmacological action goes back at least two or three decades (Ariens *et al.*, 1957; Paton, 1961), and theories of gustation based on this (Dzendolet, 1967) have invoked the dynamic concept of receptor vacation, rather than occupation, as the determinant of taste intensity. Ariens *et*

Biochemistry of Taste and Olfaction

ISBN 0-12-154450-8

al. (1957) original theory postulates the separate features of *affinity* and *intrinsic activity* in a pharmacon-receptor complex. Both features depend on chemical structure of the stimulus, but for a given structural change they may vary more or less independently. In the context of a modern outlook on chemoreception, we might more accurately describe the separate features of the stimulus/receptor complex as *accession efficiency* and *ionophor trigger efficiency*. Both may be anticipated to contribute to the observed temporal properties of a particular sweetener as well as the subjective intensity of response.

In comparing the structure/activity relationships of different sweet molecules, we must use an index of sweetening power, and this has led to the misleading concept of "degree of sweetness." Whether absolute or relative (to sucrose), the concept is inherently erroneous (Moskowitz, 1977), because the index of sweetening power depends on the conditions of measurement and hence on the different functions of molecular efficiency. Normally, we taste a suitably diluted aqueous solution of a sweetener at a similar sweetness intensity to a sucrose solution and degree of sweetness can be calculated from the ratios of concentrations of isosweet solutions. Thus for intense sweeteners, very dilute solutions are efficient for eliciting the sweet response. We do not know whether this is due to their greater accession efficiency or their ionophor trigger efficiency. However, for intense lipophilic sweetness, such as with the aminonitrobenzenes, the former seems likely by analogy with drug activity, and this may indeed be true for all intense sweeteners. Furthermore, in the related field of olfaction, Wolkowski *et al.* (1977) have pointed out that lipophilicity (as determined by the octanol: water partition coefficient) is the essential parameter in determining transport of stimuli across the mucous/lipid interface and the membrane of the olfactory organ.

It seems logical that the approach of molecules to receptors, their accession efficiencies and ionophor trigger efficiencies, might be manifested in the intensity/time curves of their responses. Whether such psychophysically determined parameters really represent molecular events at the periphery is of course open to question, but there is neurophysiological evidence in support of this idea. Bealer (1978) has recently pointed out in examining the work of Diamant *et al.* (1965), Borg *et al.* (1967), and Diamant and Zotterman (1969) that in human subjects undergoing surgery the amplitude of neural response (chorda tympani) to a gustatory stimulus declines to baseline with a time course closely parallelling loss of psychophysical sensation determined preoperatively in the same subjects. A deeper study of temporal effects should therefore lead to a better understanding of the receptors and the sweet pharmacophores.

II. MEASUREMENTS AND OBSERVATIONS IN INTENSITY/TIME RELATIONSHIPS

For psychophysical studies, the intensity/time response can be studied with any suitable chronometer, but a useful technique employs the moving chart recorder (Birch and Mylvaganam, 1976; Larson-Powers and Pangborn, 1978) upon which a magnitude estimation scale is imposed. The subject holds a pen on this scale according to his subjective intensity of response, and the time/intensity plot is thus obtained. A recent improvement of this in our laboratories utilizes a potentiometer dial box connected by a cable to the chart recorder so that the subject does nothing but taste the sample and move the dial from 1–10 according to subjective intensity. The device is referred to as a Sensory Measuring Unit for Recording Flux (SMURF) on the basis that the time course of response reflects the flux of stimuli at the receptor.

A typical time/intensity curve determined by the taste and swallow method is shown in Fig. 1. This shows the plateau of protracted maximum intensity, which is a common feature of the response if the chart speed is high enough. The curves are elevated either with constant temperature and increasing concentration or with constant concentration and increasing temperature, but the time courses are in these cases different, because sugars are essentially hydrophilic molecules and hence unable to accede better to receptors at higher temperatures. The opposite is generally true of noncarbohydrate sweeteners.

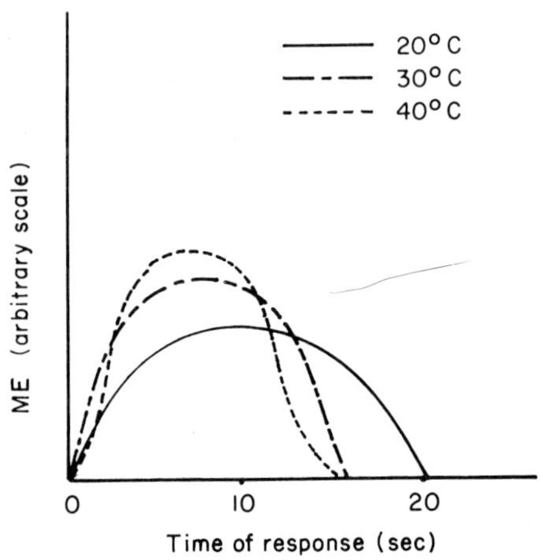

Fig. 1. Time intensity plot for sugar sweetness. Magnitude estimation (ME).

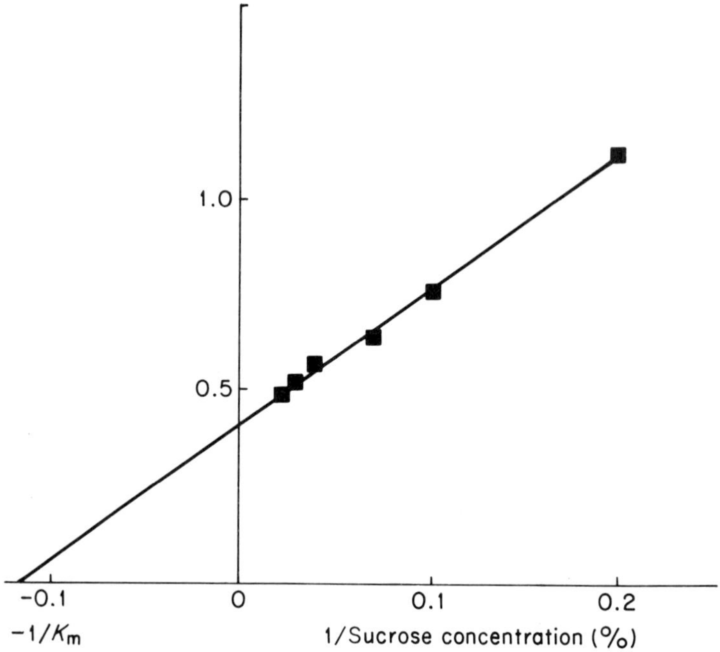

Fig. 2. Reciprocal plot of MER versus concentration of sweet stimulus.

If a line is drawn from the beginning of perception (reaction time) to the highest point of the plateau, its slope is the magnitude estimation rate (MER). A plot of MER against concentration is sigmoid in shape and MER reaches a constant value at the same concentration as magnitude estimation of response (ME) or subjective intensity (SI). With reciprocals (Fig. 2), Lineweaver-Burk types of plot are obtained from which a low affinity of sugar for receptor (approximately 10^{-3} *M*) is calculable (Ray, 1980). This agrees with the work of Cagan (1971) using radioactively labeled sucrose binding with isolated bovine receptor protein and with that of Lee *et al.* (1975) using model sugar/amino acid systems and is consistent with the concept of weak hydrogen-bonding forces between sugar and receptor.

Another study of temporal properties of the human taste system has been undertaken by McBurney (1976) by varying percentage and frequency of concentration modulation. Subjects were able to detect frequency changes up to 5 Hz, which seems to agree with our own observations of reaction time at concentrations of sucrose that elicit maximum subjective intensity of response. Meiselman and Bose (1977) used a flow-chamber device to show that subjective intensity of response increases with flow rate of stimulus.

They interpret their observations by assuming that flow rate controls the rate at which stimulus is made available to the receptor sites and thus mimics concentration, presumably by prolonging duration of stimulus. Both McBurney's (1976) and Meiselman and Bose's (1977) results seem more relevant to the explanation of reaction time than total persistence time of the sweet response. A very important related neurophysiological study of chorda tympani response to sodium chloride in rats (Marowitz and Halpern, 1977) showed that the latency period (reaction time) was about 30 msec, following which the integrated response during several seconds was measured. This was divisible into discrete phases. In many ways, this intensity/time response resembled human intensity/time plots of the type depicted in Fig. 1 and exhibited related concentration dependence. Even more details of intensity/time response may be accrued with insect models (Hansen, 1978; Kaissling, 1976) along with some accurate recording of sweet receptor kinetics. Obviously all of these data now demand evaluation in comparison with human psychophysical data to elucidate a satisfactory gustatory model.

III. SIGNIFICANCE OF TIME IN MODELS OF CHEMORECEPTION AND TRANSDUCTION

The separate stages in the time course of the sweet response suggest that attention should be paid to the *approach*, *localized concentration*, and *alignment* of the stimulus molecules at the receptors. That diffusion properties and localized concentration of stimuli near the receptor should be taken into consideration is obvious from the insect work of Hansen (1978; see also this volume, Chapter 8) and the several models of the sweet receptor reviewed by Price and DeSimone (1977). Hansen (1978) indeed points out that molecules diffusing to the dendritic outer segment of the fly taste hair tip are subject to an orifice no wider than 1000 Å. Thus bearing in mind the average length of a sugar molecule (e.g., 5–10 Å) and the considerably larger protein sweeteners (e.g., thaumatin, molecular weight 20,000; monellin, molecular weight 11,000) the idea of a channelling effect of stimulus molecules immediately presents itself. Solvation of sugar molecules also probably occurs and thus the interaction of these hydrated molecular species with organized boundary layers of water molecules at the receptor should be considered, as recently pointed out by Miller and Mooser (1979). As stimulus molecules pass from the aqueous/lipid interface to the lipid receptor membrane, some sort of stepwise molecular interaction might be envisaged as suggested previously by Kafka (1974, 1976).

A model of taste chemoreception already hypothesized (Birch *et al.*, 1978, 1980) explains the approach and alignment of stimulus molecules as an orderly queue (Fig. 3), which supplies stimuli irreversibly to the ionophor

Fig. 3. The orderly queue model of taste chemoreception. (From Birch and Lee, 1978, with permission.)

trigger mechanism. Reaction time is explained in this model as the time required to travel the length of the queue, whereas persistence is a function of queue emptying time. The ionophor trigger mechanism is envisaged as a cyclic open–shut process governed by the "fit" of the sweet pharmacophore (i.e., the AH,B system) (Shallenberger and Acree, 1967) and the supply of stimuli from queues. Intensity of response is thus the result of numbers of queues occupied ("accession efficiency") and the efficiency of the AH,B system ("ionophor trigger efficiency"). The concepts embodied in the model are not new because a lipid-dependent preliminary transport stage of chemoreception has been suggested by Wolkowski *et al.* (1977), and Hansen (1978) has discussed two stages of the process, the first involving initial binding and the second initiation of response. Ariens *et al.* (1957) also distinguished between affinity and intrinsic activity in the kinetics of drug/receptor interaction.

Accession of sweet stimuli to receptors will vary with concentrations and might be regarded as a competitive process between stimuli and other molecules, notably water. Even at maximum intensity, water molecules still vastly outnumber stimulus molecules, and the accession efficiency is therefore probably more meaningful at concentrations near threshold when the stimuli accede to the receptors in the more natural state of the receptors. Moskowitz (1973) showed that sweetness intensity (SI) obeys power functions of the form

$$SI = Kc^n$$

where c is the concentration of stimulus, n is the exponent, and K is a constant. Logarithmic plots of SI against c therefore gave straight lines of intercept K and slope n. Moskowitz (1973) pointed out that K increased with temperature, whereas n did not change, and he suggested that K (the intercept at 1% sucrose, i.e., near threshold) might represent the accession of stimuli to receptors, whereas n reflected a more profound interaction of stimulus with receptor dependent on chemical structure.

Accession efficiences (K') calculated on the basis of Moskowitz's assumptions and ionophor trigger efficiencies (ITE) were calculated on the basis of a hypothetical relationship:

$$ITE = n \frac{Tr}{K'}$$

where Tr = reaction time (assumed equal to queue crossing time). The results are listed in Table I, which shows that the intense sweetness of molecules such as thaumatin can be accounted for in this way purely on the basis of extremely high accession efficiency. Furthermore, these same molecules have low ionophor trigger efficiencies.

Careful studies of reaction times, total persistence times and intensity plateau times on the basis of the queue model (Birch *et al.*, 1980) demonstrate the same differences between sugars and intense sweeteners. The thaumatin molecules exhibit very high "accession efficiencies" whereby they occupy large numbers of queues near threshold and 1000-fold the threshold at maximum intensity. Thaumatin also exhibits a slow reaction time (Tr = 1.0 sec at maximum intensity), which may reflect its poor ionophor trigger efficiency.

Table I also shows that when accession efficiencies (K', i.e., assumed to be moles acceding divided by moles presented) are multiplied by moles of stimulus presented, the value obtained is a constant for all sugars except galactose. This suggests that near threshold the same number of molecules of each of the sugars accedes to queues and supports the molecular basis of the

TABLE I Accession Efficiencies and Ionophor Trigger Efficiencies of Sweet Molecules Calculated from Subjective Intensity/Time Observations

Sugar	Accession efficiency (K') = f (log K)	$\frac{(K')\ C_T{}^a}{MW}$	Ionophor trigger efficiency
Sucrose	1.22	0.022	0.036
D-Glucose	0.76	0.024	0.67
D-Xylose	0.75	0.024	—
D-Galactose	1.03	0.050	—
Maltose	0.89	0.022	—
Lactose	0.67	0.022	—
Sorbitol	0.93	0.034	—
Xylitol	0.98	0.032	1.32
Saccharin	0.46×10^3	—	0.003
Thaumatin	7.77×10^3	—	0.00014

[a] C_T = Percentage concentration (w/v) of stimulus solution at threshold.

sweetness hypothesis (Shallenberger and Acree, 1967; S. Munton, unpublished 1980). A higher constant is obtained for the sugar alcohols, sorbitol and xylitol. The conclusion that the accession of similar molecules is the same seems logical if the acceding molecules exist predominantly as one conformational species. The anomaly in Table I is the sugar galactose, which has an element of conformational strain due to the axial OH-4 substituent and which therefore consists of at least 4.1% furanose forms at normal tasting temperatures (Shallenberger and Birch, 1975).

IV. INTENSITY/TIME AND THE SWEET PHARMACOPHORE

The greatest contribution to our understanding of the sweet pharmacophore during the past 13 years has stemmed from the AH,B theory of Shallenberger and Acree (1967). This was subsequently modified by Kier (1972) to include a third lipophilic (γ) binding site. Although the tripartite (AH,B,γ) sweet pharmacophore was then rationalized by Shallenberger and Lindley (1977) as a scalene triangular representation, there is little reason to believe that the γ-site contributes much to the sweetness of sugars (Birch *et al.*, 1980). On the other hand, the high accession efficiency of the artificial sweeteners may indeed be ascribable to their lipid character, and this accords with the idea of a preliminary lipid-dependent transport stage of chemoreception as suggested by Wolkowski *et al.* (1977). On this basis, Kier's (1972) γ-site may be viewed as an approaching function and possibly also as a directing and aligning influence.

Previous work on model sugar homologues (Birch *et al.*, 1972; Kearsley *et al.*, 1980; Birch and Lee, 1978) has demonstrated not only the strict molar proportionality of sweetness intensity, but also the complete gustatory parallel of sweet and bitter response in a series of analogously modified, conformationally defined sugar derivatives. The results of this work allow tentative locations of AH,B systems and emphasize the alignment of the stimulus on the receptor. The results with maltodextrins (Kearsley *et al.*, 1980) and probably also the intense protein sweetener, thaumatin, show that in certain longitudinal directions there are no spatial restrictions at the receptor. On the other hand, certain 3,6-anhydro sugars (Lee and Birch, 1975) and other sugars possessing axial hydroxyl groups are tasteless, which suggests that spatial restriction does occur in certain directions. This is also well-illustrated in the simple amino acids (Shallenberger *et al.*, 1969) and has recently been theoretically reviewed by Temussi *et al.* (1978). Tastelessness may therefore result from steric exclusion at the receptor or absence of alignment. Certain small analogues of sugars, such as 1,2:4,5 cyclohexane tetrol have potential AH,B systems, and although tasteless they are not excluded sterically. Presumably their intrinsic symmetry prevents their alignment on the receptor and hence nullifies their ionophor trigger efficiency.

It is reasonable to conclude that the lipid character of sweet molecules may govern their accession efficiency and hence their localized concentration at or near receptors. In the queue model hypothesised above, this may lead to increased queue filling and hence persistence of response. However, queue emptying time may also be a function of ionophor trigger efficiency, because molecules such as thaumatin with high accession efficiencies have low ionophor trigger efficiencies and extremely long persistence times (5–50 min).

V. CONCLUSIONS AND FUTURE PROSPECTS

All attempts to measure relative sweetness have traditionally ignored the time of response and simply measured the effectiveness of molecules at eliciting a sweet response at a given dilution. It therefore now seems appropriate to seek more detailed measures of the activities of stimulus molecules in regard to both their accession efficiencies and their ionophor trigger efficiencies at the sweet receptors.

Much of the work already carried out by Bartoshuk (1977) on taste modifiers and synergism is important for the understanding of taste chemoreception, and it now demands reevaluation in terms of the time/intensity approach. For example, although the modifier, gymnemic acid, is highly time-dependent in its ability to abolish sweet taste (Meiselman and Halpern, 1970), no one has yet evaluated its separate effects on time and intensity of response. This is all the more surprising because the only sweet molecule that is reported to be unaffected by gymnemic acid is chloroform (Kurihara, 1969), a highly lipophilic molecule with a remarkably short reaction time.

The orderly queue hypothesis, as outlined above, serves to focus attention on the approach, alignment, and localized concentration of stimulus molecules and the separate mechanisms that may be responsible for the time and intensity of response. If such mechanisms do exist, this could modify all future attempts to isolate receptors and the approach to a more objective measurement of sweet taste.

Although it is arguably inappropriate to deduce molecular models of the receptors from subjective psychophysical data, the observations and conclusions reported in this chapter derive from studies with conformationally defined substrates. The queue model is proffered as a more flexible approach to understanding taste chemoreception and taste modification and to explain the temporal data that are a significant feature of the taste response.

REFERENCES

Ariens, E. J., van Rossum, J. M., and Simonis, A. M. (1957). Affinity, intrinsic activity and drug interactions. *Pharmacol. Rev.* **9**, 218–236.

Bartoshuk, L. M. (1977). Modification of taste quality. *In* "Sensory Properties of Foods" (G. G. Birch, J. G. Brennan, and K. J. Parker, eds.), pp. 71–89. Applied Science Publ., London.

Bealer, S. L. (1978). Intensity coding in the transient portion of the rat chorda tympani response. *J. Comp. Physiol. Psychol.* **92**, 185–195.

Birch, G. G., and Lee, C. K. (1978). Theory of sweetness. *In* "Developments in Sweeteners" (C. A. M. Hough, K. J. Parker, and A. J. Vlitos, eds.), pp. 165–186. Applied Science Publ., London.

Brich, G. G., and Mylvaganam, A. R. (1976). Evidence for the proximity of sweet and bitter receptor sites. *Nature* (*London*) **260**, 632–634.

Birch, G. G., Cowell, N. D., and Eyton, D. (1972). A quantitative investigation of Shallenberger's sweetness hypothesis. *J. Food Technol.* **5**, 277–280.

Birch, G. G., Latymer, Z., and Hollaway, M. R. (1978). An orderly queue hypothesis in sugar taste chemoreception. *IRCS Med. Sci. J.* **6**, 214.

Birch, G. G., Latymer, Z., and Hollaway, M. R. (1980). Intensity/time relationships in sugar sweetness: evidence for a queue hypothesis in sugar taste chemoreception. *Chem. Senses Flavor* **5**, 63–78.

Borg, G., Diamant, H., Oakley, B., Ström, L., and Zotterman, Y. (1967). A comparative study of neural and psychophysical responses to gustatory stimuli. *In* "Olfaction and Taste" (T. Hayashi, eds.), Vol. 2, pp. 253–264. Pergamon, Oxford.

Cagan, R. H. (1971). Biochemical studies of taste sensation I. Binding of 14_C-labelled sugars to bovine taste papillae. *Biochim. Biophys. Acta* **252**, 199–206.

Diamant, H., and Zotterman, Y. (1969). A comparison of neural and psychophysical responses to taste stimuli. *In* "Olfaction and Taste" (C. Pfaffmann, ed.), Vol. 3, pp. 428–435. Rockefeller Univ. Press, New York.

Diamant, H., Oakley, B., Wells, C., and Zotterman, Y. (1965). A comparison of neural and psychophysical responses to taste stimuli in man. *Acta Physiol. Scand.* **64**, 67–74.

Dzendolet, E. (1967). Beidler's taste equation: a correction. *Percept. Psychophys.* **2**, 519–520.

Hansen, K. (1978). Insect chemoreception. *In* "Receptors and Recognition" (G. L. Hazelbauer, ed.), Ser. B, Vol. 5 pp. 233–292. Chapman & Hall, London.

Kafka, W. A. (1974). A formalism on selective molecular interactions. *Colloq. Ges. Biol. Chem.* **25**, 275–278.

Kafka, W. A. (1976). Energy transfer and odour recognition. *In* "Structure-Activity Relationships in Chemoreception" (G. Benz, ed.), pp. 123–131. Information Retrieval, London.

Kaissling, K. E. (1976). The problem of specificity in olfactory cells. *In* "Structure-Activity Relationships in Chemoreception" (G. Benz, ed.), pp. 137–148. Information Retrieval, London.

Kearsley, M. W., Birch, G. G., Dziedzic, S. Z., and Smith, P. D. (1980). The production and properties of glucose syrups. III. The sweetness of glucose syrups. *Stärke* **32**, 244–247.

Kier, L. B. (1972). A molecular theory of sweet taste. *J. Pharm. Sci.* **6**, 1394–1397.

Kurihara, Y. (1969). Antisweet activity of Gymnemic acid A and its derivatives. *Life Sci.* **8**, 537–543.

Larson-Powers, N., and Pangborn, R. M. (1978). Paired comparison and time-intensity measurements of the sensory properties of beverages and gelatins containing sucrose or synthetic sweeteners. *J. Food Sci.* **43**, 41–46.

Lee, C. K., and Birch, G. G. (1975). Structural functions of taste in the sugar series: gustatory properties of anhydro sugars. *J. Food Sci.* **40**, 784–787.

Lee, C. K., Mattai, S. E., and Birch, G. G. (1975). Structural functions of taste in the sugar series: sugar-amino acid interaction as an index of sugar sweetness. *J. Food Sci.* **40**, 390–393.

McBurney, D. H. (1976). Temporal properties of the human taste system. *Sens. Processes* **1**, 150–162.

Marowitz, L. A., and Halpern, B. P. (1977). Gustatory neural response of the chorda tympani to lick duration stimuli. *Chem. Senses Flavor* **2**, 457–485.

Meiselman, H. L., and Bose, H. E. (1977). Effects of flow rate on taste intensity ratings. *Chem. Senses Flavor* **2**, 515–522.

Meiselman, H., and Halpern, B. P. (1970). Human judgements of Gymnema sylvestre and sucrose mixtures. *Physiol. Behav.* **5**, 945–948.

Miller, I. J., Jr., and Mooser, G. (1979). Taste responses to deuterium oxide. *Physiol. Behav.* **23**, 69–74.

Moskowitz, H. R. (1973). Effects of solution temperature on taste intensity in humans. *Physiol. Behav.* **10**, 289–292.

Moskowitz, H. R. (1977). *In* "Sensory Properties of Foods" (G. G. Birch, J. G. Brennan, and K. R. Parker, eds.), pp. 148–149. Applied Science Publ., London.

Paton, W. (1961). A theory of drug action based on the rate of drug/receptor combination. *Proc. R. Soc. London, Ser. B* **154**, 21–69.

Price, S., and DeSimone, J. A. (1977). Models of taste receptor cell stimulation. *Chem. Senses Flavor* **2**, 427–456.

Ray, A. (1980). Some chemical and physiological aspects of the sense of taste. Ph.D. Thesis, Reading Univ., Surrey, England.

Shallenberger, R. S., and Acree, T. E. (1967). Molecular theory of sweet taste. *Nature (London)* **216**, 480–482.

Shallenberger, R. S., and Birch, G. G. (1975). "Sugar Chemistry." Avi, Westport, Connecticut.

Shallenberger, R. S., and Lindley, M. G. (1977). A lipophilic hydrophobic attribute in the stereochemistry of sweetness. *Food Chem.* **2**, 145–154.

Shallenberger, R. S., Acree, T. E., and Lee, C. Y. (1969). Sweet taste of D- and L-sugars and amino acids and the steric nature of their receptor site. *Nature (London)* **221**, 555–556.

Temussi, P. A., Lelj, E., and Tanoredi, T. (1978). Three-dimensional mapping of the sweet taste receptor site. *J. Med. Chem.* **21**, 1154–1158.

Wolkowski, Z. W., Moccatti, D., Heymans, F., and Godfroid, J. J. (1977). A quantitative structure-activity approach to chemoreception: importance of lipophilic properties. *J. Theor. Biol.* **66**, 181–193.

10

Recognition of Taste Stimuli at the Initial Binding Interaction

ROBERT H. CAGAN

I. INTRODUCTION

Recognition of taste stimuli by receptors occurs at the initial binding interaction between the stimulus molecule and sites on the taste-receptor cell surface. Evidence supporting this hypothesis for several types of stimulus compounds is presented in this chapter (see also Chapter 12, this volume). Renqvist (1919) proposed that taste stimuli adsorb to the surface of taste receptors. Although knowledge of the supramolecular and molecular structure of cells was scant at that time, Renqvist discussed in detail the hypothesis of adsorption and the specific importance of the surface membrane of the taste receptor cell as the adsorbent. He emphasized the kinetics of the association reaction in his mathematical description. He reasoned that

Biochemistry of Taste and Olfaction
ISBN 0-12-154450-8

taste threshold determinations could be used to verify the hypothesis and consequently determined his own taste thresholds for many stimuli, including those for cane sugar (sucrose) and milk sugar (lactose), for several simple electrolytes, for inorganic acids and bases, and for homologous series of fatty acids and alcohols, thereby providing evidence in support of the hypothesis. The remarkable conceptual insight by Renqvist, over 60 years ago, appears to be fundamentally correct in his assigning adsorption as the initial event in generation of a taste sensation. The best available evidence supports the hypothesis of adsorption.

An alternative mathematical description of the binding interaction was proposed by Beidler (1954), who derived an equation based on simple mass–action principles at equilibrium. Using the adsorption isotherm, he calculated the apparent equilibrium constant of the interaction, stimulus + receptor $\rightleftharpoons$ S-R complex. The data were obtained from electrophysiological recordings of the chorda tympani nerve of animals. His formulation, therefore, placed the electrophysiological recordings from whole taste nerves into a physicochemical context, thereby departing from the traditional psychological approaches that had long dominated the field of taste. Because the data for some stimuli fit the equation (Beidler, 1954, 1962), this prompted its continued use in many electrophysiological studies to extrapolate the nerve recording data to the initial receptor interaction. The use of such recordings in this manner is, of course, not a rigorous description of the binding interaction, because the measurement is made following transduction of the information in what are undoubtedly many steps after the initial binding interaction. At the molecular level, Beidler's hypothesis assumed noninteracting binding sites, an assumption that represents the simplest case. Although it may hold for some interactions, it does not provide for the cooperative interactions revealed in studies described in this chapter and elsewhere in this volume (see Chapter 12).

In reviewing the Renqvist hypothesis, Beidler (1962) critiqued the mathematical approach of Renqvist (1919), in which the kinetics of the association reaction were postulated to underlie taste stimulation. Rather than the mathematics, perhaps the most important aspect of Renqvist's hypothesis is the clear emphasis on the adsorption process and on the role of the surface membrane of the taste-receptor cell as the adsorbent. The hypothesis provided a fundamentally important conceptual advance. It was later elaborated upon, although in a different mathematical form, by the hypothesis of Beidler (1954). Increased experimental data obtained since the time Renqvist (1919) proposed his hypothesis suggested to Beidler (1962) that the kinetics of the association reaction were not in accord with some of the data. Beidler (1962) pointed to the observation that a steady-state level of activity is achieved when recording the response to some stimuli, rather than a decrease to a zero level as the kinetic hypothesis would predict.

A study by Faull and Halpern (1972) tested predictions of both the Beidler (1954) hypothesis and the kinetic hypothesis of Paton (1961). The Beidler hypothesis did not predict the phasic electrophysiologic response. The kinetic hypothesis of Paton (1961) predicted the phasic response but did not show a good correspondence with the data over the full range of concentrations studied. The details of the adsorption interaction remain incompletely defined, but present researchers, over 60 years later, should not lose sight of the major advance of the Renqvist hypothesis. The mounting evidence supports the essentially correct conceptual view of Renqvist and makes the hypothesis worthy of classification as a theory; it is subsequently referred to as the "Renqvist Adsorption Theory."

Not until the past decade were attempts made to measure the actual binding interaction in taste (and olfactory) receptors. The basic principles and experimental approaches are similar to those used successfully in studies of several other vertebrate receptor systems, including hormone receptors (Kahn, 1976), neurotransmitter receptors (see this volume, Chapters 15, 17, and 18), acetylcholine receptors (Chapter 23), and opiate receptors (Chapter 24). Like the receptor interactions in those systems, the interactions of taste (and olfactory) stimulants appears to occur at the level of the plasma membrane of the receptor cell.

The concept of the cell membrane was derived originally from physiological experiments suggesting the presence of a diffusion barrier between the extracellular and intracellular compartments. Our understanding of the complexity of cell membranes advanced greatly with the advent of electron microscopy and the development of sophisticated preparative and analytical biochemical techniques (DePierre and Karnovsky, 1973; Korn, 1966; Singer and Nicolson, 1972). Although diversity exists in specific structural features, the plasma membrane of a cell generally consists of a lipid–protein matrix in which substantial regions contain lipid bilayer. Protein components may be present in the plasma membrane either as peripheral components, which are bound relatively weakly, or as integral components, which are strongly associated as part of the membrane structure (Singer and Nicolson, 1972). Specific proteins have in some cases been shown to be responsible for recognition of ligands (e.g., see this volume, Chapter 23 on acetylcholine receptors), and the hypothesis that taste receptor proteins exist in the membranes of taste receptor cells appears tenable. Hydrophobic interactions among hydrocarbon side chains of lipids and between lipids and nonpolar regions of proteins play a role in the structural forces maintaining the membrane. The increasingly detailed studies in recent years of the structure of plasma membranes has led to recognition not only of the complexity of molecular constituents and their geometry within the membrane, but also of their diversity and complexity of functions. Among these functions appears to be recognition of external chemical stimuli.

II. SWEET TASTE RECEPTORS

From a historical perspective, sweet-taste receptor research is a logical starting point to discuss the biochemistry of taste. In a widely cited paper published 15 years ago, Dastoli and Price (1966) claimed to have isolated a "sweet-sensitive protein" from bovine tongue epithelium. That report and its subsequent companion paper (Dastoli *et al.*, 1968) represent the first biochemical attempt to study the initial taste interaction. This work provided a focus for discussion but prompted only a relatively modest level of research activity. Although more recent research has not substantiated their claims, the idea of approaching the question biochemically deserves notice, particularly in view of the overwhelming emphasis of research activity up to that time using nerve recordings to extrapolate to peripheral receptor events. Dastoli and Price (1966) used refractive index changes and ultraviolet difference-spectral changes to measure interactions of sweet compounds with the protein extract from bovine tongue epithelium. Although one group (Hiji *et al.*, 1971; Sato *et al.*, 1977) apparently continues to use difference spectra with extracts of rat and monkey tongue epithelium, studies in other laboratories provide compelling evidence that the "sweet-sensitive protein," although an interesting hypothesis, is not supported by experimental data as a taste receptor protein (Koyama and Kurihara, 1971; Lum and Henkin, 1976b; Nofre and Sabadie, 1972; Ostretsova *et al.*, 1975).

Nofre and Sabadie (1972) confirmed that the protein fraction prepared from bovine tongue does undergo changes in refractive index on adding sugars, as had been reported by Dastoli and Price (1966). In control experiments, Nofre and Sabadie (1972) showed that nonsweet compounds cause the same changes in refractive index of the lingual protein and also that γ-globulin shows similar results. This demonstrated that the changes are nonspecific. Nofre and Sabadie (1972) concluded that the protein is *not* a sweet taste receptor protein.

Amino acid interactions with the protein were measured in a later study. Based on purported taste responses of human subjects to the amino acids, Price (1972) reported that the spectral data supported its role as a sweet receptor. If the human taste results are extrapolated to the bovine, as Price (1972) did, the data actually provide additional evidence against the hypothesis. For example, D-alanine is not sweet to humans, but L-alanine does taste sweet (Solms, 1969; Solms *et al.*, 1965; see also Schiffman, 1976). In the study by Price (1972), the binding parameters for D-alanine and L-alanine were approximately the same. D-Threonine, which does not taste sweet (Solms, 1969; Solms *et al.*, 1965), has a more favorable K_D for binding to the protein than does sucrose. Many other discrepancies in the claimed relationship between binding to the sweet-sensitive protein and taste effectiveness are evident in the data (Dastoli and Price, 1966; Price, 1972; cf. Price and Desimone, 1977).

^{14}C-Labeled sugars were shown, using a centrifugation assay (Cagan, 1971), to bind *in vitro* to taste tissue preparations; binding occurred to a lesser extent in a control tongue-tissue preparation devoid of taste buds. The circumvallate papillae on the bovine tongue contain 90% of the lingual taste buds (Davies *et al.*, 1979). The binding, however, is extremely weak, in agreement with behavioral and electrophysiological findings (Beidler, 1962; Pfaffmann, 1964). Because of the weak interactions of sugars, their binding parameters are difficult to determine accurately (Cagan, 1971, 1974). Moreover, it is difficult to interpret the report that a rat tongue protein retains a small amount of bound [^{14}C]fructose during electrophoresis (Hiji and Sato, 1973) with such weak binding. Ostretsova *et al.* (1975), using difference spectra with the protein extract (Dastoli and Price, 1966) from bovine taste papillae (fungiform), found only insignificant spectral changes. Using equilibrium dialysis with [^{14}C]glucose as the ligand, they showed binding activity present, but it was in a sedimentable fraction rather than in the Dastoli and Price (1966) protein fraction. They detected binding in the sedimentable fraction from both circumvallate and fungiform papillae but not in tongue epithelium devoid of taste buds. Their observation of binding activity in a sedimentable fraction from bovine taste tissue (Ostretsova *et al.*, 1975) extended the earlier findings (Cagan, 1971). Subsequently, further studies with bovine taste papillae by Lum and co-workers (1976; Lum and Henkin, 1976a,b), who used the centrifugation assay, showed binding activity for sugars and several other sweet compounds in a plasma membrane fraction. Furthermore, their evidence indicated competition for binding among several sweet-tasting compounds. Studies with purified plasma membranes from catfish taste tissue are described later in this chapter.

Monellin is a protein that elicits an intense sweet taste sensation at concentrations considerably below that of sucrose (by 10^{-4}–10^{-5}) (Cagan, 1973; Morris and Cagan, 1972; van der Wel and Loeve, 1973). Monellin is a protein of 11,000 molecular weight (Morris *et al.*, 1973) consisting of two polypeptide chains that are tightly, but noncovalently associated (Bohak and Li, 1976; Brand and Cagan, 1977; Cagan *et al.*, 1978; Hudson and Biemann, 1976). Because monellin can be perceived as sweet at comparatively low concentrations and it has a lingering sweet taste, it was suggested (Cagan,

1973, 1974) that monellin might be used as a ligand for studies of sweet taste receptors. For monellin to be effective at these relatively low concentrations, it was presumed that it might have a higher affinity for its receptors than does sucrose; sucrose must be at a concentration in the range of 10^{-2} M to elicit a sweet taste sensation. Moreover, the molecular size of monellin makes it unlikely that it will readily pass through the cell membrane.

To measure the binding of monellin necessitated preparation of a radiolabeled derivative. Derivatization of monellin by reductive methylation under mild conditions (Means and Feeney, 1968) results in partially methylated monellin that retains the sweet taste (Morris *et al.*, 1978). This derivative could therefore be used if the reaction were carried out in such a way as to introduce a radioactive group. Use of the methylation reaction allows preparation of ^{3}H-labeled methylated monellin, using [^{3}H]formaldehyde as the methyl group donor. Binding of this ligand was demonstrated using a membrane filtration assay with circumvallate papillae preparations from bovine (Table I) and human tongue tissues (Fig. 1) (Cagan and Morris, 1979). Taste responses of bovines to monellin are not reported, and it is possible that monellin in the bovine is an antagonist to the sweet receptor sites rather than an agonist. Binding is greater to the taste tissue than to the nontaste epithelia, although significant levels of nonspecific binding occur under the conditions used. Binding of labeled monellin appears to saturate at high ligand concentrations, and has a higher affinity ($K_D \sim 10^{-5}$ M) for binding than do sugars ($K_D \sim 10^{-1}$–10^{-3} M). Furthermore, binding data (Cagan and Morris, 1979) suggest that competition occurs between certain other sweet compounds and [^{3}H]methylmonellin at the monellin binding sites.

Sweet receptors provide an experimental model with which investigators

TABLE I **Binding of [^{3}H]Methylmonellin to Taste and Nontaste Tissues from Bovine Tongue**[a]

Tissue	Taste buds	Binding (pmole/mg protein)
Circumvallate sidewall	Yes	196 ± 51
Tongue epithelium	No	70 ± 28

[a] Each value is the mean ± SEM of 11 experiments, in each of which duplicate determinations were made. The circumvallate sidewall and tongue epithelium preparations are derived from homogenates of the respective bovine tissues which are centrifuged to obtain the tissue pellets. Binding is measured using a membrane filtration technique (Schleicher and Schuell, BA85 membranes) with [^{3}H]methylmonellin at 6 μM. Statistical analyses show the circumvallate and epithelial values to be significantly different ($p < 0.01$). (Taken from Cagan and Morris, 1979.)

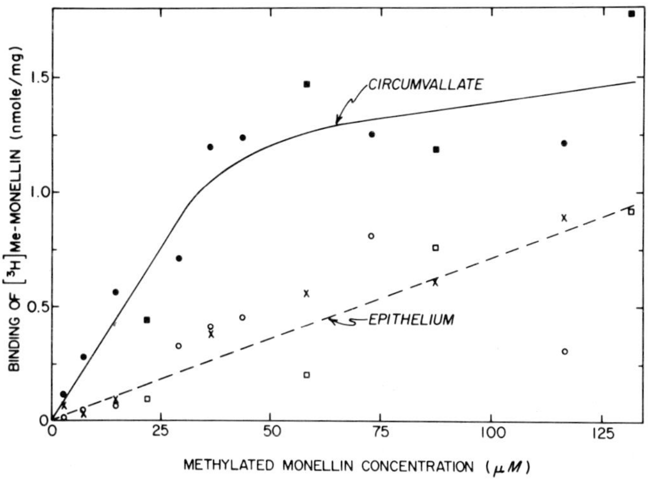

Fig. 1. Binding of [^{3}H]methylmonellin to preparations derived from human circumvallate (taste) papillae and control epithelia. Samples of circumvallate sidewall and of control tongue epithelium and pectoral skin were prepared in parallel fashion. Binding was measured using a membrane filtration assay (Schleicher and Schuell, type BA85 membrane). The data are from five experiments. Each point represents either the mean of 2–5 values (○,●) or a single value (□,■). Each point (×) is a single sample for the pectoral skin. Statistical measures show an effect of monellin concentration ($p<0.01$) and significantly greater binding to circumvallate than to epithelium ($p<0.001$). (Taken from Cagan and Morris, 1979.)

have begun to provide evidence showing that the initial binding interaction is measurable biochemically. Although the current data are limited, they agree in a general way with behavioral and electrophysiological findings that the interaction with sugars appears to be very weak (Beidler, 1962; Pfaffmann, 1964; see also this volume, Chapters 7 and 9). The data support the hypothesis that the interaction with monellin is of a higher affinity. Furthermore, the molecular size of monellin (11,000 molecular weight) as a sweet stimulus is indirect evidence pointing to the surface membrane as the location of its receptor sites.

III. GLUTAMATE TASTE RECEPTORS

Recognition of a stimulus is an important process for which the mechanisms need to be better defined. In addition to recognition per se, modulation of receptor activity is conceivable, with concomitant effects on the information content of the initial interaction. A biochemical model experimentally demonstrates this hitherto unrecognized principle for the taste system.

The taste of monosodium L-glutamate (MSG) has long been described as unique by psychologists and food scientists (Amerine *et al.*, 1965; Solms, 1969; Solms *et al.*, 1965; reviewed in Cagan, 1977a). In combination with certain 5′-ribonucleotides, the taste intensity of MSG is enhanced above that predicted by the sum of the taste intensities of the two components (Kuninaka, 1967; Kuninaka *et al.*, 1964; Yamaguchi, 1967; Yamaguchi *et al.*, 1971). This effect, called taste synergism, appears to form the basis for the Japanese culinary tradition of combining certain foodstuffs as condiments. For example, the seaweed *Laminaria* contains L-glutamate as its main taste-active constituent, while dried bonito contains 5′-IMP and black mushroom contains 5′-GMP as the characteristic taste-active components, respectively.

It was postulated (Cagan, 1977a) that the mechanism underlying the unusual taste effect of L-glutamate depends primarily on its synergism in combination with certain 5′-ribonucleotides. Accordingly, experiments to test this hypothesis were carried out. Initially, binding of L-[^{3}H]glutamate was measured to a preparation of bovine circumvallate (taste) papillae using a

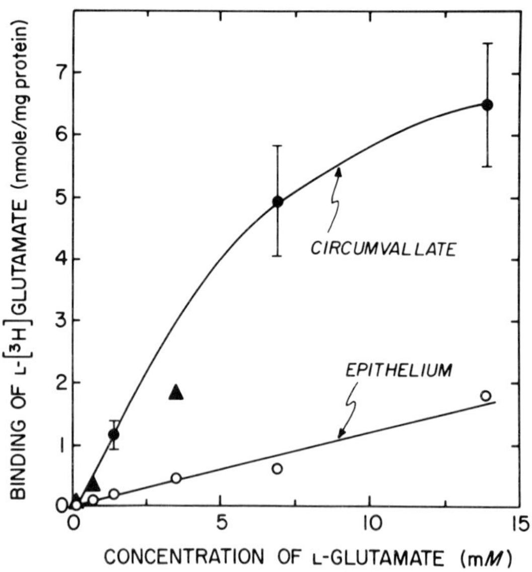

Fig. 2. Binding of L-[^{3}H]glutamate to bovine circumvallate (taste) papillae and tongue epithelium controls. Binding was measured with a Millipore filtration assay on duplicate samples. The circumvallate data are means ± SEM of three experiments, except for the values indicated (▲), which are single values. Epithelial values shown are from a single experiment. (Taken from Torii and Cagan, 1980, with permission.)

membrane filtration assay. As a control, a tongue epithelium preparation devoid of taste receptors was used. Binding to the circumvallate papillae is considerably greater than to the epithelium (Fig. 2) (Cagan *et al.*, 1979; Torii and Cagan, 1980). The apparent K_D for L-glutamate binding to the circumvallate is estimated to be 20–30 m*M*. When certain 5′-ribonucleotides (unlabeled) are included in the assay system with L-[^{3}H]glutamate as the ligand, severalfold increases in binding of L-[^{3}H]glutamate occur (Torii and Cagan, 1980). The effective nucleotides are 5′-GMP, 5′-IMP, and 5′-UMP, but 5′-XMP, 5′-AMP, and 5′-CMP do not increase L-glutamate binding (Fig. 3). The specificity with respect to the nucleotides shows a marked similarity to their specificity in evoking the synergistic taste effect in human psychophysical experiments (Kuninaka, 1967; Kuninaka *et al.*, 1964; Yamaguchi, 1967; Yamaguchi *et al.*, 1971). None of the nucleotides altered the low level of binding to the control epithelium (Fig. 3), thus the stimulatory effect is specific to the taste receptor preparation. Structurally related compounds were studied, including free bases, their nucleosides, and their di- and triphosphonucleotides, but none were effective in increasing binding of L-glutamate to the circumvallate preparation (Torii and Cagan, 1980).

The increased taste intensity of L-glutamate in the presence of a low level

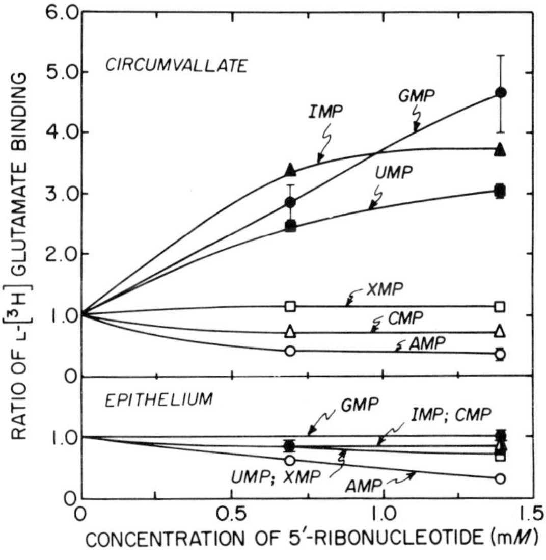

Fig. 3. Enhancement of L-[^{3}H]glutamate binding to bovine taste papillae by 5′-ribonucleotides. A Millipore filtration assay was used with 1.4 m*M* L-[^{3}H]glutamate as the ligand. The data are compiled from several experiments; in each experiment 5′-GMP was included as an internal control (see also Fig. 2). (Taken from Torii and Cagan, 1980, with permission.)

of 5′-ribonucleotide is postulated to result from increased binding of L-glutamate to the taste receptors. Although the detailed mechanism is not well defined, the limited data suggest that the major effect is an increase in maximal binding of L-glutamate without a marked change in the apparent K_D (Torii and Cagan, 1980). The possibility is suggested that the 5′-ribonucleotide unmasks previously hidden or buried receptor sites (see Cagan, 1979) in the cell membrane. The L-glutamate/5′-ribonucleotide phenomenon could be a useful model to better understand mechanisms of cooperative effects in taste receptors. The phenomenon provides additional evidence showing the initial binding interaction to be a critical recognition event in taste and in addition demonstrates that it is subject to modulation at the receptor level.

IV. CATFISH TASTE RECEPTORS AND THE ROLE OF THE PLASMA MEMBRANE

Most of the ligands identified as taste stimuli in mammals have relatively high taste thresholds and consequently are presumed to have relatively weak binding affinities. Unlike mammals, the channel catfish, *Ictalurus punctatus*, has taste receptors that are sensitive to amino acid taste stimuli at considerably lower concentrations (10^{-8}–10^{-9} *M*). Consequently, their binding affinities might be correspondingly greater. Moreover, from a practical perspective, high specific radioactivity L-amino acids are available in ^3H-labeled form.

The extensive taste system of the channel catfish is readily accessible, being located on its body surface and barbels. Originally described by Herrick (1904) and Landacre (1907), the catfish taste system was shown to subserve food location. More recent anatomical, behavioral (Atema, 1971; Bardach and Atema, 1971; Bardach *et al.*, 1967), and electrophysiological (Caprio, 1975) studies demonstrate the functional role of the catfish taste system in locating food and its sensitivity to amino acids as stimulus compounds.

Biochemical studies of catfish taste receptors (Krueger and Cagan, 1976), using a membrane filtration assay (Cuatrecasas, 1971a,b), showed the initial binding interaction to be important in taste sensation. First, the localization of binding activity was similar to the known localization of taste buds on the body surface (Atema, 1971; Bardach and Atema, 1971). Greater binding activity occurs in the sedimentable preparation from barbels than from ventral skin (Table II), in agreement with the distribution of taste buds in these regions. Second, denervation of the taste receptors established that binding is involved in taste. It has long been known (May, 1925; Olmsted, 1920; Torrey, 1934) that the taste receptors of catfish degenerate upon denervating the barbels. De-

TABLE II Binding Activity of Sedimentable Fraction from Taste Tissue of Catfish Barbel and Ventral Skin[a]

Tissue	L-[^{3}H]Alanine concentration (μM)	Binding (cpm/100 μg protein)	Percentage
Barbel	0.03	996 ± 17	
Ventral skin	0.03	657 ± 5 ($p < 0.001$)	66.0
Barbel	0.33	6594 ± 319	
Ventral skin	0.33	4565 ± 113 ($p < 0.01$)	69.2

[a] The sedimentable preparation (fraction P2) was prepared by differential centrifugation of the tissue homogenate. Binding was measured using a Millipore filtration assay in which the values are corrected for nonspecific binding (presence of large excess of unlabeled ligand). The values are means ± SEM of four replicates. (Taken from Krueger and Cagan, 1976, with permission.)

nervation results in a decline of binding activity of the membrane preparation from barbels at 2, 5, and 10 days following denervation (Krueger and Cagan, 1976) (Table III). Third, several amino acids that are taste stimuli measured electrophysiologically in the intact catfish (Caprio, 1975) exhibit binding activity with the sedimentable preparation (Krueger and Cagan, 1976) (Table IV). The relative responses were measured electrophysiologically at 10^{-4} M and binding was measured biochemically at 2×10^{-5} M. The relative ordering of responses are not in accord in every case, but there is no necessary direct relationship between the intensities of these two measures, particularly

TABLE III Decrease in Binding Activity Following Denervation of Catfish Barbel[a]

Treatment	Binding (cpm/100 mg protein) 33 nM L-[^{3}H]Alanine	% of control	0.33 μM L-[^{3}H]Alanine	% of control
Control	499 ± 17		3901 ± 107	
Denervated, 2 days	400 ± 36 ($p < 0.01$)	80.2	3187 ± 64 ($p < 0.01$)	81.7
Control	586 ± 7		5271 ± 129	
Denervated, 5 days	371 ± 10 ($p < 0.001$)	63.3	3500 ± 171 ($p < 0.01$)	66.4
Control	879 ± 7		6523 ± 42	
Denervated, 5 days	459 ± 90 ($p < 0.02$)	52.2	3429 ± 38 ($p < 0.001$)	52.6
Control	837 ± 12		6292 ± 50	
Denervated, 10 days	722 ± 40 ($p < 0.1$)	86.2	5308 ± 113 ($p < 0.01$)	84.4

[a] The right maxillary barbels of groups of four to six fish were denervated for each time period shown; the left-side remained intact as a control. The sedimentable fraction (fraction P2) was prepared from each group, and binding was determined with a Millipore filtration assay in which the values are corrected for nonspecific binding. The values are means ± SEM of four replicates. (Taken from Krueger and Cagan, 1976, with permission.)

TABLE IV Binding Activity of Sedimentable Fraction From Catfish Taste Tissue with ^{3}H-Labeled Amino Acids[a]

Experiment number	Binding (pmole/mg protein) L-Arginine	L-Serine	L-Threonine	L-Alanine	β-Alanine	Glycine
1	304.7 ± 24.8	149.8 ± 4.6	66.5 ± 4.4	54.0 ± 0.6	37.9 ± 1.3	1.1 ± 0.4
2	320.0 ± 7.9	202.2 ± 2.2	83.4 ± 24.8	81.3 ± 2.1	51.9 ± 1.5	10.2 ± 3.1
3	393.9 ± 19.6	223.2 ± 1.7	38.6 ± 2.7	55.9 ± 4.3	31.1 ± 0.9	27.2 ± 2.4

[a] Three separate preparations of the sedimentable fraction (fraction P2) were studied for binding activity using a Millipore filtration assay. Each of the ligands was present at 20 μM. The values are corrected for nonspecific binding. Each value is the mean ± SEM of four replicates. (Taken from Krueger and Cagan, 1976, with permission.)

when compared at different concentrations. Recent studies (see below) show an interesting correspondence between the electrophysiological and biochemical findings. Furthermore, recent studies of Holland (1980) show a good correspondence between behavioral sensitivity of the catfish and binding. For example, L-arginine, which binds to a high degree, is a very potent stimulus behaviorally, whereas β-alanine and glycine, which bind less, are less effective behavioral stimuli.

Binding of L-[^{3}H]alanine has been studied in considerable detail (Krueger and Cagan, 1976; Cagan, 1979). The preparation, fraction P2, is obtained by differential centrifugation (Neville, 1960; Ray, 1970) of a homogenate of catfish taste epithelium. Binding of L-[^{3}H]alanine shows saturation as the ligand concentration increases (Fig. 4); the apparent K_D calculated from linear transformations of the data show it to be 5×10^{-6} M (Krueger and Cagan, 1976; Cagan, 1979). Furthermore, binding is reversible, as demonstrated both by dissociation of bound ligand into ligand-free medium and by displacement of bound radioactive ligand upon adding a pulse of unlabeled L-alanine.

An unusual phenomenon was observed upon storing the sedimentable fraction (−65°C) in a high concentration (10 mM) of unlabeled ligand. Following removal of the unlabeled L-alanine by centrifugation and extensive washing, the binding activity was enhanced by severalfold. The data (Cagan, 1979) do not support a transport hypothesis, but the possibility of some entrapment in vesicles was raised. Binding is not increased by Na^+, thus indicating that a Na^+-coupled transport mechanism is not operative. Preloading experiments with unlabeled L-alanine failed to cause increased uptake of label, as would occur with a countertransport mechanism (Christensen, 1975); instead, binding was decreased as expected when true binding is

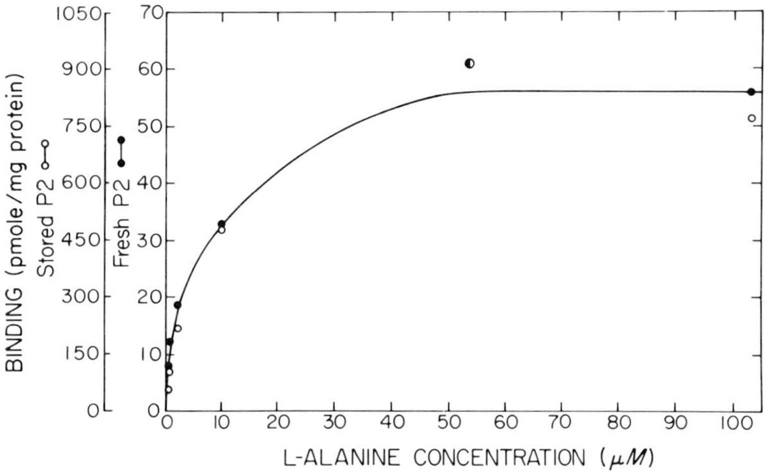

Fig. 4. Saturation curve for binding of L-[³H]alanine to catfish taste tissue sedimentable fraction P2. The points (●) are for freshly prepared fraction P2. The points (○) are for the same preparation following its storage (−65°C) in 10 m*M* L-alanine (unlabeled), with removal of the storage L-alanine by washing prior to the binding assay. (Taken from Cagan, 1979, with permission.)

measured. Osmotic perturbation, however, suggests that a significant portion of the ligand might be contained within vesicles even though transport does not occur. Following treatment of fraction P2 with unlabeled ligand and its removal, the binding data show the apparent K_D to remain unchanged (Fig. 4), whereas B_{max} increases. This suggests that the treatment unmasks "hidden" or "buried" receptor sites for L-alanine (Cagan, 1979).

An important question in taste receptor function is the localization of the taste receptor molecules in the cell. The Renqvist Adsorption Theory (Renqvist, 1919) postulates the outer (plasma) membrane of the receptor cell as the adsorbent. This localization has several lines of indirect evidence as support. Two observations were suggested by Beidler (1962) as indicating that taste compounds need not enter cells to stimulate a response. He noted that (1) the speed of the response (within milliseconds) was too fast to allow transport into the cells and (2) the ability of toxic substances to stimulate taste receptors without apparently damaging their responsiveness supported this hypothesis. Evidence that the taste-modifying principle of miracle fruit, miraculin, is a glycoprotein of about 44,000 molecular weight (Brouwer *et al.*, 1968; Kurihara and Beidler, 1968) provides additional indirect evidence that taste effects occur at the cell surface. Treating the tongue with miraculin causes sour-tasting acids to taste sweet. Kurihara and Beidler (1969) proposed that miraculin binds to the taste cell surface. They proposed further that exposure of the taste receptors to acids causes a conformational change in the receptor membrane such that the sugar moieties of miraculin are able to stimulate sweet receptor sites. Interaction of miraculin with taste receptors has not been studied biochemically, but such studies could provide interesting information. Additional indirect evidence for a role of the plasma membrane is provided by the discovery of chemostimulatory proteins (see Cagan, 1973), which are sweet taste stimuli. It was postulated (Cagan, 1973, 1974) as unlikely that molecules of this size (11,000 and 18,000–21,000 daltons) could readily pass through the cell membrane to elicit a taste sensation.

Evidence cited in this chapter and elsewhere in this volume point to the plasma membrane as the locus of receptor specificity in taste and olfactory receptors. An experimental test of this hypothesis appeared possible with the channel catfish preparation. Because of the extensive taste system of the catfish, a relatively greater quantity of membrane fraction is obtained than with other experimental models available; the binding affinities for taste stimuli are in a range ($K_D \sim 10^{-6}\ M$) to be readily and reproducibly measured, and ^{3}H-labeled ligands that are taste stimuli are available to measure the binding.

Advantage was therefore taken of this system to determine the subcellular localization of taste receptor sites. Isolation of plasma membranes and determination of their content of binding activity for taste ligands would provide important evidence in support of the hypothesis. Plasma membranes from catfish taste tissue were isolated in a reasonable state of purity. The sedimentable fraction (fraction P2) described by Krueger and Cagan (1976) and Cagan (1979) was further fractionated on sucrose density gradients (Cagan and Boyle, 1978). A discontinuous gradient was prepared of sucrose concentrations (w/w) 48, 45, 41, and 37% (Ray, 1970), with the sample being introduced in the bottom layer of the gradient. The samples were centrifuged at 23,000 rpm for 17 hr on a Spinco Model L2-65B ultracentrifuge (4°C) in a swinging bucket rotor (Model SW-41). The resulting bands and the pellet were analyzed for marker enzymes characteristic of plasma membranes, of mitochondria, and of microsomes and for binding activity towards L-[^{3}H]alanine. Under the conditions employed, Ray (1970) found that the plasma membranes from rat liver were recovered at the 37%/41% sucrose interface. With the catfish taste tissue, the plasma membranes are recovered at the uppermost surface of the gradient, indicating that they have a lighter buoyant density.

The binding activity is localized in the uppermost band, B1, which floats on the surface of the gradient (Fig. 5). Of the enzymatic markers used to characterize the preparation, only the 5′-nucleotidase activity (Heppel and Hilmoe, 1955; Chen *et al.*, 1956), which is often associated with plasma membranes (DePierre and Karnovsky, 1973; Emmelot *et al.*, 1974), is preferentially localized to band B1. The other marker enzymes, NADH cytochrome *c* reductase (Mackler, 1967) and succinate cytochrome *c* reductase

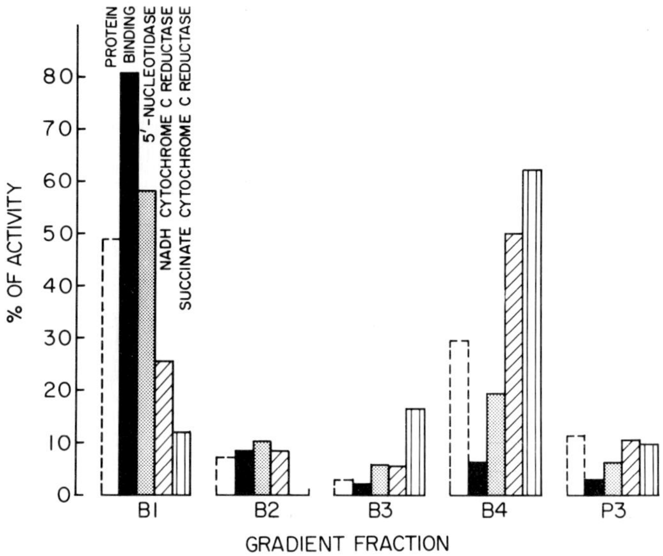

Fig. 5. Distributions of protein, binding activity, and marker enzymes following sucrose density-gradient fractionation. Catfish fraction P2 was subjected to ultracentrifugation as described in the text. Migration of the sample was in the upward direction through the gradient. The fractions are indicated starting with the band at the top of the gradient (B1) and continuing to the pellet at the bottom of the tube (P3). Binding was assayed with L-[^{3}H]alanine as described (Krueger and Cagan, 1976; Cagan, 1979). Marker enzymes were 5′-nucleotidase (Heppel and Hilmoe, 1955; Chen *et al.*, 1956) for plasma membranes, NADH cytochrome *c* reductase (Mackler, 1967) for microsomes, and succinate cytochrome *c* reductase (Tisdale, 1967) for mitochondria. Of the total recovered on the gradient, the percentage of each component in each fraction is shown.

(Tisdale, 1967), for microsomes and mitochrondia, respectively, were localized in band B4, with relatively little of these activities appearing in the plasma membrane fraction.

Modification of the gradient, based on these studies, allowed somewhat better purification of band B1. Centrifugation of fraction P2 under similar conditions but in a continuous sucrose gradient of 41→26% (w/w) resulted in flotation of band B1 to the surface of the gradient. Thus the density of the taste receptor membranes is no greater than 1.11 gm/ml (sucrose density), which is considerably lighter than the rat liver plasma membranes (>1.16 gm/ml sucrose density) prepared by Ray (1970) with this method. With the continuous gradient, the binding activity and 5′-nucleotidase levels in B1 remain high, whereas the levels of the reductase enzymes are barely detectable. Electron microscopic examination of B1 from the continuous gradient reveals it to contain membrane profiles (Fig. 6). Furthermore, binding activity for several taste ligands (L-alanine, L-arginine, L-serine, L-threonine, β-alanine, glycine, and L-lysine) occurs in this fraction; these experiments are being extended. Therefore, the catfish experimental model makes it

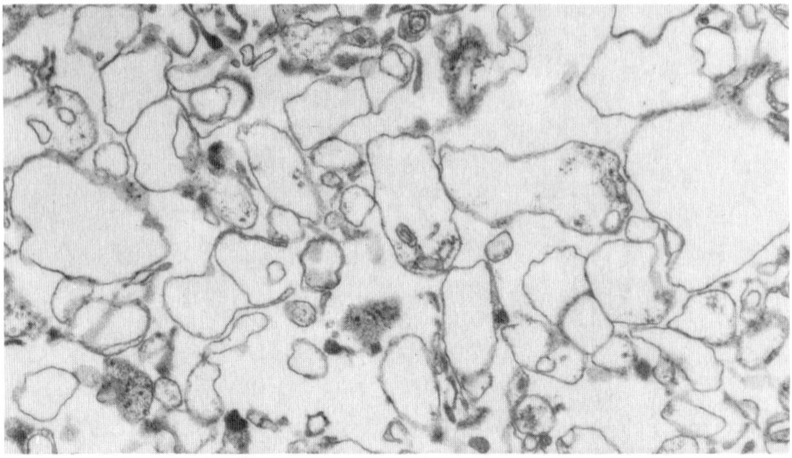

Fig. 6. Plasma membrane fraction derived from catfish taste tissue. The sample of fraction P2 was fractionated by ultracentrifugation in a continuous sucrose density gradient [41→26% (w/w) sucrose]; see also legend to Fig. 5. Band B1 was fixed, stained, sectioned, and examined by transmission electron microscopy (×23,800). (Courtesy of Dr. Paula M. Orkand.)

possible to isolate plasma membranes and to demonstrate binding with taste ligands of reasonably high affinity (for a taste receptor). This has thereby enabled localization of the binding activity to the plasma membranes. The catfish system thus provides direct experimental support for the Renqvist Adsorption Theory that taste receptor sites are located in the plasma membrane.

V. TASTE RECEPTOR SITE ANTAGONIST

To define more clearly the specificity of receptor sites, a selective inhibitor would be highly useful. None have been available, but recent studies in our laboratory have resulted in identification of such an inhibitor.

Studies of the binding of L-[^{3}H]alanine to taste receptor membrane preparations (fractions P2 and B1) from the catfish (Cagan, 1979; Cagan and Boyle, 1978; Krueger and Cagan, 1976) (see above) provide fundamental information on the interaction of this taste ligand with its receptor sites. L-Alanine is an effective stimulus to the catfish when measured electrophysiologically (Caprio, 1975). Recordings from the taste nerve in the maxillary barbel of the catfish show responses to a large number of amino acids. Recently, structure–activity relationships were also reported (Caprio, 1978) using the electrophysiological recording technique. A preliminary report by Caprio and Tucker (1976) suggests that the catfish has at least two types of amino acid receptors. A population of single nerve fibers could be classified into two major groups: one population was maximally sensitive to L-arginine but showed little or no responses to other amino acids; the other population consisted of fibers responsive to L-alanine and also to several other amino acids tested but was minimally stimulated by L-arginine. They also reported that cross-adaptation experiments (Caprio and Robinson, 1978) support these data. They postulate (at least) two types of receptors, a specific L-arginine receptor and an L-alanine receptor with broad specificity.

In survey experiments (P. R. Zelson and R. H. Cagan, unpublished), binding competition studies showed that L-alanine and L-arginine did not compete with each other for binding. The ^{3}H-labeled ligand was present at 1 μM and the potential competitor at 1 mM. Further studies show no competitive binding interactions between L-alanine and L-arginine (Table V) under conditions where an analog of L-alanine does inhibit. The fact that neither amino acid inhibits binding of the other demonstrates that L-alanine and L-arginine interact with different binding sites. This appears somewhat similar to the olfactory receptors of the rainbow trout (Cagan and Zeiger, 1978; Rhein and Cagan, 1980; see also this volume, Chapter 3) in which L-alanine and L-lysine interact with separate sites. The presence of (at least) two types of taste receptor sites allowed demonstration of the selective inhibition of

TABLE V Effects of L-Alanine, L-Arginine, and β-Chloro-L-Alanine on Binding to Sedimentable Fraction from Catfish Taste Tissue[a]

Ligand	Addition	Inhibition (%)
0.4 μM L-[^{3}H]Arginine	+ 4 μM L-Alanine	0
	+ 40 μM L-Alanine	0
	+ 4 μM β-Chloro-L-alanine	0
	+ 40 μM β-Chloro-L-alanine	0
0.4 μM L-[^{3}H]Alanine	+ 4 μM L-Arginine	0
	+ 40 μM L-Arginine	0
	+ 4 μM β-Chloro-L-alanine	59
	+ 40 μM β-Chloro-L-alanine	97

[a] The sedimentable fraction (fraction P2) was prepared from catfish taste tissue (Krueger and Cagan, 1976; Cagan, 1979). It was assayed for L-[^{3}H]alanine binding activity using a Millipore filtration assay with the correction for nonspecific binding evaluated using a large excess of unlabeled ligand in parallel samples. Because of the high degree of actual binding of L-[^{3}H]arginine to the Millipore filters, the values of total L-[^{3}H]-arginine bound were corrected using parallel samples with ^{3}H-labeled ligand under exactly the same radioactivity conditions but in the absence of fraction P2. The data are a compilation of four separate preparations, in each of which a single ligand was used (either L-[^{3}H]arginine or L-[^{3}H]alanine) and additions at either tenfold or 100-fold in excess, respectively. In each case, four replicates were run.

one type of site. The data in Table V show that β-chloro-L-alanine inhibits L-[^{3}H]alanine binding but does not inhibit L-[^{3}H]arginine binding to the catfish receptor preparation. For example, at a concentration tenfold above that of the ligand L-[^{3}H]alanine, β-chloro-L-alanine inhibits binding by 60% and essentially completely at 100-fold above the ligand concentration.

In preliminary experiments, β-chloro-D,L-alanine was used as the inhibitor, although it was later found to inhibit to a lesser extent than the L-form itself. In these experiments, reversibility of the inhibition was demonstrated. Following incubation of fraction P2 with L-[^{3}H]alanine in the absence or presence of β-chloro-D,L-alanine, fraction P2 was recovered by centrifugation. It was washed twice by suspending it in buffer and centrifuging to recover the fraction P2. Binding assays on the control and β-chloroalanine-treated samples showed the same specific binding activity in both samples.

Corroborative evidence has been obtained by electrophysiological recordings from nerve fibers of the catfish maxillary barbel (Cagan and Teeter, 1980). Summated recordings were made from multi-fiber preparations. First, the responsiveness of the taste receptors to L-alanine and L-arginine was demonstrated; for example, responses were readily recorded to 10^{-5} M of each of these amino acids. Upon stimulation with 10^{-4} M β-chloro-L-alanine, the responses were small and similar in magnitude to responses with control pulses of water. The critical measurements were those to mixtures of β-chloro-L-alanine + L-alanine and to β-chloro-L-alanine + L-arginine. Responses to the mixture of 10^{-5} M L-arginine + 10^{-4} M β-chloro-L-alanine were the same or slightly larger than those to 10^{-5} M L-arginine alone. Responses to the mixture of 10^{-5} M L-alanine + 10^{-4} M β-chloro-L-alanine were decreased by 40–60% compared with 10^{-5} M L-alanine alone. The electrophysiological evidence therefore provides strong support for the hypothesis that β-chloro-L-alanine is a receptor site antagonist of the L-alanine taste receptors.

The analogue β-chloro-L-alanine therefore appears to be an inhibitor of the taste receptor sites for L-alanine but not those for L-arginine. Further studies are in progress to more completely define its inhibition characteristics.

VI. COVALENT LABELING AND ISOLATION OF A TASTE-RECEPTOR MACROMOLECULE

An important question is the chemical character of receptor molecules. The hypothesis of "receptor proteins" in taste has been addressed experimentally by only a few laboratories and remains poorly understood. Although likely to be correct, the hypothesis of receptor proteins in the taste cell membranes requires careful documentation. Other hypotheses, for example, have been proposed ascribing recognition of taste stimuli to the lipid portion of the membrane or to the charge distribution at the surface (see this volume, Chapters 11 and 13). Recent evidence (Zelson and Cagan, 1979) suggests that both phospholipids and proteins are important in the amino acid binding activity of catfish taste receptors. Treatment of fraction P2 with phospholipase C caused a decrease in binding activity; treatment with trypsin surprisingly caused an increase in binding. The data do not, however, support the hypothesis that —SH groups are involved in the alanine receptor binding interaction because several —SH reactive reagents do not inhibit binding (Zelson and Cagan, 1979).

Better definition of the properties of the putative receptor macromolecules could be achieved by unambiguously identifying isolated molecules as specific binding components. One approach is to solubilize the receptor membranes with detergent and then to isolate the putative receptor protein in soluble form. Its identity as a receptor is based upon its binding properties with relevant ligands. Attempts to use this approach in our laboratory, in which catfish fraction P2 was solubilized with detergents, resulted in preparations with relatively little or no ligand binding activity. An alternative approach is to label the receptor molecules prior to solubilization and isolation.

Physiologically it is important to an organism that the interaction of a stimulus compound with a taste receptor be reversible. The ready reversibility of specific binding of the ligands studied to date (Cagan, 1971, 1974, 1977b, 1979; Cagan and Morris, 1979; Krueger and Cagan, 1976; Lum and Henkin, 1976b; Ostretsova *et al.*, 1975) suggests that a mechanism of taste reception involving an irreversible binding interaction is unlikely. What is a desirable functional and adaptive biological feature, however, can be an experimental disadvantage to the biochemist, for the reversibility of the interaction and the relatively low affinity of the receptor for the ligand (i.e., high K_D) makes identification of a reversibly bound ligand difficult, if not impossible, while the receptor is being isolated.

Receptor studies in other areas have sometimes benefited from the ability to covalently attach to the receptor site either an agonist or an antagonist. An attached radiolabeled ligand can be unambigously identified throughout the receptor isolation procedure. For example, an acetylcholine receptor was radiolabeled with the alkylating agent 4-(*N*-maleimido)-α-benzyltrimethylammonium iodide (MBTA). The acetylcholine receptor in the electroplax organ of *Electrophorus electricus* was first reduced by dithioerythritol and then alkylated with [^{3}H]MBTA. Dithioerythritol is believed to reduce a disulfide group near the receptor binding site; the —SH thus produced becomes alkylated by [^{3}H]MBTA (Karlin *et al.*, 1971; Karlin, 1973). The labeling approach, in conjunction with protection experiments, helped to establish MBTA as an affinity label of the acetylcholine receptor site.

A labeling approach for isolating taste receptors appears promising. By first covalently labeling the receptor preparation with an ^{3}H-labeled taste ligand, subsequent identification of the isolated receptor can be more clearly made. The approach taken in our laboratory is to utilize a cross-linking reaction to label taste receptor sites with L-[^{3}H]alanine. 3-Unsubstituted isoxazolium salts react readily with carboxylates, converting the carboxylate into a reactive ester (Woodward and Olofson, 1966). The reaction was described by Woodward *et al.* (1961) to synthesize peptides and was used subsequently to determine the functional role of carboxyl groups in proteins and to couple ligands to insoluble matrices. For example, *N*-alkyl-5-phenylisoxazolium salts were used to modify the carboxyl groups and study their functional role in trypsin (Bodlaender *et al.*, 1969; Feinstein *et al.*, 1969), in staphylococcal nuclease (Dunn *et al.*, 1974), and in yeast phosphoglycerate kinase (Brake and Weber, 1974). Enzymes were coupled to collagen films using *N*-ethyl-5-phenylisoxazolium-3′-sulfonate (Lee *et al.*, 1976), and the reagent was used to polymerize α-chymotrypsin (Patel and Price, 1967) and to couple α-chymotrypsin to other polymers (Patel *et al.*, 1967).

Woodward's Reagent K is *N*-ethyl-5-phenylisoxazolium-3′-sulfonate. When used as the coupling agent, the following reaction scheme could occur:

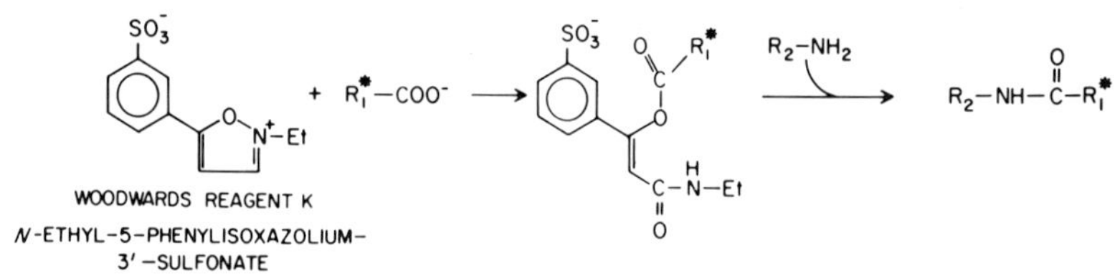

Two possible routes can be postulated. In the route shown, R_1 is that of L-[^{3}H]alanine; the carboxyl group of the ligand is activated, which reacts with an amino group in the protein. Alternatively, a carboxyl group of the protein might be activated, which would react with the primary amino group of the ligand.

Preliminary experiments in our laboratory suggested that Woodward's Reagent K might be useful to cross-link taste ligands to their receptors. In these experiments (R. W. Morris and R. H. Cagan, unpublished), treatment of the catfish fraction P2 in the presence of L-[^{3}H]alanine with Woodward's Reagent K indicated that some of the ligand was incorporated irreversibly. In these preliminary experiments, quantification of incorporation was not well-defined, because the unbound radioactivity constituted a major portion of the counts recovered in fraction P2 following labeling and an added pulse of unlabeled L-alanine. Also, because the experiments had been carried out at a higher temperature (17°C) than is routinely used for binding assays (0°C) (Krueger and Cagan, 1976), the possibility of metabolic interconversions of L-alanine also posed a question. The preliminary findings (R. W. Morris and R. H. Cagan, unpublished) suggested, however, that Woodward's Reagent K could be useful for taste receptor labeling.

Experiments have been carried out with catfish fraction P2 demonstrating irreversible incorporation of label from the ligand L-[^{3}H[alanine. Aliquots of fraction P2 (3.4-mg protein) were incubated on ice with 52 μM L-[^{3}H]alanine (5.3×10^7 cpm) as the ligand, in 0.01 M Tris-HCl, pH 7.8 + 1 mM $CaCl_2$ (Krueger and Cagan, 1976; Cagan, 1979). Samples were run in parallel either with ligand alone or with addition of 500 μM β-chloro-L-alanine or 500 μM L-arginine, respectively. Following incubation on ice for 55 min, 100 mM Woodward's Reagent K (in a small volume of 1 mM HCl) was added and allowed to react for 5 min. At that time, a pulse of unlabeled L-alanine (0.02 M final concentration) was added and the samples were immediately centrifuged at 100,000 g for 10 min to recover the labeled fraction P2. The pellet was suspended in a small volume of buffer, sonicated for 1 min, diluted to 10 ml with buffer, and centrifuged as above. The suspension, sonication, and centrifugation steps were repeated once more, and the pellet was then suspended for determination of radioactivity and protein (see Cagan, 1979). The specific binding activities of the samples at this point were as follows: ligand alone, 7.74 pmole/mg protein; ligand + β-chloro-L-alanine, 6.82 pmole/mg protein; ligand + L-arginine, 8.20 pmole/mg. The labeled fraction P2 samples were suspended in buffered sodium dodecyl sulfate and chromatographed in this medium on a column of Sephadex G-200 (Fig. 7).

Following reaction with Woodward's Reagent K, removal of most of the unbound ligand is achieved with the pulse of unlabeled L-alanine plus the two sonication and wash steps. Chromatography of the labeled preparation on Sephadex G-200 in the presence of sodium dodecyl sulfate reveals three peaks of radioactivity (Fig. 7). The first peak (I) is small (4%–7% of radioactivity) and occurs at the void volume. Peak II (33%–39% of radioactivity) and Peak III (55–63% of radioactivity) occur during elution. Peak III appears to be free L-alanine by comparison with chromatography of a standard on the same column. Therefore, despite the added pulse of unlabeled ligand and the sonication and wash steps in buffer, unbound label remains in the preparation.

The label recovered in the macromolecular peaks (I and II) during chromatography can be safely ascribed to covalently bound radiolabel. The major portion of the bound radioactivity appears in peak II; this fraction is

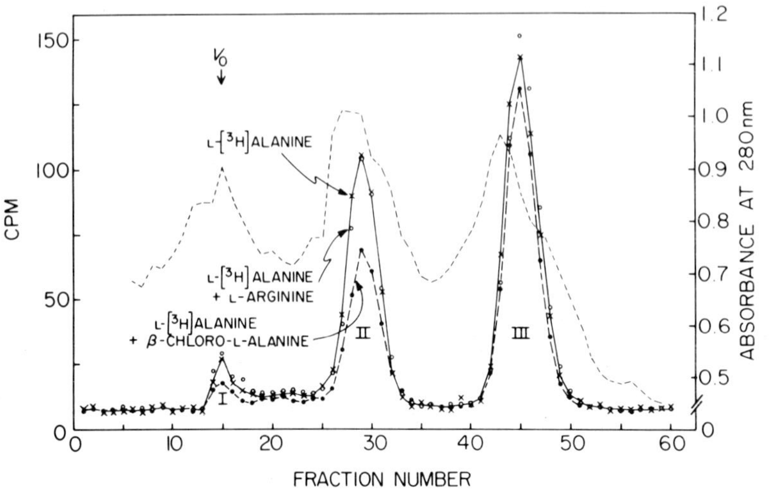

Fig. 7. Chromatography on Sephadex G-200 of catfish sedimentable fraction cross-linked with taste ligand L-[^{3}H]alanine by reacting with Woodward's Reagent K. The experimental details for labeling are described in the text. Chromatography was carried out in buffered sodium dodecyl sulfate (1% SDS, 1% 2-mercaptoethanol, 0.01 *M* sodium phosphate, pH 7.0) on a column of Sephadex G-200 (1.1 × 56 cm) at room temperature. The column flow rate was 6 ml/hr; 1 ml fractions were collected. Elution was monitored by A_{280} (an example is shown by the dashed line) and by counts on each fraction. It is noted that the elution pattern of counts was highly reproducible for that for A_{280} was highly variable, undoubtedly due to the presence of SDS. The void volume of the column is indicated (V_0). In the absence of direct protein determinations on the peak fractions (insufficient material remained), the data must be regarded as preliminary.

postulated to contain the L-alanine taste receptor macromolecule(s). It is further postulated that peak I represents label bound to membranes that become cross-linked by the reagent, such as by reaction of activated carboxyl side chains with ε-amino groups of lysine. Further experiments are needed to establish quantitatively the specific binding activity (covalent) in peak II. In the experiment shown in Fig. 7, insufficient material remained for this analysis.

The postulate is that peak II either is the receptor macromolecule or contains the receptor macromolecule (receptor protein?) along with nearest-neighbor molecules possibly cross-linked to it. Studies of its physicochemical characteristics are needed.

Advantage was taken of the findings (Table V) that β-chloro-L-alanine inhibits binding of L-[^{3}H]alanine to the taste receptor preparation and that the receptor sites for L-alanine and L-arginine are different. In aliquots of the same fraction P2 preparation (Fig. 7), β-chloro-L-alanine was included as a potential competitor of L-[^{3}H]alanine binding; L-arginine was included in other aliquots as a control. β-Chloro-L-alanine and L-arginine, respectively, were each added to the samples at tenfold higher concentration than the ligand L-[^{3}H]alanine. Cross-linking was carried out. It is encouraging to observe (Fig. 7) that β-chloro-L-alanine decreases incorporation of label from L-[^{3}H]alanine into peak II by 35%, whereas L-arginine has no effect. This initial evidence therefore supports the hypothesis that peak II contains the macromolecule(s) that determine the specificity of binding of L-alanine.

VII. RESEARCH NEEDS

The Renqvist Adsorption Theory, proposed in 1919, appears to be fundamentally correct in its concept that many taste stimuli act by adsorbing to the surface membrane of receptor cells. The importance of the kinetic aspects of the interaction, which Renqvist emphasized, remains to be studied biochemically (see this volume, Chapter 9). The apparent importance of receptor occupancy, which Renqvist did not postulate, should not detract from the conceptual advance of his theory.

During the past decade, it has been possible to demonstrate biochemically the initial interaction of taste and olfactory ligands with their respective receptors. In 1971, the common biochemical technique of radioisotopic tracers was introduced to taste receptor binding studies. This has led to its use in several systems. Binding measurements using radioactively labeled ligands and attempts to relate binding properties to physiological and behavioral effects of the stimuli have been used in studies of sweet taste receptors and glutamate taste receptors in mammals and of amino acid taste receptors in fish. These biological models provide experimental systems for detailed study of site specificities in taste and of the physicochemical nature of taste receptor macromolecules. During the coming decade, the membrane mechanisms involved in recognition of various types of stimulus compounds should become better understood.

The recent discovery that β-chloro-L-alanine is a site-specific antagonist compound that interacts with the L-alanine taste receptor sites of the catfish allows another biochemical approach to study site specificity. Along with binding competition studies of taste ligands, being carried out in our laboratory, the antagonist approach should yield valuable insights into the molecular basis of receptor specificity.

Relationships between binding measured biochemically and the activity of the taste system in the same species measured at higher organizational levels, such as electrophysiological or behavioral, will need to be more carefully examined. This approach is being used here for the channel catfish.

Basic biological principles of chemoreceptor function discovered in the vertebrate model systems should be generally applicable across vertebrate species. Questions of basic operating principles should not, however, be confused with the large differences that can be expected in specificity characteristics. For example, a principle that is emerging is the importance of the outer cell membrane (plasma membrane) in taste and in olfactory recognition (see this volume, Chapter 3). This principle will undoubtedly become better developed during the coming decade. At the same time the differences that are expected among stimulus compound specificities (Chapter 7) should not detract from the generality of the operating principle; indeed, specificity differences can, and are, being used to more clearly document the involvement of binding in taste and olfaction.

Promising results have been obtained in preliminary attempts to label a taste receptor site. Further development of this approach, especially with affinity reagents, will have a significant impact on the field. Use of reactive ligands that specifically interact with particular taste receptor sites and are capable of covalently bonding to the sites, will be of great value in enabling isolation and identification of receptor macromolecules.

Studies of the complex train of molecular sequelae following binding that occur within the cell membrane and possibly intracellularly could easily comprise a decade or more of research. The transduction mechanisms (this volume, Part III) are only beginning to be understood. How the recognition event is coupled to its molecular and cellular consequences remains a challenging problem.

The ability to identify and characterize the specificities of particular taste receptor sites lies at the very heart of the classical idea, derived from psychology, of "primary" tastes. "Primary" or "fundamental" tastes will ultimately need to be defined on the basis of receptor site specificity.

ACKNOWLEDGMENT

The author thanks his students and postdoctorals who have collaborated in various aspects of this research on taste receptors: Joseph G. Brand, James M. Krueger, Robert W. Morris, Kunio Torii (visiting scientist), and Philip R. Zelson, and the technicians who have contributed outstanding work during the past several years: Ardithanne G. Boyle, Linda Graham, Pamela Burke, and Ellen Van Auken. The research from our laboratory described in this chapter was supported by the Veterans Administration and by NIH Research Grant NS-08775/NS-15740 from the National Institute of Neurological and Communicative Disorders and Stroke.

REFERENCES

Amerine, M. A., Pangborn, R. M., and Roessler, E. B. (1965). "Principles of Sensory Evaluation of Food," pp. 115-120. Academic Press, New York.

Atema, J. (1971). Structures and functions of the sense of taste in the catfish (*Ictalurus natalis*). *Brain Behav. Evol.* **4**, 273-294.

Bardach, J. E., and Atema, J. (1971). The sense of taste in fishes. *In* "Handbook of Sensory Physiology" (L. M. Beidler, ed.), Vol. 4, Part 2, pp. 293-336. Springer-Verlag, Berlin and New York.

Bardach, J. E., Todd, J. H., and Crickmer, R. (1967). Orientation by taste in fish of the genus *Ictalurus*. *Science* **155**, 1276-1278.

Beidler, L. M. (1954). A theory of taste stimulation. *J. Gen. Physiol.* **38**, 133-139.

Beidler, L. M. (1962). Taste receptor stimulation. *Prog. Biophys. Biophys. Chem.* **12**, 107-151.

Bodlaender, P., Feinstein, G., and Shaw, E. (1969). The use of isoxazolium salts for carboxyl group modification in proteins. Trypsin. *Biochemistry* **8**, 4941-4948.

Bohak, Z., and Li, S.-L. (1976). The structure of monellin and its relation to the sweetness of the protein. *Biochim. Biophys. Acta* **427**, 153-170.

Brake, A. J., and Weber, B. H. (1974). Evidence for an essential carboxyl group for yeast phosphoglycerate kinase. Reaction with Woodward's Reagent K. *J. Biol. Chem.* **249**, 5452-5457.

Brand, J. G., and Cagan, R. H. (1977). Fluorescence characteristics of native and denatured monellin. *Biochim. Biophys. Acta* **493**, 178-187.

Brouwer, J. N., van der Wel, H., Francke, A., and Henning, G. J. (1968). Miraculin, the sweetness-inducing protein from miracle fruit. *Nature (London)* **220**, 373-374.

Cagan, R. H. (1971). Biochemical studies of taste sensation I. Binding of ^{14}C-labeled sugars to bovine taste papillae. *Biochim. Biophys. Acta* **252**, 199-206.

Cagan, R. H. (1973). Chemostimulatory protein: A new type of taste stimulus. Science **181**, 32-35.

Cagan, R. H. (1974). Biochemistry of sweet sensation. *In* "Sugars in Nutrition" (H. L. Sipple and K. W. McNutt, eds.), pp. 19-36. Academic Press, New York.

Cagan, R. H. (1977a). A framework for the mechanisms of action of special taste substances: The example of monosodium glutamate. *In* "The Chemical Senses and Nutrition" (M. R. Kare and O. Maller, eds.), pp. 343-359. Academic Press, New York.

Cagan, R. H. (1977b). Recognition of gustatory and olfactory stimulus molecules at receptor sites. *In* "Food Intake and Chemical Senses" (Y. Katsuki, M. Sato, S. F. Takagi, and Y. Oomura, eds.), pp. 131-138. Univ. of Tokyo Press, Tokyo.

Cagan, R. H. (1979). Biochemical studies of taste sensation VII. Enhancement of taste stimulus binding to a catfish taste receptor preparation by prior exposure to the stimulus. *J. Neurobiol.* **10**, 207-220.

Cagan, R. H., and Boyle, A. G. (1978). Plasma membranes from taste receptor tissue: Isolation and binding activity. *Fed. Proc., Fed. Am. Soc. Exp. Biol.* **37**, 1818. (Abstr.)

Cagan, R. H., and Morris, R. W. (1979). Biochemical studies of taste sensation [6.] Binding to taste tissue of ^{3}H-labeled monellin, a sweet-tasting protein. *Proc. Natl. Acad. Sci. U.S.A.* **76**, 1692-1696.

Cagan, R. H., and Teeter, J. H. (1980). Selective inhibition of binding to a taste receptor site. *Soc. Neurosci. Abstr.* **6**, 304.

Cagan, R. H., and Zeiger, W. N. (1978). Biochemical studies of olfaction: Binding specificity of radioactively labeled stimuli to an isolated olfactory preparation from rainbow trout (*Salmo gairdneri*). *Proc. Natl. Acad. Sci. U.S.A.* **75**, 4679-4683.

Cagan, R. H., Brand, J. G., Morris, J. A., and Morris, R. W. (1978). Monellin, a sweet-tasting protein. *In* "Sweeteners and Dental Caries" (J. H. Shaw and G. G. Roussos, eds.), pp. 311–323. Information Retrieval, London.

Cagan, R. H., Torii, K., and Kare, M. R. (1979). Biochemical studies of glutamate taste receptors: The synergistic taste effect of L-glutamate and 5′-ribonucleotides. *In* "Glutamic Acid: Advances in Biochemistry and Physiology" (L. J. Filer, Jr., S. Garattini, M. R. Kare, W. A. Reynolds, and R. J. Wurtman, eds.), pp. 1–9. Raven, New York.

Caprio, J. (1975). High sensitivity of catfish taste receptors to amino acids. *Comp. Biochem. Physiol. A* **52**, 247–251.

Caprio, J. (1978). Olfaction and taste in the channel catfish: An electrophysiological study of the responses to amino acids and derivatives. *J. Comp. Physiol.* **123**, 357–371.

Caprio, J., and Robinson, J. J. (1978). Adaptation and cross adaptation of the peripheral olfactory and gustatory systems of the catfish to amino acids. *Soc. Neurosci. Abstr.* **4**, 86.

Caprio, J., and Tucker, D. (1976). Specialist and generalist taste fibers in the catfish. *Soc. Neurosci. Abstr.* **2**, 152.

Chen, P. S., Jr., Toribara, T. Y., and Warner, H. (1956). Microdetermination of phosphorus. *Anal. Chem.* **28**, 1756–1758.

Christensen, H. N. (1975). "Biological Transport," 2nd ed., pp. 144–150. Benjamin, New York.

Cuatrecasas, P. (1971a). Insulin-receptor interactions in adipose tissue cells: Direct measurement and properties. *Proc. Natl. Acad. Sci. U.S.A.* **68**, 1264–1268.

Cuatrecasas, P. (1971b). Perturbation of the insulin receptor of isolated fat cells with proteolytic enzymes. Direct measurement of insulin-receptor interactions. *J. Biol. Chem.* **246**, 6522–6531.

Dastoli, F. R., and Price, S. (1966). Sweet-sensitive protein from bovine taste buds: Isolation and assay. *Science* **154**, 905–907.

Dastoli, F. R., Lopiekes, D. V., and Price, S. (1968). A sweet-sensitive protein from bovine taste buds. Purification and partial characterization. *Biochemistry* **7**, 1160–1164.

Davies, R. O., Kare, M. R., and Cagan, R. H. (1979). Distribution of taste buds on fungiform and circumvallate papillae of bovine tongue. *Anat. Rec.* **195**, 443–446.

DePierre, J. W., and Karnovsky, M. L. (1973). Plasma membranes of mammalian cells. A review of methods for their characterization and isolation. *J. Cell Biol.* **56**, 275–303.

Dunn, B. M., Anfinsen, C. B., and Shrager, R. I. (1974). Kinetics of Woodward's Reagent K hydrolysis and reaction with staphylococcal nuclease. *J. Biol. Chem.* **249**, 3717–3723.

Emmelot, P., Bos, C. J., van Hoeven, R. P., and van Blitterswijk, W. J. (1974). Isolation of plasma membranes from rat and mouse livers and hepatomas. *In* "Biomembranes," Part A (S. Fleischer and L. Packer, eds.), Methods in Enzymology, Vol. 31, pp. 75–90. Academic Press, New York.

Faull, J. R., and Halpern, B. P. (1972). Taste stimuli: Time course of peripheral nerve response and theoretical models. *Science* **178**, 73–75.

Feinstein, G., Bodlaender, P., and Shaw, E. (1969). The modification of essential carboxylic acid side chains of trypsin. *Biochemistry* **8**, 4949–4955.

Heppel, L. A., and Hilmoe, R. J. (1955). 5′-Nucleotidases. *In* "Preparation and Assay of Enzymes" (S. P. Colowick and N. O. Kaplan, eds.), Methods in Enzymology, Vol. 2, pp. 546–550. Academic Press, New York.

Herrick, C. J. (1904). The organ and sense of taste in fishes. *Bull. U.S. Fish. Comm.* **22**, 237–272.

Hiji, Y., and Sato, M. (1973). Isolation of the sugar-binding protein from rat taste buds. *Nature (London)* **244**, 91–93.

Hiji, Y., Kobayashi, N., and Sato, M. (1971). Sweet-sensitive protein from the rat tongue: Its interaction with various sugars. *Comp. Biochem. Physiol. B* **39**, 367–375.

Holland, K. N. (1980). Conditioned cardiac reflex and behavioral assays of the amino acid sensitivities of channel catfish *Ictalurus punctatus*. Ph.D. Thesis, Graduate Group in Biology, Univ. of Pennsylvania, Philadelphia.

Hudson, G., and Biemann, K. (1976). Mass spectrometric sequencing of proteins. The structure of subunit I of monellin. *Biochem. Biophys. Res. Commun.* **71**, 212–220.

Kahn, C. R. (1976). Membrane receptors for hormones and neurotransmitters. *J. Cell Biol.* **70**, 261–286.

Karlin, A. (1973). Molecular interactions of the acetylcholine receptor. *Fed. Proc., Fed. Am. Soc. Exp. Biol.* **32**, 1847–1853.

Karlin, A., Prives, J., Deal, W., and Winnik, M. (1971). Affinity labeling of the acetylcholine receptor in the electroplax. *J. Mol. Biol.* **61**, 175–188.

Korn, E. D. (1966). Structure of biological membranes. *Science* **153**, 1491–1498.

Koyama, N., and Kurihara, K. (1971). Do unique proteins exist in taste buds? *J. Gen. Physiol.* **57**, 297–302.

Krueger, J. M., and Cagan, R. H. (1976). Biochemical studies of taste sensation [4.] Binding of L-[^{3}H]alanine to a sedimentable fraction from catfish barbel epithelium. *J. Biol. Chem.* **251**, 88–97.

Kuninaka, A. (1967). Flavor potentiators. *In* "Symposium on Foods: The Chemistry and Physiology of Flavors" (H. W. Schultz, E. A. Day, and L. M. Libbey, eds.), pp. 515–535. Avi, Westport, Connecticut.

Kuninaka, A., Kibi, M., and Sakaguchi, K. (1964). History and development of flavor nucleotides. *Food Technol.* **18**, 287–293.

Kurihara, K., and Beidler, L. M. (1968). Taste-modifying protein from miracle fruit. *Science* **161**, 1241–1243.

Kurihara, K., and Beidler, L. M. (1969). Mechanism of the action of taste-modifying protein. *Nature (London)* **222**, 1176–1179.

Landacre, F. L. (1907). On the place of origin and method of distribution of taste buds in *Ameiurus melas*. *J. Comp. Neurol. Psychol.* **17**, 1–66.

Lee, K. H.-K., Coulet, P. R., and Gautheron, D. C. (1976). Grafting of enzymes on collagen films using Woodward's reagent "K" and a water-soluble carbodiimide derivative. *Biochimie* **58**, 489–497.

Lum, C. K. L., and Henkin, R. I. (1976a). Characterization of fractions from taste bud and non-taste bud-enriched filtrates from and around bovine circumvallate papillae. *Biochim. Biophys. Acta* **421**, 362–379.

Lum, C. K. L., and Henkin, R. I. (1976b). Sugar binding to purified fractions from bovine taste buds and epithelial tissue. Relationships to bioactivity. *Biochim. Biophys. Acta* **421**, 380–394.

Lum, C. K. L., Whittaker, N. F., and Henkin, R. I. (1976). Preparation and isolation of a taste-bud derived fraction from bovine circumvallate papillae. *Biochim. Biophys. Acta* **421**, 353–361.

Mackler, B. (1967). Microsomal DPNH[NADH]-cytochrome *c* reductase. *In* "Oxidation and Phosphorylation" (R. W. Estabrook and M. E. Pullman, eds.), Methods in Enzymology, Vol. 10, pp. 551–553. Academic Press, New York.

May, R. M. (1925). The relation of nerves to degenerating and regenerating taste buds. *J. Exp. Zool.* **42**, 371–410.

Means, G. E., and Feeney, R. E. (1968). Reductive alkylation of amino groups in proteins. *Biochemistry* **7**, 2192–2201.

Morris, J. A., and Cagan, R. H. (1972). Purification of monellin, the sweet principle of *Dioscoreophyllum cumminsii*. *Biochim. Biophys. Acta* **261**, 114–122.

Morriș, J. A., Martenson, R., Deibler, G., and Cagan, R. H. (1973). Characterization of monellin, a protein that tastes sweet. *J. Biol. Chem.* **248**, 534–539.

Morris, R. W., Cagan, R. H., Martenson, R. E., and Deibler, G. (1978). Methylation of the lysine residues of monellin. *Proc. Soc. Exp. Biol. Med.* **157**, 194–199.

Neville, D. M., Jr. (1960). The isolation of a cell membrane fraction from rat liver. *J. Biophys. Biochem. Cytol.* **8**, 413–422.

Nofre, C., and Sabadie, J. (1972). A propos de la protéine linguale dite "sensible aux sucres." *C. R. Acad. Sci., Ser. D* **274**, 2913–2915.

Olmsted, J. M. D. (1920). The results of cutting the seventh cranial nerve in *Amiurus nebulosus* (LeSueur), *J. Exp. Zool.* **31**, 369–401.

Ostretsova, I. B., Safarian, E. K., and Etingof, R. N. (1975). On the presence and localization of glucose-binding proteins in the tongue. *Dokl. Akad. Nauk USSR* **223**, 1484–1487.

Patel, R. P., and Price, S. (1967). Derivatives of proteins I. Polymerization of α-chymotrypsin by use of *N*-ethyl-5-phenylisoxazolium-3′-sulfonate. *Biopolymers* **5**, 583–585.

Patel, R. P., Lopiekes, D. V., Brown, S. P., and Price, S. (1967). Derivatives of proteins II. Coupling of α-chymotrypsin to carboxyl-containing polymers by use of *N*-ethyl-5-phenylisoxazolium-3′-sulfonate. *Biopolymers* **5**, 577–582.

Paton, W. D. M. (1961). A theory of drug action based on the rate of drug-receptor combination. *Proc. R. Soc. London, Ser. B* **154**, 21–69.

Pfaffmann, C. (1964). Taste, its sensory and motivating properties. *Am. Sci.* **52**, 187–206.

Price, S. (1972). "Sweet-sensitive protein" from bovine tongues: Stereospecific interactions with amino acids. *In* "Olfaction and Taste" (D. Schneider, ed.), Vol. 4, pp. 214–220. Wiss. Verlagsges., Stuttgart.

Price, S., and DeSimone, J. A. (1977). Models of taste receptor cell stimulation. *Chem. Sens. Flav.* **2**, 427–456.

Ray, T. K. (1970). A modified method for the isolation of the plasma membrane from rat liver. *Biochim. Biophys. Acta* **196**, 1–9.

Renqvist, Y. (1919). Über den geschmack. *Skand. Arch. Physiol.* **38**, 97–201.

Rhein, L. D., and Cagan, R. H. (1980). Biochemical studies of olfaction [2.] Isolation, characterization and odorant binding activity of cilia from rainbow trout olfactory rosettes. *Proc. Natl. Acad. Sci. U.S.A.* **77**, 4412–4416.

Sato, M., Hiji, Y., Ito, H., and Imoto, T. (1977). Sweet taste sensitivity in Japanese macaques. *In* "The Chemical Senses and Nutrition" (M. R. Kare and O. Maller, eds.), pp. 327–341. Academic Press, New York.

Schiffman, S. S. (1976). Taste of dipeptides. *Physiol. Behav.* **17**, 523–535.

Singer, S. J., and Nicolson, G. L. (1972). The fluid mosaic model of the structure of cell membranes. *Science* **175**, 720–731.

Solms, J. (1969). The taste of amino acids, peptides and proteins. *J. Agric. Food Chem.* **17**, 686–688.

Solms, J., Vuataz, L., and Egli, R. H. (1965). The taste of L- and D-amino acids. *Experientia* **21**, 692–694.

Tisdale, H. D. (1967). Preparation and properties of succinic-cytochrome *c* reductase (complex II-III). *In* "Oxidation and Phosphorylation" (R. W. Estabrook and M. E. Pullman, eds.), Methods in Enzymology, Vol. 10, pp. 213–215. Academic Press, New York.

Torii, K., and Cagan, R. H. (1980). Biochemical studies of taste sensation IX. Enhancement of L-[^{3}H]glutamate binding to bovine taste papillae by 5′-ribonucleotides. *Biochim. Biophys. Acta* **627**, 313–323.

Torrey, T. W. (1934). The relation of taste buds to their nerve fibers. *J. Comp. Neurol.* **59**, 203–220.

van der Wel, H., and Loeve, K. (1973). Characterization of the sweet-tasting protein from *Dioscoreophyllum cumminsii* (Stapf) Diels. *FEBS Lett.* **29**, 181–184.

Woodward, R. B., and Olofson, R. A. (1966). The reaction of isoxazolium salts with nucleophiles. *Tetrahedron, Suppl.* No. 7, 415–440.

Woodward, R. B., Olofson, R. A., and Mayer, H. (1961). A new synthesis of peptides. *J. Am. Chem. Soc.* **83**, 1010–1012.

Yamaguchi, S. (1967). The synergistic taste effect of monosodium glutamate and disodium 5′-inosinate. *J. Food Sci.* **32**, 473–478.

Yamaguchi, S., Yoshikawa, T., Ikeda, S., and Ninomiya, T. (1971). Measurement of the relative taste intensity of some L-α-amino acids and 5′-nucleotides. *J. Food Sci.* **36**, 846–849.

Zelson, P. R., and Cagan, R. H. (1979). Biochemical studies of taste sensation VIII. Partial characterization of alanine-binding taste receptor sites of catfish *Ictalurus punctatus* using mercurials, sulfhydryl reagents, trypsin and phospholipase C. *Comp. Biochem. Physiol. B* **64**, 141–147.

PART II

Discussion

DeSimone: The energetics, or the lack thereof, in the orderly queue mechanism are troubling. How can an isotropic solution become very organized in the vicinity of the receptor? This would require at least the equivalent of the free energy of mixing in the solution, which would be a great energy investiture. What might the source of the energy be?

Birch: I am not at all sure whether this could possibly be some new loosely associated structural component that was lost in the association with the membrane or whether it is simply a channeling effect. For example, in Hansen's diagrams of insect taste hairs, the stimuli pass through a 1000 Å diameter pore before reaching the dendrite containing the alleged receptors. With a large molecule, for example thaumatin, there would almost be a single file channeling through such a pore.

DeSimone: You envisage the pore as a morphological structure?

Birch: That's a possibility.

Dodd: Max Delbruck and colleagues have worked on reduction of dimensionality, where they envisage that ligands can adsorb to membranes and then diffuse in two dimensions. That would be part of the queueing, and is easily envisaged for many hydrophobic ligands such as typical odorants, but perhaps less likely with sugars.

DeSimone: That would reduce it by one dimension.

Karnovsky: Is there any possibility that furanosides and pyranosides, loosely ordered in some way, could open and form hemiacetals with the next sugar in the queue, at the C-4 or C-6 hydroxyls? Are there reports on properties such as unusual increased viscosities of highly concentrated sugar solutions? Possibly the energy of the ring system could provide the energy for queueing.

Brand: Did you use a whole-mouth taste test for the sweet solutions?

Birch: The procedure was simply sip, taste, and swallow.

Brand: Then you are confounding many sites in the oral cavity, and the time-intensity functions are actually a summation of many lingual receptor areas in the mouth.

Karnovsky: In the electrical recording in which there is increased amplitude, what about the temporal aspect? Are there aspects of the electrical recordings that tell the duration of the effect in similar terms to those of Dr. Birch?

Jakinovich: I don't know if there are temporal effects in our data.

Getchell: How would the "water spike" fit with the interpretation of the impulse discharges with regard to sugar specificity? How can one distinguish in an impulse discharge a "water spike"?

Jakinovich: The histogram analysis in which I measured the spike size shows it to be a different size. In this particular fly, it shows clearly in the records.

Karnovsky: In the structure by Dr. Birch of the hexose pyranoside, C-1 and the oxygen bridge were labeled the "bitter end" and C-3 and C-4 and their hydroxyls were labeled the "sweet end." And yet, everything seems to focus on the anomeric carbon as the sweet end.

Jakinovich: I think the anomeric carbon plays a role in bitter and sweet tastes. In the fly, of course, we don't think of it as a bitter taste, but in other animals and humans, β-methylglucopyranoside is bitter and α-methylglucoside is sweet. The electrophysiological response correlates well with the behavior. The same compounds that stimulate the cell also produce feeding behavior in the fly.

Karnovsky: What do compounds that are bitter in the human do in the fly?

Jakinovich: Quinine does not stimulate the sugar receptors, but rather inhibits them.

Karnovsky: But aren't there some bitter sugars?

Jakinovich: β-Methylglucoside. It doesn't stimulate flies very well.

Birch: Has anything been done on the intact catfish with chloroalanine? A colleague used chlorosucrose in frogs and it appears to bind for many hours to the taste receptors, which then recover.

Cagan: Our studies *in vitro* show β-chloroalanine to be reversible. Also, J. H. Teeter carried out electrophysiological experiments in which the barbel nerve responses recovered following treatment with the reagent. Therefore we see no persistent effect in the time scales used (min).

Karnovsky: Does β-chloroalanine have any behavioral effect?

Cagan: It has not been studied behaviorally.

Margolis: Isn't it surprising that there is only about a 30% reduction of L-[^{3}H]alanine binding in the macromolecular peak?

Cagan: The inhibitor was present at tenfold above that of the ligand, but the actual concentrations were higher than in the quantitative inhibition studies. The relative shapes of the alanine binding and chloroalanine binding curves are not known. The reduction is 35%, compared with a 60% decrease in the other studies. One possibility is that the remaining portion represents nonspecifically coupled alanine. If that is true then we should interpret the difference in incorporation as the specific component rather than the absolute amount of counts.

Price: Are there protein values for the column fractions?

Cagan: Specific binding activities, based on protein, were calculated for the samples applied to the column. Proteins were not determined for the column fractions since the samples were used up for the radioactivity measurements. We clearly need to determine the actual specific binding activities of the peaks from the column.

Price: The binding activity was associated with the major absorbance peaks. Using Woodward's K, alanine could bind to any primary amine.

Karnovsky: It is possible that the avidity of binding gives this unexpected specificity. I expected a great smear across the board. You could use an irrelevant amino acid. Also, the chromatographic peak may not represent a single entity.

Cagan: I think peak II may represent a complex entity from the membrane, which was possibly polymerized by Woodward's Reagent K.

Karnovsky: On running it on electrophoretic gels you would hope to see an effect of the chloroalanine that eliminates the labeling of one band.

Cagan: Correct. Another possibility is denervation. That will be a very demanding but informative experiment.

Karnovsky: The plasma membrane demonstration was good because everything else went away in the continuous gradient, which is the best kind of isolation to get.

Mooser: Does Woodward's Reagent K cause inhibition of binding by itself?

Cagan: Preliminary experiments with millimolar levels showed no effect.

Dodd: The phenomenon called the "ligand-retention effect" was reported for the periplasmic proteins in bacterial reception. The dissociation of ligand from receptor does not follow unimolecular kinetics when the receptor concentration is in great excess over the free ligand concentration. There is a common supposition that when a ligand, which is not covalently bound to a protein, is passed down a column with the protein, the procedure will easily remove all of the ligand. In fact, it is very difficult to remove the last 1%.

Cagan: In our case the ligand is covalently bound to the receptor, and furthermore it is in SDS. The final radioactive peak emerged at the position of free alanine, but we have not rigorously identified it as L-alanine.

Karnovsky: In the glycosidases is there any evidence as to whether those are ecto-enzymes; that is, are they in the membrane with the active sites facing outward, or are they internal enzymes? You reported data from homogenates. Have you used any nonpenetrating inhibitor? Is it not possible, since those enzymes can be assayed, to treat the tissue first with an anionic covalent linking nonpenetrating reagent to eliminate the activity?

Hansen: In homogenates of taste hair-rich labella or tarsi of flies, 20–40% of the α-glucosidase activity (sucrose as substrate) is bound to membranes. For solubilization, treatment with Triton X-100 is necessary. The enzymes have been fractionated chromatographically and partially characterized by Kijima and co-workers. However, it is not clear at which type of membrane these glucosidases are localized. Kijima and co-workers and Koizumi and co-workers discuss the dendritic membranes of the sugar receptor cells containing an α-glucosidase with its active site outward. Experiments with nonpenetrating inhibitors have not yet been done.

Margolis: At that level of activity how does one deal with possible bacterial contamination on the hairs and differential distribution of bacteria?

Hansen: We have two arguments against bacterial contamination. (1) The cuticle of flies is covered by a waxy hydrocarbon layer which appears to be rather clean and free of bacteria under the scanning electron microscope, despite the numerous cuticular hairs. (2) After preparation and before homogenization the labellae or tarsi are washed in isotonic buffer solution. The measured activities and the chromotographic enzyme pattern are quantitatively reproducible. This would not be expected if the activities originate from bacteria attached to the cuticle.

Dodd: Regarding the kinetics of the cyclic nucleotide effects, you mentioned the extra step of diffusion through the membrane. How long does it take for the material to diffuse through the membrane to give the effects?

Hansen: Upon application of the micropipette containing dibutyryl-cGMP (or aminophylline) together with the stimulating sugar the number of impulses during the counting interval (300 millisec, beginning 50 millisec after onset of stimulation) is decreased.

Dodd: In view of a time constant in the range of 100 millisec, what do you envisage as the mechanism of action of dibutyryl-cGMP? Does it act on the external face of the dendrite or the internal face?

Hansen: This is not known. But possibly our results might be discussed in terms of the model proposed for the rod outer segments of visual cells. According to this model the cGMP level falls upon stimulation (illumination). The presence of externally applied dibutyryl-cGMP or of the phosphodiesterase blocker aminophylline during stimulation prevents the lowering of the internal cGMP level and this leads to an inhibition of the response.

Matschinsky: Do you know whether the disaccharides enter the intracellular space or stay outside in the extracellular space? In most other cells the disaccharides are excluded from the intracellular water.

Hansen: There is general agreement that sensory recognition of chemical stimuli occurs at the outer surface of the receptor membrane. Therefore, penetration of disaccharides through the membrane is functionally not necessary. However, the complete decrease of the receptor potential within 50 millisec after the end of stimulating with molar concentrations of sugar might indicate that dilution of this high stimulus concentration merely by diffusion within the extracellular space of the dendrite-containing canal of the hairshaft is not sufficient to explain the fast potential decrease.

Jakinovich: Sometimes when recording from the hairs you can remove a mucus-like material when the electrode is pulled away after stimulating for a long time. Perhaps the sugar can be drawn back into the mucus and trapped.

Hansen: Maybe. This mucus-like material was mentioned first by Stuerckow in 1967.

Menco: The receptive portions of the dendrites are surrounded by a condensed layer of mucus (glycocalyx). Before reaching the lipid–protein moieties of the receptor membrane, the sugar has to pass this glycocalyx. Is it possible that the stimuli interact with this material?

Hansen: We do not believe that in flies the mucus layer has a marked influence on the electrophysiological response under our standardized experimental conditions. The layer has a thickness of less than 1 μm. Therefore, it seems improbable that the polysaccharides, if they react with the sugars, decrease their concentration substantially. Furthermore, there is no indication that the polysaccharides themselves are involved in the transduction process.

Karnovsky: Dr. Birch and I have quietly disposed of my so-called theory because of his data for methylglucosides.

Dodd: Dr. Birch raises questions about the receptor occupancy theory and the rate theory and opposes the two concepts. Monellin and thaumatin dissociate very slowly from the tongue. Doesn't that suggest an extremely tight affinity which perhaps favors occupancy rather than rate processes?

Birch: We must distinguish between the initial affinity and the persistence of the localized concentration. The movement and the approach of the molecules to the ionophore trigger mechanism must be geared to the rate at which they interact with the ionophore trigger mechanism. A molecule like thaumatin, which is intensely sweet, may in fact have a poor ionophore trigger efficiency, as our initial calculations indicate. The reason it is intensely sweet is because of its extremely high accession efficiency. The approach of the molecule to the ionophore trigger mechanism may be very, very slow because of its poor ionophore trigger efficiency. In other words, it is very slow to be removed as well.

Price: Wouldn't a molecule of thaumatins size approach a receptor slowly because it diffuses slowly? The fact of its persistence is most easily interpreted as reflecting a high affinity, which means that it dissociates much more slowly than it associates.

Birch: The reaction time for thaumatin at maximum concentration, the limiting reaction time, is about 1 sec. This is not very accurate. For sucrose it is about 0.3 sec. Both figures are considerably in excess of the neurophysiological data, with limiting values of 30–50 millisec.

Kurihara: We have performed electrophysiological experiments and obtained a result similar to that observed by Dr. Birch. Recording the response from the frog glossopharyngeal nerve, temperature changes lead to alterations in the gustatory response. At low temperature, e.g., 2°C, the dynamic portion of the response to 1 mM $CaCl_2$ becomes very small and an increase of temperature greatly enhances the dynamic portion. We have also examined the effect of flow rate of the stimulating solution on the response. A small flow rate decreases the dynamic portion and a large flow rate increases it. Thus the dynamic portion of the response greatly depends on the flow rate. These data can be explained by rate theory, but we don't like rate theory because it does not have a reasonable basis. We postulate, instead, a binary receptor conformation where an active form of receptor–stimulus complex can be transformed into an inactive form. The theoretical results based on this reaction scheme, which belongs to occupation theory, well explain behaviors of the dynamic portion of the response under various conditions.

Jakinovich: Because my electrophysiological work with the gerbil using sugars seems to show the same sort of specificity as with insects, is it possible that there are glycosidase receptors in mammalian taste receptor cells?

Price: I am not aware of studies of glycosidases in taste buds. We earlier observed glucose dehydrogenase activity in bovine tongue extract. We looked for α-glucosidase and didn't observe any.

Jakinovich: I collected the exudate after pouring sucrose solution over the tongue. In the fluids that came off I looked for hydrolysis of the sucrose by thin-layer and gas chromatography. I did not observe any hydrolysis, but it may not have been sensitive enough.

PART III

Physicochemistry and Transduction

11

Physicochemical Principles in Taste and Olfaction

JOHN A. DESIMONE

I. INTRODUCTION

Chemical stimuli are transported to the gustatory and olfactory organs by convection and diffusion. Accordingly one or more barriers to mass transfer must be traversed by the stimulus before encountering the limiting membranes of the chemoreceptor cells. In the case of olfaction in man and other terrestial vertebrates, odorants are conveyed as gases to the mucus layer overlying the olfactory epithelium. The gases then partition across the air/mucus interface and diffuse an unspecified distance before receptor recognition and subsequent transduction can occur (Bostock, 1974; Getchell *et al.*, 1980; Van Drongelen *et al.*, 1980). Under laboratory conditions, taste receptors are usually stimulated by substances conveyed to the tongue in flowing liquid streams. This flow must transfer mass and momentum to the stagnant adapting solution bathing the tongue. Mass transfer is completed by diffu-

Biochemistry of Taste and Olfaction

ISBN 0-12-154450-8

sion of the stimulus into the pore region of the taste bud. During normal mastication convective circulation of saliva is ensured by mechanical deformation of the tongue. Such movements probably enable tastants to gain access to some taste buds such as those of the circumvallate papillae, which are located in the deep grooves along their lateral edges (Beidler, 1962).

The importance of mass transfer in the overall chemosensory process can be appreciated from a rough calculation of diffusion times for small molecules in an aqueous milieu. The diffusion distance varies, of course, but it can be expected to be of the order of 10 μm in the case of controlled stimulation of the taste buds of the anterior region of the tongue (Beidler, 1962). Typical diffusion distances in olfaction are likely to be about 50 μm, the approximate thickness of the olfactory mucus layer (Getchell *et al.*, 1980). For a substance with diffusion coefficient D (approximately 10^{-5} cm^2/sec), diffusing through a distance ℓ, diffusion times are given by $t \simeq \ell^2/2D$ (Reif, 1965). We would therefore expect transport times of 50–1250 msec. In gustation and olfaction, the period between the introduction of the stimulus and the onset of the response (the latency) is typically in this range (Marowitz and Halpern, 1977; Getchell *et al.*, 1980). Accordingly, we might expect the time course of the initial phases of the neural response to depend largely on the dynamics of stimulus transport. In that neural latencies in olfaction and taste are generally concentration and flow rate dependent (Marowitz and Halpern, 1977; Getchell *et al.*, 1980) and that early phasic components of the neural response reflect flow rate and stimulus concentration (Smith and Bealer, 1975; Sato, 1976), these expectations seem reasonable. If mass transport is a significant rate-limiting process, latency can be interpreted as the time required for the olfactory or gustatory stimulus to reach threshold levels at the receptor loci by means of diffusion and convection. Getchell *et al.* (1980) and DeSimone and Heck (1980) have analyzed latency-concentration responses in olfaction and taste, respectively, using a diffusion paradigm. The analyses permit estimates of receptor location and threshold concentrations.

A mechanistic analysis of the post-threshold response pattern in the peripheral olfactory or taste nerves is presently impossible. In the case of the rat, gustatory responses are characterized by a rapidly changing phasic burst of neural activity followed by a quasi-steady, slowly adapting response. There is every indication that the phasic part of the neural response contains a significant portion of the total gustatory information that a given stimulus, delivered at a fixed velocity and concentration, is capable of imparting (Smith *et al.*, 1975; Marowitz and Halpern, 1977). If the early phasic period is limited by mass transfer, then for a time interval after threshold is reached the neural response should be governed solely by the departure of the local stimulus concentration from threshold. One immediate consequence of such a response function is the feasibility of an analysis of the early neural response characteristics without requiring a detailed model of transduction.

The reason is that at concentrations sufficiently close to threshold, all responses are linear functions of the local stimulus concentration. Accordingly, the rate of change of the response with concentration at threshold is a concentration-independent characteristic of a given receptor–stimulus pair. DeSimone and Heck (1980) have called this parameter the *response compliance* and have verified its concentration independence.

Although the mechanisms of transduction in taste are unknown, there is ample evidence that receptor cell membranes play a critical role in the recognition of tastants (Kimura and Beidler, 1961; Eyzaguirre *et al.*, 1972; Ozeki and Sato, 1972; Mooser and Lambuth, 1977; see also this volume, Chapter 10). To account for salt and acid reception, it is generally assumed that the receptor membrane carries fixed surface charge (Beidler, 1967; Kamo *et al.*, 1974). Recent work by DeSimone *et al.* (1980b) has shown that the taste suppressor, gymnemic acid, is about as surface active as the well-known surfactant, sodium lauryl sulfate, and that very dilute solutions of both surfactants easily penetrate insoluble monolayers of phospholipids. Furthermore sodium lauryl sulfate is itself a taste modifier and suppressor in man (DeSimone *et al.*, 1980). Since surfactants primarily affect cell function by altering membrane lipoproteins (Kagawa, 1974), it seems reasonable to conclude that surface active agents exert their influence on taste by altering a membrane-centered process, possibly an event in transduction. A number of studies have stressed the role of charged phospholipids in taste (Kamo *et al.*, 1974; Aiuchi *et al.*, 1976; DeSimone *et al.*, 1980a; DeSimone and Heck, 1980; see also this volume, Chapter 13). The possibility that cell surface potentials may function as receptor potentials has been discussed by Kamo *et al.* (1974) and by DeSimone and Price (1976). The possibility that cell membrane surface pressure may play a role in salt, acid, and "water" taste has been considered by DeSimone *et al.* (1977, 1978) and by DeSimone and Heck (1980).

In the following sections, the sequence of physicochemical events that occur in the early phases of stimulation of olfactory and taste receptors will be discussed. Naturally, it is possible to be more precise about preneural transport processes than events surrounding transduction. However, even in the latter case a few generalizations, supported by existing data, are possible.

II. MASS TRANSPORT IN CHEMORECEPTION

A. Olfaction

Mass transport occurs in both the gas and liquid phases before a given odorant encounters olfactory receptors. The key processes can be summarized as follows: (1) convection of the odorant in a gas stream from a source to the region of the olfactory epithelium, (2) diffusion from the bulk gas to the gas/mucus interface, (3) partition across the interface, (4) diffusion through the aqueous mucus, (5) recognition of the odorant by the receptors and transduction, and, finally, (6) clearance of the odorant. In addition to these processes, Bannister (1974) and Getchell and Getchell (1977) have considered the possibility of chemical alteration of the odorant in the mucus bathing the epithelium. The receptors are bipolar neurons consisting of a dendrite, soma, and axon. From the apical knob of each receptor, cilia project into the mucus. Some cilia are up to 200 μm in length. Because the mucus layer itself is estimated to be about 30-μm deep (Reese, 1965; Holley and MacLeod, 1977), many cilia must lie parallel to the epithelium over most of their length.

The location of the olfactory receptor sites is the subject of controversy. It has been suggested that the cilia contain the receptor sites (Vinnikov and Titova, 1949; Ottoson, 1956), but attempts to confirm this have left the question unsettled (Shibuya, 1969; Shibuya and Tucker, 1967; Bronshtein and Minor, 1977; see also this volume, Chapter 3). For example, Bronshtein and Minor (1977) removed the cilia from the olfactory epithelium of the frog by treating it with Triton X-100. This resulted in a loss of the normal electroolfactogram (EOG) in response to butyl acetate. The complete recovery of the EOG did not, however, coincide with complete regeneration of the cilia. When the EOG returned to control values, the cilia had only recovered 25% of their pretreatment length. This may mean that receptors are normally confined to the proximal cilia and the dendritic knob. It is also possible that receptor sites are uniformly distributed over the cilia, but sites on distal segments do not contribute significantly to transduction. The results do suggest, however, that receptors are not confined to the distal segments of the frog cilia. Characterization of the various receptor types *in situ* would, of course, settle the question. Goldberg *et al.* (1979) have developed antisera against proteins having some of the characteristics of olfactory receptors. These techniques may in the future prove useful in localizing receptors. Getchell *et al.* (1980) approached the question indirectly. Their methods assume that the dynamics of the response itself contain information bearing on certain receptor characteristics such as relative location and threshold concentration of stimulus. More specifically the model is based on the assumption that the variation of the latency with stimulus concentration represents the time required for the stimulus to reach threshold by diffusion through the olfactory mucus. In this view, the latency is essentially a diffusion time. It therefore follows that the diffusion path length estimates the average location of the first functional plane of receptors.

This approach assumes that, of all the rate processes enumerated above, process (4) is rate determining. This can be justified in the case of the experiments reported by Getchell *et al.* (1980). In these studies, odorant is conveyed as a flowing gas stream from a nozzle placed directly over the olfactory epithelium. As the gas leaves the source, it travels initially normal to the surface but is then deflected by the surface causing a parallel velocity component to develop. The process can be modeled hydrodynamically as flow near a stagnation point (Jones and Watson, 1963). The velocity of the gas decreases as it approaches the gas/mucus interface tending to zero directly at the interface. The narrow thickness, δ_0, over which the parallel component of the velocity decreases from its stream value toward zero is called the *hydrodynamic boundary layer*. Clearly, the efficacy of convection as a mode of mass transfer diminishes as the gas velocity approaches zero. As this occurs, diffusion becomes increasingly important in the overall transport process. The distance, δ, over which diffusion dominates is called the *diffusion boundary layer*. Figure 1 shows the stream lines developed for this type of flow field.

Solving the Navier-Stokes equation (Jones and Watson, 1963) we can estimate the hydrodynamic boundary layer thickness as a function of the kinematic viscosity, ν, the diameter of the area of epithelium stimulated, ℓ, and the fluid velocity, U:

$$\delta_0 \simeq 2.4\sqrt{\frac{\nu\ell}{U}} \tag{1}$$

If we define the diffusion boundary layer thickness, δ, as the place where diffusive and convective transport are equal, we obtain the following estimate of δ (Levich, 1962; Getchell *et al.*, 1980):

$$\delta \simeq 0.5\, \delta_0 \left[\frac{D_g}{\nu}\right]^{1/3} \tag{2}$$

where D_g is the diffusion coefficient of the stimulus in the gas phase. Assuming that diffusion dominates stimulus transport within a distance, δ, from the epithelium, the time, t, required for transport is of the order, δ^2/D_g, or using Eqs. 1 and 2 we obtain:

$$t \simeq \frac{1.44\ell}{U}\left[\frac{\nu}{D_g}\right]^{1/3} \tag{3}$$

In the gas phase ν and D_g are of the same order. For typical parameter values of ℓ = 0.5 cm, and a volumetric flow rate of 2 ml/sec delivered through a 0.1-cm-diameter orifice, t is about 3 msec. At the flow rates used by Getchell *et al.* (1980), gas phase transport is far too fast to account for the measured latencies of 50–2000 msec. However, at lower flow rates, gas phase transport could clearly become a factor in the response time course.

Most odorants for terrestrial vertebrates are only slightly water soluble.

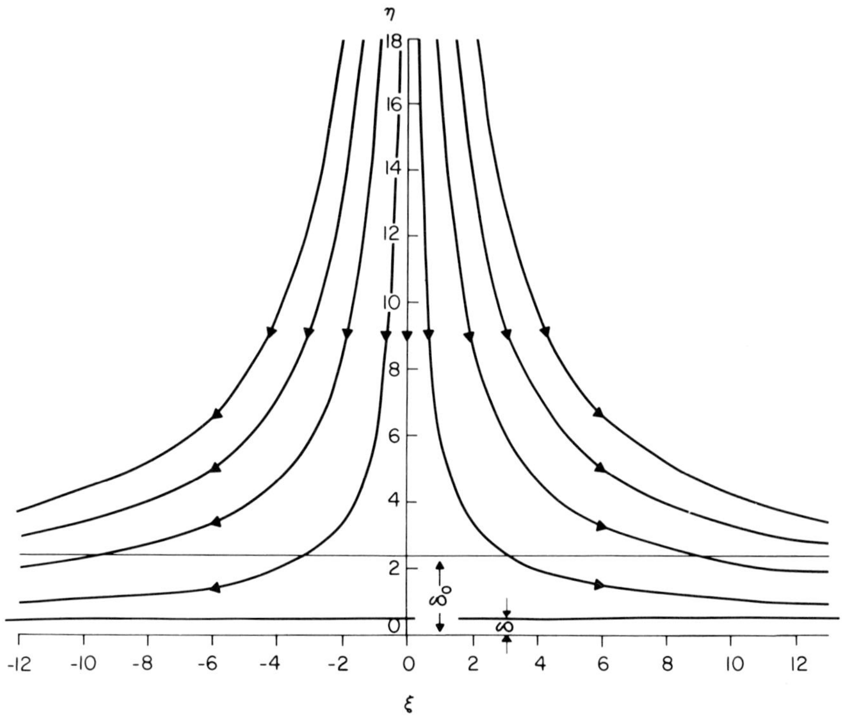

Fig. 1. The stream lines describing the stimulus path against the olfactory epithelium, or in the case of gustation, the lingual surface. η is a dimensionless normal coordinate, and ξ is a coordinate in the plane of the stimulated surface. δ_0 denotes the hydrodynamic boundary layer thickness (in units of η). For $\eta < \delta_0$, the parallel component of the velocity rapidly approaches zero. In the region δ (not drawn to scale), convection plays a minor role in mass transfer, and diffusion dominates.

Therefore in most cases saturation equilibrium across the gas/mucus interface should be reached rapidly so that interfacial kinetics are generally not expected to be rate limiting. We are left with diffusion in the aqueous mucus or transduction itself as the most plausible rate-limiting processes. Most sensory receptor cells respond to suprathreshold stimuli in times far shorter than the typical latencies observed in olfaction. If olfactory receptors are typical, then diffusion through the aqueous phase should be considered as a possible rate-limiting process. It is all the more plausible considering that diffusion coefficients in the aqueous phase are at least four orders of magnitude smaller than in the gas phase.

The concentration profile of odorant at any position x in the mucus and at any time, t, is, according to diffusion laws:

$$c(x,t) = c_0 \operatorname{erfc}\left[\frac{x}{2\sqrt{Dt}}\right] \tag{4}$$

where c_0 is the gas phase stimulus concentration, $c(x,t)$ is the local concentration, D is the aqueous diffusion coefficient, and erfc refers to the error function complement (Carslaw and Jaeger, 1959). The equivalent aqueous phase concentrations can be found by multiplying both sides of the equation by the gas/mucus partition coefficient. In using Eq. 4, it is assumed that diffusion of the stimulus can proceed well beyond the apical membrane of the receptor cell body. This is justified by the observation of Hornung and Mozell (1977) that odorants can be detected in the oral cavity and in the general circulation after contact with the olfactory epithelium (see this volume, Chapter 2).

We can define threshold, c_T, as that concentration just sufficient to cause a response at the receptor plane $x = h$ (relative to the gas/mucus interface, $x = 0$). Threshold will be achieved at $x = h$ at various times t_L (the latencies) depending on the concentration of the stimulus in the gas phase,

$$c_T = c(c_0,\ t_L,\ h) \tag{5}$$

With Eq. 4, we have

$$t_L = \frac{h^2}{4D\left[\operatorname{erfc}^{-1}\left(\frac{c_T}{c_0}\right)\right]^2} \tag{6}$$

This relation defines a two parameter model for the interpretation of latency-concentration relations. The parameters are receptor plane location, h, and apparent threshold concentration, c_T. Data obtained from extracellular recordings of two CO_2-sensitive units and two safrole-sensitive units of the tiger salamander were analyzed according to Eq. 6 (Gethcell *et al.*, 1980). The results are summarized in Fig. 2 and Table I. The parameters h and c_T are obtained using a nonlinear least squares method (Metzler *et al.*, 1974). The two safrole-sensitive units show similar thresholds and appear to be located at about the same level. The CO_2-sensitive units also have similar thresholds, but their apparent locations differ. In all cases analyzed, including those using data published elsewhere (Shibuya, 1969; Holley *et al.*, 1974), the functional receptors appear to be located well below the gas/mucus interface. An examination of the latency-concentration data shows that in general: (1) latencies for a given substance vary among units and (2) latencies for a given cell vary among substances. The mass transport model accounts for the first fact in terms of variations in the location of receptors with respect to the air/mucus interface and the second in terms of variations

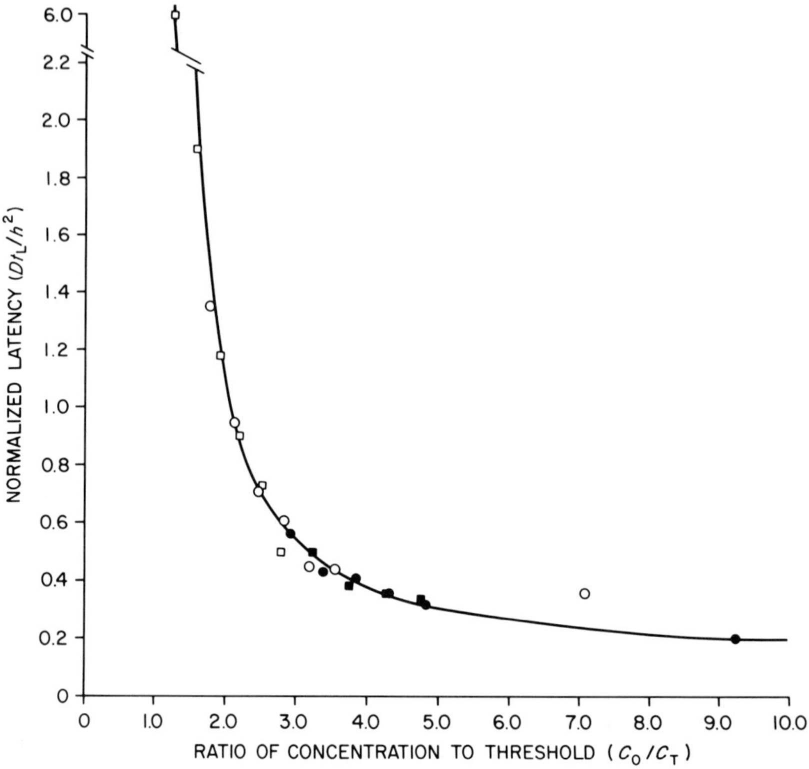

Fig. 2. Normalized latency is shown as a function of stimulus concentration relative to the calculated threshold. The parameters h and c_T are chosen according to a least squares criterion. Each data point has been normalized using these parameters. The line is the theoretical curve using diffusion theory. Circles represent two safrole sensitive olfactory units of the tiger salamander, and squares are CO_2 units. (Taken from Getchell *et al.*, 1980, with permission.)

in the physical constants of each molecule (e.g., diffusivities and partition coefficients). An analysis of olfaction in insects based upon convection-diffusion theory has recently been made by Murray (1977).

B. Taste

Considerations similar to those outlined above apply in the case of a gustatory stimulus. Here, however, mass transport occurs in a single phase. For taste buds of fungiform papillae, the flowing stimulus must mix with and displace a stagnant adapting solution (saliva or some other liquid). As in the case of olfaction, there will be a diffusion boundary layer. In this instance, it consists of an adapting solution through which the stimulus must pass before

TABLE I Parameters for Olfactory Units Analyzed[a]

Stimulus	Threshold (c_T)[b]	Receptor depth*(h)[b]
Safrole	21 n*M*	30 μm
	15 n*M*	21 μm
CO_2	0.05%	11 μm
	0.03%	64 μm

[a] Taken from Getchell *et al.* (1980), with permission.
[b] Standard errors do not exceed 7% of the reported values.

entering the pore region of the taste bud. The diffusion path length is therefore not constant, but in general it varies with flow rate. The only part of the total diffusion path that is probably independent of flow rate, and therefore likely to be constant, is the distance a stimulus must travel after entering the pore region of the taste bud. The total diffusion path, h, is therefore the sum of a flow rate-dependent component, δ, and an independent component, h_p. Using Eqs. 1 and 2 and specializing to the case of NaCl ($D = 1.2 \times 10^{-5}$ cm^2/sec), we have

$$h = h_p + 0.13\sqrt{\frac{\ell \nu}{U}} \tag{7}$$

Increasing the velocity of the stimulus therefore reduces the diffusion path length.

To a first approximation we can treat mass transfer in the diffusion boundary layer as being governed by the diffusion equation, but with a flow rate-dependent path length. If we assume that the taste receptors coincide with the first impermeable membrane encountered, then reflecting boundary conditions are appropriate (Carslaw and Jaeger, 1959). The local concentration at the reflecting surface ($x = 0$) is

$$c\,(0,t) = c_0\left[1 - \frac{4}{\pi}\sum_{n=0}^{\infty}\frac{(-1)^n}{2n+1}\;e^{-\lambda_n^2 Dt}\right] \tag{8}$$

where $\lambda_n = (2n + 1)\,\pi/2h$, and the adapting solution is assumed to be free of stimulus. For any bulk concentration, c_0, the effect of increasing stimulus velocity is to reduce h and thus sharpen the rate of rise in the local concentration.

Marowitz and Halpern (1977) have studied the neural time course in the rat for four NaCl concentrations delivered to the anterior part of the tongue at two flow rates. They provide sufficient data to calculate the diffusion boundary layer thickness for each velocity. At a velocity of 18 cm/sec (0.091 ml/sec through a 0.8 mm orifice), the diameter of tongue covered is 2.8 mm.

At 65 cm/sec (0.324 ml/sec through a 0.8-mm orifice), the diameter covered is 6.2 mm. Taking the kinematic viscosity as 0.01 cm^2/sec, and using the second term in Eq. 7, we obtain δ values of 16 μm and 13μm for the lower and higher velocities, respectively. Although the flow rates differ by a factor of nearly 4, the boundary layer thicknesses are nearly the same. This would explain why both the latencies and the rising part of the phasic response appear flow rate-independent in these studies (Marowitz and Halpern, 1977). The outcome here is somewhat fortuitous, in that the increasing velocity is almost completely compensated by the increasing area of coverage. This will not always be the case and a flow rate-dependent rise time is predicted and observed (Sato, 1976). The flow rate-independent component of the diffusion path can be estimated as 5–10 μm from the electron-microscopic studies of fungiform papillae reported by Murray (1971).

If we define latency as the time required to achieve threshold at the reflecting boundary, we can readily account for the latencies reported by Marowitz and Halpern (1977) as diffusion times. Furthermore, a threshold concentration of about 0.6 mM is required to produce agreement. Concentrations of this order have been reported for NaCl thresholds in the rat (Bartoshuk, 1974). Table II compares the calculated latencies for several path lengths with the observed values (Marowitz and Halpern, 1977). The total path length can be considered the sum of a flow-dependent contribution of about 15 μm, and a pore path length that accounts for the remaining distance. It is clear that mass transport can account for the observed latencies. This result is further supported by a prediction of threshold in accord with

TABLE II A Comparison of Calculated Latencies For Three Path Lengths with Observed Latencies[a]

	Calculated latencies			
Stimulus NaCl (mM)	t_L (20 μm) (msec)	t_L (21 μm) (msec)	t_L (22 μm) (msec)	Reported latencies[b] t_L (msec)
5	47	52	57	41
10	35	39	43	—
20	28	31	34	32
50	22	24	27	29
200	17	18	20	26

[a] The calculated latencies (t_L) are made with the aid of Eq. 8, assuming a threshold concentration of 0.6mM and the indicated diffusion path lengths (in parentheses).

[b] Taken from Marowitz and Halpern (1977), with permission.

electrophysiological measurements (Oakley, 1962; Pfaffmann, 1955). On the basis of these analyses and the data presented here and elsewhere (Getchell *et al.*, 1980; Marowitz and Halpern, 1977), we conclude, therefore, that the relative slowness of the neural responses in olfaction and taste reflect more the mode of stimulus transport than an intrinsically sluggish cellular transduction process.

III. A GENERALIZED RESPONSE FUNCTION

Marowitz and Halpern (1977) report that a rat can make a behavioral decision concerning the taste of 500 m*M* NaCl within 160 msec. It is likely that the NaCl concentration is still changing rapidly during the decision-making process. Since the slope of the early phasic response is concentration dependent, it seems reasonable to assume that for a concentration range beyond threshold the response is governed solely by the deviation of the concentration from threshold. If beyond threshold the response is an analytic function of concentration, we can represent it as a Taylor series expansion:

$$R(c) = R(c_T) + \left(\frac{\partial R}{\partial c}\right)_{c_T}(c - c_T) + \frac{1}{2}\left(\frac{\partial^2 R}{\partial c^2}\right)_{c_T}(c - c_T)^2 + \ldots + \quad (9)$$

where R is the response expressed in arbitrary units, c is the concentration of stimulus at the reflecting plane, and c_T is threshold. Since the response is zero at threshold, the first term on the right is zero and if the deviations from threshold are small, only the linear term need be retained:

$$R(c) = \left(\frac{\partial R}{\partial c}\right)_{c_T}(c - c_T) \quad (10)$$

If Eq. 10 represents the early response behavior, the coefficient $(\partial R/\partial c)_{c_T}$ should be a constant characteristic of the stimulus–receptor pair. DeSimone and Heck (1980) have designated it the *response compliance* and have calculated it for the case of NaCl stimulation of the rat. The method of calculation is as follows. Rearranging Eq. (10) gives:

$$\frac{R}{mc_T} = \frac{c}{c_T} - 1 \simeq at \quad (11)$$

where we have used m to denote the response compliance. It is easily verified from the concentration time course described by Eq. 8 that the change in concentration at the reflecting surface (the assumed receptor locus) is quasi-linear for a time period beyond latency, i.e., the second

TABLE III Calculation of the Response Compliance of NaCl

Stimulus NaCl (mM)	Theoretical slope a (msec^{-1})[a]	Observed slope b (10^5 mV/msec)[b]	Response compliance $(\partial R/\partial c)c_{\text{T}}$ (10^5 mV/mM)[c]
5	0.040	2.3 ± 0.11	95.3 ± 4.6
20	0.145	10.2 ± 0.67	117.5 ± 7.7
50	0.265	15.9 ± 0.71	99.8 ± 4.5
200	0.412	15.1 ± 0.46	61.0 ± 1.9

[a] Calculated from Eqs. 8 and 11 with $c_{\text{T}} = 0.6$ mM. The term a appears in Eq. 11.
[b] Reported by Marowitz and Halpern (1977). The term b appears in Eq. 12.
[c] Response compliance is calculated using Eq. 12.

equality in Eq. 11 holds where the slope a increases with the stimulating concentration. Marowitz and Halpern (1977) report that the early phasic responses are quasilinear in time, with slope b depending on the concentration of stimulus. With this fact and Eq. 11, we arrive at

$$m = \left(\frac{\partial R}{\partial c} \right)_{c_{\text{T}}} = \frac{b}{a c_{\text{T}}} \tag{12}$$

If the model is correct, the response compliance should be independent of concentration even though both a and b are concentration dependent (Table III). The constancy of the response compliance is clear at the three lower concentrations. At 200 mM, its value is lower. The reason for this is the narrowness of the true quasilinear region at increasing salt concentrations. The observed slope includes time periods beyond the region of validity of the model. During these later phases a linear model fails and higher order terms in the Taylor series must be considered.

Although the concept of response compliance is independent of any one particular molecular model, a calculation based upon a rate theory of transduction (Heck and Erickson, 1973) shows that

$$\left(\frac{\partial R}{\partial c} \right)_{c_{\text{T}}} \propto k_1 S_{\text{T}} \tag{13}$$

where k_1 is the rate constant governing the binding of the stimulus to the receptor site, and S_{T} is the concentration of unbound receptors at threshold.

The response compliance is, therefore, an index of the efficacy of any tastant-receptor pair. A comparison among the response compliances of several tastants delivered under the same conditions could provide a quantitative basis for the study of taste responses in a given species.

IV. SURFACE ACTIVITY AND TASTE

The fact that transduction in chemoreception is fast compared to mass transport suggests that taste stimuli need not penetrate the taste cell in order to cause depolarization of the cell and subsequent nervous excitation. Beidler (1967) points out that the rapid response of the receptor potential to changes in tastant concentration implies a rapid transfer of information from the microvilli of the taste cell to regions below the tight junctions where the cells must undergo an ion permeability change. Thus the site of stimulus recognition and cell depolarization may be spatially separated by as much as 30 μm. It is conceivable that for salt and acid responses depolarization may occur at the microvilli as suggested by the models of Kamo *et al.* (1974) and DeSimone and Price (1976). However, not all tastants are charged and some are large proteins whose membrane permeability is probably very low (Cagan, 1973). In addition taste responses can occur after receptors adapted to salts such as NaCl and KCl are treated with water (Bartoshuk, 1965, 1974). Since the oral cavity is subject to sudden changes in ion composition, any hypothesized mechanism that relies on its ion composition as the sole source of depolarizing ion current has inherent difficulty in explaining depolarization induced by sugars and other nonelectrolytes. From the standpoint of providing a reliable source of depolarizing ions, the homeostatically controlled extracellular fluid below the tight junctions seems better suited. Thus it seems most plausible that recognition of the tastant by the receptor is a membrane-centered event which occurs at the apical border of the taste cell and depolarization occurs along the basolateral parts of the cell.

By what mechanism is information propagated from recognition sites to depolarization sites? Beidler and Gross (1971) postulate a mechanochemical model in which conformational changes at recognition sites may be propagated in the membrane. Are highly specific receptor proteins necessary aspects of such conformational changes? Certainly in cases where taste reflects subtle differences among sterioisomers as in the taste of some of the D and L isomers of amino acids (Solms, 1969; Price, 1972), such entities must be present. In the case of salts and acids, however, conformational changes can be produced by less specific means. It is well-known that structural protein, such as collagen, can swell or shrink in response to ionic strength and pH changes (Ciferri, 1971). Charged phospholipid monolayers can undergo phase transitions induced by pH and ionic strength (DeSimone *et al.*, 1980a; DeSimone and Heck, 1980). In both cases, structural protein and phospholipid monolayers, events in the microenvironment control these conformational changes.

The surface pressure of various charged phospholipids is largely determined by protons and other cations in the lipid microenvironment. This is because phosphate and carboxyl residues create a region of electronegativity near the membrane. The net result is increased concentrations of cations near the surface relative to the bulk solution, i.e., the creation of an electrical double layer. Mobile ions may electrostatically screen fixed ions, i.e., change the electrical state of the surface without actually binding to a group, or a chemical bond might be formed. In either case the surface pressure would be altered.

These changes in bulk ionic strength or pH could be expected to alter the surface pressure in the fluid mosaic membrane of the taste cell. The constraints of electroneutrality in the double layer also suggest an explanation of the water response. Removal of sodium ions from the solution requires that hydrogen ions take their place in the microenvironment. Thus, water responses may be local acid responses. This would explain the close correlation between water-sensitive and acid-sensitive taste nerves in the cat and other mammals (Cohen *et al.*, 1955; Bartoshuk, 1965; Bartoshuk and Frank, 1972; Boudreau and Nelson, 1977).

If altered states of surface pressure play a role in transduction, one would expect taste to be very sensitive to even small amounts of surface agents, i.e., substances capable of inserting themselves into phospholipid structures and altering the state of surface-free energy. A recent study by DeSimone *et al.* (1979, 1980b) suggests this is true. It shows that (1) the well-known taste suppressor gymnemic acid has surface activity comparable with that of sodium lauryl sulfate, an anionic detergent and (2) the latter substance is itself a taste modifier and suppressor. Ionic and neutral surface active agents may provide the chemical probes for further investigation of the role of surface forces in transduction.

V. RESEARCH NEEDS

We have attempted to illustrate some of the insights that physicochemical analysis can bring to problems in chemoreception. Although a clearer picture of preneural events is emerging and a generalized interpretation of early phasic neural responses is possible, the events surrounding transduction are still obscure. The possibility that surface mechanical forces may play a role in transduction is, however, suggested by the *reversible* effects of surface active agents, including gymnemic acid on taste responses. Clearly more work must be done in each of these areas, particularly in exploring the fascinating connection between taste modification and surface activity.

ACKNOWLEDGMENT

This work was supported by NIH research grant NS-13767. The author wishes to thank Ms. Wanda Houston for her careful typing of the manuscript.

REFERENCES

Aiuchi, T., Kamo, N., Kurihara, K., and Kobatake. Y. (1976). Physicochemical studies of taste reception VI. *Chem. Senses Flavor* **2**, 107–119.

Bannister, L. H. (1974). Possible functions of mucus at gustatory and olfactory surfaces. *In* "Transduction Mechanisms in Chemoreception" (T. M. Poynder, ed.), pp. 39–46. Information Retrieval, London.

Bartoshuk, L. M. (1965). Effects of adaptation on responses to water in cat and rat. Ph.D. Thesis, Brown Univ., Providence, Rhode Island.

Bartoshuk, L. M. (1974). NaCl Thresholds in man: Thresholds for water taste or NaCl taste. *J. Comp. Physiol. Psychol.* **87**, 310–325.

Bartoshuk, L. M., and Frank, M. (1972). Taste of water: Neural recordings from rat, hamster, cat and squirrel. *Soc. Neurosci., Houston, Tex.*

Beidler, L. M. (1962). Taste receptor stimulation. *Prog. Biophys. Biophys. Chem.* **12**, 107–151.

Beidler, L. M. (1967). Anion influences in taste receptor response. *In* "Olfaction and Taste" (T. Hayashi, ed.), Vol. 2, pp. 509–534. Pergamon, Oxford.

Beidler, L. M., and Gross, G. W. (1971). The nature of taste receptor sites. *Contrib. Sens. Physiol.* **5**, 97–127.

Bostock, H. (1974). Diffusion and the frog EOG. *In* "Transduction Mechanisms in Chemoreception" (T. M. Poynder, ed.), pp. 27–39. Information Retrieval, London.

Boudreau, J. C., and Nelson, T. E. (1977). Chemical stimulus determinants of cat geniculate ganglion chemoresponsive Group I unit discharge. *Chem. Senses Flavor* **2**, 353–374.

Bronshtein, A. A., and Minor, A. V. (1977). The regeneration of olfactory flagella and restoration of the electroolfactogram after treatment of the olfactory mucosa with Triton X-100. (In Russ.) *Tsitologiya* **19**, 33–38.

Cagan, R. H. (1973). Chemostimulatory protein: a new type of stimulus. *Science* **181**, 32–35.

Carslaw, H. S., and Jaeger, J. C. (1959). "Conduction of Heat in Solids." Oxford Univ. Press, London and New York.

Ciferri, A. (1971). Swelling and phase transition of insoluble collagen. *In* "Biophysical Properties of Skin" (H. R. Elden, ed.), Vol. 1, pp. 101–151. Wiley (Interscience), New York.

Cohen, M. J., Hagiwara, S., and Zotterman, Y. (1955). The response spectrum of taste fibers in the cat: A single fiber analysis. *Acta Physiol Scand.* **33**, 316–332.

DeSimone, J. A., and Heck, G. L. (1980). An analysis of the effects of stimulus transport and membrane charge on the salt, acid, and water-response of mammals. *Chem. Senses* **5** (in press).

DeSimone, J. A., Heck, G. L., and DeSimone, S. K. (1980a). A surface chemical model of salt, *Biophys. J.* **16**, 869–881.

DeSimone, J. A., Bond, J. D., and Price, S. (1977). A model for transduction in gustatory receptors. *Int. Symp. Olfact. Taste, 6th, Gif-sur-Yvette, Fr.*

DeSimone, J. A., Heck, G. L., and Bartoshuk, L. M. (1979). Surface activity: Its influence on taste. *Annu. Meet. Assoc. Chemorecept. Sci., 1st, Sarasota, Fla.*

DeSimone, J. A., Heck, G. L., and DeSimone, S. K. (1980a). A surface chemical model of salt, acid and "water" taste. *In* "Bioelectrochemistry" (M. Blank, ed.), Advances in Chemistry Series, No. 188, pp. 239–260. Am. Chem. Soc., Washington, D.C.

DeSimone, J. A., Heck, G. L., and Bartoshuk, L. M. (1980b). Surface active taste modifiers: A comparison of the physical and psychophysical properties of gymnemic acid and sodium lauryl sulfate. *Chem. Senses* **5** (in press).

Eyzaguirre, C., Fidone, S., and Zapata, P. (1972). Membrane potentials recorded from the mucosa of the toad's tongue during chemical stimulation. *J. Physiol.* (*London*) **221**, 515–532.

Getchell, T. V., and Getchell, M. L. (1977). Early events in vertebrate olfaction. *Chem. Senses Flavor* **2**, 313–326.

Getchell, T. V., Heck, G. L., DeSimone, J. A., and Price, S. (1980). The location of olfactory receptor sites: Inferences from latency measurements. *Biophys. J.* **29**, 397–411.

Goldberg, S. J., Turpin, J. A., and Price, S. (1979). Anisole binding protein from olfactory epithelium: evidence for a role in transduction. *Chem. Senses Flavor* **4**, 207–214.

Heck, G. L., and Erickson, R. P. (1973). A rate theory of gustatory stimulation. *Behav. Biol.* **8**, 687–712.

Holley, A., and MacLeod, P. (1977). Transduction et codage des informations olfactives chez les vertébrés. *J. Physiol.* (*Paris*) **73**, 725–848.

Holley, A., Duchamp, A., Revial, M. F., Juge, A., and MacLeod, P. (1974). Qualitative and quantitative discrimination in the frog olfactory receptors: analysis from electrophysiological data. *Ann. N.Y. Acad. Sci.* **237**, 102–114.

Hornung, D. E., and Mozell, M. M. (1977). Factors influencing the differential absorption of odorant molecules across the olfactory mucosa. *J. Gen. Physiol.* **69**, 343–361.

Jones, C. W., and Watson, E. J. (1963). Two-dimensional boundary layers. *In* "Laminar Boundary Layers" (L. Rosenhead, ed.), pp. 198–257. Oxford Univ. Press, London and New York.

Kagawa, Y. (1974). Dissociation and reassembly of the inner mitochondrial membrane. *In* "Methods in Membrane Biology" (E. D. Korn, ed.), Vol. 1, pp. 201–269. Plenum, New York.

Kamo, N., Miyake, M., Kurihara, K., and Kobatake, K. (1974). Physiochemical studies of taste receptor II. Possible mechanism of generation of taste receptor potential induced by salt stimuli. *Biochim. Biophys. Acta* **367**, 11–23.

Kimura, K., and Beidler, L. M. (1961). Microelectrode study of taste receptors of rat and hamster. *J. Cell. Comp. Physiol.* **58**, 131–140.

Levich, V. G. (1962). "Physicochemical Hydrodynamics." Prentice-Hall, Englewood Cliffs, New Jersey.

Marowitz, L. A., and Halpern, B. P. (1977). Gustatory neural response of the chorda tympani to lick-duration stimuli. *Chem. Senses Flavor* **2**, 457–485.

Metzler, C. M., Elfring, C. M., and McEwen, A. J. (1974). A package of computer programs for pharmacokinetic modeling. *Biometrics* **30**, 562.

Mooser, G., and Lambuth, N. (1977). Inactivation of taste receptor cell function by two cationic protein modification reagents. *J. Neurobiol.* **8**, 193–206.

Murray, J. D. (1977). "Lectures on Nonlinear- Differential-Equation Models in Biology," p. 83. Oxford Univ. Press (Clarendon), London and New York.

Murray, R. G. (1971). Ultrastructure of taste receptors. *In* "Handbook of Sensory Physiology" (L. M. Beidler, ed.), Vol. 4, Part 2, pp. 31–50. Springer-Verlag, Berlin and New York.

Oakley, B. (1962). Microelectrode analysis of second order gustatory neurons in the albino rat. Ph.D. Thesis, Brown Univ., Providence, Rhode Island.

Ottoson, D. (1956). Analysis of the electrical activity of the olfactory epithelium. *Acta Physiol. Scand, Suppl.* No. 122, 1–83.

Ozeki, M., and Sato, M. (1972). Responses of gustatory cells in the tongue of the rat to stimuli representing four taste qualities. *Comp. Biochem. Physiol.* A **41**, 391–407.

Pfaffmann, C. (1955). Gustatory nerve impulses in rat, cat and rabbit. *J. Neurophysiol.* **18**, 429–440.

Price, S. (1972). Sweet-sensitive protein from bovine tongues: sterospecific interactions with amino acids. *In* "Olfaction and Taste" (D. Schneider, ed.), Vol. 4, pp. 214–220. Wiss. Verlagsges., Stuttgart.

Reese, T. S. (1965). Olfactory cilia in the frog. *J. Cell Biol.* **25**, 209–230.

Reif, F. (1965). "Fundamentals of Statistical and Thermal Physics." McGraw-Hill, New York.

Sato, T. (1976). Does an initial phasic response exist in the receptor potential of taste cells? *Experientia* **32**, 1426–1427.

Shibuya, T. (1969). Activities of single olfactory receptor cells. *In* "Olfaction and Taste" (C. Pfaffmann, ed.), Vol. 3, pp. 109–116. Rockefeller Univ. Press, New York.

Shibuya, T., and Tucker, D. (1967). Single unit responses of olfactory receptors in vultures. *In* "Olfaction and Taste" (T. Hayashi, ed.), Vol. 2, pp. 219–233. Pergamon, Oxford.

Smith, D. V., and Bealer, J. (1975). Sensitivity of the rat gustatory system to the rate of stimulus onset. *Physiol. Behav.* **15**, 303–314.

Smith, D. V., Steadman, J. W., and Rhodine, C. N. (1975). An analysis of the time course of gustatory neural adaptation in the rat. *Am. J. Physiol.* **229**, 1134–1140.

Solms, J. (1969). The taste of amino acids, peptides and proteins. *Agric. Food Chem.* **17**, 686–688.

Van Drongelen, W., Pagnotti, Y., and Hendricks, M. H. (1980). Primary events in odour detection. *Bull. Math. Biol.* (in press).

Vinnikov, Y. A., and Titova, L. K. (1949). Histophysiological characteristics of the olfactory receptor. (In Russ.) *Dokl. Akad. Nauk SSSR* **65**, 903–906.

12

Transduction Through Receptor State Transitions

GREGORY MOOSER

I. INTRODUCTION

The study of biochemical events associated with taste reception in mammals suffers from major limitations in availability of tissue, isolation of homogeneous preparations of taste-cell derived membranes or membrane macromolecules, and availability of high-specificity, high-affinity receptor ligands. Nonetheless, there have been several courageous attempts and reports on isolation of enriched preparations from mammalian taste receptor cells (for review, see Price and De Simone, 1977). Some look particularly promising, but none has been studied sufficiently to meet the generally accepted major criteria used to establish authenticity of preparations. This certainly does not result from any lack of perseverance, but given the above three limitations, meeting all of the criteria using isolated mammalian taste receptor tissue presents a formidable task. The criteria, however, cannot be

Biochemistry of Taste and Olfaction

ISBN 0-12-154450-8

taken lightly since many cases of artifactual results have been reported on experiments with other membrane receptor systems. This point is perhaps best emphasized by experiments simulating studies on ligand binding to isolated receptor preparations, where nonbiological rather than biological material is used. Nonspecific binding can appear "specific" based on criteria such as specificity, saturability, and reversibility, which are often used to define receptors. For example, [^{125}I]insulin binds to talc, alumina, and microsilica in a manner that closely mimics receptor characteristics even to the point of displaying positive cooperativity (Cuatrecasas *et al.*, 1975). Thus, relying on one or a few criteria to establish authenticity can be hazardous even though some of the most difficult criteria to meet experimentally may seem redundant, particularly when dealing with a difficult system such as mammalian taste tissue. There are some minor differences in accepted criteria suggested by various researchers, but those enumerated by Kahn (1976) are as comprehensive as any.

II. STUDY OF BIOCHEMICAL EVENTS IN TASTE RECEPTION WITHOUT ISOLATING RECEPTOR TISSUE

Although the problems mentioned above are limiting, a reasonable approach can be designed to fractionate receptor cells. For example, high-specificity, high-affinity ligands are potentially available for application to probing and isolating sweet receptor macromolecules (Cagan, 1973), and, if attention is turned toward nonmammalian experimental animals, a richer source of taste receptor cells can be found (Kreuger and Cagan, 1976). However, dealing with some aspects of taste reception, such as salt stimulation, is further constrained by the fact that a large number of biological macromolecules not associated with taste cells bind salt stimuli with an affinity of the order of that expected with taste cells. This does not exclude the possibility that the membrane events in salt reception are common to taste as well as nontaste cells, but this cannot be finally established until the discrete events operating in taste cells themselves are more fully understood.

There remains a reasonable method to study biochemical events in mammalian salt reception without isolating the tissue. This takes advantage of the observation that some animals, such as the rat, show a concentration-dependent, steady-state level of neural activity when exposed to salt stimuli (Beidler, 1953). The possibility that the magnitude of whole-nerve chorda tympani activity in the rat is directly proportional to the magnitude of the receptor events on the taste cell membrane presents a great potential to explore the dynamics of receptor activation without isolating the tissue.

Several lines of evidence justify the assumption that a linear, proportionate relationship exists. The strongest support comes from the observation that modification of taste receptor cell membranes in rats, by treating the surface of the tongue with irreversible protein modification reagents, causes a decrease in neural activity which follows the kinetics of a psuedo first-order decay for several half-lives (Mooser, 1976; Mooser and Lambuth, 1977). In addition, measurement of the rates of inactivation at several concentrations of the inhibitor is consistent with a purely second-order reaction between the reagent and receptor cell. This means that the protein modification reagents do nothing more than react with the receptor cell membrane and, by a mechanism which interferes with membrane changes necessary for receptor cell activation, causes a decrease in neural activity. The simple scheme shown below

$$R \xrightarrow[\text{reagent}]{\text{protein modification}} R'$$

(where R is the unmodified receptor cell and R' is the inhibited, modified cell) is all that is required to explain the kinetics of the inhibition. The fact that a change induced at the cell membrane causes a pattern of inhibition of whole nerve summated activity, which is predicted by the most simple of kinetic schemes, offers strong support for a linear relationship between the magnitude of inhibition (or activation) at the receptor cell and the magnitude of whole nerve activity. The possibility is remote that several counteracting

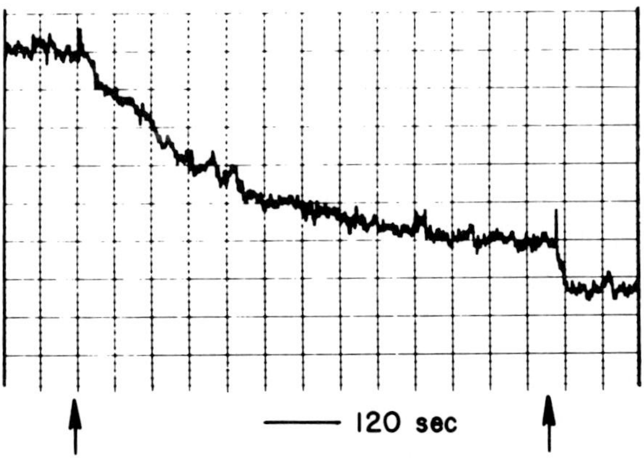

Fig. 1. Inactivation of the 0.3 *M* NaCl-stimulated, whole-nerve chorda tympani response by 0.035 *M* HNB-dmS at pH 4.0. The preparation was stimulated with NaCl followed by application of inhibitor in the presence of 0.3 *M* NaCl (left arrow). The right arrow indicates the time of application of a water rinse. (Reprinted from Mooser and Lambuth, 1977, with permission).

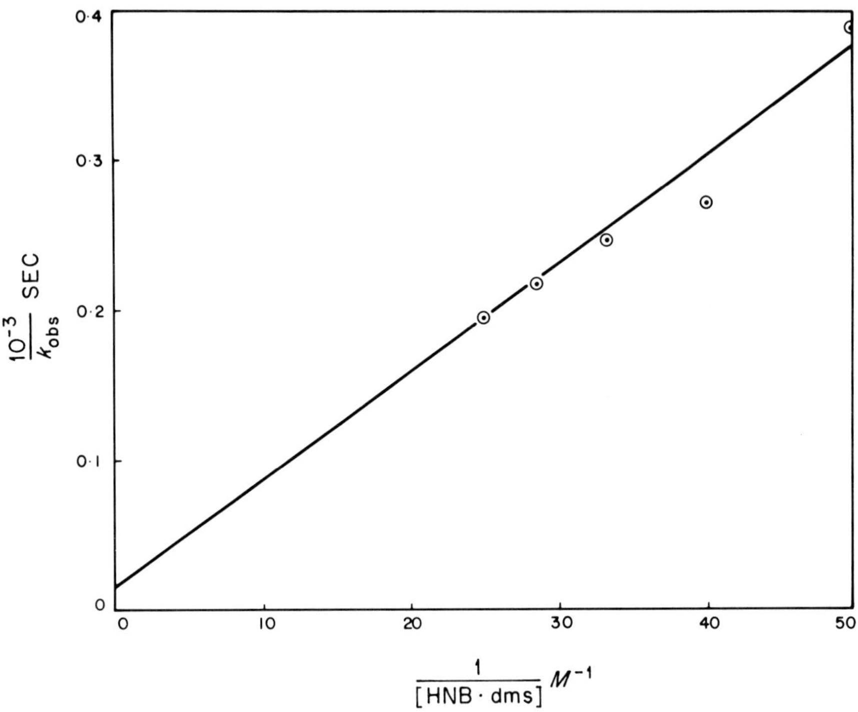

Fig. 2. Double reciprocal plot of the psuedo first-order rates of inactivation of the 0.3 *M* NaCl response as a function of HNB-dmS concentration. (Reprinted from Mooser and Lambuth, 1977, with permission.)

nonlinear events give an overall linear relationship between the points of stimulus interaction and whole nerve response.

A typical graph of the decay using the potein modification reagent dimethyl(2-hydroxy-5- nitrobenzyl)sulfonium bromide (HNB-dmS) is shown in Fig. 1, and the respective plot of the psuedo first-order rate constants as a function of inhibitor concentration is shown in double reciprocal form in Fig. 2.

III. DYNAMICS OF RECEPTOR ACTIVATION BY SODIUM AND POTASSIUM SALTS

If the degree of receptor cell activation can be judged from the magnitude of tonic chorda tympani activity in the rat, it should be possible to evaluate the dynamics of receptor-stimulus interactions. The effects of stimulus concentration on activity can be analyzed using an approach analogous to the study of the effects of substrate concentration on enzyme activity or the effects of ligand concentration on some spectral property of an isolated receptor. These data provide quantitative information on complexes that contribute to changes in the parameters observed in the assay whether the measured parameter is enzyme activity, protein absorption, or neural activity, provided that the measured parameter is linearly related to the event.

We found that only certain salt stimuli, such as sodium and potassium salts, give a steady-state level of whole nerve chorda tympani activity in rats that is of sufficient duration and stability to analyze quantitatively. The characteristic response after stimulation with sucrose (and other sweet stimuli), quinine, or HCl is not acceptable because of too rapid a decrease in the steady-state level.

The concentration of NaCl required to give half-maximal activity is of the order of 0.1–0.2 *M*. At least three reasonable explanations can account for the ability of this relatively small binding energy to induce conformational changes in the receptor cell membrane. First, the true affinity could be very high with a major part of the energy coupled to conformational changes resulting in a greatly reduced observed affinity; second, multiple weak binding events could collectively provide sufficient energy to drive conformational changes; and third, the system could be poised between active and inactive states so that only a small amount of energy is required for activation.

These three possibilities cannot be distinguished with current methods available to study taste receptor mechanisms. The third possibility, however, has particular merit because of major precedents established for the mechanism of other membrane receptor systems (Edelstein, 1972; DeMeyts, 1976; Limbird and Lefkowitz, 1976; Catterall, 1977). For example, two-state allosteric models, which involve an equilibrium between an active and inactive state of the receptor, can be activated or inactivated with very little energy if the intrinsic equilibrium between the two states does not greatly favor one receptor form or the other.

Since allosteric mechanisms, like any kinetic mechanisms, follow a defined set of mathematical functions, they can be tested for consistency with taste-receptor response data. In the following section, the theory of two-state mechanisms is discussed as a foundation for analyzing data on taste receptor activation by sodium and potassium salts.

A. Theory of Two-State Allosteric Mechanisms

The theory of two-state mechanisms was first developed in detail by Monod *et al.* (1965) as a new model that could explain the behavior of a large class of enzymes that do not follow classical Michaelis-Menten kinetics.

These were allosteric enzymes, many of which serve a central role in controlling metabolism. The kinetic behavior of allosteric enzymes demanded incorporating into the kinetic scheme a mechanism for activation or inhibition by effectors (activators or inhibitors) which alter the conformation at the enzyme active site through binding at a site remote from the active site. Of course when applied to receptor activation and inhibition, the conformational change involves activation or inhibition of the events initiating the transduction mechanism rather than enzyme activity.

A two-state mechanism is based on an equilibrium between active (R) and inactive (T) receptor forms.

$$R \overset{L}{\rightleftharpoons} T \tag{1}$$

The position of the equilibrium in the absence of ligands is defined solely by an equilibrium or allosteric constant (L) according to Eq. (2):

$$L = \frac{T}{R} \tag{2}$$

such that the higher the value of L, the further the equilibrium lies in the direction of the inactive state. L must be significantly greater than 1 for the system to be inactive in its resting state. However, the energy required to displace the equilibrium toward the active state is directly dependent on the magnitude of L. When L is very large, a greater amount of energy (in this case energy in the form of ligand binding) is required than if L were smaller.

It is important to emphasize that even systems with a very high allosteric constant can be further displaced toward the inactive state since this is an equilibrium. In the absence of stimulus, the *fraction* of total receptors in the active state, $\bar{R}$, is

$$\bar{R} = \frac{R}{R + T} = \frac{1}{1 + L} \tag{3}$$

As L gets large, R approaches but never equals zero. In our analysis of taste receptors, L had the relatively low value of 9, which predicts that approximately 10% of the receptors are active in the resting state (Mooser, 1980a).

Ligands displace the equilibrium toward the active or inactive state according to basic laws of mass action. The equilibrium is displaced toward the active state if an activator binds to that state but not to the inactive state, and the converse holds for inhibitors. However, it is not mandatory for ligands to bind exclusively to one state; all that is required is that the receptor-ligand affinity be higher for one state relative to the other. This situation is referred to as *nonexclusive binding* and has been developed theoretically and applied to several systems (Rubin and Changeux, 1966).

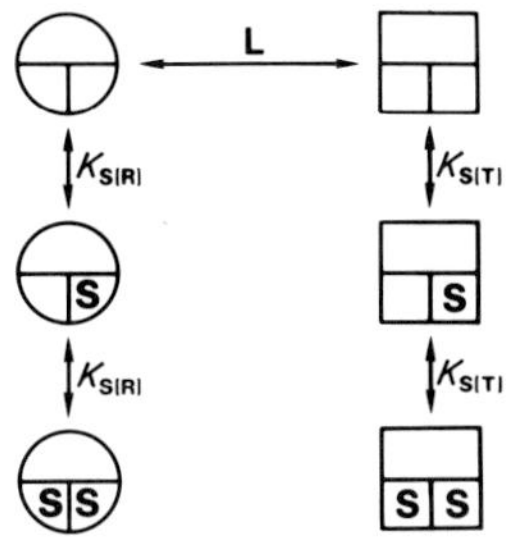

Fig. 3. Schematic representation of nonexclusive ligand binding in a two-state model. The circles represent the active conformation of the receptor and the squares are the inactive conformation. Two identical stimulus sites per receptor are shown. All active forms have equivalent activity regardless of the state of ligand occupancy. The equilibrium is displaced toward activation when the stimulus dissociation constant for the active state [$K_{S(R)}$] is less than that for the inactive state [$K_{S(T)}$].

Nonexclusive ligand binding in a two-state model is shown schematically in Fig. 3. The squares represent the inactive conformation and the circles are the active conformation. A stimulus, S, can bind to either state and will displace the equilibrium toward the active state only if the receptor-stimulus dissociation constant for the active state [$K_{S(R)}$] is less than that for the inactive state [$K_{S(T)}$].* Two identical sites per receptor are shown. Any number greater than 1 is possible, although the data for sodium ion activation of taste receptors is best resolved with two sites as is used for illustration. All R state receptors have equal activity regardless of whether stimulus is bound or not. The fraction of total receptors in the R state in the presence of S is given in Eq. 4.

$$\bar{R} = \frac{(1 + \alpha)^n}{(1 + \alpha)^n + L(1 + c\alpha)^n} \quad (4)$$

The function is based on the ratio of stimulus concentration to the dissociation constants where $\alpha = S/K_{S(R)}$ and c, referred to as the nonexclusive binding coefficient, equals $K_{S(R)}/K_{S(T)}$. The parameter n is the number of sites per receptor and equals 2 for the pattern in Fig. 3.

Note that if the affinities of the stimulus for both states of the receptor are identical [$K_{S(R)} = K_{S(T)}$ and $c = 1$], Eq. 4 reduces to $1/(1+L)$ and is independent of stimulus concentration; the addition of stimulus will not change the state of the equilibrium relative to that of the resting state. S causes receptor

*Dissociation constants for the receptor states are indicated with a subscript denoting stimulus, S, or inhibitor, I, binding to the R state, R, or T state, T; for example, inhibitor binding to the T state is $K_{I(T)}$.

activation only when $K_{S(R)} < K_{S(T)}$; conversely, S would be an inhibitor if $K_{S(R)} > K_{S(T)}$.

Some of the taste receptor activation data can be explained only if certain salt solutions are considered to contain both activators and inhibitors. For example, data for sodium salts of organic anions, such as sodium acetate and sodium propionate, are most easily resolved by considering sodium ion as an activator and the organic anion as an inhibitor, with the two effectors binding to separate and distinct sites. This situation (referred to as heterotropic inhibition since the activator and inhibitor are different ions which bind to different sites), follows Eq. 5:

$$\bar{R} = \frac{(1 + \alpha)^n (1 + \beta)^m}{(1 + \alpha)^n(1 + \beta)^m + L(1 + c\alpha)^n(1 + d\beta)^m} \tag{5}$$

in which β is the normalized inhibitor concentration $[I/K_{I(R)}]$, d is the nonexclusive binding coefficient for the inhibitor $[K_{I(R)}/K_{I(T)}]$, and m is the number of equivalent sites that bind inhibitor. For organic anions, $m = 1$ gave the best fit to our data, although $m = 2$ was also acceptable.

One final point requires explanation. The magnitude of taste receptor activation is measured relative to a baseline of chorda tympani activity determined with water on the surface of the tongue. It was noted earlier that even in the absence of stimulus, a finite portion of receptors are in the R state. Thus the observed magnitude of activation (observed activity–baseline activity) is representative of only a portion of the total R state receptors. However, those receptors initially in the R state must also be considered, which, according to Eq. 3, always equals $1/(1+L)$. If the observed activity is a function only of receptor activation after stimulus addition, the observed fraction of R state receptors is

$$\bar{R}_{\text{obs}} = \bar{R} - \frac{1}{1 + L} \tag{6}$$

Thus, Eq. (4) becomes

$$\bar{R}_{\text{obs}} = \frac{(1 + \alpha)^n}{(1 + \alpha)^n + L(1 + c\alpha)^n} - \frac{1}{1 + L} \tag{7}$$

and Eq. (5) becomes

$$\bar{R}_{\text{obs}} = \frac{(1 + \alpha)^n(1 + \beta)^m}{(1 + \alpha)^n(1 + \beta)^m + L(1 + c\alpha)^n(1 + d\beta)^m} - \frac{1}{1 + L} \tag{8}$$

All of the equations presented up to this point have been based on the *fraction* of total R state receptors (indicated by $\bar{R}$). For convenience, the taste data analyzed by this approach were normalized to a value of 100 for activation by 0.3 *M* NaCl. Regression analysis of several sets of data made it possible to estimate the value of $R + T$ relative to the normalized NaCl response. This was found to be 202 so that the observed magnitude of chorda tympani activity, R_{obs}, can be expressed as

$$R_{obs} = \bar{R}_{obs} \times 202 = \text{observed} - \text{baseline activity when } 0.3\ M\ \text{NaCl} = 100.$$

B. Two-State Mechanisms and Taste Receptor Activation

Analyses of this type only establish consistency with a mechanism and can never prove a mechanism. However, the validity of a model increases when the response from variations in the system and/or stimulus parameters can be predicted by the model. Some of the results obtained through application of this analysis are discussed below with examples of varying the stimulus by using different anion-cation combinations alone and in mixtures as well as examples where the system itself is modified with an irreversible inhibitor.

1. Sodium Chloride, Sodium Acetate, and Sodium Propionate Stimulation with and without Inhibiton by HNB-dmS

It has been well established that sodium salts become less active stimuli as the hydrophobic character of the anion increases (Beidler, 1954). Typical profiles of the concentration-response curve for NaCl, sodium acetate (NaAc), and sodium propionate (NaPr) are shown in Fig. 4. Two-state mechanisms can potentially account for this characteristic if sodium chloride is treated purely as a receptor activator and any organic anions replacing chloride are treated as heterotropic inhibitors that interact with the receptor membrane at a site distinct from sodium ions. Under these conditions, chloride has no effect on the magnitude of receptor activation, but it is not excluded from interacting at the same site as organic anions. This is not contradictory since, as was noted above, an ion must fulfill two requirements to either activate or inhibit the system. First, there must be binding, and, second, the affinity must be different for the two states. If the affinities are the same, there will be no influence on the state of the equilibrium and no functional effect. Note that it is not possible to distinguish between the absence of binding by a particular ion and binding with equal affinity for both states with these data alone. Rather, competition experiments are required, some of which are discussed as part of the studies using salt mixtures.

The NaCl activation profile can be well resolved by nonlinear regression based on Eq. 7, which treats sodium ion as a receptor activator and disregards chloride ion. The sodium ion dissociation constant for the *R* state [$K_{S(R)}$] was 0.098 *M* and 1.63 *M* for the *T* state [$K_{S(T)}$] (Mooser, 1980a). These same affinities, with small variation, can be incorporated into the regression analysis of NaAc and NaPr profiles with the inclusion of acetate

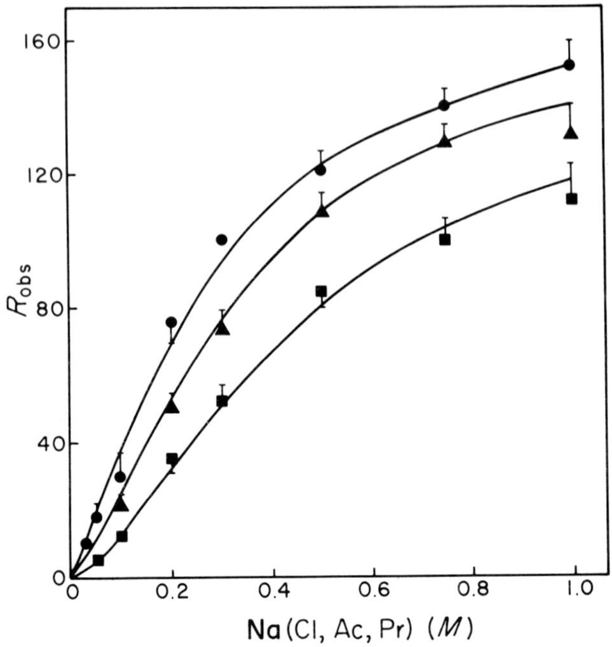

Fig. 4. Concentration-response profiles for NaCl (circles), NaAc (triangles), and NaPr (squares). The lines are nonlinear least-squares regressions based on Eq. 7 for the NaCl profile and Eq. 8 for the NaAc and NaPr profiles. Equation 8 was constrained by the sodium ion dissociation constant values determined in Eq. 7 with the NaCl data. (Reprinted from Mooser, 1980b, with permission.)

and propionate as heterotropic inhibitors according to Eq. 8. As Fig. 4 shows, the results are fully consistent with these considerations.

This is a limited analysis, and the data can be resolved by a number of suitable mechanisms as has been shown using a classical Michaelis-Menten scheme (Beidler, 1954). In our laboratory, we tested several related models all of which were biochemically feasible; six of them fit the data with reasonable accuracy.

Altering the system can aid in narrowing the possibilities. One approach involves treating the surface of a rat's tongue with HNB-dmS prior to measuring the response to sodium salt stimuli. The value of this compound lies in the observation that it is an irreversible inhibitor of the taste response but does not totally destroy receptor activity (Mooser and Lambuth, 1977). HNB-dmS inhibits by reacting with the receptor cell membrane and altering an event that occurs subsequent to stimulus interaction. The two-state mechanism can account for this if HNB-dmS acts by stabilizing the inactive state of the receptor with a resultant increase in the $R \rightleftharpoons T$ equilibrium constant (L). The stability of the inactive state increases and more energy (higher stimulus concentration) is required to displace the equilibrium toward activation. No change in anion-receptor or cation-receptor dissociation constants is involved.

Since measurable activity remains after inhibition, the hypothesis can be tested on sodium salt concentration-response profiles of the HNB-dmS-inhibited preparation. The data were found to be consistent and were explained on the basis of an approximately tenfold increase in L without invoking an effect on salt ion dissociation constants (Mooser, 1980a).

2. *Potassium Chloride, Potassium Acetate, and Potassium Propionate*

Analysis of potassium salt profiles offers the potential to confirm these assumptions. In addition, potassium propionate has a characteristic that must be explained by the model. Salts of certain hydrophobic organic anions, including KPr, depress the whole nerve response below baseline (Mooser, 1980b). The most simple explanation for this phenomenon, consistent with a two-state mechanism, involves displacing the resting state $R \rightleftharpoons T$ equilibrium toward the T state on addition of KPr at appropriate concentrations. The potential for this exists because, as noted earlier, the allosteric constant is about 9, resulting in some 10% of the receptors in the active state in the absence of all stimuli.

Characteristic profiles for KCl, KAc, and KPr are shown in Fig. 5 and resolved by nonlinear regression using the dissociation constants for organic anions determined from the sodium salt profile. Clearly, inhibitory properties

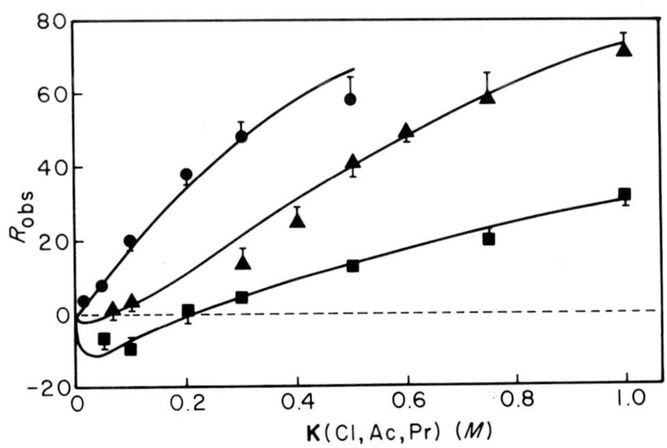

Fig. 5. Concentration-response profiles for KCl (circles), KAc (triangles), and KPr (squares). The dissociation constants for the organic anions are the same as those used in Fig. 4. (Reprinted from Mooser, 1980b, with permission.)

of propionate ion are observed that can be considered responsible for the decreased activity of NaPr relative to NaCl and can also adequately resolve the KPr data relative to KCl.

Inclusion of two potassium sites was required to resolve these data: an inhibitor site and an activator site. Two sites have been proposed previously (Beidler, 1962). Of these two, it is the inhibitor site that accounts for the decreased activity of potassium relative to sodium salts and the reduction of activity below baseline in part of the KPr profile.

3. *Salt Mixtures*

Since all of the sodium and potassium salt profiles can be resolved with given values for R state and T state dissociation constants for each ion plus a single $R \rightleftharpoons T$ allosteric constant, it should be possible to predict profiles for ion mixtures. Two series of mixtures have been analyzed: one in which anions vary but not cations (NaCl/NaPr) and the converse in which cations vary but not anions (NaCl/KCl).

The response activity from mixtures is consistent with predicted values (Mooser, 1980a,b) and provides further information on competition between ions with the same charge. For example, when the ratio of chloride to priopionate was varied at constant sodium ion concentration, it was clear that chloride ion affects propionate ion binding, even though chloride in the absence of propionate does not influence activation. Thus, chloride apparently binds with near identical affinity to both the R and T receptor states and has no effect in a NaCl solution, but in the presence of NaPr, chloride competes with the inhibition properties of propionate. This suggests that interaction at an anion site is dependent on charge, but the potential to alter the equilibrium state is a function of the hydrophobic character of the anion.

Analysis of mixtures of NaCl and KCl at constant ionic strength but with different Na^+/K^+ ratios is also predictable based on the parameter values determined from data generated with NaCl and KCl alone. In addition, it was possible to determine whether sodium competes with the potassium inhibition site, with the potassium activation site, or with both. As might be expected, the data were consistent with sodium–potassium competition at the activation site only (Mooser, 1980b).

4. *Conclusions*

The two-state model in the presence of stimuli can be reduced to interacting equilibria, which determine the total fraction of R state receptors and the magnitude of the response. Figures 6 and 7 are schematic representations of these equilibria for the sodium and potassium salts discussed in this section. For NaCl, the scheme is relatively simple, but as other ions are added the number of steps increases proportionately. Subtle differences in dissociation

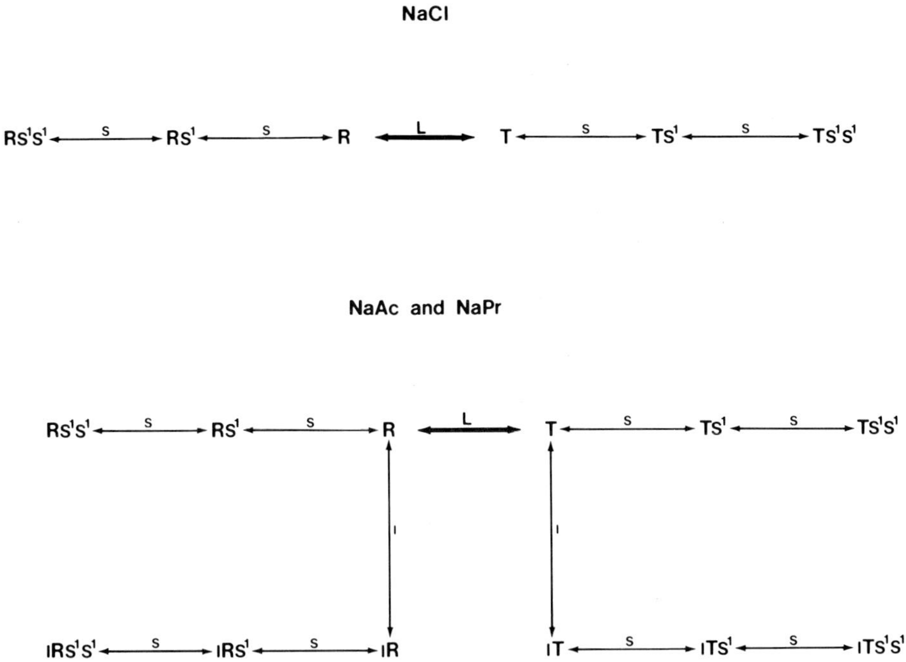

Fig. 6. Models for NaCl, NaAc, and NaPr binding in a two-state receptor system. *R* and *T* are the active and inactive receptor states, respectively. S^1 represents sodium ion binding to one of two identical activation sites; *I* represents acetate or propionate ion binding to an inhibition site.

constants or nonexclusive binding coefficients can have significant effects on the equilibrium state. For example, acetate and propionate have a ratio of R state to T state dissociation constants that is close to 1. Yet a slightly higher affinity for the *T* state, possibly due to a small change in hydrophobic character near the binding site, accounts for the reduction in activity.

Since inorganic and organic salts have a wide variety of taste properties in man, the transduction of the response requires a mechanism to generate information interpretable in terms of taste quality. The results of the analysis summarized above provides an insight into an aspect of one possible mechanism.

Of those salts analyzed, only sodium chloride has a purely salty taste at moderate to high concentrations. Replacing sodium with potassium or chloride with organic anions adds a taste component that is most closely associated with bitterness (Schiffman, 1980). Of interest is the observation that only NaCl is a pure receptor activator in the two-state model. Potassium is both a receptor activator and inhibitor, and organic anions are receptor

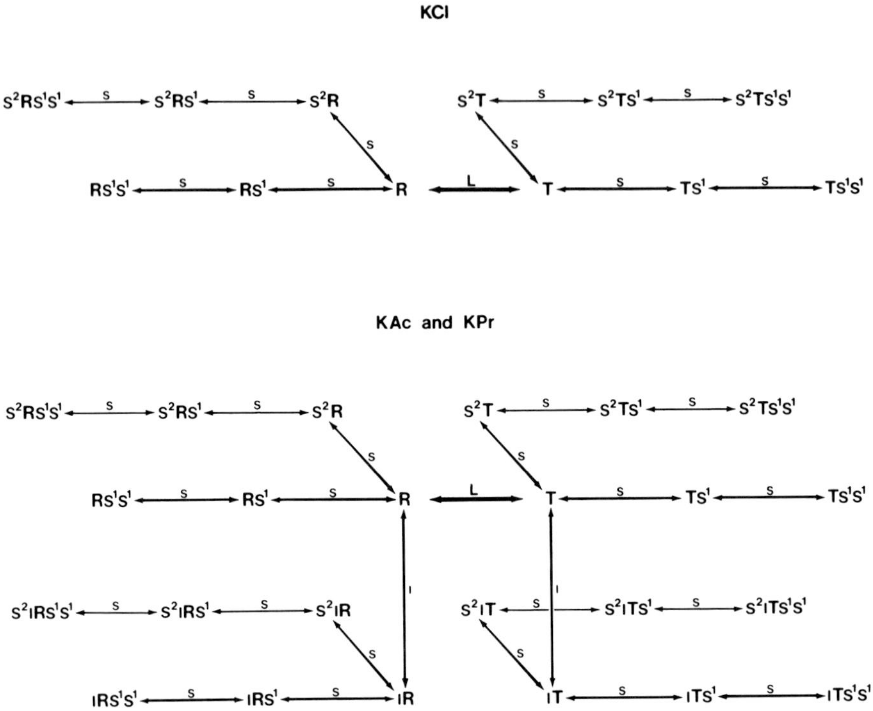

Fig. 7. Models of KCl, KAc, and KPr binding in a two-state receptor system. S^1 represents potassium ion binding to one of two identical activation sites; S^2 represents potassium ion binding to an inhibition site; and I represents acetate or propionate ion binding to an inhibition site.

inhibitors. Thus, there exists a relationship between the effects of discrete ions on the receptor state and the taste quality of a stimulus, which may be relevant to the mechanism involved in distinguishing taste qualities.

This chapter was intended to review the work on salt activation of taste receptors in the context of a two-state model and to provide an example of an approach to explore discrete events in receptor activation where isolating the pertinent macromolecule is impractical. This indirect approach carries advantages and disadvantages. It is a convenient method to generate data for testing kinetic models, but it lacks the simplicity of isolated systems. In contrast, there is concrete value in characterizing the intact system in as much detail as possible in order to provide a framework to test isolated preparations as they become available. Furthermore, since receptor characteristics can change during isolation due to loss of critical components or partial denaturation as a consequence of removing receptors from their usual environment (Popot *et al.*, 1976), knowledge of the intact system provides a necessary reference on function.

IV. RESEARCH NEEDS

One of the major values of biochemical studies in taste reception lies in gaining insight into the relationship between molecular events and taste function. Understanding taste and placing it in the context of other membrane receptor systems requires a thorough study of all areas. Only when disciplines extend sufficiently to overlap concepts is it possible to explain fundamental properties.

Receptor biochemistry is a relatively young field and taste biochemistry is younger still. Today it is easier to draw relationships between the biochemistry of taste receptors and the biochemistry of other membrane receptors than it is to identify those biochemical properties of taste that make it unique. There remain numerous gaps to fill, many of which will require experiments at the front of biochemical technology.

At the beginning of this chapter, it was noted that several limitations pose significant problems in exploring taste receptors at the molecular level. One in particular, the puacity of high affinity, high specificity receptor ligands, seems critical for distinguishing isolated receptor membranes from membranes associated with adjacent cells or intracellular organelles. Some potential receptor probes exist, such as the chemostimulatory proteins (Cagan, 1973), and preliminary experiments using one of these, monellin, look promising (Cagan and Morris, 1979).

There can be great advantage in using receptor probes that bind covalently to receptors. The disadvantage of labeling with reversible probes lies in the inability to perform preliminary protection experiments where nonspecific sites are blocked prior to labeling the specific site. Protection experiments used with covalent probes amplifies the labeling specificity so that even relatively nonspecific covalent probes can be used to selectively tag receptors as was done with *N*-ethylmaleimide and frog olfactory tissue (Getchell and Gesteland, 1972).

For functional reasons, taste buds are relatively insensitive to covalent modification. Taste is used, in part, to determine whether ingested substances should be permitted to proceed down the alimentary tract, and sensitivity to chemical disruption would compromise this function. The added stability of taste cells may account for the observation that even when separated from taste buds, the cells maintain their characteristic shape

(Brand and Cagan, 1976). The strength and resistence to chemical modification makes it difficult to design covalent probes directed at reacting with specific functional groups on the receptor, and it may be necessary to use techniques such as photoaffinity labeling that eliminates the requirement for a reactive group at the labeling site (Bayley and Knowles, 1977).

REFERENCES

Bayley, H., and Knowles, J. R. (1977). Photoaffinity labeling. *In* "Methods in Enzymology" (W. B. Jakoby and M. Wilchek, eds.), Vol. 46, pp. 69–114. Academic Press, New York.

Beidler, L. M. (1953). Properties of chemoreceptors of tongue of rat. *J. Neurophysiol.* **16,** 595–607.

Beidler, L. M. (1954). Taste receptor stimulation. *J. Gen. Physiol.* **38,** 133–139.

Beidler, L. M. (1962). Taste receptor stimulation. *Prog. Biophys. Biophys. Chem.* **12,** 107–151.

Brand, J. G., and Cagan, R. H. (1976). Biochemical studies of taste sensation. III. Preparation of a suspension of bovine taste bud cells and their labeling with a fluorescent probe. *J. Neurobiol.* **7,** 205–220.

Cagan, R. H. (1973). Chemostimulatory proteins: A new type of taste stimulus. *Science* **181,** 32–35.

Cagan, R. H., and Morris, R. W. (1979). Biochemical studies of taste sensation: binding to taste tissue of ^{3}H-labeled monellin, a sweet tasting protein. *Proc. Natl. Acad. Sci. U.S.A.* **76,** 1692–1696.

Catterall, W. A. (1977). Activation of the action potential Na^{+} ionophore by neurotoxins: an allosteric model. *J. Biol. Chem.* **252,** 8669–8676.

Cuatrecasas, P., Hollenberg, M. D., Chang, K.-J., and Bennett, V. (1975). Hormone receptor complexes and their modulation of membrane function. *Recent Prog. Horm. Res.* **31,** 37–94.

DeMeyts, P. (1976). Cooperative properties of hormone receptors in cell membranes. *J. Supramol. Struct.* **4,** 241–248.

Edelstein, S. J. (1972). An allosteric mechanism for the acetylcholine receptor. *Biochem. Biophys. Res. Commun.* **48,** 1160–1165.

Getchell, T. V., and Gesteland, R. C. (1972). The chemistry of olfactory reception: Stimulus specific protection from SH inhibition. *Proc. Natl. Acad. Sci. U.S.A.* **69,** 1494–1498.

Kahn, C. R. (1976). Membrane receptors for hormones and neurotransmitters. *J. Cell Biol.* **70,** 261–286.

Krueger, J. M., and Cagan, R. H. (1976). Biochemical studies of taste sensation: Binding of L-[^{3}H]alanine to a sedimentable fraction from catfish barbel epithelium. *J. Biol. Chem.* **251,** 88–97.

Limbird, L. E., and Lefkowitz, R. J. (1976). Negative cooperativity among adrenergic receptors in frog erthyrocyte membranes. *J. Biol. Chem.* **251,** 5007–5014.

Monod, J., Wyman, J., and Changeux, J.-P. (1965). On the nature of allosteric transitions: a plausable model. *J. Mol. Biol.* **12,** 88–118.

Mooser, G. (1976). *N*-Substituted maleimide inactivation of the response to taste cell stimulation. *J. Neurobiol.* **7,** 457–468.

Mooser, G. (1980a). Membrane transitions in taste receptor cell activation by sodium salts. *In* "Biological and Behavioral Aspects of Salt Intake" (M. R. Kare, M. J. Fregly, and R. A. Bernard, eds.), pp. 275–287. Academic Press, New York.

Mooser, G. (1980b). Sodium and potassium salt stimulation of taste receptor cells: An allosteric model. *Proc. Natl. Acad. Sci. U.S.A.* **77**, 1686–1690.

Mooser, G., and Lambuth, N. (1977). Inactivation of taste receptor cell function by two cationic protein modification reagents. *J. Neurobiol.* **8**, 193–206.

Popot, J.-L., Sugiyama, H., and Changuex, J.-P. (1976). Studies on the electrogenic action of acetylcholine with *Torpedo marmorata* electric organ. II. The permeability response of the receptor-rich membrane fragments to choline agonists *in vitro*. *J. Mol. Biol.* **106**, 469–483.

Price, S., and DeSimone, J. A. (1977). Models of taste receptor cell stimulation. *Chem. Senses Flavor* **2**, 427–456.

Rubin, M. M., and Changeux, J.-P. (1966). On the nature of allosteric transitions: Implications of nonexclusive ligand binding. *J. Mol. Biol.* **21**, 265–274.

Schiffman, S. S. (1980). Contribution of the anion to the taste quality of sodium salts. *In* "Biological and Behavioral Aspects of Salt Intake" (M. R. Kare, M. J. Fregly, and R. A. Bernard, eds.), pp. 99–111. Academic Press, New York.

13

Molecular Mechanisms of Transduction in Chemoreception

KENZO KURIHARA, MICHIHISA MIYAKE,
AND KIYONORI YOSHII

I. INTRODUCTION

Living organisms have the ability to recognize chemical stimuli in external environments. The most primitive type of chemoreception is seen in

Biochemistry of Taste and Olfaction

ISBN 0-12-154450-8

chemotaxis by bacteria or protozoa. In higher vertebrates, chemical stimuli in external environments are received at gustatory or olfactory receptor cells. The transduction process of chemoreception in lower organisms and in higher vertebrates is composed of the following processes: (a) Chemical stimuli are adsorbed on the receptor membrane. (b) The adsorption induces the receptor potential. (c) In higher vertebrates, the receptor potential is transduced to nerve impulses. In lower organisms, the receptor potential triggers the chemotactic response, although the role of the receptor potential in chemotaxis is still unclear at present in organisms except for ciliated protozoa.

Output of chemoreception is quite different between lower organisms and higher vertebrates, but there are many common mechanisms in the transduction process in both types of organisms (Kurihara *et al.*, 1978). In this chapter, the transduction mechanisms in lower organisms (true slime mold *Physarum polycephalum* and *Tetrahymena*) and in higher vertebrates are reviewed.

II. COMPARISON OF CHEMORECEPTIVE FUNCTION IN VARIOUS ORGANISMS

A variety of chemicals (salts, acids, sugars, amino acids, "bitter substances" and "odorants") stimulate gustatory or olfactory receptors. Among these chemical stimuli, amino acids and sugars stimulate only membranes having specific receptor proteins (see Chapters 4, 8, and 10). On the other hand, other stimuli interact with most cell membranes. In the following sections it is shown that cell membranes of various organisms respond similarly to odorants and bitter substances.

A. Reception of "Odorants"

In higher vertebrates, olfactory reception takes place at the receptor cells in the olfactory epithelium (Chapters 1–3). However, the elicitation of neural responses by "odorants" is not restricted to the olfactory organ; other excitable systems such as the trigeminal nerve and vomeronasal organ of the tortoise also respond to odorants (Tucker, 1971). Tucker (1971) pointed out that all of the odorants that stimulate the olfactory cells of the tortoise also stimulate the trigeminal nerves.

Kashiwagura *et al.* (1977b) found that frog gustatory cells, which are secondary sensory cells, also respond to various species of so-called odorants. The cross-adaptation experiments between various pairs of "odorants" indicated that the frog gustatory receptors can discriminate various odorants, although the receptors would not be expected to have specific receptor molecules for odorants.

Lower organisms also respond to various odorants. For example, *Tetrahymena* and the true slime mold *Physarum polycephalum* exhibited negative chemotaxis against all of the odorants examined (Ueda and Kobatake, 1977). The internodal cell of *Nitella*, which is a large excitable cell, also responded to various odorants, which depolarized the internodal cell (Ueda *et al.*, 1975a,b). In Fig. 1A, the threshold concentrations for odorants to stimulate various organisms (C_{th}) are compared with human olfactory thresholds (T) (Koyama and Kurihara, 1972a). A plot of log T versus log C_{th} gives a straight line with all of the organisms examined.

In general, odorants are lipid-soluble, and hence odorants applied to cells seem to be adsorbed on the hydrophobic region of the cell membranes. As a model for biological membranes, lipid monolayers were used and the interaction of odorants with the monolayers was examined by measuring surface pressure increases in the monolayers (Koyama and Kurihara, 1972a). A plot of log T versus log $C_{1.0}$ ($C_{1.0}$ is the concentration of odorant in the subphase giving a surface pressure increase of 1.0 dyn/cm) also gives a straight line as shown in Fig. 1A. A linear relationship observed with various organisms can be understood provided that the responses of the surface

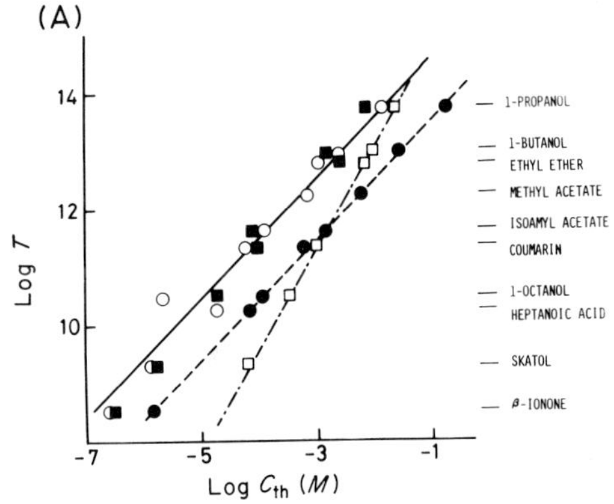

Fig. 1A. Linear relationship between log T and log C_{th}, where T and C_{th} stand for the olfactory threshold in human and the threshold of the responses to odorants in the frog gustatory nerve (■), *Nitella* internodal cell (●), the slime mold, *Physarum*, (□), and lipid monolayer (○), respectively. For the lipid monolayer, the abscissa value is actually log $C_{1.0}$ as defined in the text. The values of T are represented as the number of molecules of odorant per milliliter of air. (Taken from Kurihara *et al.*, 1978, with permission.)

membranes in these organisms to odorants are induced by their adsorption on the hydrophobic region of the membranes.

B. Reception of "Bitter Substances"

A variety of chemicals induce a bitter taste in humans. All of the bitter substances that we tested induced responses in the frog gustatory nerve Ataka *et al.*, 1978). The bitter substances acted not only on the gustatory receptor membranes but also on other membrane systems. *Physarum* and *Tetrahymena* were repelled by all of the "bitter substances" examined. In addition, the bitter substances induced depolarization in *Nitella* internodal cells (Ataka *et al.*, 1978). Figure 1B represents the threshold concentrations eliciting the responses in various organisms. The sequences of the thresholds of the bitter substances are similar among various organisms. As is similar to odorants, bitter substances showed strong affinities to the lipid monolayers (Koyama and Kurihara, 1972b). Hence bitter substances also seem to induce the responses in various organisms by adsorption on the hydrophobic region of the membranes. Moreover, Brand *et al.* (1976) found a similar amount of quinine binding to isolated bovine taste bud cells and to control tongue epithelial cells. Thus biological membranes recognize both odorants and bitter substances as hydrophobic stimuli. In this connection, it is interesting to note that bitter substances elicited responses in the frog EOG and in the olfactory bulb when aqueous solutions of the substances were applied to the olfactory epithelium (Getchell, 1969; Takagi *et al.*, 1978). Some bitter substances have a positive or negative charge. The reason why a relationship

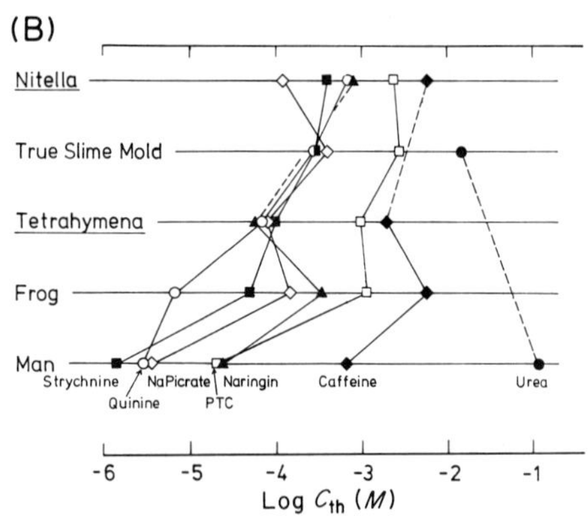

Fig. 1B. Threshold concentrations of various bitter substances to induce responses in man, frog (neural activity), slime mold (chemotaxis), *Tetrahymena* (chemotaxis), and *Nitella* (membrane potential). (Taken from Ataka *et al.*, 1978, with permission.)

between the thresholds of bitter substances in various organisms is not as simple as that with odorants seems to stem from the variation in electrostatic interaction of the substances with various membranes, which will be discussed below.

III. REACTION SCHEME FOR INITIAL PROCESS OF CHEMORECEPTION

A. Dynamic and Steady Responses

The initial event of chemoreception is adsorption of chemical stimuli on the receptor membrane (Beidler, 1954) (Chapter 10). This appears to be true for chemoreception in lower organisms as well as higher vertebrates. A mathematical treatment of the initial process of chemoreception has been carried out for taste reception in higher vertebrates. In 1954, Beidler proposed the following reaction scheme for the initial process of taste reception.

$$S + Re \rightleftharpoons SRe \tag{1}$$

Here S and *Re* are stimulus chemicals and receptor sites unbound, respectively. It is assumed that the taste response is proportional to the number of receptor sites occupied (*SRe*).

The neural responses or receptor potentials of sensory systems often consist of an initial, rapidly changing (dynamic) component and a subsequent prolonged, stable (steady) component (Ottoson, 1973). The gustatory neural response in the frog typically shows such a dynamic time course (Fig. 2). This dynamic time course cannot be explained simply by Eq. 1, since it predicts that the response increases monotonically with time to reach a steady level. A similar dynamic time course was observed with the drug action on guinea pig ileum (Paton, 1961). In order to explain the dynamic time course, Paton (1961) proposed a rate theory in which the response is postulated to be proportional to the rate of stimulus adsorption to the receptor sites. Although the rate theory could explain the dynamic time course, the assumption that the response is proportional to the rate of adsorption seemed to have no reasonable basis. In order to explain the dynamic time course based on occupation theory, Kamo *et al.* (1980) proposed the following model for chemical stimulation.

$$S + Re \underset{k_{-1}}{\overset{k_1}{\rightleftharpoons}} (SRe)_{\text{active}} \underset{k_{-2}}{\overset{k_2}{\rightleftharpoons}} (SRe)_{\text{inactive}} \tag{2}$$

Here $(SRe)_{\text{active}}$ and $(SRe)_{\text{inactive}}$ are the receptor domains with bound stimulus molecules or ions in the active and the inactive conformations, respectively. The response is assumed to be proportional to $(SRe)_{\text{active}}$. Differential equations of the law of mass action were set up in accordance with Eq. 2 to obtain the following equation.

$$p = \alpha_1 e^{-w_1 t} + \alpha_2 e^{-w_2 t} + \frac{C}{k_{-1}/k_1 + (1 + k_2/k_{-2})\,C} \tag{3}$$

where

$$w(w_1 < w_2) = \frac{k_1C + k_{-1} + k_2 + k_{-2} \pm \sqrt{(k_1C + k_{-1} - k_2 - k^2_{-2}) + 4\,k_{-1}\,k_2}}{2}$$

$$\alpha_1 = \frac{(w_1 - k_{-2})k_1\,C}{w_1(w_2 - w_1)} \quad \text{and} \quad \alpha_2 = \frac{(k_{-2} - w_2)k_1C}{w_2(w_2 - w_1)}$$

Here, p is the ratio of the number of active receptor domains occupied by S to the total number of receptor sites, and C is the concentration of stimulus molecules or ions.

Kashiwagura *et al.* (1980) compared the time courses of the frog gustatory responses obtained under varying conditions with the theoretical curves derived from Eq. 3. Figure 2A illustrates typical responses of the frog to 1 mM $CaCl_2$ under varying conditions. The response under the control condition (18°C, pH 5.7; flow rate, 1.7 ml/sec) increases rapidly after onset of stimulation and declines gradually to a steady level. Hereafter the initial large portion and the steady portion are referred to as the dynamic and steady components, respectively. The dynamic component becomes very small at low temperature, in the presence of procaine or at pH 6.7. The solid curves in Fig. 2A represent the corresponding theoretical curves obtained from Eq. 3. The theoretical curves fit well with the respective experimental data. Comparison of the rate constant used indicated that the value of k_1 is decreased at low temperature, in the presence of procaine and at pH 6.7, which implied that the binding of S to *Re* did not easily lead to a conformational change into $(SRe)_{\text{active}}$ under these conditions.

The dynamic properties of the response greatly depend on the flow rate of the stimulating solution. When the flow rate is small, the dynamic response is greatly reduced, whereas the steady response is unchanged (see Fig. 2B). The effect of the flow rate is also simulated by the theoretical curves obtained from Eq. 2 (Fig. 2C).

B. Concentration–Response Relationship

Based on psychological and physiological data on sensory intensities, Weber and Fechner (Eq. 4) (Fechner, 1860) and Stevens (1971) (Eq. 5) pro-

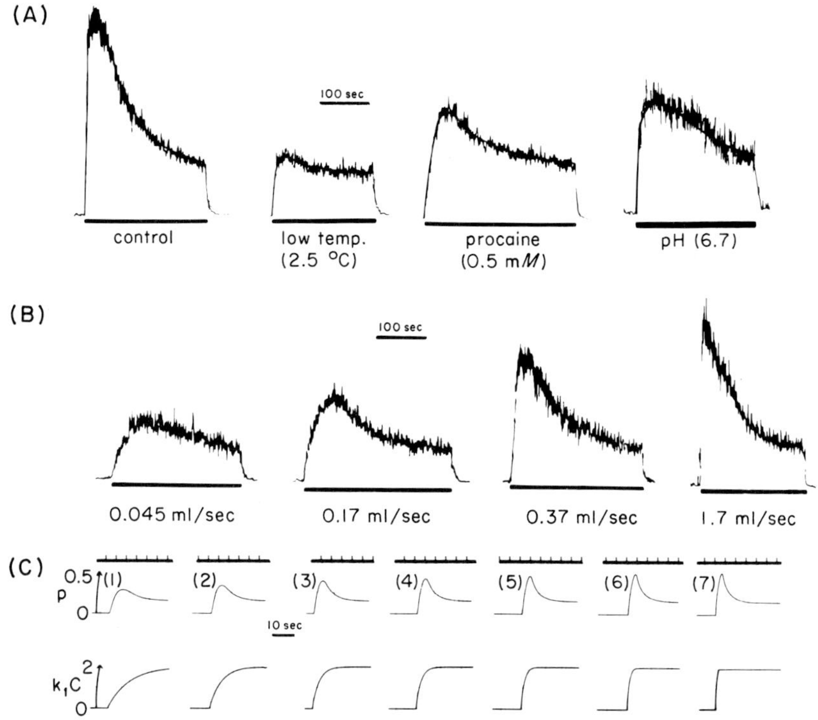

Fig. 2. (A) Responses of the frog gustatory nerve to 1 mM $CaCl_2$ under various conditions and theoretical curves (solid curves) calculated from Eq. 3 (taken from Kashiwagura *et al.*, 1980, with permission). (B) Effect of flow rate on the gustatory nerve responses to 1 mM $CaCl_2$ (taken from Kashiwagura *et al.*, 1980). (C) Analog computer simulation for the time course of response (upper graph) when the stimulus is applied at different flow rates (lower graph). The flow rate is increased from curve (1) to (7) (taken from Kamo *et al.*, 1980).

posed the following laws for the relationship between the magnitude of a perceived sensation (Q) and that of the stimulus intensity (I).

$$Q = A \log I \tag{4}$$

$$\log Q = B \log I + \log D \tag{5}$$

where A, B, and D are constants.

As described above, Beidler proposed Eq. 1, which explained the concentration–response relationship in many gustatory systems. In Fig.3A, the hyperbolic function of Eq. 1 is plotted with the magnitude of response (Q) as a function of the logarithm of stimulus concentration ($\log C$) (Beidler, 1971). A linear relationship between R and $\log C$ holds only within two log

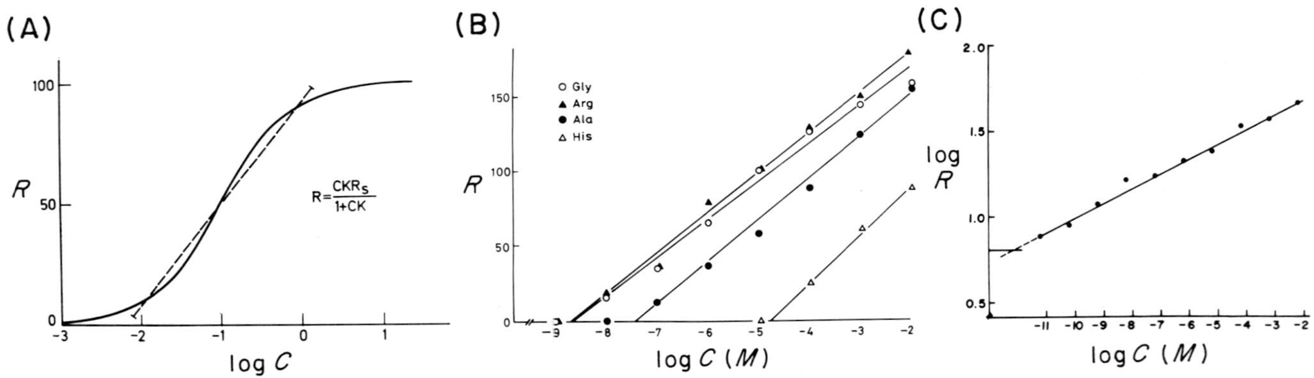

Fig. 3. (A) A semi-log plot of hyperbolic function derived from Eq. 1 (Beidler, 1971). (B) Relationship between the gustatory nerve response (R) of the eel and log concentration (log C) of amino acid stimuli; ○, Gly; ▲, Arg; ●, Ala; △, His (Yoshii *et al.*, 1979). (C) Relationship between log response (log R) of the catfish gustatory nerve and log concentration (log C) of L-alanine (taken from Caprio, 1975).

units of stimulus concentration. Weber and Fechner's law, therefore, holds only within two log units when Eq. 1 holds for nerve responses.

Recently it was shown that the gustatory receptors of the eel and of the catfish electrophysiologically respond extremely well to amino acids. Fig. 3B shows the concentration–response relationship for the eel to amino acids, where the response (R) is plotted against logarithmic concentration of amino acids (Yoshii *et al.*, 1979). In the cases of glycine and L-arginine, a linear relationship between R and log C holds over seven log units. That is, Weber and Fechner's law holds over seven log units. In Fig.3C, logarithmic response (log R) of the catfish is plotted against logarithmic concentration (log C) of L-alanine (Caprio, 1975). The figure shows that Stevens' law holds over nine log units of stimulus concentration. Thus the function of the gustatory receptors in the eel and in the catfish have a great advantage in recognizing concentration gradients of chemical stimuli over a wide concentration range. In fact, it was pointed out (Yoshii *et al.*, 1979; Caprio, 1975) that the gustatory receptors of these animals serve for sensing foods at a distance. Other chemoreceptor systems with high sensitivities also seem to have a similar function to recognize chemical stimuli over a wide concentration range. The molecular mechanisms by which stimuli are recognized over a wide concentration range is unknown. Binding of a chemical stimulus to a receptor site may decrease the affinity of the surrounding sites to the stimulus due to a negative cooperativity between receptor sites (Chapter 12).

IV. STRUCTURAL CHANGES OF RECEPTOR MEMBRANES

A. Effect of Changes in Membrane Lipid Composition on Receptor Function

The lipids in receptor membranes are considered to play an important role in chemoreception. To examine directly the role of the membrane lipids in chemoreception, *Tetrahymena* is a suitable organism since the composition of membrane lipids can be easily changed by the exogeneous supplementation of lipids or lipid analogues. Nozawa *et al.* (1975) replaced tetrahymanol, which is localized principally in the surface membrane of *Tetrahymena*, with ergosterol by exogeneous supplementation. Tanabe *et al.* (1979) examined chemotaxis of the ergosterol-replaced cells with inorganic salts and hydrophobic compounds such as odorants or bitter substances.

Fig. 4A shows the chemotactic responses of the native and the ergosterol-replaced cells to various inorganic salts. The thresholds for inorganic salts in the replaced cells are lower than in the native cells by a factor of 1/20–1/35;

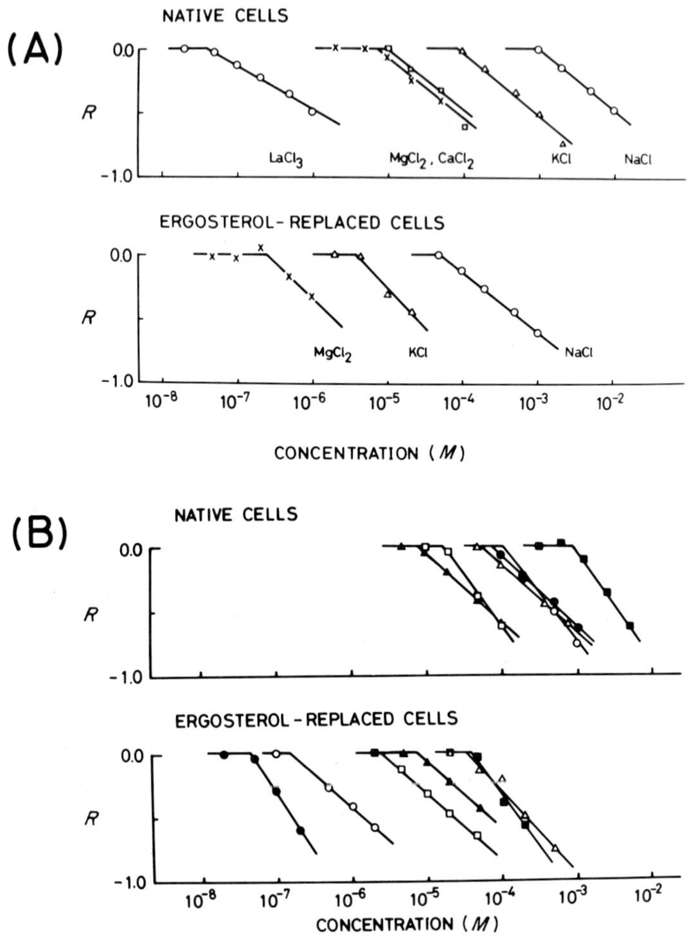

Fig. 4A and B. The magnitude of chemotactic responses (R) of native and ergosterol replaced *Tetrahymena* to inorganic salts (A) and hydrophobic compounds (B) as a function of concentration of the chemical stimuli. A negative sign of R represents negative chemotaxis. For (B): ○, strychnine; ●, nicotine; □, skatole; ■, PTC; △, picrate; ▲, m-NBS (taken from Tanabe *et al.*, 1979).

the replaced cells respond to the salts 20–35 times more sensitively than to the native cells.

Figure 4B shows the chemotactic responses of the native and the replaced cells to hydrophobic compounds. The responses to the positively charged compounds (strychnine and nicotine) are sensitized about 1000-fold. Those to the neutral compounds (phenylthiocarbamide and skatole) are sensitized about tenfold. Those to the negatively charged compounds (sodium

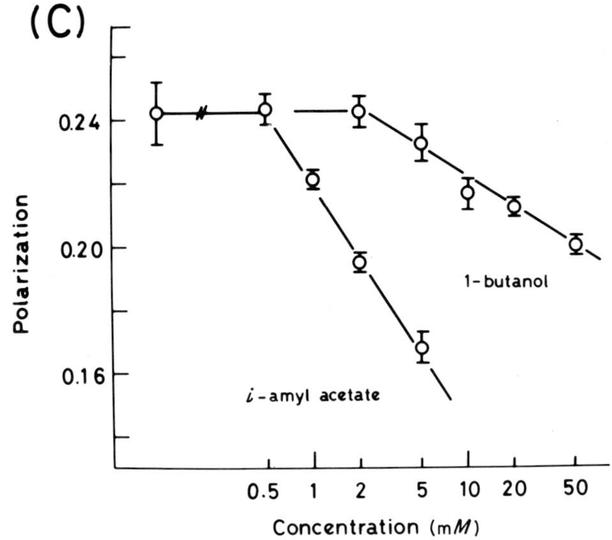

Fig. 4C. The degree of fluorescence polarization of DPH added to a *Tetrahymena* suspension as a function of concentration of the added chemical stimuli shown (Tanabe *et al.*, 1980).

m-nitrobenzenesulfonate and sodium picrate) are not or only slightly sensitized.

The sensitizing effect of ergosterol replacement was also observed in our laboratory with interactions between liposomes and stimulus chemicals; liposomes made from the lipids of the replaced cells had higher affinities for the chemicals than those of the native cells, although the degree of increases in the affinities were not as dramatic as seen in intact *Tetrahymena*.

B. Changes in Membrane Fluidity in Response to Chemical Stimuli

It would be interesting to know how the membrane fluidity is changed with reception of chemical stimuli. Tanabe *et al.* (1980) measured changes in the fluidity of *Tetrahymena* surface membranes in response to chemical stimuli with use of DPH (1,6-diphenyl-1,3,5-hexatriene).

Figure 4C illustrates the degree of fluorescence polarization of DPH as a function of concentration of isoamyl acetate and 1-butanol. The polarization value decreases linearly with an increase in concentration of these chemical stimuli. Concentrations of the chemicals at which the fluorescence polarization starts to decrease are practically identical to the respective thresholds for chemotaxis. All of the hydrophobic compounds examined, which are "odorants" for humans, decreased the fluorescence polarization. On the other hand, inorganic salts such as NaCl, KCl, and $CaCl_2$ did not bring about any change in fluorescence polarization, although these inorganic salts induced changes in the membrane potential and in chemotactic responses in *Tetrahymena*. The above results indicate that the membrane fluidity is decreased associated with reception of odorants, whereas reception of inorganic salts does not affect the fluidity. Probably odorants are adsorbed on the hydrophobic region of the lipid layer of *Tetrahymena* surface membrane and increase the fluidity. These structural changes seem to play an essential role in generation of the receptor potential induced by the neutral compounds. On the other hand, inorganic salts interact with the hydrophilic region of the membrane and hence do not affect the structure of the hydrophobic region of the lipid layer.

C. Role of Membrane-Bound Ca^{2+}

1. *True Slime Mold*

The true slime mold *Physarum* exhibits negative chemotaxis for various salts of monovalent cations (Ueda *et al.*, 1975c). Associated with chemotaxis, the membrane potential changes in response to these salts. The threshold concentrations (C_{th}) of these salts are greatly influenced by Ca^{2+} in the

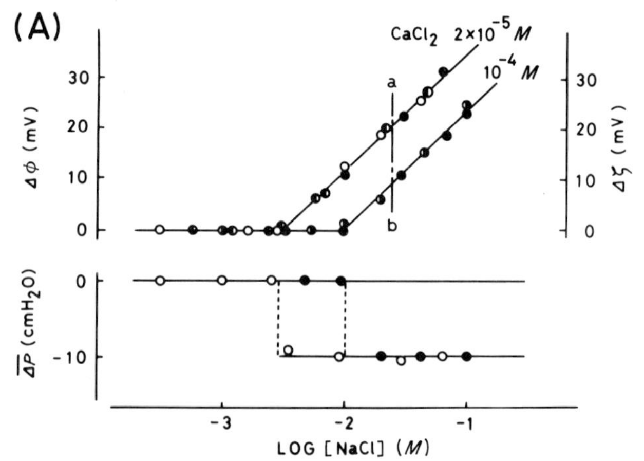

Fig. 5A. Changes in membrane potential ($\Delta\phi$), zeta potential ($\Delta\zeta$), and chemotactic motive force ($\overline{\Delta P}$) of *Physarum* as a function of NaCl concentration in the presence and absence of $CaCl_2$. ○, $\Delta\phi$, $\overline{\Delta P}$ in the absence of $CaCl_2$ in medium; ⊕, $\Delta\phi$, ◐ $\Delta\zeta$ in the presence of 2×10^{-5} *M* $CaCl_2$; ●, $\Delta\phi$, $\overline{\Delta P}$, ◑, $\Delta\zeta$ in the presence of 10^{-4} *M* $CaCl_2$ (Terayama *et al.*, 1977).

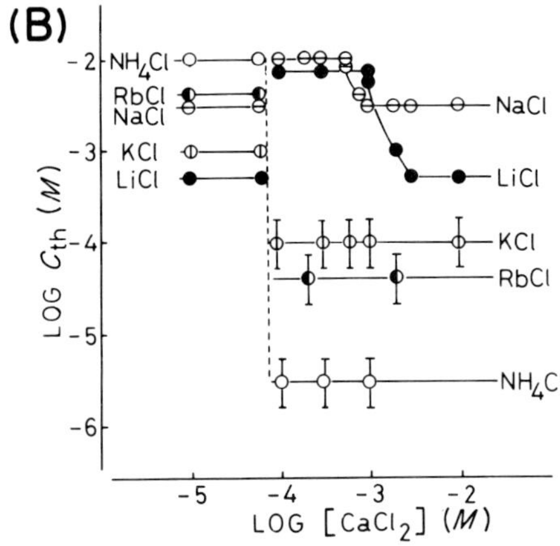

Fig. 5B. The threshold concentrations (C_{th}) of *Physarum* for monovalent cations as a function of $CaCl_2$ concentration added in the external solution (Terayama *et al.*, 1977).

external medium (Terayama *et al.*, 1977). As shown in Fig. 5A, the presence of 0.1 mM $CaCl_2$ increased the threshold for the membrane potential change in response to NaCl. The threshold for chemotaxis was also shifted in accordance with changes in C_{th}. In Fig. 5B, C_{th} for various 1:1 type salts is plotted as a function of $CaCl_2$ concentration added to the medium. The sequence of thresholds of monovalent cations in the presence of $CaCl_2$ becomes NH_4 > Rb > K > Li > Na, whereas that in the absence of $CaCl_2$ is Li > K > Na > Rb > NH_4. Mg^{2+}, La^{3+}, and Th^{4+} showed an effect similar to Ca^{2+} on the thresholds of monovalent cations.

Fig. 5A also shows that the threshold for the zeta potential change is increased by addition of 0.1 mM $CaCl_2$. The value of the zeta potential of microplasmodia of the slime mold at a given ionic strength is shifted to the negative direction by addition of 0.1 mM $CaCl_2$ to the medium (see broken line a–b in the figure), which implies that the negative surface charge density increases by Ca^{2+} reception. Probably the binding of Ca^{2+} causes a conformational change of the surface membrane so that the negative charge increases in the surface.

2. *Frog Gustatory Membrane*

The divalent cations in the gustatory receptor are so tightly bound to the receptor membrane that decreasing the concentration of divalent cations in the external medium does not lead to removal of the cations from the receptor membranes. However, treatment of the frog tongue with an alkaline solution (Kamo *et al.*, 1978) or 1-anilinonaphthalene-8-sulfonate (ANS) solution (Kashiwagura *et al.*, 1977a) led to removal of Ca^{2+} from the gustatory receptor membranes.

The frog gustatory responses to various salt stimuli and distilled water were greatly enhanced after the tongue had been treated with an alkaline solution (pH 10) for a short time; the responses to galactose and amino acids were enhanced by a longer treatment. On the other hand, the treatment did not affect the response to quinine. Figure 6A represents the responses to salts of various monovalent cations as a function of their concentrations be-

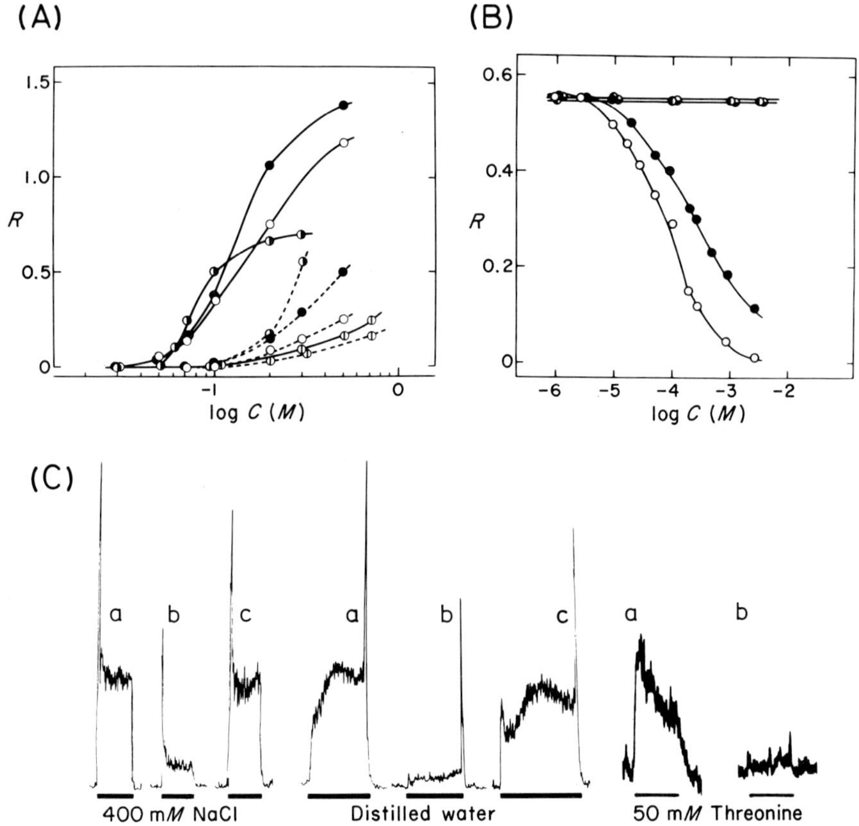

Fig. 6. Role of membrane-bound Ca^{2+} on the frog gustatory nerve responses. (A) Relative magnitude of responses (R) to various salts as a function of concentration. The dashed and continuous lines represent responses before and after Ca^{2+} removal, respectively. ○, NaCl; ●, KCl; ◑, NH_4Cl; ⦶, LiCl (taken from Kamo *et al.*, 1978). (B) Effect of divalent cations on the enhanced response. After the ANS treatment, 100 mM NaCl solutions containing divalent cations of various concentrations were applied to the tongue. ○, $CaCl_2$; ●, $SrCl_2$; ◒, $MgCl_2$; ◐, $BaCl_2$ (Kashiwagura *et al.*, 1977a). (C) The integrated responses to various chemical stimuli before (a) and after (b) the tongue was incubated in Ringer solution of pH 5.3 for 15 min. The records (c) represent the responses after the tongue was incubated in the acidic Ringer solution and was then incubated in Ringer solution of pH 7.0 for 1 hr (Kamo *et al.*, 1978; Yoshii *et al.*, 1981, in press).

fore and after the alkali treatment. The degree of enhancement varies among the species of cations. The magnitude of the enhanced response to 200 m*M* NaCl, for example, is about 9 times that before the treatment, whereas the response to LiCl remains small after the treatment.

The enhanced responses after removal of Ca^{2+} are suppressed by addition of $CaCl_2$ or $SrCl_2$ to the stimulating solution; addition of $CaCl_2$ or $SrCl_2$ to 100 m*M* NaCl restores the enhanced response to the original level before the treatment (Fig. 6B). On the other hand, $MgCl_2$ and $BaCl_2$ exhibit practically no effect on the enhanced responses.

The responses to salt stimuli, distilled water, sugars, and amino acids were greatly reduced after the tongue had been incubated in Ringer solution containing Ca^{2+} at pH 5.3. Experiments using ^{45}Ca indicated that this treatment led to binding of extra Ca^{2+} to the tongue. The records *a* and *b* in Fig. 6C represent the responses before and after incubation in the acidic Ringer, respectively. The responses are greatly reduced by the treatment, and are recovered under appropriate conditions. These results imply that the binding of extra Ca^{2+} to the receptor membrane leads to reduction of the responses.

Much evidence has accumulated that Ca^{2+} stabilizes the structure of the biological membranes. Hence the binding of Ca^{2+} to the gustatory receptor membranes seems to stabilize the receptor domains so that a conformational change of the domains is hardly induced by adsorption or desorption of chemicals and the removal of Ca^{2+} destabilizes the domains. Here it is emphasized that a conformational change of the receptor domain plays an important role in the transduction process of the gustatory response.

V. MEMBRANE POTENTIAL CHANGES IN RESPONSE TO CHEMICAL STIMULI

A. Membrane Potential Changes in Lower Organisms

As shown in Fig.4, *Tetrahymena* exhibits negative chemotaxis with various chemical stimuli. Aiuchi *et al.* (1980) monitored the changes in the membrane potential of *Tetrahymena* associated with chemotaxis using fluorescence changes of rhodamine 6G. Figure 7A shows changes in the fluorescence intensity of the dye–*Tetrahymena* suspension as a function of the concentration of chemical stimuli. The fluorescence starts to change at respective threshold concentrations for chemotaxis and increases with increasing stimulus concentration, which implies that *Tetrahymena* is depolarized by chemical stimuli.

As shown in Fig.4A the effectiveness of inorganic salts to induce the

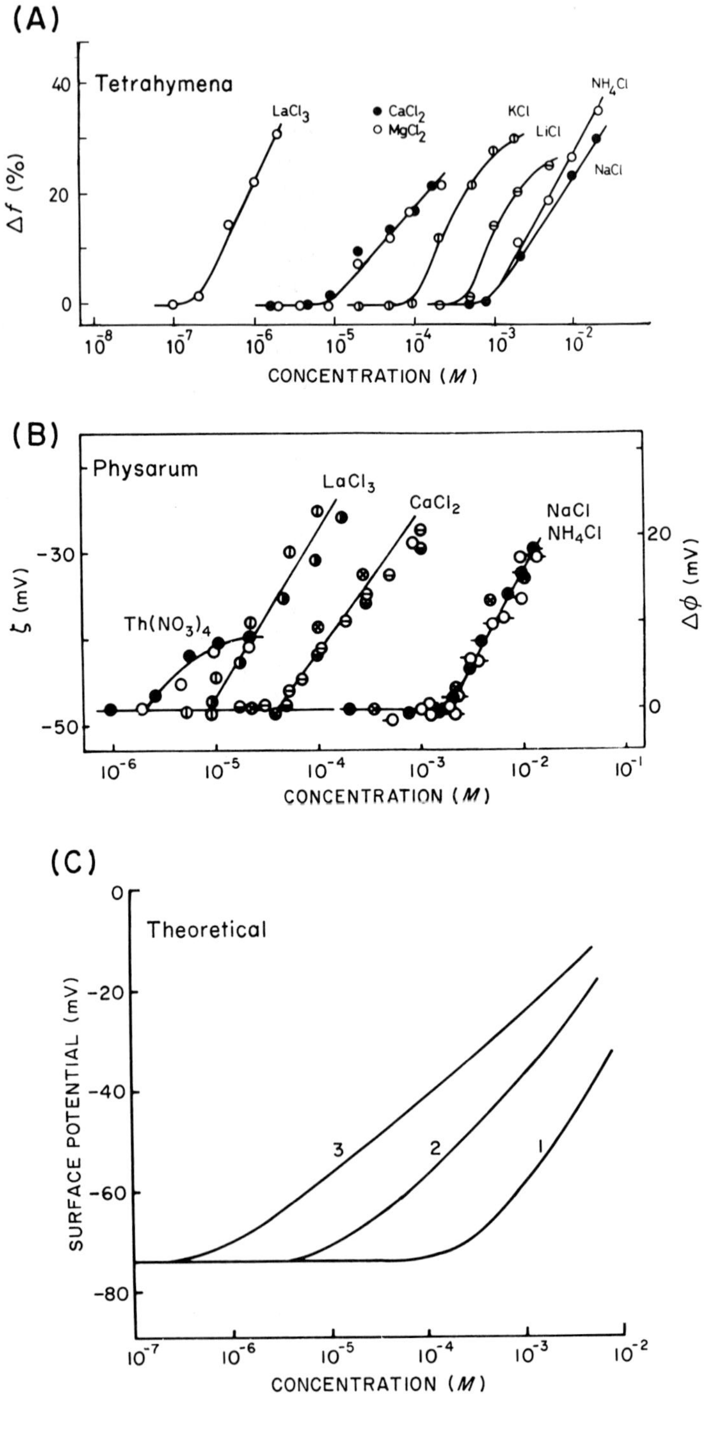

(A)
Tetrahymena
Δf (%)
40
20
0
LaCl3
● CaCl2
○ MgCl2
KCl
LiCl
NH4Cl
NaCl
10^-8
10^-7
10^-6
10^-5
10^-4
10^-3
10^-2
CONCENTRATION (M)
(B)
Physarum
ζ (mV)
-30
-50
Δφ (mV)
20
0
Th(NO3)4
LaCl3
CaCl2
NaCl
NH4Cl
10^-6
10^-5
10^-4
10^-3
10^-2
10^-1
CONCENTRATION (M)
(C)
Theoretical
SURFACE POTENTIAL (mV)
0
-20
-40
-60
-80
3
2
1
10^-7
10^-6
10^-5
10^-4
10^-3
10^-2
CONCENTRATION (M)

membrane potential changes in *Tetrahymena* depended on the valence of the cation. The order of effectiveness was trivalent > divalent > monovalent. Similar results were observed with *Physarum*. The changes in membrane potential of the slime mold were measured directly with a microelectrode or by the double-chamber method (Kamiya, 1942). Figure 7B shows changes in membrane potential of the slime mold as a function of concentration of inorganic salts. The membrane potential changes are closely associated with the chemotactic response. The order of effectiveness of the cations was the same as that observed with *Tetrahymena*.

In general, the permeabilities of membranes to cations decreases with an increase of their valence if the membranes have no specific channels for particular ions. Therefore the above results of the effectiveness of cations cannot be explained simply by assuming that the membrane potential stems only from the diffusion of ions across the membrane. According to Teorell (1935) and Meyer and Sievers (1936), the membrane potential (ΔE) is the algebraic sum of two surface potentials (or phase boundary potentials) at both sides of the membrane (ΔE_i and ΔE_0) and a diffusion potential within the membrane (ΔE_d); $\Delta E = \Delta E_i + \Delta E_d + \Delta E_0$. In living organisms, ΔE_i (surface potential inside cells) seems to be constant during chemical stimulation. If ΔE_d is not significantly changed in response to chemical stimulation, ΔE_0 (surface potential outside cells) will contribute significantly to changes in the total membrane potential. This was true with *Physarum* (see below). The changes in ΔE_0 of the slime mold were estimated from measurements of the zeta potential (an experimental approximation for ΔE_0) of microplasmodia in the presence of various salts. As seen in Fig. 7B, the changes in the zeta potential are approximately equal to the changes in the membrane potential (Hato *et al.*, 1976). Similar results were obtained when D-glucose, cAMP, and ATP (attractants for the slime mold) were used as chemical stimuli. Therefore it was concluded that changes in the membrane potential of the true slime mold in response to chemical stimuli stem mainly from changes in the surface potential outside the cells.

Aiuchi *et al.* (1977) showed that fluorescence changes of ANS applied to the membrane systems reflect the surface potential changes. H. Tanabe *et*

Fig. 7. Effect of various inorganic salts on the membrane potential and the surface potential (A) Depolarization of *Tetrahymena* as a function of salt concentration monitored by fluorescence of rhodamine 6G (Aiuchi *et al.*, 1980). (B) Changes in the membrane potential, $\Delta\phi$ (also ΔE) (●, ◑, ◒, ●, ●) and zeta potential, ζ (○, ⦶, ⊖, ○, ○) of *Physarum* as a function of salt concentration (Hato *et al.*, 1976). (C) Theoretical changes in the surface potential of a membrane with addition of various mono-(1), di-(2), and trivalent (3) cations. The surface potential was calculated according to Gouy-Chapman equation (Verwey and Overbeek, 1948) using the assumption that the surface charge density of the membrane is 0.75 $\mu C/cm^2$. The various cations were added to a control solution containing 1 mM Tris-HCl, pH 7.0.

al. (unpublished) monitored the surface potential changes of *Tetrahymena* in response to chemical stimuli with ANS fluorescence and showed that the surface potential changed in response to various inorganic salts except for KCl.

As shown above, the effectiveness of salts on the membrane potential changes in *Physarum* and *Tetrahymena* depended greatly on the valence of the cation. The order of this effectiveness can be explained in terms of the surface potential. Figure 7C illustrates the surface potential of negatively charged membranes on addition of various monovalent, divalent, and trivalent cations calculated with use of the Gouy-Chapman theory (Verwey and Overbeek, 1948). The order of effectiveness is trivalent > divalent > monovalent, which is the same as the order observed with *Physarum* and *Tetrahymena.* Thus the changes in the membrane potential of these organisms in response to inorganic salts were interpreted in terms of the surface potential but it should be noted that the intermembrane diffusion potential of particular ions, such as K^+ in *Tetrahymena*, also contributes to the total membrane potential.

B. Taste Receptor Potentials

Application of chemical stimuli to taste cells changes their membrane potential in the direction of depolarization, which is referred to as the taste receptor potential. Although the mechanism of nerve excitation is well explained by the Na theory (Hodgkin, 1964), that of the taste receptor potential is not explained simply by the theory. For example, the receptor membranes of taste cells are nearly impermeable to taste stimuli. Not only K^+ but also many other ions such as Na^+, Li^+, NH_4^+, Ca^{2+}, or Mg^{2+} depolarize taste cells. The membrane resistance of taste cells is decreased in response to high concentrations of salts of monovalent cations but is only slightly changed or unchanged during depolarization induced by other stimuli (Akaike *et al.*, 1976).

1. Responses to Salts and Acids

As noted above, the receptor membranes of taste cells are hardly permeable to taste stimuli. Kamo *et al.* (1974a,b) prepared a model membrane, which is impermeable to ions, by impregnating a Millipore filter with a large amount of lipid and measured the membrane potential in response to added chemical stimuli. This model membrane simulated the taste receptor potential in response to salts, acids, and distilled water. This suggested that the surface potential significantly contributed to the receptor potential, since the intermembrane diffusion potential in the model membrane could be ignored.

Whether or not the surface potential actually contributes to changes in the membrane potential of the taste cell may be determined by measuring the zeta potential of the cells, but it is not possible experimentally. For this purpose the neuroblastoma cell, which is excitable, is an excellent model since the electrophoretic technique is applicable. Miyake (unpublished) measured changes in the membrane potential and changes in the zeta potential of the neuroblastoma cell in response to KCl, $CaCl_2$, and HCl. As shown in Fig. 8, the neuroblastoma cell is depolarized by these chemicals. The zeta potential is unchanged with increasing KCl concentrations. This result seems to be reasonable because the membrane potential change in response to KCl stems from the intramembrane diffusion potential of K^+. On the other hand, application of $CaCl_2$ or HCl decreases the zeta potential closely associated with the membrane potential change. These results suggest that the changes in membrane potential of the neuroblastoma cell in response to $CaCl_2$ or HCl stem mainly from those of the surface potential. A similar mechanism seems to be applicable to the taste receptor potential in response to acids and salts of polyvalent cations. It is noted that the taste receptor

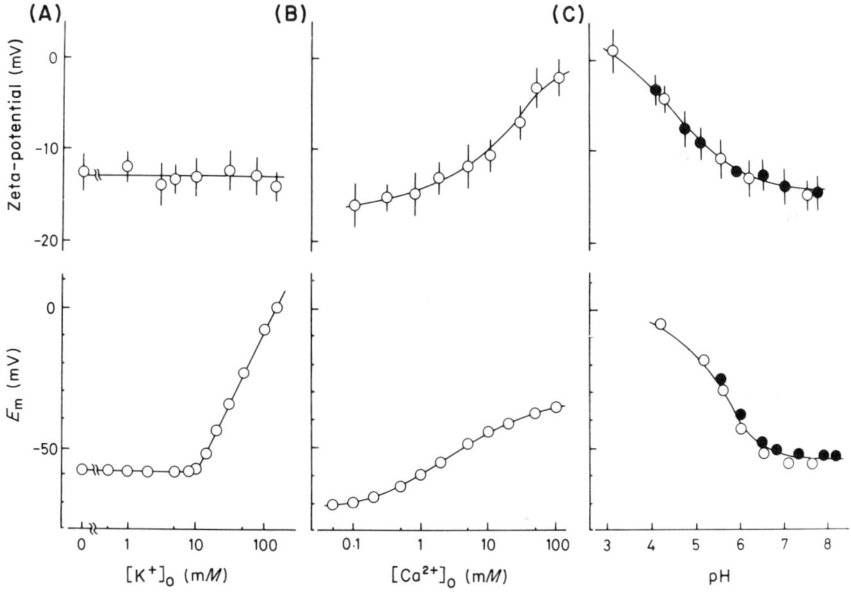

Fig. 8. The membrane potential (E_m) and the zeta potential of neuroblastoma cells as a function of concentration of KCl, $CaCl_2$, and HCl (M. Miyake, unpublished).

potentials in response to HCl or $CaCl_2$ are produced with only a slight change in the membrane resistance (Akaike *et al.*, 1976).

The taste responses to salts are largely influenced by species of anions as well as cations. Aiuchi *et al.* (1976) interpreted the anion influence in terms of the surface potential.

2. *Water Response*

Application of distilled water to the gustatory organ adapted to saliva or Ringer solution leads to depolarization of taste cells and to an increase in gustatory nerve activities. This phenomenon is observed in a variety of animal species and has been referred to as the *water response*. The Millipore–lipid membrane described above simulated the water response well (Miyake *et al.*, 1976a). Figure 9A shows the water responses in the frog (the taste nerve response) and in the Millipore–lipid membrane (the membrane potential) after adaptation to 100 m*M* of various 1:1 type salts and Ringers solution. For both the frog and the model membrane, adaptation to

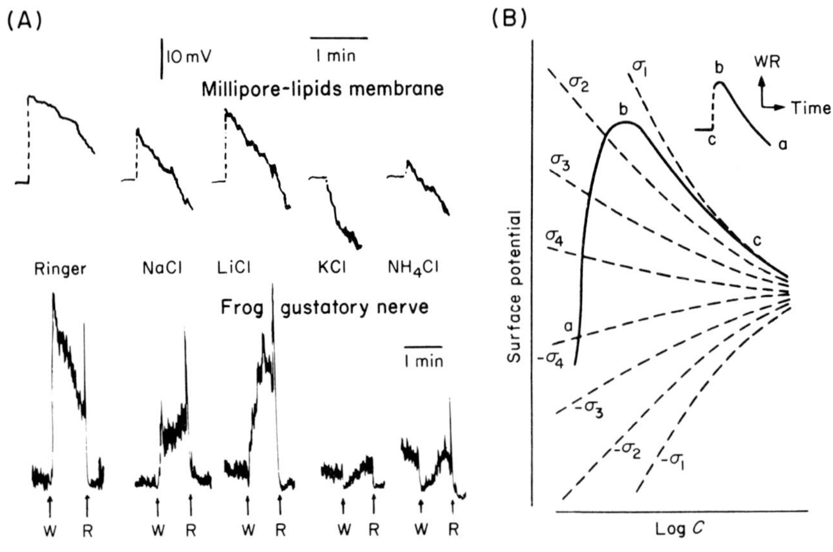

Fig. 9. Water responses in the model membrane and in the frog. (A) Upper records; the potential responses of Millipore–lipid membranes to distilled water after adaptation to Ringer and to various 1:1 type salts at 100 m*M*. Lower records: the integrated responses of the frog gustatory nerve to distilled water when the tongue was adapted to various salt solutions (Miyake *et al.*, 1976a). (B) Schematic diagram of the surface potential as a function of logarithmic concentration of salt. Dotted lines indicate the potential changes with constant charge density ($-\sigma_1 < -\sigma_2 < -\sigma_3 < -\sigma_4 < \sigma_4 < \sigma_3 < \sigma_2 < \sigma_1$). The insert is the schematic time course of the water response due to a change in the surface potential.

Ringer, NaCl, or LiCl solution led to a large water response, whereas that to KCl or NH_4Cl solution led to little or no response. The water response is suppressed by adding an electrolyte to the distilled water. The suppression curves of the water response in the model membrane as a function of concentration of added electrolytes was similar to that in the frog. Thus the model membrane simulated the water response in the animal. The mechanism of the water response is explained as follows. When distilled water is applied to the membrane adapted to a salt solution, the salts dissolved in the membrane surface will diffuse out into the bulk solution. This produces the diffusion potential in the bulk solution outside of the membrane phase and depolarizes the taste cells or the model membrane. It is noted that LiCl and NaCl produce a large diffusion potential, and KCl and NH_4Cl produce only a small diffusion potential.

There is another origin of the water response (Miyake *et al.*, 1976a; Kurihara *et al.*, 1978). When distilled water is applied to the model membrane and the cell membrane that have been adapted to salt solutions, the screening effect of ions on the double layer potential is eliminated. If cations in the adapting solution are strongly bound to the membranes, the surface potential is changed along c→b→a in Fig. 9B, which temporarily depolarizes the membranes. This type of water response was seen after taste cells of the frog were adapted to a solution containing polyvalent cations (Miyake *et al.*, 1976a) or the cells of the carp were adapted to a solution containing movovalent cations (Konishi, 1967).

As described above, the water response has two origins. The potential changes produced by both origins occur at the membrane–solution interface, which leads to depolarization of the cell. Hence, the water response is produced by a change in the electrical potential at the membrane–solution interface induced by elimination of salts from the membrane surface.

3. *Responses to Sugars and Amino Acids*

It is known that sugar response is suppressed by the presence of salts; Andersen *et al.* (1963) reported that the response of the dog chorda tympani to sucrose was suppressed by 0.2 *M* or 0.5 *M* NaCl. Ozeki and Sato (1972) stated that the receptor potential of the rat taste cell in response to sucrose was diminished by 0.04 *M* NaCl.

The responses of the frog to sugars were also suppressed by the presence of salts (Miyake *et al.*, 1976b). The effectiveness of the suppression of salts depended strongly on the valences of either the cation or the anion involved. Figure 10 shows magnitudes of response to 0.5 *M* D-galactose and 1 *M* D-fructose as a function of ionic strength of the medium. The data fall on a single curve for the respective sugars irrespective of diversity of salt species added. These results are explained as follows. The adsorption of sugars to the

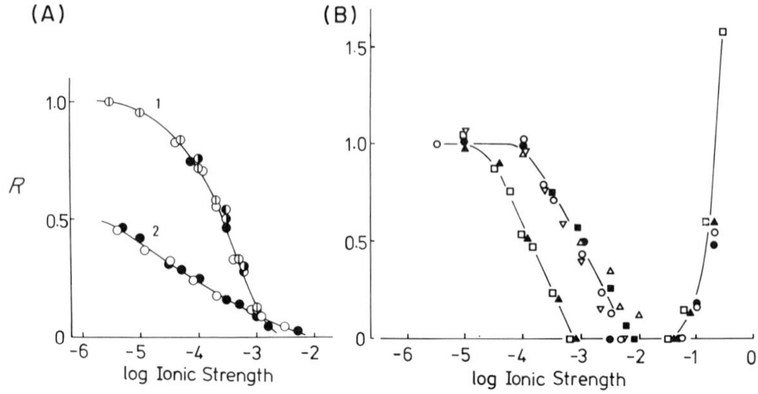

Fig. 10. Relative magnitude of the gustatory nerve response (R) in the frog to sugars and amino acids as a function of ionic strength (A) curve 1, 0.5 M galactose; curve 2, 1 M fructose: ⦶, NaCl; ●, KCl; ◐, $MgCl_2$; ○, $MgSO_4$; ◑, $K_4Fe(CN)_6$ (Miyake *et al.*, 1976b). (B) 50 mM L-threonine: ○, NaCl; ●, KCl; ▽ CH_3SO_3Na; ■ Na_2SO_4; △ $Na_4Fe(CN)_6$; □ $MgCl_2$; ▲ $MgSO_4$ (Yoshii *et al.*, 1981, in press).

receptor membrane leads to a conformational change of the receptor domain, and the surface charge of the domains becomes positive. The surface potential is a function both of the charge density of the membrane surface and of the ionic strength of the medium. An increase of ionic strength leads to a diminution of the surface potential in the domain of the sugar.

The response of the frog to amino acids was also suppressed by the presence of salts (Yoshii *et al.*, 1981, in press). Figure 10B shows the magnitude of the response to 50 mM L-threonine as a function of the ionic strength of the medium. The response decreases with increasing ionic strength. As is similar to the sugar response, the suppressive effect of the salts on the response to the amino acid is interpreted in terms of the surface potential. The fact that salts of divalent cations showed stronger effects may imply that the specific binding of divalent cations to the membrane surface contributes partly to the suppressive effect. The response to the amino acid reappears when a relatively high concentration of salt is present in the stimulating solution. One explanation for this phenomenon is that the presence of a relatively high concentration of salt decreases the membrane resistance, and the diffusion potential across the membrane contributes to the total membrane potential. However, the following results cannot be explained simply by this mechanism; the divalent cation, which is less permeable to the receptor membrane than is the monovalent cation, showed a larger effect on the generation of the response to the amino acid than the monovalent cation of equal concentration.

In physiology, considerable attention has been given to the permeablilty changes to specific ions, and the surface potential has been considered to be less important. This seems to come from the following two reasons. In nerve cells, depolarization is usually accompanied by a remarkable decrease in the membrane resistance, and hence the diffusion potential across the membrane significantly contributes to the total potential change. Salt concentrations of internal and external solutions separated by a nerve cell membrane are generally rather high, and the total ionic concentrations in two solutions are not very different from each other. Therefore, the surface potentials at both sides of the membrane are considered to cancel each other. On the other hand, depolarization of chemoreceptor systems in response to certain chemical stimuli often occurs without changes in the membrane resistance (Akaike *et al.*, 1976). In addition, ionic concentrations in the external solutions in contact with chemoreceptor systems vary over a wide range. Thus the surface potential as well as the diffusion potential significantly contribute to changes in the membrane potential of chemoreceptor systems.

VI. TRANSDUCTION MECHANISMS

A. Chemotactic Responses

In lower organisms, stimulus information received at the surface membrane is transduced to their motile systems and induces chemotactic responses. Mito *et al.* (1980) showed that the transduction mechanism for positive chemotaxis in *Physarum* is different from that for negative chemotaxis. The slime mold exhibits normal motility for a long time under anaerobic conditions as well as under aerobic conditions. Figure 11A represents variations of the chemotactic motive force as a function of concentration of an attractant (D-glucose) and a repellent (NaCl) under aerobic and anaerobic conditions. The slime mold exhibits positive chemotaxis toward glucose under aerobic conditions but does not exhibit any chemotaxis toward glucose under anaerobic conditions. On the other hand, the slime mold exhibits negative chemotaxis with NaCl under both aerobic and anaerobic conditions. Similar results were obtained when other attractants (D-galactose, D-mannose, KH_2PO_4) and a repellent (D-fructose) were used as chemical stimuli. Thus exposure of the slime mold to the anaerobic conditions led to a selective suppression of chemotaxis toward attractants without affecting negative chemotaxis against repellents. Treatment of the slime mold with rotenone, which is a respiratory inhibitor, also led to a selective suppression of positive chemotaxis. These results are interpreted as follows. The Ca-pump of mitochondria of the slime mold is suppressed by anaerobic condi-

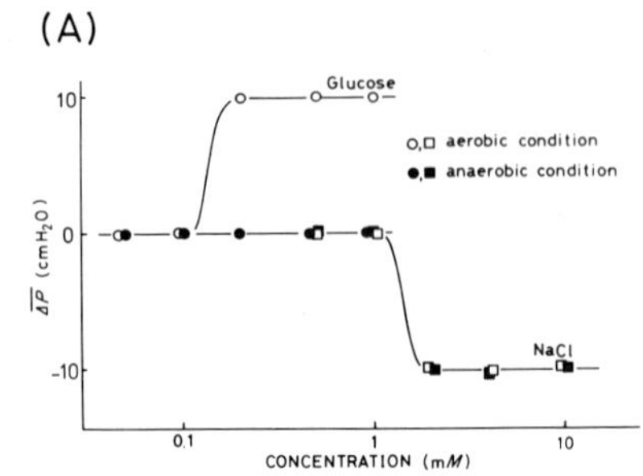

Fig. 11A. Chemotactic motive force ($\overline{\Delta P}$) of *Physarum* as a function of concentration of glucose (○, ●) and NaCl (□, ■) under aerobic and anaerobic conditions (Mito *et al.*, 1980).

tions or rotenone treatment, and hence the concentration of free Ca^{2+} in the cytoplasm is increased. This will lead to suppression of the positive chemotaxis.

The chemotactic responses of *Tetrahymena* to repellents depend on the $CaCl_2$ concentration in the external medium (H. Tanabe *et al.*, unpublished). Figure 11B represents negative chemotactic responses to phenylthiocarbamide as a function of $CaCl_2$ concentration in the external medium. *Tetrahymena* exhibits fully the chemotactic response above 1 μM, but becomes less responsive with decreasing $CaCl_2$ concentration, although the cells exhibit normal motility under this condition. As described above, inorganic salts, odorants, and bitter substances, which are repellents for *Tetrahymena,* depolarized the cells above the respective threshold concentrations for chemotaxis. Probably the depolarization opens the voltage-dependent Ca-channel of *Tetrahymena* and induces Ca influx from the external medium into the cell, which leads to generation of a chemotactic response.

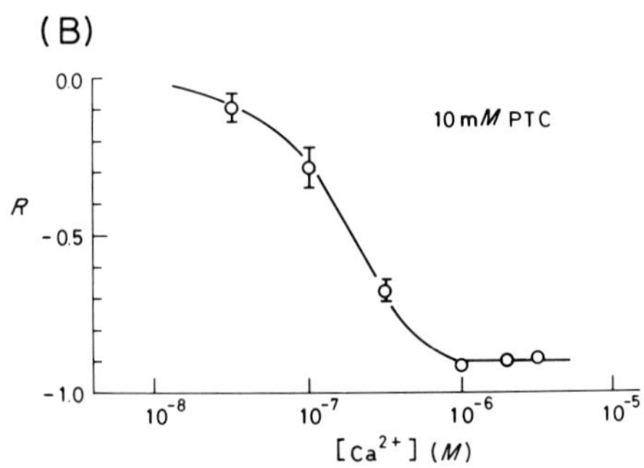

Fig. 11B. Magnitude of chemotactic responses (*R*) of *Tetrahymena* to 10 m*M* phenylthiocarbamide (PTC) as a function of Ca^{2+} concentration in the external medium (H. Tanabe *et al.*, unpublished).

B. Taste Nerve Responses

1. *Role of Ca^{2+} in Transduction Process*

K. Morimoto (unpublished) and Nagahama *et al.* (unpublished) perfused the frog lingual artery with Ringers solution and examined dependence of the gustatory responses on $CaCl_2$ concentration in the perfusing solution. Figure 12A shows gustatory responses to various chemical stimuli as a function of $CaCl_2$ concentration in the perfusing solution. The responses to chemical stimuli of group 1 ($CaCl_2$, NaCl, distilled water, D-galactose, and L-threonine) are greatly decreased at low $CaCl_2$ concentration and increased with increasing $CaCl_2$ concentration. The responses to group 2 (quinine, HCl, acetic acid, and ethanol) are practically independent of $CaCl_2$ concentration below 0.2 m*M*, and they decreased above 0.2 m*M*. The responses to group 1 were suppressed reversibly by addition of 0.1 m*M* Ca-channel blockers ($MnCl_2$ and verapamil) to the perfusing solution, but those to group 2 were practically unaffected by 0.1 m*M* of the blockers (Fig. 12B). The above results suggest that Ca-influx from intercellular medium into a taste cell is needed for generation of the gustatory nerve responses to group 1 and that the chemical stimuli of group 2 produce neural responses without accompanying Ca-influx into a taste cell. It is not known whether the response to group 2 is produced without accompanying a release of a chemical transmitter from a taste cell or whether chemical stimulation by group 2 leads to a release of Ca^{2+} into the cytosol of a taste cell from intracellular store.

2. *Spreading of Depolarization at Microvillus Membrane to Synaptic Area*

It is well known that electrical stimulation of the tongue elicits taste sensations (Bujas, 1971) but its mechanism is unknown. Since the main part of taste transduction is electrical phenomena, the means of electrical stimulation must be a useful tool for elucidating the transduction mechanisms.

K. Kashiwayanagi *et al.* (unpublished) stimulated the frog tongue with electrical current and recorded the responses from the glossopharyngeal nerve. Figure 13A compares the responses to chemical stimuli and those induced by anodal current under various conditions. ANS treatment of the tongue leads to a great enhancement of both the response to the chemical stimulus (NaCl) and that to anodal current. Similar enhancement was observed when 1 m*M* $NiCl_2$ was present in the stimulating solution (chemical stimulation) or in the adapting solution (electrical stimulation). The records in Fig. 13A also represent the effect of uranyl acetate of varying concentrations on the responses to chemical and electrical stimulations. Both responses to 0.4 *M* NaCl and to electrical stimulation in 10 m*M* NaCl solution increase in the presence of

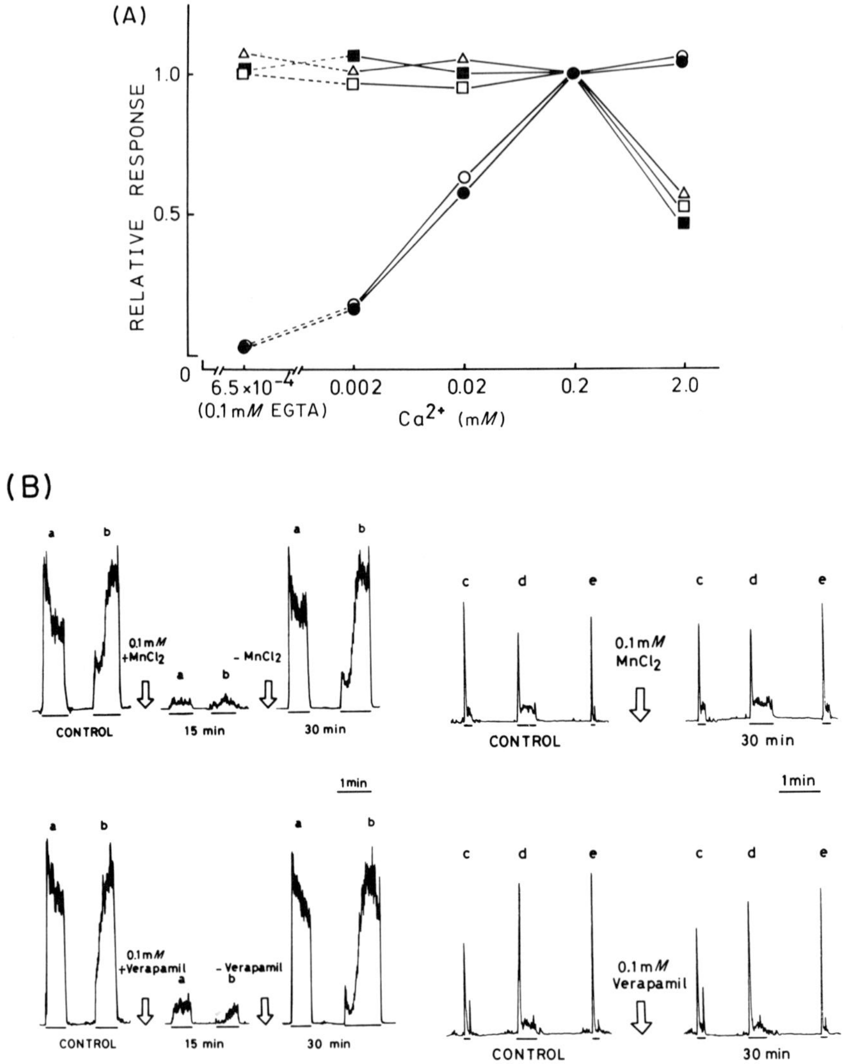

Fig. 12. (A) The responses of the frog gustatory nerve to various chemical stimuli as a function of Ca^{2+} concentration in a perfusing solution for the lingual artery. A response to each stimulus at 0.2 m*M* Ca^{2+} is taken as unity. ●, 1 m*M* $CaCl_2$; ○, distilled water; △, 0.1 m*M* quinine; □, 0.1 m*M* HCl; ■, 0.4 *M* ethanol. (B) The integrated responses of the frog gustatory nerve to various stimuli before (control) and after Ca-channel blockers are added to a perfusing solution containing 0.2 m*M* Ca^{2+}. a, 1m*M* $CaCl_2$; b, distilled water; c, 0.1 m*M* quinine; d, 0.4 *M* ethanol; e, 0.1 m*M* HCl. (Nagahama *et al.*, in preparation.)

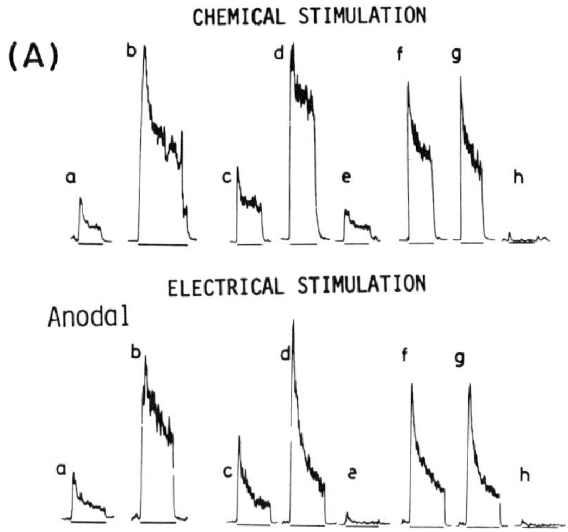

Fig. 13A. Effect of ANS treatment and uranyl acetate on the responses of the frog gustatory nerve to chemicals and anodal current. Solutions for chemical stimulation: 0.4 *M* NaCl (a-e), 0.2 *M* $MgCl_2$ (f-h). Adapting solutions for the tongue in electrical stimulation: 10 m*M* NaCl (a-e), 5 m*M* $MgCl_2$ (f-h). (a) Before ANS treatment; (b) after ANS treatment; (c) and (f) in the absence of UO_2^{2+}; (d) and (g), in the presence of $10^{-6}M$ UO_2^{2+}; (e) and (h), in the presence of $10^{-4}M$ UO_2^{2+}.

10^{-6} *M* uranyl acetate and decrease in that of 10^{-4} *M*. On the other hand, both the response to 0.2 *M* $MgCl_2$ and that to electrical stimulation in 5 m*M* $MgCl_2$ solution do not change at 10^{-6} *M* uranyl acetate and decrease at 10^{-4} *M*. Thus the behavior of the response to electrical stimulation is quite similar to that to chemical stimulation.

Figure 13B represents the effect of cathodal current on the responses

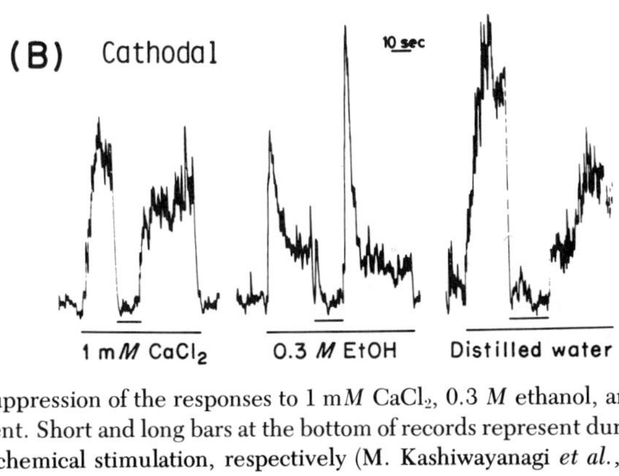

Fig. 13B. Suppression of the responses to 1 m*M* $CaCl_2$, 0.3 *M* ethanol, and distilled water by cathodal current. Short and long bars at the bottom of records represent duration of electrical stimulation and chemical stimulation, respectively (M. Kashiwayanagi *et al.*, unpublished).

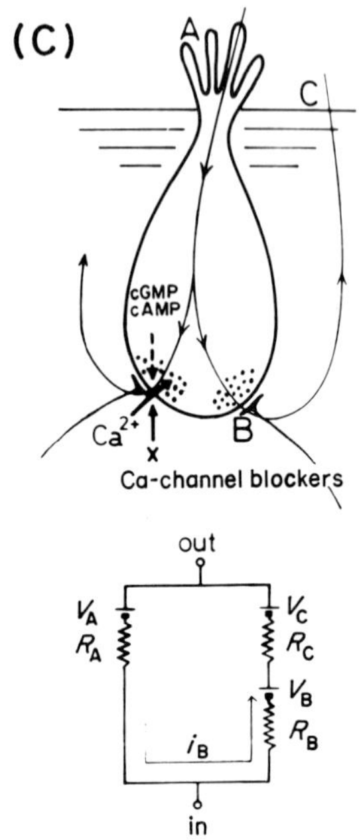

$$i_B = \frac{V_B + V_C - V_A}{R_B + R_C + R_A}$$

Fig. 13C. A schematic diagram illustrating a mechanism of the responses to chemical stimuli and anodal current. V_A, V_B, V_C, R_A, R_B, and R_C represent the potential difference and the resistance at the sites indicated by the subscripts; i_B represents the circulating current.

induced by chemical stimuli. The responses produced by $CaCl_2$, ethanol, or distilled water are suppressed by cathodal current.

As shown in Fig. 12, the responses to chemical stimuli of group 1 are suppressed greatly with decreasing $CaCl_2$ concentration in the perfusing solution or by addition of Ca-channel blockers to the solution. The response to electrical stimulation is also suppressed under these conditions; the response of the tongue adapted to 10 m*M* NaCl to anodal current is suppressed reversibly by elimination of Ca^{2+} from the perfusing solution or by addition of 0.1 m*M* verapamil or 0.1 m*M* $MnCl_2$ to the solution. This indicates that the response to electrical stimulation is not produced by direct stimulation of the gustatory nerve but that the release of the chemical transmitter is concerned with elicitation of the electrical response as well as the chemical response.

The above results indicate that the response to electrical stimulation is

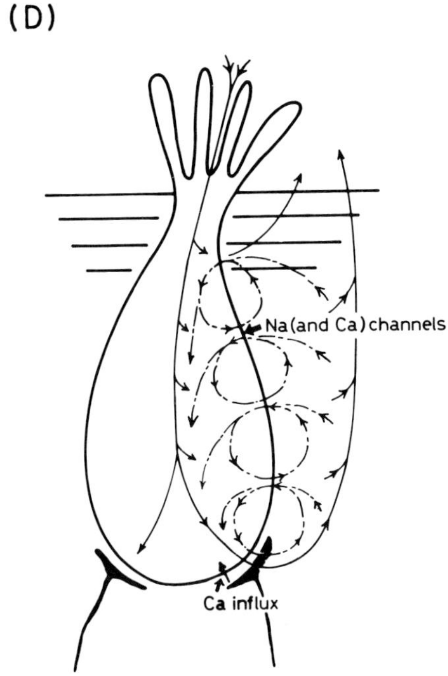

Fig. 13D. Contribution of voltage-dependent Na-channel to electrotonic spreading of depolarization at the microvilli to the synaptic area.

produced by a mechanism similar to that for chemical stimulation. Based on these results, the following mechanism is proposed for chemical and electrical stimulations (see Fig. 13C), although a more detailed mechanism is shown by Fig. 13D. Application of taste stimuli to the tongue brings about a decrease of V_A, which produces circulating electrical current, i_B. Since i_B is a function of R_A as well as V_A, a decrease in the membrane resistance of the microvillus membrane will also increase i_B. The circulating current depolarizes the synaptic area of taste cells, which opens the voltage-dependent Ca-channel and induces Ca influx from the intercellular medium into taste cells. This Ca influx will lead to a release of the chemical transmitter. Instead of chemical stimulation, the circulating current, i_B, can be applied to the tongue by anodal current, which induces the responses similar to those induced by chemical stimulation. Cathodal current cancels the circulating current produced by chemical stimuli applied to the tongue surface.

The above experiments suggested that the potential changes at the microvilli are conducted to the synaptic area by electrical current. However, the potential changes may decline during the electrotonic spreading to the synaptic area. The decline of the potential changes is considered to be prevented by the mechanism described below. M. Kashiwayanagi *et al.* (unpublished) found that a frog taste cell produces regenerative potentials in response to anode break stimulation. The regenerative potentials are voltage-dependent and appear when a taste cell is hyperpolarized (hyperpolarization releases the inactivation of ionic channels). Application of tetrodotoxin (TTX) and cobalt chloride to the tongue surface does not affect the regenerative potentials, but the potentials are decreased by addition of TTX or cobalt chloride to the perfusing solution for the lingual artery: the potentials are composed of TTX-sensitive component (Na-component) and cobalt-sensitive component (Ca-component). Note that TTX generally blocks voltage-dependent Na-channels. Sato (1980) noted that when the frog taste cells are soaked in Ringer containing TTX, the taste receptor potentials still appear, but the magnitudes of the potentials are reduced to about 50% of the control. The above results can be explained by the following mechanism (see Fig. 13D). The depolarization which occurred at the microvillus membrane spreads electrotonically along a taste cell membrane and activates voltage-dependent Na-channels. The Ca-channel may also be activated if it exists on the taste cell membrane beside the synaptic area. The number of Na-channels (and Ca-channels) on the taste cell membrane is probably not so large as on a nerve membrane, and the channels are mostly inactivated at the normal resting potential level. Hence the depolarization, which spreads electrotonically, does not produce all-or-none action potentials, but produces a graded response stemming from the activation of the voltage-dependent channels. This graded depolarization conducts along the taste cell membrane by the cycle reaction (depolarization at the microvilli → electrotonic spreading → graded response of voltage-dependent channels → electrotonic spreading). This type of conduction can prevent a decline of electrotonic spreading of depolarization, and the depolarization at the microvilli is conducted effectively to the synaptic area, which leads to activation of voltage-dependent Ca-channels at the area.

3. *Modulation of Gustatory Responses by cGMP and cAMP*

A role of cyclic nucleotides in the chemosensory systems was initially suggested by Kurihara and Koyama (1972); they showed that homogenates of bovine taste bud-bearing papillae and those of the rabbit olfactory epithelium exhibited high adenylate cyclase activities comparable to the activity found in the brain. Cagan (1976) showed that labeling of cAMP of taste papillae using those from bovine tongue prelabeled with [8-^{14}C]-adenine was increased only slightly in the presence of stimuli. Bovine taste papillae were also found to exhibit high phosphodiesterase activities (Kurihara, 1972; Price, 1973). Nomura (1978, 1980) examined the localization of adenylate cyclase, guanylate cyclase, and cAMP phosphodiesterase activities in the taste bud-bearing papillae of various animals by means of histochemistry and demonstrated that the activities of these enzymes were limited only to the apex of taste buds. Since chemical stimuli are received at the apex of taste buds, the above results suggest that the cyclic nucleotides may play an essential role in gustatory reception.

The effect of added cyclic nucleotides on taste responses was examined with the sugar receptor of the blowfly. Daley and Vande Berg (1976) applied the cyclic nucleotides to the receptor with stimulant sucrose and observed that cAMP decreased neural firing of the sugar sensitive receptor and dibutyryl cGMP increased the firing. These results were quite interesting, but the extent of the changes in the firing response by the cyclic nucleotides was not pronounced. It is generally considered that the gustatory receptor membranes are not very permeable and hence the external application of the cyclic nucleotides to the gustatory receptor would seem to hardly increase the cyclic nucleotide level in the taste cell. In fact, the cyclic nucleotides or their dibutyryl derivatives added to the tongues of the frog did not cause any effect on the gustatory response.

Nagahama *et al.* (1981, in preparation) perfused the lingual artery of the bullfrog with an artificial solution and examined the effect of the cyclic nucleotides in the perfusate on the gustatory response in the glossopharyngeal nerve. Figure 14 represents the effect of 0.5 m*M* cGMP added to the perfusing solution on the responses to various chemical stimuli. As seen from the figure, the responses to chemical stimuli of group 1 ($CaCl_2$, NaCl, distilled water, D-galactose, and L-threonine) are greatly enhanced immediately after addition of cGMP, but the responses to group 2 (quinine, HCl, acetic acid, and ethanol) are practically unaffected by cGMP. As described before, the gustatory responses to group 1 were decreased by decreasing the $CaCl_2$ concentration in the perfusing solution or by addition of Ca-channel blockers to the solution. Under these conditions, cGMP did not bring about any enhancement of the gustatory responses.

Figure 14 also shows the effect of 0.5 m*M* cAMP on the responses to various stimuli. The responses to group 1 are decreased by addition of cAMP to the perfusing solution. Contrary to the above stimuli, the response to group 2 is not affected by addition of cAMP. Application of chemical stimuli to the tongue is postulated to affect the activities of cyclases or phosphodiesterase located at the microvillus membrane and to change cyclic nucleotides level in the taste cell. However, the cyclic nucleotides themselves seem not to contribute to the changes in the membrane potential of the taste cell since the nucleotides did not affect the spontaneous activities of the gustatory nerve.

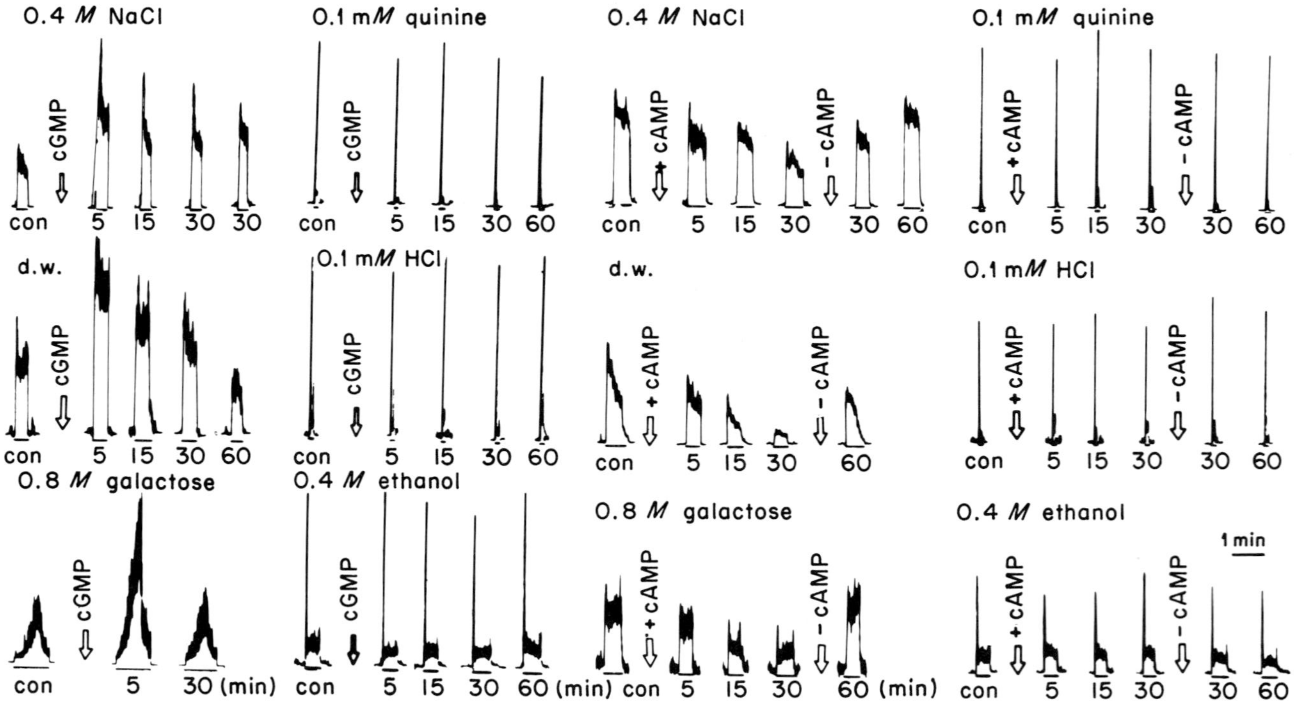

Fig. 14. The effect of 0.5 m*M* cGMP and 0.5 m*M* cAMP on the responses of the frog gustatory nerve to various chemical stimuli where the lingual artery was perfused with Ringer solution and the nucleotides were added to the perfusing solution. The number indicated at the bottom of each record represents time (min) after cyclic nucleotides were added or eliminated. (From Nagahama *et al.*, 1981, in preparation.)

Probably the cyclic nucleotides modulate Ca influx triggered by depolarization.

VII. RESEARCH NEEDS

Recently a number of research groups have explored the possible protein nature of receptor molecules for sugars and amino acids on chemoreceptor membranes (Chapters 8, 10). On the other hand, receptor molecules for other species of chemical stimuli are still unknown. Hydrophobic compounds such as bitter substances and odorants stimulate various membrane systems. It may be unlikely that reception of these stimuli occurs via adsorption on receptor molecules unique to gustatory or olfactory cells. The question arises how these membranes, which may not carry specific receptor proteins, can discriminate various hydrophobic compounds. In this connection, it would be important to study by what mechanism the frog gustatory receptors or the trigeminal nerves discriminate various species of hydrophobic compounds.

The responses of the frog gustatory receptors to various stimuli are enhanced in a quite different way by removal of Ca^{2+} from the receptor membrane, although the chemical composition of the membrane was unchanged. This suggests that an assembly of chemical components of the receptor membrane determines the specificity to ions. Various combinations of proteins, lipids, and water bound to the membrane may give a variety of receptor sites for ions, but this idea must be confirmed by future study.

Variation of the time course of the frog gustatory responses under various conditions were explained by the reaction scheme including $(SRe)_{\text{active}}$ and $(SRe)_{\text{inactive}}$. However, molecular features of $(SRe)_{\text{active}}$ and $(SRe)_{\text{inactive}}$ are completely unknown at present.

Receptor membranes are heterogeneous and contain various domains with different functions. Hence changes in the surface potential or in the membrane resistance occur nonuniformly at the receptor membrane. The changes in these electrical properties at each functional domain, not the averaged changes at the whole area of the gustatory receptor membrane seem to be effective in producing a circulating current to release a chemical transmitter.

The responses of the frog gustatory nerve to chemical stimuli of group 2 (quinine, HCl, acetic acid, and ethanol) were practically independent of $CaCl_2$ concentration in the perfusing solution for the lingual artery. It should be clarified whether these responses are produced without accompanying a release of a chemical transmitter from a taste cell or whether the chemical stimulation by these stimuli releases Ca^{2+} into the cytosol of a taste cell from an intracellular store.

The effects of cGMP and cAMP on the gustatory responses were examined by an electrophysiological method. In future studies, the biochemical reactions by which the nucleotides modulate Ca-channels should be explored.

Morimoto and Sato (1977) studied extensively to determine a chemical transmitter from frog taste cells to nerve terminals. Although they could not reach a definite conclusion, studies along this line will offer important information on taste transduction mechanisms.

REFERENCES

Aiuchi, T., Kamo, N., Kurihara, K., and Kobatake, Y. (1976). Physicochemical studies of taste reception: VI. interpretation of anion influences on taste reception. *Chem. Senses Flavor* **2**, 107–119.

Aiuchi, T., Kamo, N., Kurihara, K., and Kobatake, Y. (1977). Significance of surface potential in interaction of ANS with mitochondria fluorescence intensity and zeta-potential. *Biochemistry* **16**, 1625–1630.

Aiuchi, T., Tanabe, H., Kurihara, K., and Kobatake, Y. (1980). Fluorescence changes of rhodamine 6G associated with chemotactic responses in *Tetrahymena pyriformis*. *Biochim. Biophys. Acta* **628**, 355–364.

Akaike, N., Noma, A., and Sato, M. (1976). Electrical responses of frog taste cells to chemical stimuli. *J. Physiol. (London)* **254**, 87–107.

Andersen, H. T., Funakoshi, M., and Zotterman, Y. (1963). Electrophysiological responses to sugars and their depression by salt. *In* "Olfaction and Taste I" (Y. Zotterman, ed.), pp. 177–192. Pergamon, Oxford.

Ataka, M., Tsuchii, A., Ueda, T., Kurihara, K., and Kobatake, Y. (1978). Comparative studies on the reception of bitter stimulation in the frog, *Tetrahymena*, slime mold and *Nitella*. Comp. Biochem. Physiol. A **61**, 109–115.

Beidler, L. M. (1954). A theory of taste stimulation. *J. Gen. Physiol.* **38**, 133–139.

Beidler, L. M. (1971). Taste receptor stimulation with salts and acids. *In* "Handbook of Sensory Physiology, Vol. IV/2, Chemical Senses" (L. M. Beidler, ed.), pp. 200–220. Springer-Verlag, Berlin and New York.

Brand, J. G., Zeeberg, B. R., and Cagan, R. H. (1976). Biochemical studies of taste sensation V. Binding of quinine to bovine taste papillae and taste bud cells. *Int. J. Neurosci.* **7**, 37–43.

Bujas, Z. (1971). Electrical taste. *In* "Handbook of Sensory Physiology, Vol. IV/2, Chemical Senses" (L. M. Beidler, ed.), pp. 180–199. Springer-Verlag, Berlin and New York.

Cagan, R. H. (1976). Biochemical studies of taste sensation II. Labeling of cyclic AMP of bovine taste palillae in response to sweet and bitter stimuli. *J. Neurosci. Res.* **2**, 363–371.

Caprio, J. (1975). High sensitivity of catfish taste receptors to amino acids. *Comp. Biochem. Physiol. A* **53**, 247–251.

Daley, D. L., and Vande Berg, J. S. (1976). Apparent opposing effects of cyclic AMP and dibutyryl cyclic GMP on the neuronal firing of the blowfly chemoreceptors. *Biochim. Biophys. Acta* **437**, 211–220.

Fechner, G. T. (1860). "Elements of Psychophysic." (Engl. Transl., Holt, New York, 1966.)

Getchell, T. V. (1969). The interaction of the peripheral olfactory system with nonodorous stimuli. *In* "Olfaction and Taste III" (C. Pfaffmann, ed.), pp. 117–126. Rockeffeller Univ. Press, New York.

Hato, M., Ueda, T., Kurihara, K., and Kobatake, Y. (1976). Change in zeta-potential and membrane potential of slime mold *Physaurm polycephalum*. *Biochim. Biophys. Acta* **426**, 73–80.

Hodgkin, A. L. (1964). "The Conduction of the Nerve Impulses." Liverpool Univ. Press, Liverpool.

Kamiya, N. (1942). *In* "The Structure of Protoplasm: Physical Aspects of Protoplasmic Streaming" (W. Seifriz, ed.), pp. 199–224. Iowa State Univ. Press, Ames.

Kamo, N., Miyake, M., Kurihara, K., and Kobatake, Y. (1974a). Physicochemical studies of taste reception: I. model membrane stimulating taste receptor potential in response to stimuli of salts, acids and water. *Biochim. Biophys. Acta* **367**, 1–10.

Kamo, N., Miyake, M., Kurihara, K., and Kobatake, Y. (1974b). Physicochemical studies of taste reception: II. possible mechanism of generation of taste receptor potential induced by salt stimuli. *Biochim. Biophys. Acta* **367**, 11–23.

Kamo, N., Kashiwagura, T., Kobatake, Y., and Kurihara, K. (1978). Role of membrane-bound Ca^{2+} in frog gustatory receptors. *J. Physiol. (London)* **282**, 115–129.

Kamo, N., Kashiwagura, T., Kurihara, K., and Kobatake, Y. (1980). A theory of dynamic and steady responses in chemoreception. *J. Theor. Biol.* **83**, 111–130.

Kashiwagura, T., Kamo, N., Kurihara, K., and Kobatake, Y. (1977a). Enhancement of salt responses in frog gustatory nerve by removal of Ca^{2+} from the receptor membrane treated with ANS. *J. Membr. Biol.* **35**, 205–217.

Kashiwagura, T., Kamo, N., Kurihara, K., and Kobatake, Y. (1977b). Responses of the frog gustatory receptors to various odorants. *Comp. Biochem. Physiol. C.* **56**, 105–108.

Kashiwagura, T., Kamo, N., Kurihara, K., and Kobatake, Y. (1980). Nature of dynamic and steady components in the frog gustatory nerve response and interpretation by a theoretical model. *Am. J. Physiol.* **238**, G445–G452.

Konishi, J. (1967). Studies on the stimulation of chemoreceptors of freshwater fish by dilute solutions of electrolytes. *In* "Olfaction and Taste II" (T. Hayashi, ed.), pp. 667–692. Pergamon, Oxford.

Koyama, N., and Kurihara, K. (1972a). Effect of odorants on lipid monolayers from bovine olfactory epithelium. *Nature (London)* **236**, 402–404.

Koyama, N., and Kurihara, K. (1972b). Mechanism of bitter taste reception: interaction of lipid monolayers from bovine circumvallate papillae with bitter compounds. *Biochim. Biophys. Acta* **288**, 22–26.

Kurihara, K. (1972). Inhibition of cycle 3′, 5′-nucleotide phosphodiesterase in bovine taste papillae by bitter taste stimuli. *FEBS Lett.* **27**, 270–281.

Kurihara, K., and Koyama, N. (1972). High activity of adenyl cyclase in olfactory and gustatory organs. *Biochim. Biophys. Acta* **288**, 22–26.

Kurihara K., Kamo, N., and Kobatake, Y. (1978). Transduction mechanism in chemoreception. *Adv. Biophys.* **10**, 27–95.

Meyer, K. H., and Sievers, J. F. (1936). La perméabilite des membranes. I. Theorie de la perméabilite ionique. *Helv. Chim. Acta* **19**, 649–664.

Mito, Y., Kurihara, K., and Kobatake, Y. (1980). Selective suppression of positive chemotaxis in *Physarum polycephalum* by treatment with rotenone or anaerobic condition. *Eur. J. Cell Biol.* **21**, 48–52.

Miyake, M., Kamo, N., Kurihara, K., and Kobatake, Y. (1976a). Physicochemical studies of taste reception: III. interpretation of the water response in taste reception. *Biochim. Biophys. Acta* **436**, 843–855.

Miyake, M., Kamo, N., Kurihara, K., and Kobatake, Y. (1976b). Physicochemical studies of taste reception: V. suppressive effect of salts on sugar response of frog. *Biochim. Biophys. Acta* **436**, 856–862.

Morimoto, K., and Sato, M. (1977). Is serotonin a chemical transmitter in the frog taste organ? *Life Sci.* **21**, 1685–1696.

Nagahama, S., Kobatake, Y., and Kurihara, K. (1981). Modulation of frog gustatory responses by cyclic GMP and cyclic AMP. In preparation.

Nomura, H. (1978). Histochemical localization of adenylate cyclase and phosphodiesterase activities in the folliate papillae of the rabbit. I. light microscopic observations. *Chem. Senses Flavour* **3**, 319–324.

Nomura, H. (1980). Is cyclic nucleotide involved in transduction process of mammalian taste receptor cells? *Olfaction and Taste* **7**, 219.

Nozawa, Y., Fukushima, H., and Iida, H. (1975). Studies on *Tetrahymena* membranes: modification of surface membrane lipids by replacement of tetrahymanol by exogeneous ergosterol in *Tetrahymena pyriformis. Biochim. Biophys. Acta* **406**, 248–263.

Ottoson, D. (1973). Generator potential. *In* "Transduction Mechanisms in Chemoreception" (T. M. Poynder, ed.), pp. 231–240. Information Retrieval, London.

Ozeki, M., and Sato, M. (1972). Responses of gustatory cells in the tongue of rat to stimuli representing four taste qualities. *Comp. Biochem. Physiol* A. **41**, 391–407.

Paton, W. D. (1961). A theory of drug action based on the rate of drug-receptor combination. *Proc. R. Soc. London, Ser. B* **154**, 21–69.

Price, S. (1973). Phosphodiesterase in tongue epithelium: activation by bitter taste stimuli. *Nature (London)* **241**, 54–55.

Sato, T. (1980). Recent advances in the physiology of taste cells. *Progr. Neurobiol.* **14**, 25–67.

Stevens, S. S. (1971). Sensory power functions and neural events. *In* "Handbook of Sensory Physiology" (W. R. Loewenstein, ed.), Vol. I, pp. 226–242. Springer-Verlag, Berlin and New York.

Takagi, S. F., Iino, M., and Yarita, H. (1978). Effects of gustatory stimulants upon the olfactory epithelium of the bullfrog and the carp. *Jpn. J. Physiol.* **28**, 109–128.

Tanabe, H., Kurihara, K., and Kobatake, Y. (1979). Sensitization of chemotactic response by changing lipid composition of surface membrane *in Tetrahymena pyriformis. Biochim. Biophys. Acta* **553**, 396–403.

Tanabe, H., Kurihara, K., and Kobatake, Y. (1980). Changes in membrane potential and membrane fluidity in *Tetrahymena pyriformis* associated with chemoreception of hydrophobic stimuli: fluorescence studies. *Biochemistry*, 5339–5344.

Teorell, T. (1935). An attempt to formulate quantitative theory of membrane permeability. *Proc. Soc. Exp. Biol. Med.* **33**, 282–285.

Terayama, K., Kurihara, K., and Kobatake, Y. (1977). Variation in selectivity of univalent cations in slime mold *Physarum polycephalum* caused by reception of polyvalent cations. *J. Membr. Biol.* **37**, 1–12.

Tucker, D. (1971). Nonolfactory responses from the nasal cavity: Jacobson's organ and the trigeminal system. *In* "Handbook of Sensory Physiology" (L. M. Beidler, ed.), Vol. IV/1, pp. 151–181. Springer-Verlag, Berlin and New York.

Ueda, T., and Kobatake, K. (1977). Hydrophobicity of biosurface as shown by chemoreceptive thresholds in *Tetrahymena, Physarum* and *Nitella. J. Membr. Biol.* **34**, 351–368.

Ueda, T., Kurihara, K., and Kobatake, Y. (1975a). Responses of *Nitella* internodal cell to chemical stimuli: a model for olfactory receptor system. *J. Membr. Biol.* **23**, 271–284.

Ueda, T., Terayama, K., Kurihara, K., and Kobatake, Y. (1975b). Threshold phenomena in chemoreception and taxis by slime mold *Physarum polycephalum. J. Gen. Physiol.* **65**, 223–234.

Ueda, T., Muratsugu, M., Kurihara, K., and Kobatake, Y. (1975c). Olfactory response in excitable protoplasmic droplet and internodal cell of *Nitella*. *Nature (London)* **235**, 629–631.

Verwey, E., and Overbeek, J. T. G. (1948). "Theory of the Stability of Lyophobic Colloids." Elsevier, Amsterdam.

Yoshii, K., Kamo, N., Kurihara, K., and Kobatake, Y. (1979). Gustatory responses of eel palatine receptors to amino acids and carboxylic acids. *J. Gen. Physiol.* **74**, 301–317.

Yoshii, K., Kobatake, Y., and Kurihara, K. (1981). Selective enhancement and suppression of frog gustatory responses to amino acids. *J. Gen. Physiol.*, in press.

14

Intracellular Calcium and Taste Cell Transduction

JOSEPH G. BRAND, STEPHEN J. KRON, AND DAVID M. SENSEMAN

I. INTRODUCTION

In comparing the physiological and biochemical properties of the vertebrate taste and olfactory systems, a number of similarities and differences can be drawn. One of the central themes of this volume is the remarkable similarity of the biochemical mechanisms that apparently underlie the initial interaction of taste and olfactory stimuli with the receptor cell membrane. There is, however, an important fundamental difference between these systems with respect to the processes of sensory transduction.

Before taste or olfactory information is received by the central nervous

Biochemistry of Taste and Olfaction

ISBN 0-12-154450-8

system, the energy of the receptor–stimulus pair must be transformed into a propogated action potential in an afferent nerve fiber. The sequence of events between the initial stimulus–receptor interaction and generation of the action potential is collectively termed the *transduction process.* In the olfactory system, this initial interaction occurs on the primary afferent neuron. Transduction in this case involves processes that are largely, if not entirely, *intracellular.* In the taste system, the initial interaction occurs on a nonneural receptor cell. Transduction in this case is *intercellular* since excitation must pass from the receptor cell to an innervating afferent taste fiber before a neural impulse can be generated.

The physiological and morphological evidence discussed below suggests that communication between the taste receptor cell and the afferent fiber is by a chemical synapse, although a rigorous demonstration for this hypothesis has yet to be made. By analogy with other systems in which synaptic release has been investigated in some detail, we would expect calcium ion (Ca^{2+}) to play an essential role (Katz and Miledi, 1967, 1970). The work by Kurihara *et al.* (this volume, Chapter 13) is consistent with an hypothesis that calcium ion influx into taste cells is necessary for sensory transduction in the vertebrate (frog) taste system.

Because of the importance of Ca^{2+} in promoting transmitter release, as well as its involvement in other metabolic processes, it is not surprising that several subcellular mechanisms are active in the regulation and control of cytosolic Ca^{2+} concentrations. Of particular importance is the uptake, sequestration, and release of Ca^{2+} by mitochondria. Because very large (so-called "swollen") mitochondria are typically observed in close proximity to the synaptic release site in the vertebrate taste cell (Murrary, 1971, 1973), mitochondrial Ca^{2+} fluxes could be involved in the control of transmitter release and therefore in the taste transduction process.

It is of obvious importance that Ca^{2+}, as with other effectors of cellular processes, not be uniformly distributed within the intracellular environment. Both the location and effective concentration of Ca^{2+} is realized through its compartmentation in organelles, such as mitochondria. Such compartmentation is achieved only at the expense of cellular energy.

Some pertinent morphological features related to Ca^{2+} control in taste cells will be reviewed first. Second, since much of what is known about intracellular Ca^{2+} regulation is relatively recent in origin and is derived from areas of research other than chemosensory physiology and biochemistry, we will discuss some of the Ca^{2+} buffering processes that might influence transmitter release. Next, this information will be incorporated into a model for taste cell transduction. Finally, we will present initial successes (and failures) of investigations into the bioenergetics of taste cell transduction.

II. STRUCTURAL ASPECTS OF TASTE CELLS

The vertebrate taste bud is comprised of specialized epithelial cells and innervating afferent nerve fibers. On ultrastructural grounds, the specialized epithelial cells can be divided into different cell types. In the rabbit foliate taste bud, Murray (1971) has recognized four distinct cell types: type I, a "classical dark cell;" type II, a "classical light cell;" type III, a synaptic cell; and type IV, a basal cell.

Although the functional role of each cell type has yet to be determined, it is possible to postulate plausible functions from structural considerations. It is generally assumed that the initial reception of taste stimuli occurs on the apical end of the receptor cell that extends into the taste pore. In the rabbit, two cell types have a specialized microvillous structure on the apical end, which implies a receptive function: the type I and type II cells. The microvilli of the type II cell do not extend as far into the pore as those of the type I cell. The type III cell is very similar to the type II, except that it has a nonmicrovillous apex extending far up into the pore, and it contains small vesicles in the basal region in close proximity to a nerve. This basal area of the type III cell makes synaptic contact with the innervating sensory nerve fiber (Murray, 1971, 1973; Akisaka and Oda, 1977). Very few of these synaptic cells have been identified. Murray (1973) suggests that about 5–15% of the cells within the taste bud are type III cells. The synaptic region of the type III cell is usually closely associated with several large mitochondria. On the other hand, the type I cell possesses small dense mitochondria in the basal region (Murray, 1973). Type I cells surround both type II and type III cells with a continuous sheet of cytoplasm (Murray, 1971, 1973) so that cell–cell contact is very limited between or among cell types II and III.

Communication between type I and type III cells could occur via gap junctions. Evidence from freeze-fracture studies of rat papillae show that gap junctions do occur between taste bud cells (Akisaka and Oda, 1978), but the cell types involved were not specified. West and Bernard (1978) reported electrical coupling between dark cells of the mudpuppy taste bud, demonstrating the possibility of intercellular communication between two taste bud cells. However, coupling was not found between light cells, or between light and dark cells. It is also noted that the response of dark cells to chemical stimuli did not imply a purely receptive function (West and Bernard, 1978).

The structural differences of the cells within the vertebrate taste bud raise many questions. Of particular interest is the abundance of mitochondria in the basal regions of types I and III cells. That mitochondria are regulators of intracellular calcium (Carafoli *et al.*, 1974; Carafoli and Crompton, 1978) and that calcium is a requirement for synaptic release (Martin, 1977) leads us to speculate that calcium ion and its buffering by mitochondria are necessary for controlled release of neurotransmitter by taste receptor cells.

III. PLASMA MEMBRANE AND MITOCHONDRIA AS CALCIUM REGULATORS

The concentration of ionized calcium in excitable cells is maintained within a range of 0.1–5 μM/liter of cell water whereas total cell calcium is in the range of 0.2–10 mM/liter of cell water (Carafoli and Crompton, 1978). These estimates, suggesting that ionized calcium is only ~0.1% of the total cell calcium, emphasize the importance of calcium sequestration by the cell. In contrast with these values, total plasma calcium is about 2.5 mM, of which approximately one-half is ionized (Neuman and Neuman, 1958). The large concentration gradient of ionized calcium (10^3–10^4:1, outside:inside) across the plasma membrane presents a significant electrochemical force favoring Ca^{2+} movement into the cell. The low passive permeability of plasma membranes to Ca^{2+} as well as the ability of the cell to sequester and actively extrude Ca^{2+} permits long-term maintenance of optimal Ca^{2+} concentrations. This control may be realized by the action of several cellular agents including the plasma membrane, intracellular proteins, and intracellular organelles.

The sarcoplasmic reticulum regulates Ca^{2+} translocation in the concentration–relaxation cycle of muscle (Endo, 1977) by a process coupled to ATP hydrolysis (Froehlich and Taylor, 1976). The reticula of nonmuscle cells, such as those of rat liver (Moore *et al.*, 1975) and salivary gland (Selinger *et al.*, 1970), also participate in the control of intracellular calcium through ATP hydrolysis (for review, see Carafoli and Crompton, 1978). Intracellular calcium-binding proteins have been hypothesized in several control systems, including transmitter release at the neuromuscular (Rahamimoff *et al.*, 1978) and neural–neural synapses (Blaustein *et al.*, 1978).

Whereas many of the above systems have a marked ability to regulate calcium, the morphology of taste receptor cells implicates the plasma membrane and mitochondria as the principal systems involved in calcium regulation. The plasma membrane regulates both Ca^{2+} influx and efflux. Calcium ion channels and passive carriers are present in many cell plasma membranes. Calcium ion influx occurs across the membrane of some excitable cells during stimulation, and such influx during taste cell stimulation has been postulated (Sato and Beidler, 1975). In excitable tissues, the plasma membrane controls Ca^{2+} efflux via a Mg^{2+} –ATPase pump and/or a Na^+–Ca^{2+} exchange (Carafoli and Crompton, 1978). Transcellular transport of

Ca^{2+} (Terepka *et al.*, 1976) may also be active, particularly in cases where calcium accumulates in the cell to very high levels. Ultrastructural studies have localized membrane-associated Mg^{2+}–ATPase to the plasma membrane of taste cell types I, II, and III (Akisaka and Oda, 1977). Interestingly, Mg^{2+} –ATPase activity is not found in the region of the type III cell facing the nerve axon nor in regions of close contact (nonsynaptic) of the nerve and type II cells (Akisaka and Oda, 1977). The energy-requiring extrusion of Ca^{2+} could presumably be used by the type III cell to deplete itself of calcium after receptor-directed rises in intracellular Ca^{2+} concentration.

Although the Mg^{2+}–ATPase pump can extrude Ca^{2+} from the cell, recent evidence points to mitochondria as a high-capacity sink for Ca^{2+}. This organelle can both sequester and release Ca^{2+} in a controlled manner so that the cytosolic concentration can be maintained. Mitochondria have a large buffering capacity for Ca^{2+} at the levels considered to be present within most cells. For example, the *in vitro* buffering capacity of heart mitochondria yields K_m values of 5–15 μM in the energy-linked Ca^{2+} uptake system (Carafoli, 1978). However, *in vivo* the buffering capacity may be higher, leading to cytosolic Ca^{2+} concentrations less than 5 μM (Mela and Wrobel-Kuhl, 1978).

A schematic representation of the calcium translocation systems of a "typical" mitochondrial inner membrane is presented in Fig. 1. A large potential difference exists across the inner mitochondrial membrane, the inside being negative with respect to the cytosol. The magnitude of this potential is uncertain, but it is estimated to be on the order of 100–200 mV (Mitchell and Moyle, 1969; Padan and Rottenberg, 1973; Nicholls, 1974). This potential difference is generated, in part, by the active extrusion of protons from the mitochondria during respiration (Mitchell, 1966). Since the inside of the mitochondria is negative with respect to the cytosol, there is an electromotive force favoring movement of Ca^{2+} into the mitochondrion from the cytosol.

Three processes have been identified that transport Ca^{2+} across the inner mitochondrial membrane: a uniporter system that functions in a nonexchange fashion and two antiporter systems that involve a direct cation exchange mechanism.

The Ca^{2+} uniporter (Fig. 1, process I) is operationally defined by its sensitivity to the blocking agent ruthenium red (Crompton *et al.*, 1976; Puskin *et al.*, 1976). In contrast to ion pumps, uniporters (and antiporters) do not require expenditure of metabolic energy (e.g., hydrolysis of ATP). Though transport of Ca^{2+} across the inner membrane is a passive process, the maintenance of Ca^{2+} in relatively high concentrations within the mitochondrion (i.e., in excess of that specified by calcium's electrochemical potential) is obviously energy-dependent. This energy is supplied by mitochondrial res-

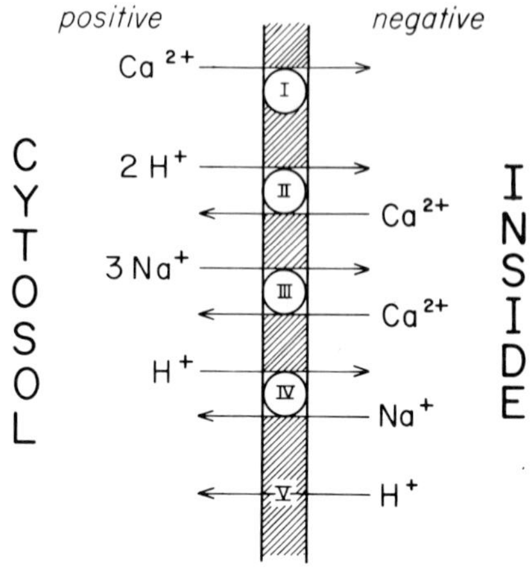

Fig. 1. Fluxes of Ca^{2+}, Na^+, and H^+ across the inner membrane of the mitochondria. The inside of the membrane is negative with respect to the cytosol. Process I represents the ruthenium red-sensitive Ca^{2+} uniporter, permitting Ca^{2+} influx. Process II represents a Ca^{2+}–H^+ antiporter described in liver mitochondria, which functions concomitant with alterations in the NADH/NAD^+ ratio. Calcium efflux occurs with pyridine nucleotide oxidation; calcium influx with pyridine nucleotide reduction. Process III represents the Na^+–Ca^{2+} antiporter operative in mitochondria of excitable cells. The stoichiometry indicated for process III is approximate. Process IV represents the rapid Na^+–H^+ antiporter. Process V is the respiration-driven efflux of H^+. See text for details. This schematum is based on Bygrave (1977), Crompton *et al.* (1976), Crompton and Heid (1978), and Fiskum and Lehninger (1979).

piration. Disruption of the electron transport chain with inhibitors (e.g., cyanide) leads to dissipation of the mitochondrial–cytosolic proton electrochemical gradient and the loss of sequestered Ca^{2+}. Because it has been found that the Ca^{2+} uniporter is unidirectional, the efflux of Ca^{2+} from the mitochondrion must occur via an alternate pathway (Crompton *et al.*, 1976; Puskin *et al.*, 1976; Bygrave, 1977).

One bidirectional process for Ca^{2+} transport across the mitochondrial inner membrane is the Ca^{2+}–H^+ antiporter (Fig. 1, process II). In liver mitochondria, the activity of the Ca^{2+}–H^+ antiporter is independent of respiration rate; unlike the Ca^{2+} uniporter, it cannot be blocked by ruthenium red (Fiskum and Lehninger, 1979). Another Ca^{2+} transport system that may be analogous to the liver mitochondrial Ca^{2+}–H^+ antiporter was observed in mitochondria isolated from cardiac muscle fibers (Lehninger *et al.*, 1978a,b). As with the Ca^{2+}–H^+ antiporter of liver mitochondria, the activity of the cardiac Ca^{2+} transport system appears to be directly related to the relative concentrations of reduced and oxidized pyridine nucleotides. Heart mitochondria can be induced to release Ca^{2+} by stimulating the oxidation of NADH with oxaloacetate. Conversely, stimulating the reduction of NAD^+

by addition of β-hydroxybutyrate promotes the uptake of Ca^{2+}. These data suggest that when the ratio of NADH–NAD^+ is high, Ca^{2+} influx is favored; when the ratio is low, Ca^{2+} efflux is favored. Because this transport process is not blocked by electron transport inhibitors (e.g., rotenone), the role of pyridine nucleotides in this antiporter process is separate and distinct from their role as proton carriers in site 1 of the electron transport chain.

The dynamic interrelationship between pyridine nucleotide metabolism and mitochondrial Ca^{2+} transport in intact cells is understandably complex and remains little understood. Whether this type of Ca^{2+} transport system operates in taste cell mitochondria and, if so, whether it plays a significant role in the control of intracellular Ca^{2+}, are open questions.

Another bidirectional process for Ca^{2+} transport across the mitochondrial inner membrane is the Na^+–Ca^{2+} antiporter (Fig. 1, process III). Stoichiometric and kinetic analyses of this antiporter process suggest that three sodium ions are exchanged for each calcium ion and that the rate of exchange is proportional to $[Na^+]^3$ (Carafoli *et al.*, 1974; Crompton *et al.*, 1976). These data have two important implications for mitochondrial Ca^{2+} transport. First, processes that result in elevation of cytosolic Na^+, such as the activation of inward Na^+ current by chemical and/or electrical stimulation of an excitable cell, will favor the efflux of mitochondrial Ca^{2+} by this antiporter process. Second, because the velocity of Ca^{2+}–Na^+ exchange is dependent upon the *third* power of the Na^+ concentration, the rate of Ca^{2+} efflux will be sensitive to relatively small changes in cytosolic Na^+ (Carafoli and Crompton, 1978).

If the Na^+–Ca^{2+} antiporter primarily functions in the extrusion of mitochondrial Ca^{2+}, then a net accumulation of Na^+ within the mitochondrion might be expected over time. Without some process to promote the efflux of Na^+, operation of the Na^+–Ca^{2+} antiporter would be self-limited by this rise in intramitochondrial Na^+. A Na^+–H^+ antiporter (Fig. 1, process IV) has been described that serves to reduce mitochondrial Na^+ levels at the expense of the proton gradient (Crompton *et al.*, 1977). Since the rate of the Na^+–H^+ antiporter (110 nmole H^+ /mg protein-min) is significantly faster than the Na^+–Ca^+ antiporter (15 nmole Ca^{2+} /mg protein-min), Na^+ loading of the mitochondrion should be prevented under normal physiologic conditions.

Operation of the Na^+–Ca^{2+} antiporter (as well as of the Ca^{2+} uniporter) would also lead to the net influx of positive charge because this transport process is not fully charge compensated. Restoration and long-term maintenance of the mitochondrial–cytosolic proton electrochemical gradient would result from the active extrustion of protons (Fig. 1, process V).

In attempting to elucidate the possible role played by the mitochondrion in control and regulation of cytosolic Ca^{2+} concentrations, it is important to remember that most of the investigations on Ca^{2+} transport systems are with isolated mitochondria. In whole cells, these processes are likely to be affected by a variety of metabolic interrelationships. For example, Ontko *et al.* (1975) demonstrated that $CaCl_2$ addition to a suspension of hepatocytes caused oxidation of pyridine nucleotides (presumably, mitochondrial), but the magnitude of the effect was greater in cells oxidizing succinate than in those oxidizing palmitate. That buffering of Ca^{2+} by mitochondria is an important physiological process in intact excitable cells is demonstrated by an elegant experiment on the squid giant axon (Brinley *et al.*, 1977). Blockage of axolemmal Ca^{2+} pumping by depletion of intracellular ATP had no immediate effect on intracellular Ca^{2+} levels following electrical stimulation. However, electrical stimulation produced a rapid rise in intracellular Ca^{2+} when mitochondrial metabolism was blocked by 2 m*M* cyanide. These data suggest that rapid Ca^{2+} buffering is a mitochondrial function requiring coupled oxidation of metabolic substrates. On the other had, the role of the axolemmal Ca^{2+} pumps appears to be in the long-term control of cytosolic Ca^{2+} under resting conditions.

IV. MODELS OF TASTE RECEPTOR TRANSDUCTION

The biochemical processes of taste reception have been reviewed elsewhere (Brand, 1978; Price and DeSimone, 1977; see also this volume, Chapter 10). On the other hand, the mechanisms of transduction in the vertebrate taste cell have received relatively little attention in the literature. A role for cyclic AMP in taste cell transduction has been proposed (Cagan, 1974; Kurihara and Koyama, 1972; Price, 1973), although Cagan's (1976) results lend only partial support for this hypothesis. A role for phosphodiesterase activity in bitter taste transduction has also been proposed (Kurihara, 1972; Price, 1973; Cagan, 1976). Stimuli that taste bitter to humans are usually phosphodiesterase inhibitors. A role for a microtuble system in the spread of excitation within olfactory receptor neurons has been postulated (Atema, 1975). However, the relative paucity of microtubles in taste receptor cells makes this mechanism less likely for gustation (Atema, 1975). It has been suggested that excitation of the primary taste fiber by the receptor cell is not the result of chemical transmission but rather of direct electrical transmission (Kurihara, 1974; Aiuchi *et al.*, 1976; Kurihara *et al.*, 1978). More recent experiments by Kurihara *et al.* (this volume, Chapter 13) seem to be inconsistent with a purely electrical transmission model. Most of these models deal with events prior to the initiation of release of chemical transmitter. We present here mechanisms that the taste cell may use to effect release of transmitter. Chemical and electrical events that undoubtedly link stimulus reception with the processes that initiate transmitter release will not be considered in this report.

Our hypothesis states that the processes that control the intracellular concentration and location of ionized calcium play a major role in transduction in taste receptor cells. As in other cellular mechanisms where Ca^{2+} plays a critical role (for reviews, see Bygrave, 1978; Blaustein, 1974), we further hypothesize Ca^{2+} movement within the cell to be under the control of various agents including the plasma membrane, intracellular proteins, and intracellular organelles. Morphological, histochemical, and biochemical considerations of the taste bud lead us to favor the plasma membrane and mitochondria in this control of Ca^{2+} location and concentration. We favor a predominent role for Ca^{2+} in taste cell transduction because transmitter release in physiologically analogous systems is normally potentiated by elevated Ca^{2+} concentrations within the presynaptic region.

If transmission between the taste receptor cell and the primary afferent nerve fiber is mediated by a conventional chemical synapse, then the requirements of the calcium hypothesis impose at least three constraints on the transduction process:

1. Transduction must be accompanied by a rise in cytosolic Ca^{2+} in the vicinity of the vesicular release site.
2. The rise in cytosolic Ca^{2+} must be sufficiently rapid to be consistent with electrophysiological measurements of stimulus–response latencies. Although the chemical stimulus–response latency is not known for certain, Sato (1976) has used electrical stimulation to estimate a gustatory neural impulse latency of around 5 msec. The increase in intracellular Ca^{2+} must be rapid enough to conform to this approximate 5 msec constraint.
3. The increased cytosolic Ca^{2+} levels must return to a lower value before synaptic release is terminated. The cell is then poised again for another excitation.

We now consider the above constraints on cytosolic Ca^{2+} fluxes in taste cell transduction under two cases. In the first case (Case I), we assume that the reception of the taste stimulus is by a cell that makes a direct synaptic contact with an afferent taste fiber. In the second case (Case II), we consider the possibility of stimulus reception occurring on both synaptic and nonsynaptic taste cells, with the former cell type serving as a transducer of the collective reception process.

A. Case I

A semidiagramatic representation of possible Ca^{2+} fluxes in a synaptic taste receptor cell (a type III cell in Murray's nomenclature (Murray, 1971, 1973)) is shown in Fig. 2. Although the apical terminus of the type III is not elaborated into a microvilli, it does extend into the taste pore region and could serve as a site of stimulus reception.

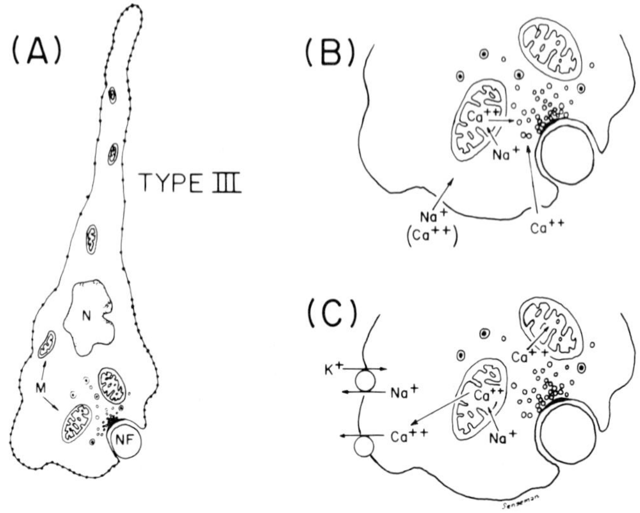

Fig. 2. Diagram of a type III taste cell and the postulated processes controlling Ca^{2+} movement during excitation and recovery. (A) illustrates the type III cell showing nucleus (N), mitochondria (M), nerve fiber (NF), and the relative concentration of Mg^{2+}-ATPase on the plasma membrane (•). The type III membrane near the nerve fiber represents the presynaptic portion of this taste cell. (B) shows Ca^{2+} fluxes during excitation. This illustration of the synaptic region of the taste cell during the excitation phase shows Ca^{2+} and Na^+ influx to the synaptic region and a possible release of calcium from the mitochondria via a Na^+–Ca^{2+} antiporter. Calcium probably enters throughout the plasma membrane via Ca^{2+}-specific channels and/or Na^+ channels, including directly across the presynaptic membrane (not shown due to space limitations). (C) shows Ca^{2+} fluxes during recovery. After stimulation begins, calcium is sequestered by the mitochondria. Intracellular Ca^{2+} levels remain high during stimulation. After stimulation ceases, Ca^{2+} and Na^+ are pumped out of the cell via their respective ATPases. Mitochondrial Ca^{2+} is slowly released via the Na^+–Ca^{2+} antiporter as long as cytosolic Na^+ levels are high. When cytosolic Ca^{2+} concentrations return to resting levels, the cell can respond fully to another stimulation. See text for details. [(A) was drawn after Murray (1973).]

At least for certain classes of stimuli, stimulus reception is associated with a depolarization of the receptor cell (i.e., generator potential). It is believed that this depolarization results from an increase in the sodium conductance of the plasma membrane (Sato and Beidler, 1975). Depolarization of the receptor cell could lead to an increase in cytosolic Ca^{2+} by at least three mechanisms (Fig. 2B). The most parsimonious mechanism assumes that the ionic channels activated by stimulus reception are not totally selective for Na^+, but also allow Ca^{2+} to carry part of the inward current. A second mechanism is a modification of the first. Depolarization of the receptor cell may lead to activation of voltage-dependent Ca^{2+} channels near the synaptic release site (e.g., Martin, 1977). A third possible mechanism that could produce an increase in cytosolic Ca^{2+} is the release of the sequestered ion by intracellular organelles such as mitochondria. For example, increased levels of cytosolic Na^+ could promote the release of mitochondrial Ca^{2+} via the Na^+–Ca^{2+} antiporter process. However, it seems unlikely that this mechanism is of major importance to synaptic release for two reasons. First, the data presented by Kurihara *et al.* (this volume, Chapter 13) argues for an external rather than an internal source of Ca^{2+}. Second, the rate of Na^+–Ca^{2+} exchange would probably be too slow to account for a rapid rise in cytosolic Ca^{2+} required for synaptic release.

As cytosolic Ca^{2+} rises during stimulation, mechanisms for controlling and removal of Ca^{2+} would be activated (Fig. 2C). Cytosolic Ca^{2+} would probably move rapidly into mitochondria via the ruthenium red-sensitive Ca^{2+} uniporter. If the rate of Ca^{2+} removal exceeded the rate of Ca^{2+} influx, a decrease in transmitter release would occur. Mitochondrial uptake of Ca^{2+} by the Ca^{2+} uniporter might therefore explain the rapid decline in nerve activity observed during the first few seconds following stimulus onset (e.g., Smith and Bealer, 1975; Smith *et al.*, 1978). Initial calcium influx may be in excess of that required for sufficient transmitter release. A large neural response will thus result. However, as the mitochondrial and other cytosolic buffer systems begin to act (probably within 100–200 msec), Ca^{2+} concentration in this synaptic region will decrease resulting in less transmitter release and a neural response that falls to a tonic level. The level of transmitter release would therefore represent a dynamic balance between Ca^{2+} entry and Ca^{2+} uptake. Additional control over the rate of Ca^{2+} uptake might be accomplished through a Na^+–Ca^{2+} antiporter. As long as cytosolic Na^+ levels remained elevated, the Na^+–Ca^{2+} antiporter will tend to reduce the Ca^{2+} buffering capacity of the mitochondrion. This would ensure that sufficiently high cytosolic Ca^{2+} levels were available for synaptic transmission.

During activation of the receptor cell, some lowering of cytosolic Ca^{2+} will occur through active extrusion via the Mg^{2+}–dependent ATPase pump. However, based on the experiments of Brinley *et al.* (1977), the total amount of Ca^{2+} extruded is expected to be very small compared with that sequestered by the mitochondria. The slow release of Ca^{2+} by this mitochondrial sink can be compensated for by the plasmalemmal Ca^{2+} extrusion pumps, ensuring little or no release of synaptic vesicular contents.

The return of the receptor cell to its resting steady-state thus involves the interaction of plasmalemmal cation pumps and cycling of Na^+ and Ca^{2+} through the mitochondria. Operation of the Mg^{2+}–ATPase and Na^+,K^+–ATPase pumps results in hydrolysis of ATP. As the cytosolic [ATP]/[ADP] [P_i] ratio declines, ATP production by the mitochondria is stimulated. Ejection of ATP from mitochondria is accompanied by calcium influx into the mitochondria. Since the Ca^{2+} uniporter system allows Ca^{2+} only to enter the mitochondria, accumulated Ca^{2+} must be extruded from the mitochondria by the Na^+–Ca^{2+} antiporter, by a slow Ca^{2+}–Ca^{2+} exchange (Crompton *et al.*, 1977) or, at least in liver, by a Ca^{2+}–H^+ antiporter. Entry of Na^+ into the mitrochrondria promotes activity of the Na^+–H^+ antiporter, rapidly returning Na^+ to the cytosol. As this cycling of cations continues, the levels of cytosolic Ca^{2+} and Na^+ are slowly lowered by their active extrusion into the extracellular space. The final resting steady state is reached when the cytosolic Ca^{2+} level is below the calcium buffering capacity of the mitochondria. At this point, the mitochondria is again ready to function as a site of Ca^{2+} sequestration following the next stimulus presentation.

This hypothetical mechanism for the participation and control of Ca^{2+} in taste cell transmitter release suggests two levels of regulation of intracellular Ca^{2+}: a rapid one, typified by the sequestration of Ca^{2+} by the ruthenium red-sensitive uniporter, and a long-term one, dominated by an integrated feedback extrusion of Ca^{2+} occurring via the Mg^{2+}–ATPase and Na^+, K^+–ATPase pumps, the Na^+–Ca^{2+} antiporters, other calcium efflux mechanisms of the mitochondria, and coupled oxidation of substrates. It should be emphasized that many of the integrated processes involved in the intracellular control of Ca^{2+} require expenditure of metabolic energy. The repletion of cytosolic ATP levels through the phosphorylation of ADP requires oxidation of NADH and reduction of flavoproteins. Restoration of the proton electrochemical gradient following operation of the Ca^{2+} uniporter and the Na^+–Ca^{2+} antiporter is respiration-dependent. We would, therefore, expect to observe changes in the levels of various co-factors following taste stimulation, specifically increases in levels of NAD^+ and reduced flavoproteins.

B. Case II

Taste buds of the mammal contain at least four types of cells. If only one type makes synaptic contact with a sensory nerve and if it is a very small percentage of the total number of cells, do the other cells have receptive properties as well? The apical morphology of the types I and II cells would seem to classify them as receptor cells, yet it is apparently the nonmicrovillous type-III cell that makes synaptic contact with the innervating nerve. Many cells of the taste bud display what have been termed *receptor potentials*, but in light of the morphology we are forced to ask how many of these potentials generate a postsynaptic response in the taste nerve fiber? This mechanism begins with a question: Could many type I cells and a type III cell all have receptive properties and be electrically coupled so that all respond as a functional unit, even though the actual transductive signal is generated only by the single type III cell?

As noted above, Murray (1971) has observed that type I cells (dark cells) surround each type III cell. Each type of cell is heavily invested with mitochondria, although the type I cell appears to have physically smaller mitochondria than those of the type III cell. Each cell contains plasma membrane-associated Mg^{2+}-dependent ATPase (Akisaka and Oda, 1977), and gap junctions have been reported between cells in the taste bud (Akisaka and Oda, 1978). Figure 3A is a digrammatic outline of these observations, assuming gap junctions between type I and type III cells. Figures 3B and 3C summarize our hypothesis for Ca^{2+}-potentiated transmitter release, sequestration, and removal assuming that type I cells participate with type III cells in reception.

Both type I and type III cells respond to stimulus receptor complexation by alterations in their plasma membranes below the tight junction levels. Inward Na^+ and Ca^{2+} movements induce a change in membrane potential in both types of cells. This Ca^{2+} movement into the type III cell potentiates transmitter release across the synaptic membrane as discussed in Case I

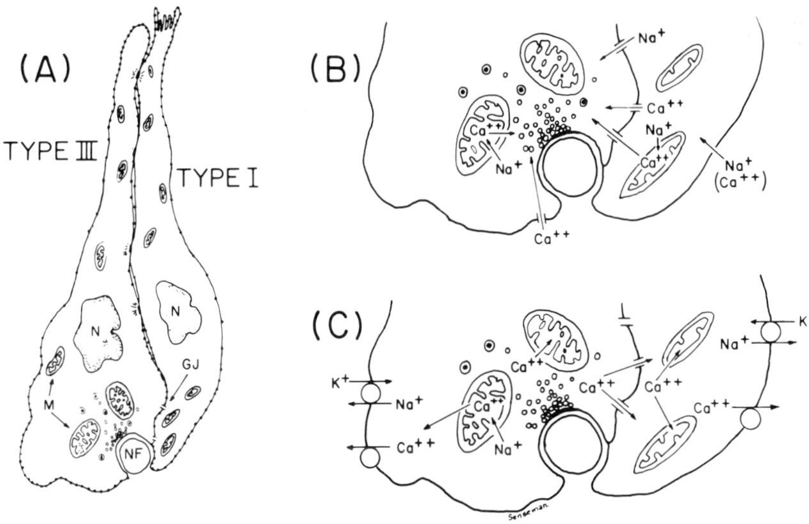

Fig. 3. Diagram of a type III taste cell and a type I taste cell and the postulated processes that might control Ca^{2+} movement both within and between these cell types. (A) illustrates a type III cell closely applied to a type I cell showing nuclei (N), mitochondria (M), nerve fiber (NF), gap junctions (GJ), and the relative density of Mg^{2+}-ATPase on the plasma membranes of the two cells. (B) shows Ca^{2+} fluxes during excitation. Calcium enters the cells via Ca^{2+} and/or Na^+ channels and is also released from type I cell mitochondria and possibly from type III cell mitochondria via the Na^+–Ca^{2+} antiporters. Calcium ions and Na^+ are driven down a concentration gradient from type I cells to type III cells where this additional Ca^{2+} further potentiates transmitter release. (C) shows Ca^{2+} fluxes during recovery. Calcium ion and Na^+ are extruded from the cells via the same mechanisms as explained for Fig. 2C. In the present case, however, we also envision a "backward" movement of Ca^{2+} from type III cells to type I cells. See text for details. [Cells in (A) are drawn after Murray (1973).]

above. In the type I cell, however, additional calcium is released from mitochondria due to increased cytosolic Na^+. This calcium ion, along with that which enters through the calcium channels during the membrane conformational change, increases type I cytosolic Ca^{2+} to a level greater than that of the adjacent type III cell. The Na^+-induced release of Ca^{2+} from mitochondrial stores in the type I cell would be expected to occur after transmitter release by the type III cell begins since the release mechanism is probably slower than Ca^{2+} influx across the synaptic membrane. Sodium ion levels in the type I cell might also be expected to be higher than those of the type III cell, leading to enhanced Na^+ and Ca^{2+} levels in type I cells compared with type III cells. Thus, the concentration gradients favor Ca^{2+} and Na^+ movement from type I cells to the type III cell. Calcium moves between type I cells and the adjacent type III cell via gap junctions down a concentration gradient. However, it has been demonstrated (Lowenstein and Rose, 1978) in salivary gland cells that free diffusion of calcium ion in the cytosol is severely restricted. This restriction has an energy component to it that leads to rapid sequestering of free Ca^{2+}. Uncoupling of mitochondria results in an increase in cytosolic Ca^{2+} concentration *in situ*, and poisoning the mitochondrial uptake system with ruthenium red leads to an increase in extramitochondrial calcium *in vitro*. Thus Ca^{2+} release toward type III cells needs to be directional. This additional increase in Ca^{2+} potentiates additional transmitter release leading to a graded neural response that now depends directly on the magnitude and duration of the collective membrane potentials evoked in the types I and III cells. Calcium is extruded from type I cells primarily via the ATPase system, and it is extruded from type III cells by the same processes outlined for Case I above. Experimentally this mechanism predicts an alteration in the NADH–NAD^+ ratio due to sequestration of Ca^{2+} by the type I and type III cell mitochondria and due to its eventual extrustion from the cells via the active pump mechanisms.

This rather cumbersome mechanism implies unique functional roles for the cell types of the mammalian taste bud. The type I cell, with its microvilli at the apex, may be a receptor cell that can communicate only via coupling with a type III cell. Many type I cells surrounding a single type III cell could be useful as an amplification mechanism for release of vesicular contents. In addition, one can consider the hypothesis that the type III cell has no receptive function at all, but it is present only as a communicator of reception for the surrounding type I cells. Coupling between type I cells and the type III cell could induce the type III cell generator potential necessary for transmitter release.

The taste bud can be envisioned as a collection of packets of type I–type III cell units and type I–type II cell units. Each type I–type III packet would have a unique receptive function, and its innervating nerve fiber would carry that sensation (sweet, sour, salty, bitter, etc.) appropriate to the unit. The exact role of the type II cell (light cell) is unknown, although a secretory function has been suggested (Farbman, 1965). Type II cells may also be useful to total calcium control in the bud region since they contain large mitochondria and swollen endoplasmic reticulum (Murray, 1973). Whether or not mitochondrial activity is involved or, indeed, whether it can even be observed in this system, provides a test of the hypotheses.

Many questions obviously arise in outlining such schemes, not the least of which are the following: How is the increase in intracellular sodium and calcium ion signaled? What is the quantitative relationship among intracellular organelles and the plasma membrane that permits calcium and sodium ion levels to return to the resting state? What is the rate-limiting step for this adaptation and recovery? While answers to these and other questions are germane to resolution of the mechanism of taste cell transduction, a more immediate question asks for evidence other than morphological for this proposed Ca^{2+} movement. A few neurophysiological studies have suggested a role for calcium in taste cell transduction. In addition, some studies have recently monitored co-factors known to be involved in mitochondrial activity. This information will be discussed in relation to the hypothetical mechanisms outlined above.

V. MONITORING OF MITOCHONDRIAL ACTIVITY

The well-known functions of mitochondrial energy metabolism involve many enzymes and enzyme systems, yet only a few co-factors. Whereas many of the enzymes cannot themselves be individually studied without physical disruption of the mitchondria, the collective process of energy metabolism can be monitored by following the fluorescence emission of two co-factor classes—the reduced pyridine nucleotides, NADH and NADPH, and the oxidized state of flavins, usually referred to collectively as flavoprotein co-factors (FP). A brief look at the mechanism whereby mitochondrial ATP is produced shows that during electron transport NADH is oxidized and FP is reduced. Because relative concentrations of both NAD(P)H and FP can be monitored fluorimetrically at different wavelengths, their ratio becomes an index of mitochondrial activity. In order to simplify this observation both mechanistically and instrumentally, either NAD(P)H or FP can be monitored individually. This practice introduces additional assumptions to the experiment, but it is justified if, for example, only preliminary data are desired or if there is reason to believe that other cellular constituents might also fluoresce in the region of interest. (For example, many of the monoamine transmitters and their metabolites fluoresce in the same spectral region as bound FMN and FAD).

Studies with isolated mitochondrial preparations have led to two interesting observations bearing on the relationship between Ca^{2+} movement and the levels of these nucleotides. First, calcium ion efflux, at least in some mitochondria, is accompanied by an alteration in the NADH/NAD ratio (Lehninger *et al.*, 1978a; Fiskum and Lehninger, 1979). Second, since calcium uptake is associated with energy dissipation and since calcium extrusion from the cell via an ATPase requires energy consumption, NADH/NAD^+ ratios as well as ratios of oxidized to reduced FP might be expected to alter during Ca^{2+} sequestration by the mitochondria and its extrusion from the cell (Carafoli and Crompton, 1978).

VI. MONITORING OF NADH AND FLAVOPROTEIN DURING TASTE STIMULATION

Investigations of the sense of taste have not focused on the effect that Ca^{2+} movement might have on the transduction of chemical information. At least one study has, however, attempted to alter the extracellular ion concentration (Sato and Beidler, 1975). This showed that sodium ion is required to generate the receptor potential and suggested that K^+ and Ca^{2+} also play a role. Kashiwagura *et al.* (1977) reported results for calcium complexing in which, presumably, calcium removal from the peripheral receptor membrane leads to enhancement of the membrane response to salt. Kurihara *et al.* (this volume, Chapter 13) also reports that low calcium ion in frog lingual artery perfusates diminishes the glossopharyngeal response. Due to the complexity of the vertebrate taste system and the fact that external Ca^{2+} concentration is around 10^3 greater than the internal concentration, detailed studies of Ca^{2+} movement in an intact system will be difficult. One possible approach would be to directly inject Ca^{2+} into the taste cell and record activity of an innervating fiber. Calcium injection into invertebrate photoreceptors causes a decrease in amplitude and latency of the photoresponse to a constant stimulus (Brown and Lisman, 1975). Although a parallel experiment has not been reported for the taste system, J. H. Teeter (1980, personal communication) has demonstrated rapid and prolonged firing of an innervating sensory fiber after calcium is iontophoretically injected into an epithelial electroreceptor cell in the glass catfish, *Kryptopterus*.

A second approach for monitoring Ca^{2+} fluxes is to use dyes or photoproteins whose spectral properties change with Ca^{2+} concentration. Prominent examples are the fluorescent protein aequorin and the dye arsenazo III. Both of these have been used to monitor calcium movements. Brown and Blinks (1974) reported experiments using aequorin in *Limulus* photoreceptors during illumination. They observed a stimulus-induced rise of intracellular calcium, which was due to calcium influx to the receptor as well as to release of calcium from intracellular stores.

Both of our hypothetical mechanisms assume that mitochondrial activity is involved in control of intracellular Ca^{2+}. A few studies have attempted to monitor co-factors of mitochondrial activity in taste-related tissue during chemical stimulation. These studies, although open to criticism, provide evidence for mitochondrial activity during taste stimulation and therefore, indirectly, for calcium movement.

Solov'yev *et al.* (1977a,b) and Samoylov *et al.* (1978) measured the ratio of FP–NADH fluorescence and concluded that in the taste cells of the frog tongue, taste stimulation is quickly followed by changes in the redox state of these nucleotides and, by inference, in the level of metabolic activity in the cells themselves. Using epi-illumination and microscopic imaging, they visually isolate a single taste disk in the frog tongue (Graziadei and Dettan, 1971) and illuminate it with monochromatic light of 365 nm. By using a monochromator in the emission mode in the fluorescence microscope, they were able to record an emission spectrum from 420 to 600 nm.

The notable features of this spectrum are emission maxima at 460 nm and 520–540 nm (Solov'yev *et al.*, 1977a), which are in agreement with those of NADH and oxidized FP, respectively. Solov'yev *et al.* (1977b) calculated the FP/NADH ratio (Chance, 1952) during stimulation of the frog tongue with acetic acid. After the acetic acid concentration reaches 10 m*M* (pH = 3.4), the ratio begins to decrease. Prior to this, the ratio remained unchanged. They interpret this decrease as an indication that metabolic activity, particularly ATP production, accompanies taste stimulation and taste cell transduction.

Additionally, Samoylov *et al.* (1978) demonstrated a reduction in oxygen tension at the taste disk of the frog tongue during caffeine stimulation. They assume that this reduction in oxygen tension is due to mobilization by the cell of substrates (presumably intracellular) into the respiratory chain. They also observed a decrease in the FP/NADH ratio after caffeine had reached a concentration of 2.5 m*M*.

The work from Solov'yev's laboratory (Solov'yev *et al.*, 1977a,b; Samoylov *et al.*, 1978) used only frogs as experimental animals. Admittedly, the frog has proved to be an important animal for taste research. However, a well-appreciated and even exploited property of the frog lingual epithelium is its permeability (Ozeki and Noma, 1972; Sato and Beidler, 1975). The fact that stimulatory agents may cross the epithelial barrier and enter not only the taste cells but also the underlying nerves makes the observations of Solov'yev and co-workers less general than if these workers employed an animal with a less permeable epithelium. The stimuli, caffeine and acetic acid, may be expected to cause pharmacological effects as well as purely taste effects. Penetration of either agent could cause changes in the level of the nucleotide and flavin co-factors in both epithelial cells and nerves. Weber and Herz (1968) noted that 2–3.5 m*M* caffeine can release calcium from sarcoplasmic reticulum. We also noted (Brand and Bayley, unpublished observations) large decreases in reduced pyridine nucleotide fluorescence from bovine taste epithelium upon stimulation of this tissue with HCl below pH 4.0.

The origin of the fluorescence that Solov'yev and co-workers classify as NADH and flavoprotein is also in question. In a report describing histofluorescence of the frog's taste organ Solov'eva *et al.* (1978) describe a weak pale green luminescence from the adrenergic nerve fibers (emission probably between 470–490 nm) and a serotonin-like fluorescence near 520 nm. Their study provided evidence that the predominant fluorescence observed in the 510–540 nm region from the frog taste apparatus is from a serotonin-like biogenic amine. This fluorescence was not reported by DeHan and Graziadei (1973) for frog. The fluorescence changes observed by Solov'yev *et al.* (1977a,b) and Samoylov *et al.* (1978) may thus be the result of a combination of mitochondrial NADH–flavoprotein and adrenergic nerve NADH, and epithelial cell biogenic amine.

At the same time as Solov'yev's reports were being published, Brand and Bayley (1978, 1980) reported that inorganic salts have the ability to decrease NADH levels in steer circumvallate papilla taste receptor epithelium. Several features of this phenomenon are notable. First, the change in NADH fluorescence was not specific to NaCl and LiCl, two salts that in humans elicit a classical salty response. Rather, changes were also observed for salts having complex tastes to humans, including KCl, NH_4Cl, and $(Na)_2SO_4$. No changes were observed, however, with the "bitter" salt, $MgSO_4$. Second, the change in NADH level was effectively antagonized by using a medium that had been deoxygenated and re-gassed with N_2. Third, the changes were reversible, but their degree of reversibility as well as the magnitude of the initial response depended on the relative ability of the papillae to respond to agents that control cellular respiration. In a report of these data, Brand and Bayley (1980) suggested that the decrease in apparently reduced pyridine nucleotide levels was a manifestation of calcium flux in the epithelial and taste cells.

This work is open to criticism for its lack of specificity, since the fluorescence signal is recorded from tissue that contains both taste and nontaste epithelial cells. There is also a question of the physiological state of the excised taste tissue. Excising the tissue results in loss of its blood supply, limiting not only nutrients but also other extracellular fluids, and the procedure thereby could greatly affect any reversal of activity that might be under metabolic control. Excision also creates a surface unprotected by an epithelial barrier. Salt ions could then diffuse into the extracellular spaces or cut ends of innervating fibers and affect NADH-dependent reactions. We reported NADH changes to Na^+, Li^+, K^+, and NH_4^+, yet only the first two ions are active in promoting Ca^{2+} release directly from isolated mitochondria (Crompton *et al.*, 1976).

We have used both epi-illumination and transverse illumination to pass 366-nm light to a catfish barbel either as an excised appendage or when attached to an anesthetized animal. The barbel epithelium is focused through a 40 × water immersion objective. Light passes through this objective, through a 435-nm long-wavelength band pass filter and then through a 471-nm interference filter. After the interference filter, light impinges on a photomultiplier tube. A continuously movable iris is placed just before the photomultiplier tube in order to observe light only from a small area of the total barbel under the objective. The filters ensure that appropriate excitation and emission wavelengths are being used to monitor reduced pyridine nucleotides. In using this technique, one must be cognizant of the possibility that the emission being monitored at 471 nm is not from NADH in the taste bud. Agents that block the electron transport chain cause an increase in NADH because further oxidation of substrates is not possible. We have observed increases in emission of 471 nm of a catfish barbel when CN^- (either as the Na^+ or K^+ salt) is flowed over the barbel. This is consistent with the known activity of CN^- to block electron transport. Interestingly, in the barbel preparations that did show this effect, it was long (~2 min) in developing, suggesting a fair degree of impenetrability of this ion into the barbel epithelium.

Using catfish barbel, we observed small decreases in 471-nm emission upon addition of stimuli such as NaCl, KCl, and caffeine. However, no change has yet been seen for L-alanine, a potent taste stimulus for the catfish, even at 10^{-2} *M*. This raises the possibility that receptor transduction could be different for different stimuli. Alternatively, the sensitivities of our techniques may still not meet those required for observation of this taste cell response.

VII. RESEARCH NEEDS

The work summarized above suggests that mitochondrial activity is enhanced during taste cell response. However, the criticisms need to be answered before specific conclusions can be reached on the control of calcium ion by metabolic activity. Most of these can be answered using a fluorescence microscopy technique on the taste bud-rich catfish barbel. With proper instrumentation, fluorescence microscopy can quantitatively monitor emission from these cells and be used to observe fluorescence from a single taste bud. The catfish barbel has a dense collection of taste buds that protrude from the epidermal surface (Ovalle and Shinn, 1977). The epidermis of this fish is expected to be relatively impermeable to chemicals and relatively resistant to osmotic shock. In the ideal, we can observe only a single taste bud without excessive epidermis either below or around it. Use of this preparation to test the model presented here must be tempered by the ultrastructure of the catfish barbel taste bud (Grover–Johnson and Farbman, 1976), a morphology only partially similar to that of the rat and rabbit.

Since axons invade the taste bud, NADH fluorescence will be observed from both the epidermal taste bud cells and the nerve axons. It would be prudent to be able to monitor NADH from one exclusive of the other. To do this, one would need to block metabolic activity in one, for example the axon, without affecting metabolic activity in the other. Tetrodotoxin (TTX) would block voltage-dependent sodium channels in the nerve membrane and thereby inhibit spike generation. It may not block taste cell sodium channels, however, since these channels are probably chemically gated. Even so, TTX poisoned nerves may still be receptive to transmitter, and interaction of the transmitter with the postsynaptic membrane may continue to initiate other metabolic alterations in the nerve that may manifest themselves as changes in NADH levels. If the chemical transmitter released by taste cells during stimulation were known, it might be possible to block its reception by the postsynaptic membrane. Then, NADH changes could be associated primarily with receptor cells. Although this is theoretically possible, a more appropriate approach for the near future might use an isolated taste cell preparation. Such a preparation has been described (Brand and Cagan, 1976), and its use in this model should be pursued.

ACKNOWLEDGMENT

This work was supported in part by National Institutes of Health Research Grant No. NS-15565 (to J. G. Brand) from National Institute of Neurological and Communicative Disorders and Stroke and by National Institutes of Health Research Grant No. DEO-5271 (to D. M. Senseman) from National Institute of Dental Research. The authors thank Dr. Thomas V. Getchell for critical review of the manuscript.

REFERENCES

Aiuchi, T., Kamo, N., Kurihara, K., and Kobatake, Y. (1976). Physicochemical studies of taste reception. VI: Interpretation of anion influences on taste response. *Chem. Senses Flavor* **2**, 107–119.

Akisaka, T., and Oda, M. (1977). The fine structural localization of adenosine triphosphatase activity on the taste bud in the fungiform papillae of the rat. *Arch. Histol. Jpn.* **40,** 63–72.

Akisaka, T., and Oda, M. (1978). Taste buds in the vallate papillae of the rat studied with freeze-fracture preparation. *Arch. Histol. Jpn.* **41,** 87–98.

Atema, J. (1975). Stimulus transmission along microtubules in sensory cells: An hypothesis. *In* "Microtubles and Microtuble Inhibitors" (M. Borgers and M. deBradander, eds.), pp. 247–257. North-Holland Publ., Amsterdam.

Blaustein, M. P. (1974). The interrelationship between sodium and calcium fluxes across cell membranes. *Rev. Physiol. Biochem. Pharmacol.* **70,** 34–82.

Blaustein, M. P., Ratzlaff, R. W., and Kendrick, N. K. (1978). The regulation of intracellular calcium in presynaptic nerve terminals. *Ann. N.Y. Acad. Sci.* **307,** 195–211.

Brand, J. G. (1978). Biophysics of taste receptors. *In* "Sweeteners and Dental Caries" (J. H. Shaw and G. G. Roussos, eds.), pp. 13–32. Information Retrieval, Washington, D.C.

Brand, J. G., and Bayley, D. L. (1978). Monitoring fluorescence from intact taste papillae. Applications to the biophysics of taste reception. *Biophys. J.* **21,** 136a.

Brand, J. G., and Bayley, D. L. (1980). Peripheral mechanisms in salty taste reception. *In* "Biological and Behavioral Aspects of Salt Intake" (M. R. Kare, M. J. Fregly, and R. A. Bernard, eds.), pp. 289–298. Academic Press, New York.

Brand, J. G., and Cagan, R. H. (1976). Biochemical studies of taste sensation. III. Preparation of a suspension of bovine taste bud cells and their labeling with a fluorescent probe. *J. Neurobiol.* **7,** 205–220.

Brinley, F. J., Tiffert, J. T., Mullins, L. J., and Scarpa, A. (1977). Kinetic measurement of Ca^{2+} transport by mitochondria in situ. *FEBS Lett.* **82,** 197–199.

Brown, J. E., and Blinks, J. R. (1974). Changes in intracellular free calcium concentration during illumination of invertebrate photoreceptors. *J. Gen. Physiol.* **64,** 643–665.

Brown, J. E., and Lisman, J. E. (1975). Intracellular Ca modulates sensitivity and time scale in *Limulus* ventral photoreceptors. *Nature (London)* **258,** 252–254.

Bygrave, F. L. (1977). Mitochondrial calcium transport. *Curr. Top. Bioenerg.* **6,** 260–318.

Bygrave, F. L. (1978). Mitochondria and the control of intracellular calcium. *Biol. Rev. Cambridge Philos. Soc.* **53,** 43–79.

Cagan, R. H. (1974). Biochemistry of sweet sensation. *In* "Sugars in Nutrition" (H. L. Sipple and K. W. McNutt, eds.), pp. 19–36. Academic Press, New York.

Cagan, R. H. (1976). Biochemical studies of taste sensation: II. Labeling of cyclic AMP of bovine taste papillae in response to sweet and bitter stimuli. *J. Neurosci. Res.* **2,** 363–371.

Carafoli, E. (1978). Discussions of the paper by Lehninger *et al. Ann. N.Y. Acad. Sci.* **307,** 175.

Carafoli, E., and Crompton, M. (1978). The regulation of intracellular calcium. *Curr. Top. Membr. Transp.* **10,** 151–216.

Carafoli, E., Tiozzo, R., Lugli, G., Crovetti, F., and Kratzing, C. (1974). The release of Ca^{2+} from heart mitochondria by sodium. *J. Mol. Cell. Cardiol.* **6,** 361–371.

Chance, B. (1952). Spectra and reaction kinetics of respiratory pigments of homogenized and intact cells. *Nature (London)* **169,** 215–221.

Crompton, M., and Heid, I. (1978). The cycling of calcium, sodium and protons across the inner membrane of cardiac mitochondria. *Eur. J. Biochem.* **91,** 599–608.

Crompton, M., Capano, M., and Carafoli, E. (1976). The sodium-induced efflux of calcium from heart mitochondria. *Eur. J. Biochem.* **69,** 453–462.

Crompton, M., Kunzi, M., and Carafoli, E. (1977). The calcium-induced and sodium-induced effluxes of calcium from heart mitochondria. *Eur. J. Biochem.* **79,** 549–558.

DeHan, R. S., and Graziadei, P. (1973). The innervation of frog's taste organ. *Life Sci.* **13,** 1435–1449.

Endo, M. (1977). Calcium release from the sarcoplasmic reticulum. *Physiol. Rev.* **57,** 71–108.

Farbman, A. I. (1965). Fine structure of the taste bud. *J. Ultrastruct. Res.* **12,** 328–350.

Fiskum, G., and Lehninger, A. L. (1979). Regulated release of Ca^{2+} from respiring mitochondria by Ca^{2+}/H^+ antiport. *J. Biol. Chem.* **254**, 6236–6239.

Froehlich, J. P., and Taylor, E. W. (1976). Transient state kinetic effects of calcium ion on sarcoplasmic reticulum adenosine triphosphatase. *J. Biol. Chem.* **251**, 2307–2315.

Graziadei, P. P. C., and DeHan, R. S. (1971). The ultrastructure of frog's taste organs. *Acta Anat.* **80**, 563–603.

Grover-Johnson, N., and Farbman, A. I. (1976). Fine structure of taste buds in the barbel of the catfish, *Ictalurus punctatus*. *Cell Tissue Res.* **169**, 395–403.

Kashiwagura, T., Kamo, N., Kurihara, K., and Kobatake, Y. (1977). Enhancement of salt responses in frog gustatory nerve by removal of Ca^{2+} from the receptor membrane treated with 1-anilinonaphthalene-8-sulfonate. *J. Membr. Biol.* **35**, 205–217.

Katz, B., and Miledi, R. (1967). A study of synaptic transmission in the absence of nerve impulses. *J. Physiol. (London)* **192**, 407–436.

Katz, B., and Miledi, R. (1970). Further study of the role of calcium in synaptic transmission. *J. Physiol. (London)* **207**, 789–801.

Kurihara, K. (1972). Inhibition of cyclic 3′, 5′-nucleotide phosphodiesterase in bovine taste papillae by bitter taste stimuli. *FEBS Lett.* **27**, 279–281.

Kurihara, K. (1974). Physico-chemical aspects of chemoreceptor mechanism: stimuli-receptor interaction and receptor potential in taste stimulation. *In* "Transduction Mechanisms in Chemoreception" (T. M. Poynder, ed.), pp. 163–172. Information Retrieval, London.

Kurihara, K., and Koyama, H. (1972). High activity of adenyl cyclase in olfactory and gustatory organs. *Biochem. Biophys. Res. Commun.* **48**, 30–34.

Kurihara, K., Kamo, N., and Kobatake, Y. (1978). Transduction mechanism in chemoreception. *Adv. Biophys.* **10**, 27–95.

Lehninger, A. L., Raynafarje, B., Vercesi, A., and Tew, W. P. (1978a). Transport and accumulation of calcium in mitochondria. *Ann. N.Y. Acad. Sci.* **307**, 160–174.

Lehninger, A. L., Vercesi, A., and Bababunmi, E. A. (1978b). Regulation of Ca^{2+} release from mitochondria by the oxidation-reduction state of pyridine nucleotides. *Proc. Natl. Acad. Sci. U.S.A.* **75**, 1690–1694.

Lowenstein, W. R., and Rose, B. (1978). Calcium in (junctional) intercellular communication and a thought on its behavior in intracellular communication. *Ann. N.Y. Acad. Sci.* **307**, 285–305.

Martin, A. R. (1977). Junctional transmission II. Presynaptic mechanisms. *In* "Handbook of Physiology, Sect. 1, The Nervous System, Vol. I/1" (E. R. Kandel, ed.), pp. 344–348. Am. Physiol. Soc. Bethesda, Maryland.

Mela, L., and Wrobel-Kuhl, K. (1978). Special characteristics of brain mitochondrial calcium accumulation. *Ann. N.Y. Acad. Sci.* **307**, 242–245.

Mitchell, P. (1966). "Chemiosmotic Coupling in Oxidative and Photosynthetic Phosphorylation." Glynn Res., Bodmin, Cornwall, England.

Mitchell, P., and Moyle, J. (1969). Estimation of membrane potential and pH difference across the cristal membrane of rat liver mitochondria. *Eur. J. Biochem.* **7**, 471–484.

Moore, L., Chen, T. S., Knapp, H. R., and Landon, E. J. (1975). Energy-dependent calcium sequestration activity in rat liver microsomes. *J. Biol. Chem.* **250**, 4562–4568.

Murray, R. G. (1971). Ultrastructure of taste receptors. *In* "Handbook of Physiology, Vol. IV/2, Chemical Senses" (L. M. Beidler, ed.), pp. 31–50. Springer-Verlag, Berlin and New York.

Murray, R. G. (1973). The ultrastructure of taste buds. *In* "The Ultrastructure of Sensory Organs" (I. Friedmann, ed.), pp. 1–81. Am. Elsevier, New York.

Neuman, W. F., and Neuman, M. W. (1958). "The Chemical Dynamics of Bone Mineral." Univ. of Chicago Press, Chicago, Illinois.

Nicholls, D. G. (1974). The influence of respiration and ATP hydrolysis on the proton—electrochemical gradient across the inner membrane of rat-liver mitochondria as determined by ion distribution. *Eur. J. Biochem.* **50,** 305–315.

Ontko, J. A., Otto, D. A., Oshino, N., and Chance, B. (1975). Calcium-mediated alterations in the oxidation-reduction state of pyridine nucleotides in isolated liver cells. *FEBS Lett.* **53,** 297–301.

Ovalle, W. K., and Shinn, S. L. (1977). Surface morphology of taste buds in catfish barbels. *Cell Tissue Res.* **178,** 375–384.

Ozeki, M., and Noma, A. (1972). The actions of tetrodotoxin, procaine, and acetylcholine on gustatory receptions in frog and rat. *Jpn. J. Physiol.* **22,** 467–475.

Padan, E., and Rottenberg, H. (1973). Respiratory control and the proton electrochemical gradient in mitochondria. *Eur. J. Biochem.* **40,** 431–437.

Price, S. (1973). Phosphodiesterase in tongue epithelium: Activation by bitter taste stimuli. *Nature (London)* **241,** 54–55.

Price, S., and DeSimone' J. A. (1977). Models of taste receptor cell stimulation. *Chem. Senses Flavor* **2,** 427–456.

Puskin, J. S., Gunter, T. E., Gunter, K. K., and Russell, P. R. (1976). Evidence for more than one Ca^{2+}-transport mechanism in mitochondria. *Biochemistry* **15,** 3834–3842.

Rahamimoff, R., Erulkar, S. D., Lev-Tov, A., and Meiri, H. (1978). Intracellular and extracellular calcium ions in transmitter release at the neuromuscular synapse. *Ann. N.Y. Acad. Sci.* **307,** 583–597.

Samoylov, V. O., Solov'yev, V. N., Kozhemyakin, L. A., and Shmarov, D. A. (1978). Role of the cAMP system in taste receptor mechanisms. *Dok. Akad. Nauk SSSR* **241,** 1478–1480.

Sato, T. (1976). Latency of taste nerve signals in frog. *Experientia* **32,** 877–879.

Sato, T., and Beidler, L. M. (1975). Membrane resistance change of the frog taste cell in response to water and NaCl. *J. Gen. Physiol.* **66,** 735–763.

Selinger, Z., Nains, E., and Lasser, M. (1970). ATP-dependent calcium uptake by microsomal preparations from rat parotid and submaxillary glands. *Biochim. Biophys. Acta* **203,** 326–334.

Smith, D. V., and Bealer, S. L. (1975). Sensitivity of the rat gustatory system to the rate of stimulus onset. *Physiol. Behav.* **15,** 303–314.

Smith, D. V., Bealer, S. L., and Van Buskirk, R. L. (1978). Adaptation and recovery of the rat chorda tympani response to NaCl. *Physiol. Behav.* **20,** 629–636.

Solov'eva, N. A, Krokhina, E. M., and Esakov, A. I. (1978). Possible nature of the biogenic amine in the dumbbell cell of the frog taste bud. *Byull. Eksp. Biol. Med.* **86,** 596–598.

Solov'yev, V. N., and Samoylov, V. O. (1977a). Investigation of the fluorescence of pyridine nucleotides and flavoproteins in the region of the receptor epithelium of the frog's tongue. *Fiziol. Zh.* **63,** 1476–1478.

Solov'yev, V. N., and Samoylov, V. O. (1977b). Change in the fluorescence spectra of the receptor epithelium of the frog's tongue under the influence of acetic acid. *Fiziol. Zh.* **63,** 1740–1742.

Terepka, A. R., Coleman, J. R., Armbrecht, H. J., and Gunter, T. E. (1976). Transcellular transport of calcium. *Symp. Soc. Exp. Biol.* **30,** 117–140.

Weber, A., and Herz, R. (1968). The relationship between caffeine contracture of intact muscle and the effect of caffeine on reticulum. *J. Gen. Physiol.* **52,** 750–759.

West, C. H. K., and Bernard, R. A. (1978). Intracellular characteristics and responses of taste bud and lingual cells of the mudpuppy. *J. Gen. Physiol.* **72,** 305–326.

15

Isolation, Separation, and Analysis of Cells from Olfactory Epithelium

JAMES D. HIRSCH AND FRANK L. MARGOLIS

I. INTRODUCTION

If we were to think of ways to investigate the biochemistry of olfaction, one approach would surely be to physically isolate and separate the major cell types from the olfactory neuroepithelium in order to study *in vitro* their individual contributions to olfaction. In mammals, the olfactory epithelium contains four major cell types (Graziadei and Graziadei, 1978). The basal cells are located in one or two layers just distal to the basal lamina and are distinguished in thin sections by their nuclear shape and position in the epithelium. The flask-shaped olfactory neurons are found in several layers distal to the basal cells and are organized into vertical columns extending from just above the basal cells to the free surface of the epithelium. Neuronal nuclei have a dense pattern of chromatin and these cells taper into a thin dendrite, which expands into a cilia-bearing olfactory vesicle at the surface of the epithelium. The nuclei of the sustentacular cells form a layer distal to the neuron layers, and these cells surround the neurons and bear many microvilli, which extend into the mucus layer that covers the whole epithelium.

Biochemistry of Taste and Olfaction

ISBN 0-12-154450-8

The fourth cell type is the secretory cell of the Bowman's glands, which are found below the basal lamina. The ducts of these glands extend through the whole epithelium and open on to the surface of the tissue. Although not part of the olfactory epithelium, ciliated respiratory epithelial cells are often closely associated with the tissue as well.

Ideally, one would like to obtain purified populations of each of these cell types for a variety of *in vitro* studies. For example, questions relating to odorant recognition by olfactory neurons could be addressed by purifying these cells, isolating their membranes, and solubilizing and characterizing the odorant-binding molecules from the membranes. Direct binding of radiolabeled odorant amino acids to partially purified olfactory cell preparations has been performed already (see below). Such *in vitro* binding techniques have been applied to neurotransmitter and drug receptors extracted from brain and electroplaque membranes, and many of these soluble receptor preparations retain the desired ligand-binding characteristics (Lowy *et al.*, 1976; Valderrama *et al.*, 1976; Simon and Hiller, 1977; Yousoufi *et al.*, 1979; Greenlee and Olsen, 1979). The role played by basal cells in neurogenesis and regeneration in the olfactory neuroepithelium (Graziadei and Graziadei, 1978) could be studied by isolating basal cells and investigating their growth and differentiation in cell culture. Development of basal cells into neurons *in vitro* has been observed in embryonic olfactory placodes maintained in culture (Farbman, 1977), but it would clearly be advantageous to have cultures of pure basal cells. In this way, the factors and signals involved in the transformation of basal cells into olfactory neurons could be more easily manipulated. Perhaps mixed cultures with olfactory bulb cells would permit study of the mechanisms by which olfactory neurons interact with their targets.

The contributions of sustentacular cells to the ionic mechanisms necessary for olfactory action-potential generation and stimulus-coupled cell secretion (Getchell, 1977) could be more easily investigated using isolated cells. If sustentacular cells could be maintained in culture one could, for example, add fractions prepared from olfactory neurons to them and study the interactions between these two cell types more directly. Similarly, the biosynthesis and secretion of the mucosubstances covering the epithelium by cells of the Bowman's glands (Cushieri and Bannister, 1974) could be investigated using isolated cells. Such cells could possibly serve as a source of mucus permitting direct studies of the solubility and metabolism of odorants.

Isolated olfactory cells could be used for studies of the fate of odorant molecules after they bind to olfactory receptors. Are odorants internalized in neurons or other cells after binding, or are they dealt with extracellularly? Which cells are involved in dealing with odorants? Mozell and co-workers studied this with the intact epithelium (Hornung and Mozell, 1977a,b; see also this volume, Chapter 2), but this aspect would be facilitated by using isolated olfactory cells. Also, interaction of environmental pollutants and noxious chemicals with the olfactory epithelium could be investigated more specifically using isolated cells. These are a few of the possible questions that can be studied with cells isolated from the olfactory epithelium.

However, few studies of isolated olfactory cells have appeared in the literature, and expectations about this approach have not yet been realized. In this chapter we will discuss these published studies and attempt to analyze why isolation of olfactory cells has proved to be so difficult. We will also present some pharmacological studies of the olfactory epithelium and suggest how such an approach might be used to isolate olfactory cells.

II. OLFACTORY CELL SEPARATION

In 1966, Ash and co-workers at Honeywell Corporation published studies on suspensions of rabbit olfactory cells (Ash *et al.*, 1966). Their aim was to isolate the various cell types and look at odorant recognition. Rabbit olfactory tissue was stripped from the turbinate bones and minced in cold buffer with scissors. The tissue was then homogenized in a loose-fitting Potter–Elvehjem-type grinder. The resulting homogenate was then filtered through gauze, and the cells were recovered by centrifugation at 700 *g* for 5 min. The cells were washed by resuspension and recentrifuged. Using this method, about 21% of the protein in the homogenate was recovered in the cell pellet.

These workers published good quality photomicrographs of an olfactory neuron, a sustentacular cell, and a respiratory epithelial cell found in their preparations, but stated that the cells were easily damaged by their manipulations. The photographs depict the cells essentially as they appear in thin sections of the olfactory epithelium with no evidence of swelling or loss of the characteristic flask-shaped morphology. Subsequent work in our laboratory failed to confirm this observation (see below).

Ash *et al.* (1966) reported cell yields of 4–7.5 $\times$ 10^5 cells per mg of epithelial protein depending on the age of the rabbits used. Older rabbits yielded more total cells and slightly more respiratory cells in relation to olfactory cells than young animals. Young rabbits were $\leq$ 2 months old and adult rabbits were $\geq$ 4 months old. These workers reported that the number of olfactory cells decreased with storage of the suspensions in the cold (50% loss after two weeks), but stated that the cells retained their morphology after freezing in liquid nitrogen. The former but not the latter observation was confirmed by work in our laboratory (J. D. Hirsch and F. L. Margolis, unpublished). They also reported that attempts to use proteolytic enzymes, such as trypsin or bromelain, to dissociate epithelia were unsuccessful because the cells were destroyed. In addition, shaking the tissue with glass beads caused extensive destruction of the cells; the beads could not be used to dissociate epithelia.

These workers used the activity of cytochrome oxidase as a marker for cellular breakage. They reasoned that since this enzyme is located in intact mitochondria and that mitochondria would not sediment at 700 *g*, any enzyme activity recovered in the cell pellet was due to intact olfactory cells. Addition of deoxycholate to the cell pellets resulted in substantial activation of cytochrome oxidase activity (tenfold), but detergent treatment of supernatant fractions led to no or minimal activation of the enzyme. About 20% of the total activatable cytochrome oxidase activity was recovered in the cell pellets. Thus, it was suggested that a portion of the cells survived the isolation procedure.

Ash and co-workers (1966) stated that their cell suspensions would be subjected to density-gradient centrifugation so that the individual cell types could be separated and studied in detail. Unfortunately, no subsequent papers on this particular aspect of olfactory research appeared from the Honeywell group, and it might be concluded that further separation of rabbit olfactory cells proved to be more difficult than anticipated.

For the next eleven years, no other publications appeared on this subject. Cancalon and Beidler (1977) and Cancalon (1978) reported on the isolation of olfactory cells from the catfish *Ictalurus punctatus*. These workers adapted a combined enzymatic–mechanical disruption procedure to dissociate olfactory tissue. Olfactory rosette tissue from the catfish was first treated with a low concentration of trypsin (0.5 mg/ml) for 20 min at 37°C. Then the tissue was treated with 2 mg/ml of DNase followed by EDTA and trypsin-inhibitor in Ca^{2+}–Mg^{2+}-free medium. After several more washes with this medium, the tissue was incubated for 30 min at 0°C in this medium. The olfactory lamellae were then freed of adhering tissue, and cell suspensions were generated by vibratory shaking. The cells were purified by centrifuging the suspensions at 100 *g* for 6 hr in discontinuous Ficoll gradients prepared in siliconized tubes. At the start of the centrifugation, each gradient contained from top to bottom: the cell suspension, a layer of 10% Ficoll, a layer of 20% Ficoll, a layer of 30% Ficoll, a layer of 33% Ficoll, a layer of 37% Ficoll, and a cushion of 50% sucrose.

The cell types concentrated at the various Ficoll interfaces. Mucus and unidentified "round" cells were found on top of the 10% Ficoll, while respiratory epithelial cells collected at the 10–20% interface. Several different cell types were found at the 20–30% interface. A mixture of 60% sustentacular cells and 40% olfactory neurons was found at the 30–33% interface. These percentages were reversed at the 33–37% interface. Connective tissue and undissociated cells were found between the 37% Ficoll and the sucrose.

The isolated olfactory cells synthesized RNA from [^{3}H]uridine and 70–80% were viable as measured by dye exculsion. These cells represented about 5% of the protein and 11% of the DNA of whole catfish olfactory tissue with an overall yield of about 20%. Cancalon (1978) also tried to dissociate catfish olfactory rosettes with pronase, high concentrations of trypsin, collagenase and hyaluronidase, and dithiothreitol, but he observed that cells produced by these methods lost their ability to specifically bind radiolabeled amino acids. The collagenase–hyaluronidase method was reported to be particularly good for dissociation, although it was very destructive to amino acid binding.

The morphology of the cells was monitored by both light and scanning electron microscopy. The scanning electron micrographs published show isolated olfactory neurons, sustentacular cells, and respiratory epithelial cells (Cancalon and Beidler, 1977; Cancalon, 1978). Cancalon and Beidler (1977) also published a phase-contrast micrograph of the cell suspensions before fractionation. This revealed that the suspensions contained predominantly single cells and small clumps, which became well-separated by their gradient technique. In all of the photographs, the morphology of the cells was excellent and their flask-shaped nature was retained *in vitro*. The trypsin–DNase–EDTA method yielded cells with the best preserved morphology. Binding of 12 radiolabeled amino acids to the cells was measured and an excellent correlation ($r = 0.9$) was obtained between the amounts of amino acids bound and the electrophysiological activity of the same amino acids in the catfish. This suggested that isolated catfish olfactory cells retained the odorant-binding capabilities they have in the intact epithelium. The correlation between the amino acid binding data and the electrophysiological results they obtained in the trout was somewhat lower ($r = 0.6$). Significantly, although isolated respiratory cells also bound amino acids, there was no correlation between amounts bound and olfactory electrophysiological response *in vivo*.

In 1976, our laboratory began research on the isolation and separation of olfactory cells from the rat olfactory neuroepithelium (Hirsch and Margolis, 1977, 1979a). As suggested in the introduction, our aims were to obtain the various cell types and to study their individual biochemical characteristics. Initially we attempted to replicate the method of Ash *et al.* (1966), but all experiments were unsuccessful. Occasionally a single neuron or a respiratory epithelial cell was seen under the phase contrast microscope, but cell destruction was too extensive to be practical. Instead a previously published method for the separation of isolated pancreatic exocrine cells (Amsterdam and Jamieson, 1972) was modified for use with rat olfactory epithelia.

Before describing the method several important points must be emphasized. First, in contrast to the report of Ash *et al.* (1966), epithelia from young rats (3–4 weeks old) yielded more cells than did adult rats; and epithelia from young rats were used throughout our work. Second, based on previous work from our laboratory showing the detrimental effects of upper respiratory disease on olfactory morphology and biochemistry (Harding *et al.*, 1977), specific pathogen-free female WKY rats obtained from Laboratory Animal Resources of Hoffmann-LaRoche, Inc., were used for this work. These animals had no significant titers to the eleven common rodent viruses, and they were free of endo- and ectoparasites, as well as any known bacterial, mycoplasmal, protozoan, fungal, or yeast pathogens. These animals were used as soon as possible after arrival in our colony and were discarded after 10 days, because they tended to develop rhinitis due to exposure to animals from commercial breeders. The animals were asphyxiated with CO_2 and thoroughly exsanguinated.

Olfactory tissue was obtained by rapidly removing the ethmoid turbinates and mucosal sheets on both surfaces of the nasal septum and placing the tissue on ice. A flow diagram of the cell separation procedure is shown in Fig. 1. The tissue was rinsed 4 times with Krebs–Ringer phosphate buffer (buffer) pre-equilibrated with 95% O_2–5% CO_2; each of the four washes was saved. The tissue was then minced into small pieces and combined with collagenase, hyaluronidase, and trypsin inhibitor in buffer. The four tissue washes were centrifuged at 50*g*, and the pellets were carefully resuspended in buffer and combined with the minced tissue and enzymes. It should be emphasized that no buffer that came in contact with the epithelia was ever discarded, because cells were found in all such fractions. The tissue–enzyme mixture was then incubated for 15 min at 37° with gentle rotary shaking. The medium was removed and saved on ice. The tissue was then incubated twice more at 37°C for 5 min in Ca^{2+}–Mg^{2+}-free buffer containing EDTA. These 2 EDTA washes were also saved on ice. The tissue was then rinsed with buffer containing Ca^{2+} and Mg^{2+}. This wash was removed and saved on ice as well.

These four incubation media were centrifuged at 50*g* to recover the cells, and the four pellets obtained were gently resuspended in buffer containing the enzymes. The mixture was then incubated for 30 min at 37°C. During this time, the tissue mass remaining after the first digestion was incubated again with enzymes for 1 hr at 37°. At the end of these incubation periods, the two digestion flasks were chilled, and the material in each was pipetted slowly up and down 5 times in glass pipettes with 2–3 mm apertures and then 5 times in pipettes with 0.5–1 mm apertures. This step was important for good cell recovery, but pipeting too rapidly led to extensive cell rupture. The bone chips from the tissue digestion were removed by filtering the contents of the flask through two layers of fine cheesecloth. The fibrous debris in the other digestion was removed by gently forcing the suspension through a small sheet of 70-μm-mesh nylon bolting cloth that was fastened

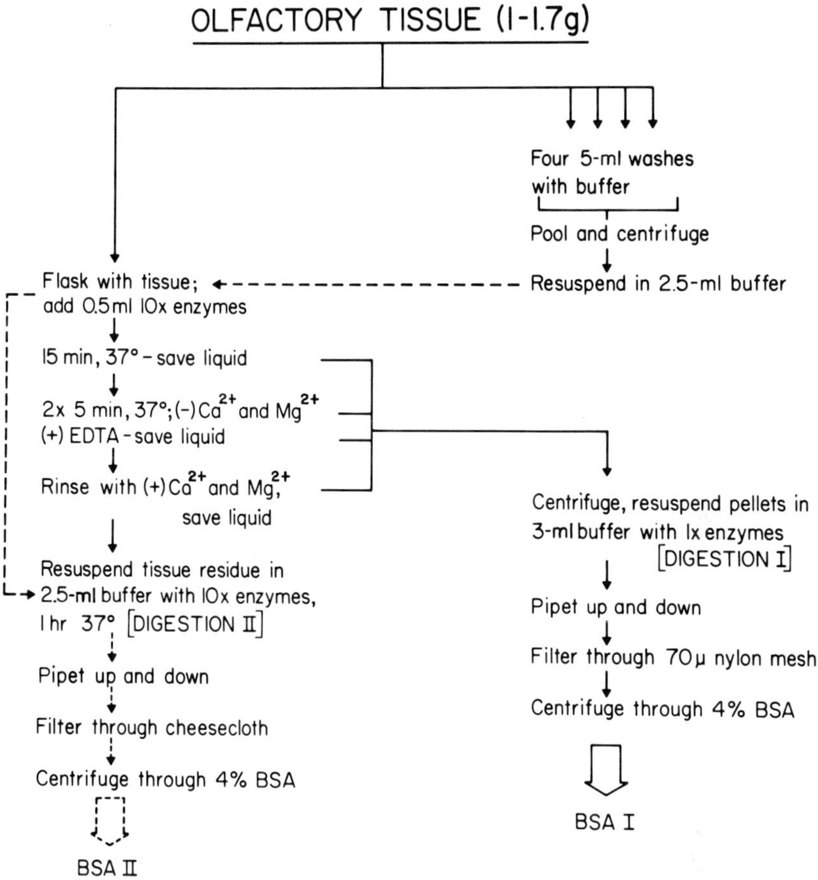

Fig. 1. Procedure for olfactory cell isolation. Rat olfactory epithelia were obtained from 4–6 female WKY rats of 3–4 weeks of age. See text for procedure. (From Hirsch and Margolis, 1979a.)

over the end of a syringe. Again, too rapid or forceful straining at this point led to increased cellular destruction. Clearly the cells are very sensitive to shear forces.

At this stage, both flasks contained suspensions of olfactory cells, but they also contained a considerable amount of debris, mucus, erythrocytes, broken cells, the enzymes, and other unwanted material. To remove these materials, the two cell suspensions were centrifuged twice through cushions of 4% bovine serum albumin in buffer. The pellets were resuspended in buffer that was equilibrated with 95% O_2–5% CO_2. At this point, the pellet obtained from the 1-hr incubation of the tissue mass was designated BSA II, and the pellet obtained from the initial digestion and washings from the tissue mass was designated BSA I.

Originally the tissue mass was not incubated with enzymes a second time. Since microscopic examination of the bone chips revealed that patches of epithelium still remained, a second incubation was included in an attempt to recover these cells. This second population of cells had a different biochemical profile than the first, more easily dissociated, population (see below).

We then characterized the cell populations by phase contrast microscopy. Cell viability, judged by the ability of the cells to exclude a 1% solution of nigrosin dye, was usually 80–90% and similar to values published by Cancalon (1978). In addition, we routinely observed synchronously beating cilia on respiratory epithelial cells. Seeing such cells "swimming" in the suspensions was usually a sign that a healthy preparation was obtained. Figure 2 shows three representative olfactory neurons in the suspensions. These cells had a few spike-like cilia. We also observed numerous smaller olfactory neurons with shorter dendrites and a large number of fine hair-like olfactory cilia (not shown). A sustentacular cell is shown in Fig. 3 and a respiratory epithelial cell in Fig. 4. Comparing the photographs of our neurons with those published by Cancalon and Beidler (1977) and Cancalon (1978), olfactory cilia appeared to be preserved much better by our procedure.

In contrast to previous reports (Ash *et al.*, 1966; Cancalon, 1978), cells in our preparations always swelled in suspension, but they retained enough of their characteristic morphology to be identified. Manipulation of both the ionic constituents and osmolarity of the buffer failed to prevent this swelling.

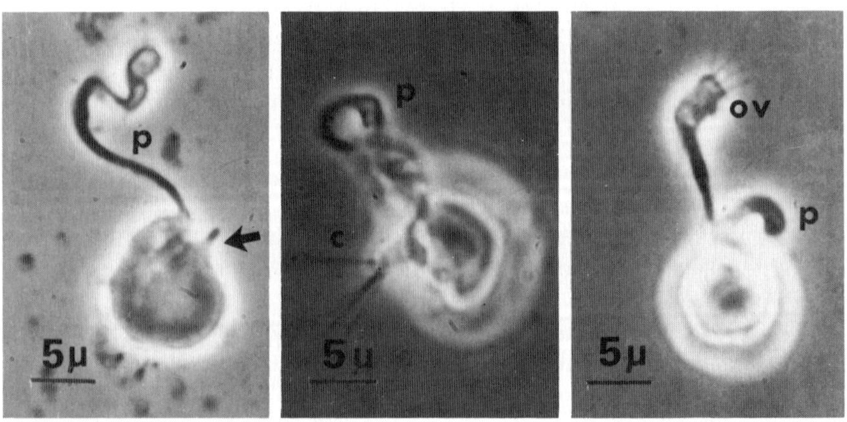

Fig. 2. Appearance of neurons from rat olfactory cell suspensions. These neurons possessed long, usually twisted dendrites (p) and coarse spike-like cilia (c) on the olfactory vesicle (ov). (From Hirsch and Margolis, 1979a.)

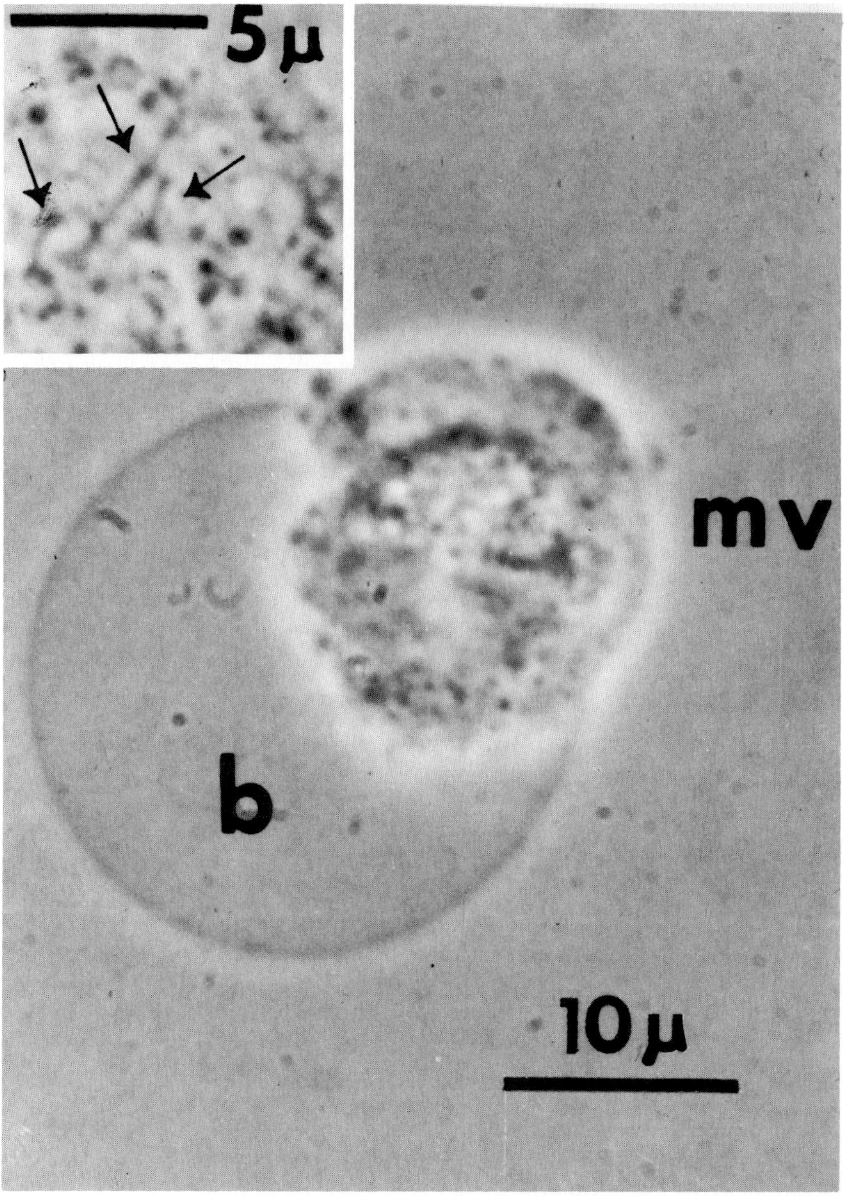

Fig. 3. Appearance of a sustentacular cell in rat olfactory cell suspensions. These cells had a spherical cell body (b) and a cap of microvilli (mv). In the inset, three prominent microvilli from another sustentacular cell are indicated by arrows. (From Hirsch and Margolis, 1979a.)

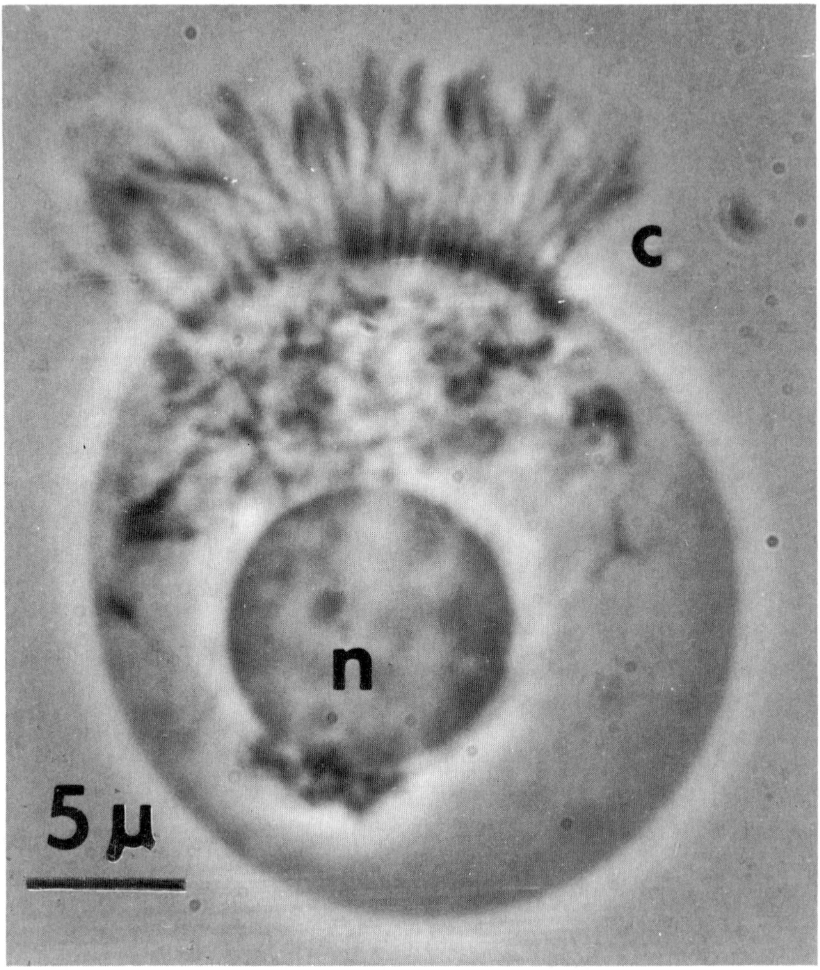

Fig. 4. Appearance of a respiratory epithelial cell in rat olfactory cell suspensions. Note the large round nucleus (n) and the array of cilia (c), which beat synchronously *in vitro*. (From Hirsch and Margolis, 1979a.)

Unfortunately, we were unable to distinguish basal cells in our suspensions, but there were many large and small spherical "phase-bright" and "phase-dark" cells in the suspensions that may have been basal cells, cells of the Bowman's glands, or immature neurons that did not yet possess a dendrite (Graziadei and Graziadei, 1978). We also found that respiratory epithelial cells and sustentacular cells readily aggregated in suspension, whereas olfactory neurons were almost always observed singly. The presence of bovine

TABLE I Olfactory Cell Neurochemical Markers in Olfactory Cell Populations[a]

	OMP (ng/mg protein)				Carnosinase (nmole/mg protein/hr)			
	1	2	3	4	1	2	3	4
BSA I	108	626	770	660	3.42	2.95	5.67	9.46
BSA II	47	397	316	398	8.67	5.49	8.72	27.80

[a] BSA I and BSA II were prepared from rat olfactory epithelia as described in the text and as shown in Fig. 1. Quantities of OMP were determined by RIA and carnosinase activity was determined as described in the text. Data from four representative experiments are shown. (From Hirsch and Margolis, 1979a.)

serum albumin promoted this aggregation, but removing Ca^{2+} and/or Mg^{2+} ions from the buffer did not. As discussed below, the aggregation-promoting property of bovine serum albumin was exploited in further purifying the cell suspensions. This feature is in direct contrast to the report of Cancalon (1978), who found that catfish cell reaggregation was reduced by bovine serum albumin. This may reflect species differences or different modifications of the cells by the two enzymatic treatments.

The two cell pellets obtained in these experiments contained between 12–25% of the DNA and 8–10% of the protein of intact olfactory epithelia, and yields of 14.4–76.8 × 10^4 cells/mg of tissue were obtained in several experiments. Our yields of protein and DNA were very similar to those obtained by Cancalon (1978). As alluded to previously, rat olfactory cells were extremely delicate and susceptible to rupture. When we determined the presence of olfactory marker protein (OMP, for review see Margolis, 1980) in the buffers and incubation media as a measure of neuronal destruction and of carnosinase (Harding and Margolis, 1976; Harding *et al.*, 1977) as a measure of nonneuronal cell rupture, substantial quantities of both markers were found in supernatants at all steps in the procedure. Because ionic and osmolarity manipulations did not alleviate this problem, we concluded that the variations in yield were due to the extremely delicate nature of the cells of the rat olfactory epithelium.

The two cell populations, BSA I and BSA II, were then biochemically characterized using the presence of olfactory marker protein (OMP) as a neuronal marker (Harding and Margolis, 1976; Harding *et al.*, 1977) and carnosinase as a nonneuronal cell marker (Harding *et al.*, 1977). Carnosinase activity was quantitated by the [^{14}C]β-alanine released from [^{14}C]carnosine by cell extracts, and OMP was determined in the extracts using the solid-phase radioimmunoassay (RIA) (Keller and Margolis, 1975). Table I shows results from four different cell isolation experiments. Although yields of markers were variable, BSA I always contained 2–2.5 times more OMP/mg of protein than did BSA II. Conversely, BSA II consistently contained 2–3 times more carnosinase activity than BSA I. In keeping with these data, it was found that cells from BSA I incorporated more [^{14}C]β-alanine into carnosine than did cells from BSA II. However, both cell populations incorporated [^{3}H]uridine into RNA and [^{14}C]histidine into protein to about the same extent.

At this stage, we began fractionating the two cell populations to obtain samples of olfactory neurons and nonneuronal cells. The cell suspensions were layered over discontinuous gradients of bovine serum albumin in a buffer prepared in cellulose nitrate ultracentrifuge tubes. The albumin was diluted from a dialyzed 20% solution. From top to bottom each gradient contained: the cell suspension in buffer, a layer of 4% albumin, a layer of 10%

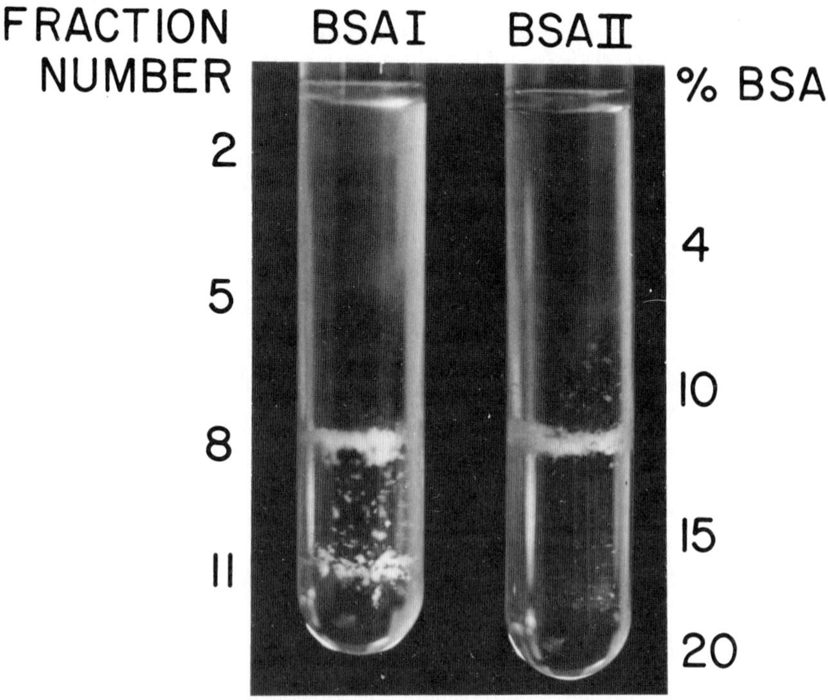

Fig. 5. Appearance of discontinuous BSA gradients after olfactory cell fractionation. The positions of the four different concentrations of BSA (bovine serum albumin) are indicated on the right, and the fraction number containing the four major cell bands are shown on the left. Cells were allowed to sediment at unit gravity into these gradients for 3 hr at 4°C, and fractions were obtained by piercing the tubes with a small bore needle and removing aliquots with a syringe.

albumin, a layer of 15% albumin, and a cushion of 20% albumin. The cells were allowed to settle for 2–3 hours at unit gravity in the cold room. The gradients were fractionated by piercing the side of the tube with a small-bore needle and removing samples of the desired volume with a syringe. At this step, it was important to draw out the cell samples carefully to avoid lysis. In these gradients, the cells tended to form distinct bands at the different albumin interfaces (Fig. 5), and each band constituted its own fraction. Each gradient fraction was then centrifuged at 50 *g* and resuspended in the appropriate buffer for use in further experiments.

Figure 6 shows the cell numbers and amounts of DNA in typical gradient-fractionated samples of BSA I and II. BSA I contained two major peaks of cells in the upper half of the gradient, whereas the cell distribution of BSA II was more homogeneous. As mentioned above the bovine serum albumin promoted aggregation of the nonneuronal cells, and cells in the lower gradient fractions occurred in clumps containing 10–15 cells. Cells in the upper gradient fractions (1–6) were found either singly or in small clumps of 2–3 cells. Morphological examination of cells in the upper gradient fractions of either gradient revealed the presence of many olfactory neurons. Neurons with long dendrites and a few spike-like cilia (see Fig. 2), as well as short dendrites with fine cilia, were seen. Toward the bottom of the gradient, a greater number of sustentacular and respiratory epithelial cells were seen. Unidentifiable cells of several types were found in all of the gradient fractions.

On the basis of morphology, we suspected that partial separation of

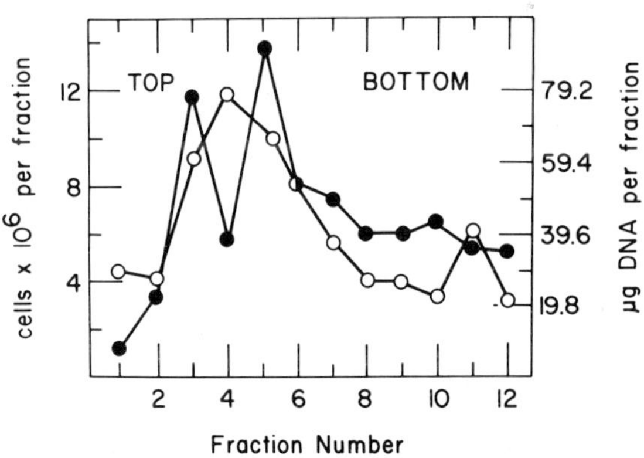

Fig. 6. Cell number per fraction in discontinuous BSA gradients allowed to sediment for 3 hr at unit gravity. BSA I and BSA II were obtained and fractionated as described in the text. Fractions were obtained by syringe through the sides of the tubes and cell number was determined by quantitation of DNA. BSA I, (●——●); BSA II, (○——○). (From Hirsch and Margolis, 1979a.)

neuronal from nonneuronal cells had been achieved, but more definitive biochemical data were required. Figure 7 shows the results of experiments designed to evaluate the gradient fractions for various olfactory neurochemical markers. BSA I contained several peaks of OMP, indicating at least three size classes of neurons present in this cell population. Carnosine synthetase activity, another neuronal marker (Harding and Margolis, 1976; Margolis and Grillo, 1977; Harding *et al.*, 1977), was assayed by measuring incorporation of [^{14}C]β-alanine into [^{14}C]carnosine. This marker was coincident with the distribution of OMP in both gradients. The nonneuronal marker carnosinase gradually increased from the top to the bottom of the gradient. A single peak of neuronal markers (OMP, carnosine synthetase) was observed at the top of the gradient in BSA II, whereas a broad peak of carnosinase activity was found in the bottom two-thirds of the gradient. Although the carnosinase activity was not localized to any one fraction, it did not substantially overlap the peaks of OMP and carnosine synthetase. These data suggested a separation of olfactory neurons from nonneuronal cells and, moreover, that BSA II contained a more homogeneous population of neurons than BSA I. As stated above, BSA II contained more carnosinase activity than BSA I, even after fractionation.

To obtain still more information about the apparent separation of olfactory

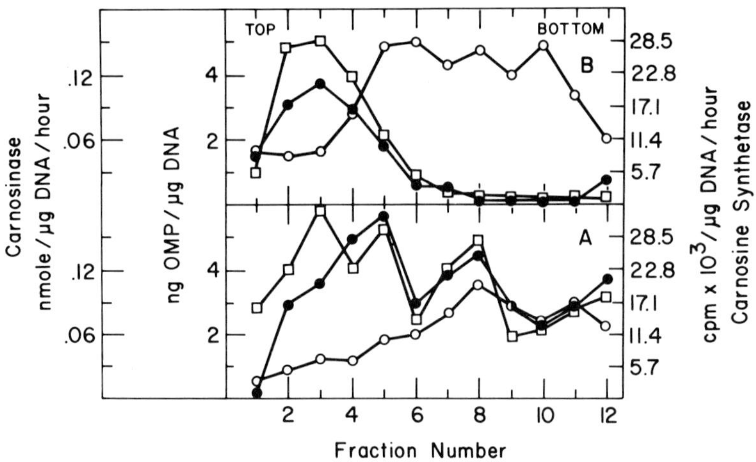

Fig. 7. Distribution of olfactory neurochemical markers in fractionated olfactory cells. BSA I and II were prepared and fractionated as described. OMP was determined by RIA; carnosinase was determined as described in the text; carnosine synthesizing activity was determined by measuring incorporation of [^{14}C]β-alanine into [^{14}C]carnosine. Results are expressed as activity per μg of DNA. OMP, (□——□); carnosine synthetase, (●——●); carnosinase, (○——○). (A) BSA I; (B) BSA II. (From Hirsch and Margolis, 1979a.)

cells, a variety of immunocytochemical and histochemical techniques were employed with the gradient fractions. The initial problem was to get the cells to adhere to glass slides for processing. This problem was solved by using poly-L-lysine coated slides (Mazia *et al.*, 1975), which resulted in rapid and irreversible adherence of the cells to the poly-L-lysine. Although in most cases the cells flattened substantially, the presence of several cell markers was unaffected by the loss of morphology. We used the immunoperoxidase staining method of Graziadei *et al.* (1977) to localize OMP and S-100 protein, a nonneuronal marker (Zomzely-Neurath and Walker, 1980), in cells in the gradient fractions. Our evaluation of the results was facilitated by the observations that OMP is restricted to mature olfactory neurons and is not found in basal, sustentacular, or respiratory epithelial cells (Graziadei *et al.*, 1977; Farbman and Margolis, 1980) and that S-100 protein is associated with nonneuronal cells (Graziadei, Neurath, and Margolis, unpublished). We observed a large number of cells staining for either antigen in both BSA I and BSA II, but in most cases an unequivocal determination of their identity could not be made due to the loss of distinctive morphology on poly-L-lysine. An exception is shown in Fig. 8, where 3 OMP-positive cells which are clearly olfactory neurons were found in a field with several negative cells. In both BSA I and BSA II, cells staining for OMP slightly outnumbered those staining for S-100. This was particularly clear in the upper fractions of both gradients and was especially striking at the top of BSA I, where OMP-positive cells outnumbered S-100-positive cells by almost fivefold. There was greater overlap of the distributions of cells staining for either of these markers in the gradient fractions than was revealed by our biochemical measurements. This was undoubtedly due to the greater sensitivity of the immunoperoxidase technique and the difference in bulk measurements versus cell identification.

The cells in the gradient fractions were also stained for carbohydrate materials. Cushieri and Bannister (1974) reported that Bowman's gland cells in the olfactory epithelium of mice were the only ones stained by Alcian Blue at pH 1.0 due to the presence of acidic sulfomucins in these cells. In our hands almost 3 times as many cells stained with Alcian Blue at pH 1.0 in BSA II than in BSA I; particularly in BSA II, the number of stained cells increased markedly from the top to the bottom of the gradient (Figure 9). A comparison of Fig. 7 with Fig. 9 shows that a clear separation of Bowman's cells from neurons was achieved. The stained cells were large and polygonal with a round centrally located nucleus, and they contained a cluster of intense blue granules at one end of the cell. When the Alcian Blue (pH 1.0) technique was followed by immunoperoxidase staining for OMP and S-100, the Bowman's gland cells contained neither of these antigens.

Even though our data demonstrated only a partial separation of neuronal

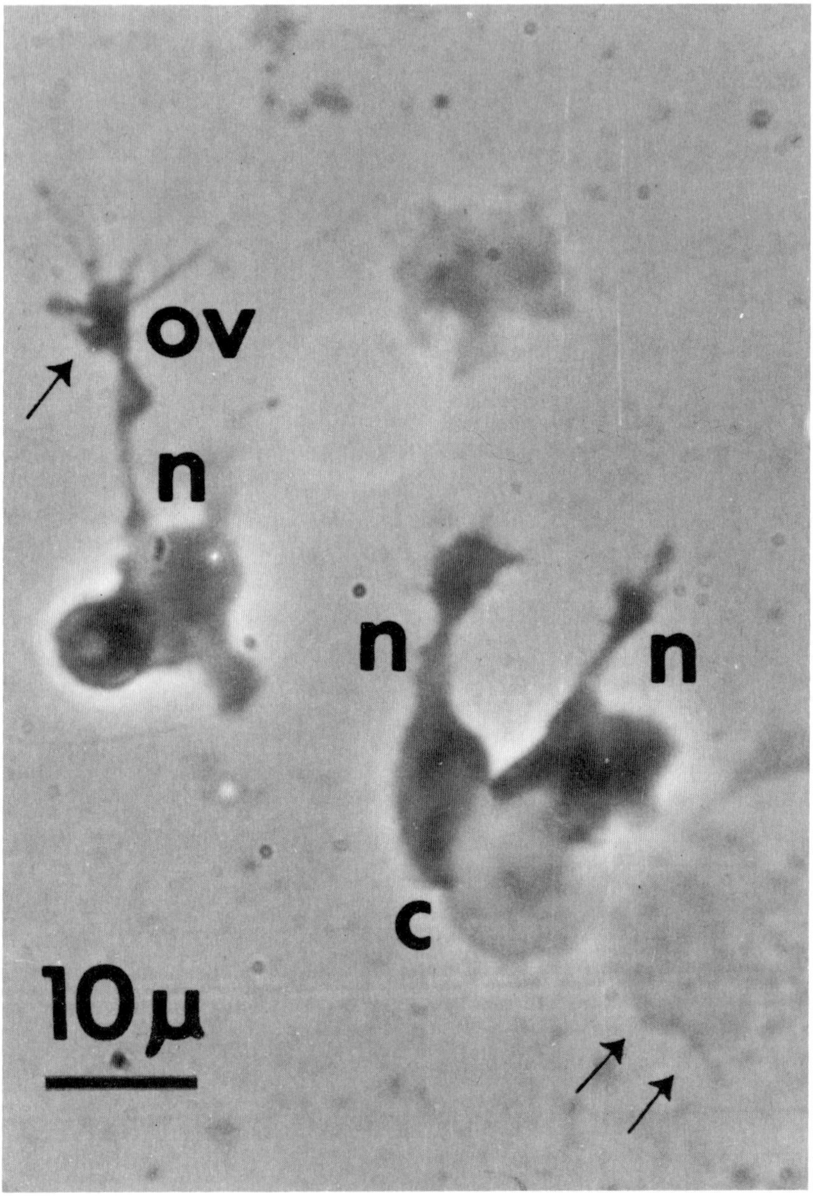

Fig. 8. Appearance of three clearly discernible olfactory neurons (n) stained for OMP by the immunoperoxidase technique. Two positive neurons are attached to a negative cell (c) at their bases. The olfactory vesicle (ov and arrow) and cilia of one cell are well-defined. One other negative cell is also present (double arrow). (From Hirsch and Margolis, 1979a.)

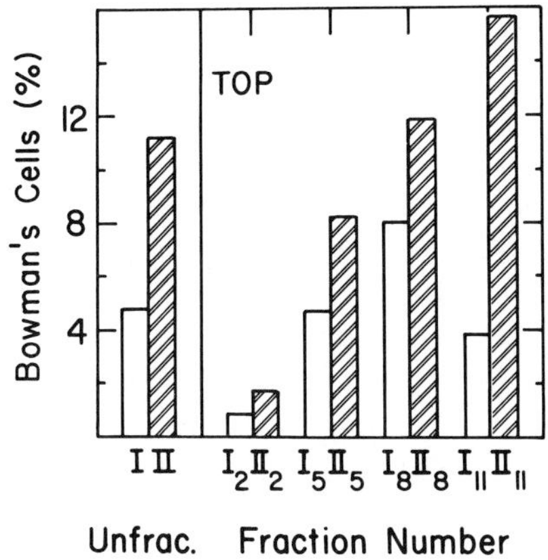

Fig. 9. Distribution of Bowman's gland cells stained with Alcian Blue at pH 1.0. Details of cell separation and other methods are given in the text. The number of blue-stained cells out of the total number counted was converted to a percentage. The left panel shows the percentage of Bowman's gland cells in unfractionated BSA I and II, and the right panel shows the percentage of Bowman's gland cells in BSA I and II fractions (indicated by subscripts). □, BSA I; ▨, BSA II. (From Hirsch and Margolis, 1979a.)

and nonneuronal cells, several additional experiments were performed. We tried to further purify neuronal and nonneuronal cell-containing fractions by both unit gravity and centrifugation gradient techniques but had no success. The extra time necessary for further purification led to a continuing problem of cell lysis and prevented further purification by these more conventional methods.

The wide variety of neurochemical markers available to us during these studies was both a blessing and a curse. On the positive side, these markers allowed us to analyze the cells from several points of view and provided information about the *in vitro* biochemistry of isolated olfactory cells. It was very encouraging to find that these markers were maintained in our cell suspensions and that such markers were useful in characterizing the fractionated cells. On the negative side, our ability to detect these markers with immunoperoxidase staining demonstrated that we had achieved only a partial separation or enrichment of the cells. Our analysis showed that olfactory neurons in the rat are not all of one size, but apparently represent a continuum of developmental stages. Thus, these data provide *in vitro* support for the work of Graziadei and Graziadei (1978), who reached this conclusion on anatomical grounds. Our work also shows that any kind of cell separation should be analyzed as thoroughly as possible. Reliance on any one criterion of purity exclusively could lead to overestimations of the homogeneity of the isolated cell fractions.

III. PHARMACOLOGICAL STUDIES OF THE OLFACTORY EPITHELIUM

Several other authors in this volume (see Gower, Chapter 1; Fesenko, Chapter 3; Price, Chapter 4) discuss the recognition and binding of odorant molecules by homogenates and other fractions of the olfactory epithelium. We have also performed ligand binding studies in the epithelium but with different aims than those of previous workers. We reasoned that if we could show that certain ligands bound to olfactory neurons, such ligands could be coupled to a solid support such as Sepharose to make an affinity column. Suspensions of olfactory cells could then be passed over the column to which the neurons would bind and the remainder of the cells would pass through. The neurons could then be eluted by washing the column with the appropriate competitor drug.

Specific binding of a number of radioligands to homogenates of mouse olfactory epithelium was investigated. The ligands were drugs known to bind to various neurotransmitter and drug receptors in the olfactory bulb and brain (Hirsch *et al.*, 1978; Hirsch and Margolis, 1979b; Nadi *et al.*, 1980). The binding assay methods were also virtually identical to those used with membrane fractions prepared with central nervous system tissue (Hirsch *et al.*, 1978; Hirsch and Margolis, 1979b; Nadi *et al.*, 1980). The mouse olfactory epithelium contained binding sites for [^{3}H]diazepam, a benzodiazepine; for [^{3}H]clonidine, WB-4101, and dihydroergocryptine, which are α-adrenergic ligands; [^{3}H]dihydroalprenolol, a β-adrenergic ligand; [^{3}H]quinuclidinyl benzilate, a muscarinic cholinergic ligand; and L-[^{3}H]carnosine, the olfactory peptide neurotransmitter candidate. Pharmacologically these binding sites had characteristics similar to sites in the brain (Snyder and Bennett, 1976), except for the [^{3}H]diazepam site, which was the "non-brain" type (Braestrup and Squires, 1977), and the L-[^{3}H]carnosine site, which was not saturable (Hirsch, unpublished). The epithelium was devoid of specific opiate binding ([^{3}H]dihydromorphine), of specific GABA binding ([^{3}H]muscimol), and of specific kainic acid binding ([^{3}H]kainic acid). We attempted to determine whether several of these binding sites were localized to olfactory neurons by performing unilateral olfactory bulbectomies on mice and measuring ligand binding in the epithelium ipsi- and contralateral to the bulbectomy. Because only olfactory neurons disappear from the epithelium after bulbectomy (Clark, 1957; Takagi, 1971), any ligand binding sites that decreased after the lesion were presumably located on the neurons. The results of these studies are shown in Table II. By 30 days after bulbectomy, there was a 75% decline in muscarinic cholinergic binding on the side ipsilateral to the bulbectomy. In contrast, [^{3}H]clonidine and [^{3}H]dihydroergocryptine binding to α-adrenergic sites in-

TABLE II Effect of Unilateral Olfactory Bulbectomy on ^{3}H-labeled Ligand Binding in Mouse Olfactory Epithelia[a]

^{3}H-labeled ligand (nM)	Exp. No.	Specific binding (fmole/mg protein)		
		Normal	Unoperated side	Bulbectomized side
[^{3}H]Quinuclidinyl benzilate (1.0)	1	54	50	12
	2	70	73	27
[^{3}H]Clonidine (3.7)	1	95	128	165
	2	52	65	98
[^{3}H]Dihydroergocryptine (5.0)	1	46	118	188

[a] Unilateral olfactory bulbectomies were performed 30 days before the binding assays were performed. Normal refers to unoperated control animals. Unoperated side refers to the portion of epithelium obtained contralaterally to the removed bulb. Bulbectomized side refers to the portion of epithelium obtained ipsilaterally to the removed bulb. Specific binding of [^{3}H]quinuclidinyl benzilate was determined with and without 1 μM atropine, and specific binding of [^{3}H]clonidine and [^{3}H]dihydroergocryptine were determined with and without 10 μM phentolamine. Results are means of triplicate determinations which varied about 15%.

creased after bulbectomy on both sides of the epithelium. A larger increase was observed on the operated side. There were no changes in [^{3}H]dihydroalprenolol binding on either side of the epithelium after bulbectomy, but there was a 30% decline in [^{3}H]diazepam binding on the bulbectomized side (not shown). It remains to be determined if the [^{3}H]quinuclidinyl benzilate was associated with olfactory neurons, but preliminary data suggest that this may be the case (J. D. Hirsch, unpublished).

IV. CONCLUSIONS AND SUGGESTIONS FOR FUTURE RESEARCH

Separation of olfactory cells for *in vitro* study has so far been a difficult and frustrating task. The olfactory epithelium is a delicate tissue, and interference with its integrity can easily damage the cells. It is significant that in all of the studies reviewed here, cell yields were low. Also, mammalian and probably fish olfactory neurons apparently do not exist as a single size class, but are found in several sizes. Thus we suspect that olfactory cell separations based on cell size or density may continue to be unsuccessful. Before ruling such techniques out, however, we recommend that the Ficoll gradient technique used by Cancalon (1978) to separate catfish olfactory cells be tried with a mammalian preparation. This method may result in separation of the cells without excessive damage and with better morphology.

Another possibility would be to obtain cell suspensions from epithelia at short times after bilateral olfactory bulbectomy. This would remove neurons from the tissue and facilitate isolation of nonneuronal cells. Later, a population of synchronized neuronal or basal cells might be obtainable. Also, further research using labeled ligands that might be coupled to affinity columns may lead to purification of olfactory cell types. Our preliminary results with [^{3}H]quinuclidinyl benzilate suggest that an affinity solid support prepared with this ligand might be a powerful tool for neuronal separation.

In the event that further purification of olfactory cell types is achieved, studies of their membrane proteins should be made. In particular, isolation and purification of cell-type-specific proteins should be undertaken. Antibodies to these proteins, if coupled to magnetic microspheres (Molday *et al.*, 1977), could be mixed with olfactory cell suspensions. Application of a magnetic field to the preparation would then allow the cells that selectively bind to the beads to be recovered. This approach has been used to label and separate lymphocytes (Molday *et al.*, 1977), and it simply represents a special case of a general approach.

Efforts should be made to maintain separated cells from olfactory epithelia in culture. In particular, it would be advantageous to have cultures of basal cells and to investigate if their differentiation into neurons can be induced in culture. Aside from the work of Farbman (1977), little is known about the factors necessary for maturation of basal cells and the olfactory placode *in vitro*. One way of tracking basal cells would be to isolate them from epithelia prelabeled *in vivo* with [^{3}H]thymidine (Graziadei and Graziadei, 1978). This could be combined with the olfactory-nerve section technique (Harding *et al.*, 1977) to stimulate the basal cells to divide before labeling and isolating them.

The olfactory-cell separation studies reviewed in this chapter might be viewed as exploratory efforts. The method of Ash *et al.* (1966) is too destructive for general use, but the method developed by Cancalon and Beidler (1977) and Cancalon (1978) seems to be promising. This is especially true since the morphology of the cells was preserved much better than with our method. If such a technique were applied to mammalian cells, in combination with analysis of the wide variety of neurochemical markers used by us, a breakthrough in olfactory cell separation might occur, and studies such as those suggested could be performed.

ACKNOWLEDGMENT

The work by J. D. Hirsch was supported by a Postdoctoral Fellowship from the Roche Institute of Molecular Biology. Dr. Claire Z. Neurath generously provided antiserum to the

S-100 protein. We thank Carole Ryan for excellent secretarial assistance. Permission to use Figures 1,2,3,4,6,7,8,9, and Table I was granted by Elsevier/North Holland Biomedical Press, Amsterdam, The Netherlands.

REFERENCES

Amsterdam, A., and Jamieson, J. D. (1972). Structural and functional characterization of isolated pancreatic exocrine cells. *Proc. Natl. Acad. Sci. U.S.A.* **69,** 3028–3032.

Ash, K. O., Bransford, J. E., and Koch, R. B. (1966). Studies on dispersion of rabbit olfactory cells. *J. Cell Biol.* **29,** 554–561.

Braestrup, C., and Squires, R. F. (1977). Specific benzodiazepine receptors in rat brain characterized by high affinity [^{3}H]diazepam binding. *Proc. Natl. Acad. Sci. U.S.A.* **74,** 3805–3809.

Cancalon, P. (1978). Isolation and characterization of olfactory epithelial cells of the catfish. *Chem. Senses Flavor* **3,** 381–395.

Cancalon, P., and Beidler, L. M. (1977). Isolation of olfactory cells and characterization of olfactory receptors. *Int. Symp. Olfaction Taste, 6th, Paris.*

Clark, W. L. G. (1957). Inquiries into the anatomical basis of olfactory discrimination, *Proc. R. Soc. London, Ser. B* **146,** 299–319.

Cushieri, A., and Bannister, L. H. (1974). Some histochemical observations on the mucosubstances of the nasal glands of the mouse. *Histochem. J.* **6,** 543–558.

Farbman, A. I. (1977). Differentiation of olfactory receptor cells in organ culture. *Anat. Rec.* **189,** 187–200.

Farbman, A. I., and Margolis, F. L. (1980). Olfactory marker protein during ontogeny: immunohistochemical localization. *Dev. Biol.* **74,** 205–215.

Getchell, T. V. (1977). Mechanisms of excitation in vertebrate olfactory receptors and sustentacular cells. *In* "Food Intake and Chemical Senses" (Y. Katsuki, M. Sato, S. Takagi, and Y. Oomura, eds.), pp. 3–12. Univ. of Tokyo Press, Toyko.

Graziadei, G. A. M., Margolis, F. L., Harding, J. W., and Graziadei, P. P. C. (1977). Immunocytochemistry of the olfactory marker protein. *J. Histochem. Cytochem.* **25,** 1311–1316.

Graziadei, P. P. C., and Graziadei, G. A. M. (1978). The olfactory system: a model for the study of neurogenesis and axon regeneration in mammals. *In* "Neuronal Plasticity" (C. W. Cotman, ed.), pp. 131–153. Raven, New York.

Greenlee, D. V., and Olsen, R. W. (1979). Solubilization of gamma-aminobutyric acid receptor protein from mammalian brain. *Biochem. Biophys. Res. Commun.* **88,** 380–387.

Harding, J., and Margolis, F. L. (1976). Denervation in the primary olfactory pathway of mice. III. Effect on enzymes of carnosine metabolism. *Brain Res.* **110,** 351–360.

Harding, J., Graziadei, P. P. C., Graziadei, G. A. M., and Margolis, F. L. (1977). Denervation in the primary olfactory pathway. IV. Biochemical and morphological evidence for neuronal replacement following nerve section. *Brain Res.* **132,** 11–28.

Hirsch, J. D., and Margolis, F. L. (1977). Preparation and characterization of cells from rat olfactory mucosa. *Neurosci. Abstr.* **3,** 79.

Hirsch, J. D., and Margolis, F. L. (1979a). Cell suspensions from rat olfactory neuroepithelium: biochemical and histochemical characterization. *Brain Res.* **161,** 277–291.

Hirsch, J. D., and Margolis, F. L. (1979b). L-[^{3}H-carnosine binding in the olfactory bulb. II. Biochemical and biological studies. *Brain Res.* **174,** 81–94.

Hirsch, J. D., Grillo, M., and Margolis, F. L. (1978). Ligand binding studies in the mouse olfactory bulb: identification and characterization of a L-[^{3}H-carnosine binding site. *Brain Res.* **158,** 407–422.

Hornung, D. E., and Mozell, M. M. (1977a). Odorant removal from the frog olfactory mucosa. *Brain Res.* **128**, 158–163.

Hornung, D. E., and Mozell, M. M. (1977b). Factors influencing the differential sorption of odorant molecules across the olfactory mucosa. *J. Gen. Physiol.* **69**, 343–361.

Keller, A., and Margolis, F. L. (1975). Immunological studies of the rat olfactory marker protein. *J. Neurochem.* **24**, 1101–1106.

Lowy, J., McGregor, J., Rosenstone, J., and Schmidt, J. (1976). Solubilization of an α-bungarotoxin-binding component from rat brain. *Biochemistry* **15**, 1522–1527.

Margolis, F. L. (1980). A marker protein for the olfactory chemoreceptor neuron. *In* "Proteins of the Nervous System" (R. A. Bradshaw and D. M. Schneider, eds.), 2nd ed., pp. 59–84. Raven, New York.

Margolis, F. L., and Grillo, M. (1977). Axoplasmic transport of carnosine (β-alanyl-L-histidine) in the mouse olfactory pathway. *Neurochem. Res.* **2**, 507–519.

Mazia, D., Schatten, G., and Sole, W. (1975). Adhesion of cells to surfaces coated with polylysine. Applications to electron microscopy. *J. Cell Biol.* **66**, 198–200.

Molday, R. D., Yen, S. P. S., and Rembaum, A. (1977). Application of magnetic microspheres in labeling and separation of cells. *Nature (London)* **268**, 437–438.

Nadi, N. S., Hirsch, J. D., and Margolis, F. L. (1980). Laminar distribution of putative neurotransmitter amino acids and ligand binding sites in the dog olfactory bulb. *J. Neurochem.* **34**, 138–148.

Simon, E. J., and Hiller, J. M. (1977). *In Vitro* studies on opiate receptors and their ligands. *Fed. Proc., Fed. Am. Soc. Exp. Biol.* **37**, 141–146.

Snyder, S. H., and Bennett, J. P., Jr. (1976). Neurotransmitter receptors in the brain: biochemical identification. *Annu. Rev. Physiol.* **38**, 153–175.

Takagi, S. F. (1971). Degeneration and regeneration of the olfactory epithelium. *In* "Handbook of Sensory Physiology" (L. M. Beidler, ed.), Vol. IV/1, pp. 76–94. Springer-Verlag, Berlin and New York.

Valderrama, R., Weill, C. L., McNamee, M., and Karlin, A. (1976). Isolation and properties of acetylcholine receptors from Electrophorus and Torpedo. *Ann. N.Y. Acad. Sci.* **274**, 108–115.

Yousoufi, M. A. K., Thomas, J. W., and Tallman, J. F. (1979). Solubilization of benzodiazepine binding site from rat cortex. *Life Sci.* **25**, 463–470.

Zomzely-Neurath, C., and Walker, W. (1980). Nervous system specific proteins: 14-3-2 protein, neuron specific enolase, and S-100 protein. *In* "Proteins of the Nervous System" (R. A. Bradshaw and D. M. Schneider eds.), 2nd ed., pp. 1–58. Raven, New York.

16

Biochemical Mechanisms in Vertebrate Primary Olfactory Neurons

GEORGE DODD AND KRISHNA PERSAUD

Practically nothing is known about the biochemical processes that take place at the receptors of the sensory nerve cells. [From G. Ohloff. "Firmenich Symposium on Taste and Smell," 1970 (Ohloff and Thomas, 1971).]

I. MOLECULAR MECHANISMS IN OLFACTION

A. Current Problems

In this chapter the biochemical steps involved in the stimulation of the olfactory primary neurons in air-breathing vertebrates have been reviewed.

Biochemistry of Taste and Olfaction

ISBN 0-12-154450-8

We believe that olfactory neurons employ some combination of the molecular mechanisms that are found in other receptor systems (Greaves, 1976).

The focus of our work has been on ideas that are biochemically realistic and that have received some experimental support (for opposing views, see Moncrieff, 1951; Beidler, 1971). Interest in the biochemical aspects of olfaction has increased throughout this past decade (Benz, 1976; Poynder, 1974), but in comparison with vision, there are few experimentalists. Much experimental work is unrepeated, and the amount of reliable data (see Chapters 1–6) is so minute that it is not possible to offer support for any particular olfactory mechanism. Understanding has advanced only a little beyond the situation outlined by Ohloff and Thomas (1971).

B. Odorants as Ligands

Some of the constraints on the biochemistry of olfactory processes can be appreciated by considering the ligands. A representative range of odorants that are particularly interesting for human olfaction are shown in Table I. In contrast with the easily discernable molecular features that are observed with the ligands of other receptor systems (Greaves, 1976), there is no discernable structural requirement for a molecule to be an odorant, provided it is sufficiently volatile to reach the olfactory epithelium. The odorants shown in Table I are all small ligands spanning less than an order of magnitude in molecular weight.

The odor type of a molecule is a subtle function of the molecular properties. The size and shape of the molecule, together with the distribution of polar groups, determine the odor type (Amoore, 1970; Dodd, 1976), but the structural requirements for a particular odor type are only partially defined (Beets, 1978; Boelens, 1974; Ohloff and Flament, 1979). Some polar groups give a distinctive odor type to an odorant series. The lower fatty acids have a sweaty note; the lower thiols have an easily recognizable type of putrid note; and the lower amines have a fishy note. For these particular odorant series, it is easy to imagine odorant binding sites with complementary polar groups. Other common polar groups have a less idiosyncratic odor impact.

Odorants differ from the ligands of many other receptor systems in respect to their stereochemical requirements for activity. For the enantiomeric odorants that have so far been examined, both enantiomers have a smell with relatively minor differences in odor type (Beets, 1978). This relatively unselective "fit" of the odorants to the binding sites may reflect the fact that most odorants have only a single polar group, which can be used for orienting the ligand at the binding site using molecular interactions such as hydro-

TABLE I Odor Properties of Typical Odorants

Molecular weight	Odorant	Odor type[a]	Olfactory threshold[b] (ppb)
48	Methanethiol CH_3SH	Putrid	0.02
59	Trimethylamine $(CH_3)_3N$	Fishy	0.47
104	Methional 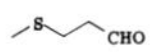	Boiled potatoes	0.2
132	Limonene 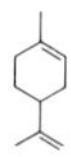	Lemons	10
166	2-isobutyl-3-methoxypyrazine 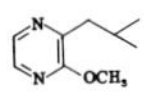	Green peppers	0.002
224	Cyclopentadecanone 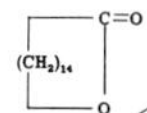	Musk	15
272	5α-Androst-16-en-3-one 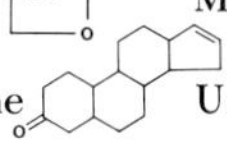	Urinous	0.18

[a] Odor description from Arctander (1969).
[b] Olfactory thresholds are for odorant solutions in water and are taken from the compilation given in Ohloff and Flament (1979).

gen bonding and electrostatic interactions that have a directional element. The recent studies using bifunctional odorants is of interest in this respect (Ohloff and Giersch, 1980).

There is no obvious correlation between olfactory thresholds and molecular size, for odorants taken as a whole (see Table I). Several molecular parameters may determine the olfactory threshold, and with increasing molecular size, there may be "compensating interactions" between these parameters analogous to enthalpy–entropy compensation found in the thermodynamics of some biological systems.

II. TRANSDUCTION AND CODING IN PRIMARY NEURONS

A. General Principles of Receptor Mechanisms

The olfactory system, in common with other major senses, processes primary stimuli through a complex interactive system of synapses. This system uses principles of information processing similar to that used in man-made pattern-recognition systems (Deutsch, 1967; Uttal, 1973). Thus, when we consider the biochemical mechanisms for quality coding at the level of the primary neurons, it must be realized that detailed knowledge of the recognition mechanisms in the primary neuron would not be sufficent for a full explanation of the complex quality coding mechanism, since this is a property of the circuits of the system (see Chapters 1–4 and 17–20).

Transduction and coding, in analogy with other receptor systems (Dodd, 1974; Greaves, 1976), are likely to be intrinsic properties of a macromolecular receptor complex. It is convenient to discuss the ideas in the context of a specific molecular mechanism that we previously proposed—the allosteric–membrane–enzyme (AME) hypothesis (Dodd *et al.*, 1977; Dodd, 1978; see also this volume Chapter 12). This model takes into account the biochemical features of the olfactory system that has been implicated in alternative mechanisms and conveniently considers all the postulated mechanisms. These general biochemical features, which may be found in olfactory cilia, are shown diagrammatically in Fig. 1.

In general, much is known about receptor mechanisms (Cuatrecasas and Hollenberg, 1976) and about the structure of membranes (Chapman, 1968; Chapman and Wallach, 1973). Universal molecular mechanisms operating through a limited number of biopolymer types are responsible for the great variety of receptor systems, and the olfactory system may not have unique biochemical mechanisms. An analysis of olfactory mechanisms can be started by considering the ciliary membranes of the receptor neurons, making the

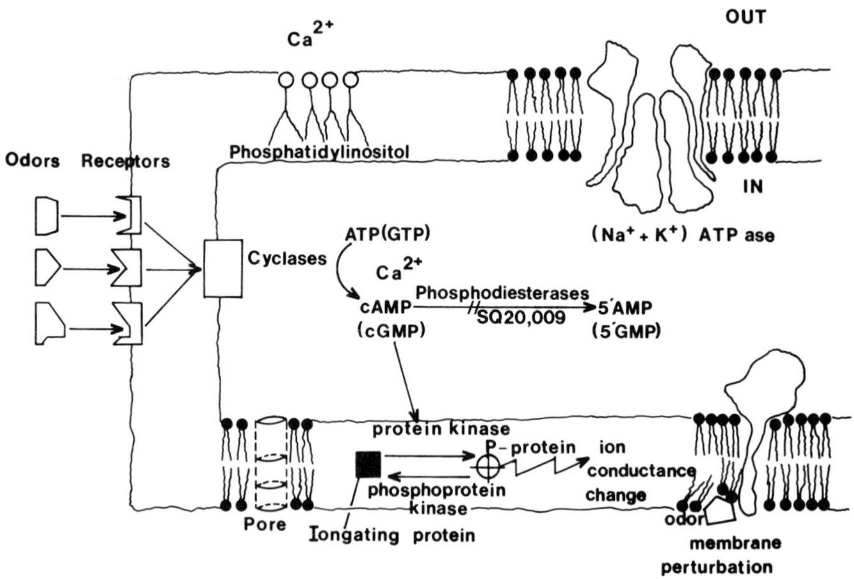

Fig. 1. Some general biochemical mechanisms suggested in olfactory cilia.

following basic assumptions: (a) The ligands interact with the external face of the ciliary membranes; and (b) the biochemical events of interest to us—coding and transduction—take place in this membrane.

Odorants reach the olfactory epithelium via the vapor phase and dissolve in the mucus layer overlying the surface of the tissue. The role of the mucus is likely to be preservation of the necessary ionic environment of the cells—the detailed structure and biochemistry of mucus remains unknown (Bannister, 1974). Calculations from diffusion coefficients and experimental measurements (Bostock, 1974) suggest that the ligand will reach the membrane rapidly (see Chapter 11). Chromatographic separation of constituents of odor mixtures, as the mixture diffuses along the surface of the mucosa and into the mucus to the cilia, has been demonstrated experimentally (Mozell and Jagodowicz, 1974; see also this volume, Chapter 2). This event, taken in conjunction with the observed differential sensitivity of odorants found for regions of the olfactory mucosa (Mustaparta, 1971; Squirrell, 1978), may indicate a contribution to olfactory coding. It is unlikely that the mucus participates directly in the transduction steps, since this would require one of the following events.

1. Specific binding of ligands to components of the mucus with consequent propagation of a conformational change or other ligand-induced event to the receptor membrane.
2. The odorant–mucus component complex could diffuse to the receptor membrane and initiate the receptor potential. Such a mechanism would require the neurons to secrete soluble receptor molecules into the external medium in analogy with prokaryotic secretory mechanisms and some immunological mechanisms. It is a general feature of receptors that receptor proteins are associated with the receptor membrane and this feature is to be expected for the olfactory system (see Chapter 4).

The portion of the neuronal plasma membrane accessible to odorants approaching from the mucus is chiefly ciliary membrane (Menco *et al.*, 1978). It is reasonable to assume that this membrane is the locus of the initial binding of the ligand (see Chapter 3). In other sensory systems, cilia—sometimes extensively modified—are the sensory organelles, e.g., the retinal rod cells (Hagins, 1979), the inner ear (De Reuck and Knight, 1968), and protozoa (Browning *et al.*, 1976). These observations strengthen the supposition that olfactory cilia are the sensory organelles.

The general structure of the ciliary membrane can be imagined as an arrangement of proteins and phospholipids having a general disposition of the fluid–mosaic model of membranes. No data are available on the composition of ciliary membranes from air-breathing vertebrates (see Chapters 2 and

6). The ratios of phospholipid species of various membranes differ considerably and reflect the function of membrane. Information on the phospholipid composition from a plasma-membrane fraction of olfactory epithelium is no guide to the composition of the ciliary membranes. From our general knowledge of other plasma membranes, including receptor membranes, we can speculate about the likely phospholipid composition. About 50% of the lipid can be expected to be the zwitterionic phosphatidylcholine, which is found universally in plasma membranes, with smaller amounts of the anionic phosphatidylserine and smaller amounts of phosphatidylethanolamine—the only lipid with a nucleophilic head group. It will be essential to know the amount of phosphotidylinositol in the sensory membrane, since this will determine the possibility of a transduction mechanism based on breakdown of this lipid (Michell *et al.*, 1976). The phospholipids can be expected to be in a lamellar phase and, like other nerve membranes, have a high degree of acyl chain unsaturation with consequent high membrane fluidity.

The presence of unsaturated fatty acid chains may lead to autoxidation and formation of lipofuschin pigments because the olfactory cilia are exposed to a high concentration of oxygen. If this is the case, it would be expected that the cilia could compensate for such damage. One feature may be the continuous growth and destruction of the ciliary membrane, which is analogous to the constant migration of the retinal rod disk membranes from the cell body to the distal end of the rod, from where they are shed and destroyed (Hagins, 1979). Some studies on centriole migration in the olfactory neurons support this concept, but further studies are necessary. The decay and regeneration of olfactory neurons is another such repair mechanism in the tissue (Moulton, 1974).

Another general feature of the phospholipids, which might be expected, are an asymmetric arrangement of the lipids across the bilayer, both on the basis of the lipid head group and the acyl chain. In an extreme case this could lead to (a) independent phase transitions in the two halves of the bilayer, (b) phase separation of the lipids in the membrane on the basis of their relative fluidity, and (c) diffusion of the lipid molecules both within the plane of the membrane and across the plane of the membrane.

The ciliary membrane will contain membrane proteins in the phospholipid matrix with a similar arrangement to that found in other membranes and with similar rotational diffusion characteristics.

B. The Allosteric Membrane Enzyme (AME) Hypothesis

The allosteric membrane enzyme hypothesis for olfaction (Dodd *et al.*, 1977; Dodd, 1978) is an example of a general hypothesis for the arrangement and function of receptor systems. (Fig. 1) (see also this volume, Chapters 3, 4, 12–14, 21, 23). Some of the features of this hypothesis are:

1. Binding sites for odorants. These may be proteins or phospholipids and may exhibit different degrees of selectivity to different odorants, with possibly high binding specificity to some odorants of biological significance, such as pheromones.

2. Receptor potential. This is brought about by ligand-initiated conformational change in an ion-gating protein, by any of the following mechanisms: (a) modulation of membrane fluidity, with consequent alteration in membrane phospholipid-protein interactions; (b) direct ligand-induced conformational change of an allosteric protomer of a membrane enzyme with intrinsic ion-gating properties, such as a Na^+, K^+-ATPase or a Ca^{2+}-ATPase; (c) modulation of a second messenger system through an enzyme cascade system or through noncovalent binding or covalent chemical modification of the protein. In the latter mechanisms it is not necessary for the binding protein to be part of an oligomeric transducer complex. Additional degrees of freedom are open if it is not, and the ligand–protein complex may diffuse on or off the transducer molecule, allowing several receptors to share the same transducer. This may allow cooperative interactions between clusters of binding proteins.

3. Intensity coding. The fraction of sites occupied will mainly determine the magnitude of the receptor potential.

4. Quality coding. The degree of response of different primary neurons to identical stimuli may be determined by either membrane phospholipid composition or by relative distribution of receptor proteins. With the former, little difference may be expected for many congeneric odorants, while with the latter, greater differences may be expected.

An important feature of the model is that some degree of signal processing is expected directly at the membrane level. Natural olfactory stimuli generally consist of complex mixtures of odorants, and these can be expected to bind to several types of receptors in the membrane. Interaction between these receptors through lipid–lipid interactions, lipid–protein interactions, or protein–protein interactions in an oligomeric complex can be expected to contribute to the establishment of a primary information pattern for a particular odor stimulus.

It is noted that the AME hypothesis considers that ciliary membranes are unlikely to have properties different from the other well-described biological membranes and membrane enzyme systems. Further elaboration of this hypothesis in terms of the established equations for describing ligand binding, enzyme kinetics, and receptor behavior is premature in the absence of reliable evidence of the membrane structure of ciliary membranes. The evidence available on olfactory mechanisms will be examined below.

III. OLFACTORY MECHANISMS—EXPERIMENTAL STRATEGIES

The experimental difficulties encountered in investigation of hormone receptors and neurotransmitters (Cuatrecasas and Hollenberg, 1976) are fewer than those encountered with the vertebrate olfactory system. A logical approach to the isolation and study of olfactory receptors follows the pattern outlined in Table II. The advantage of approaching the problem through the sequence of steps shown is particularly marked for the mammalian olfactory system. Here, hydrophobic ligands are the norm, so that considerable nonspecific binding to membranes and proteins occurs. This makes the identification of receptors through binding studies difficult in homogenates and membrane fractions. However, reversible labeling of olfactory receptors in the intact tissue, with subsequent tissue fractionation, should be a suitable strategy for isolation and purification of receptors. The study of acetylocholine receptors successfully followed this approach (Eldefrawi and Eldefrawi, 1977).

Binding studies to fish olfactory membranes using water soluble odorants, such as some α-amino acids, give little nonspecific binding, and olfactory receptors can be identified from such binding experiments (Cagan and Zeiger, 1978). In general, however, the identification of receptors requires

TABLE II Biochemistry of Olfaction: Experimental Strategies[a]

Level	Comment
Intact epithelium	Experiments can be monitored electrophysiologically
Isolated "intact" neurons	If electrophysiological methods are not applicable, spectroscopic methods should be used to monitor the experiments
Homogenates from epithelium	Electrophysiological methods not applicable at this level
Sensory membranes	The potential across membrane vesicles can be measured by spectroscopic probes
Receptor proteins	Can be "reconstituted" into membranes
Receptor phospholipids	Membrane vesicles can be used to study properties of the lipids

[a] Two key biochemical experiments—(1) odorant binding to determine the number of specific binding sites and (2) specific labeling of olfactory receptors—can be carried out at the levels suggested in the table.

information from several types of experiments other than ligand binding studies.

There is little published information on the molecular structures involved in olfactory mechanisms, so the various aspects shown in Fig. 1 cannot be related to each other. The work will be reviewed by proceeding through the steps in Table II in reverse order.

A. Role of Phospholipids

The first olfactory transduction mechanism proposed with a realistic theoretical basis and supported by experimental work involved membrane lipids. This was the "penetrating and puncturing" model (Davies, 1970). In this model, it was postulated that diffusion of an odorant molecule through the membrane left behind a hole, which healed relatively slowly so that ions could pass through, leading to depolarization of the membrane. Experimentally, a satisfactory correlation exists between the olfactory thresholds of a range of odorants and their ability to accelerate the hemolysis of erythrocytes suspended in hypotonic media. Odor quality coding at the level of the primary neurons was explained on the supposition that there were differences in the lipid composition of different classes of neurons.

This model was introduced before the unique physical chemistry of cell membranes and phospholipids was elucidated. From current knowledge, we can see that small ligands, like typical odorants, will not puncture membranes and create transient holes as envisaged by Davies. Odorants can interact with the lipid regions of cell membranes producing a variety of effects such as phase transitions or altered packing of lipid molecules. These could lead to membrane labilization or stabilization depending on odorant concentration and alteration of lipid–protein interactions. These effects are more plausible than the puncturing concept and could lead to production of a receptor potential through conformational changes in ion-gating proteins.

The response to odorants of nonsensory membranes, like that of erythrocytes, is to be expected from the physical properties of both odorants and phospholipids (see Chapter 13). Hemolysis is irreversible, so this is not a satisfactory model system to study the reversible events of olfactory transduction. Membrane stabilization of erythrocytes has been explored (Seeman and Weinstein, 1966) by using a wide range of ligands. Our studies have revealed that odorants from several classes produce stabilization of the membranes (Fig. 2A). Further investigation of this new model system awaits the isolation of the sensory membranes.

Changes in the electrical conductance of phospholipid bilayers made from highly purified phosphatidylcholine are induced by odorants. This effect is observed over the same concentration range to which the olfactory system

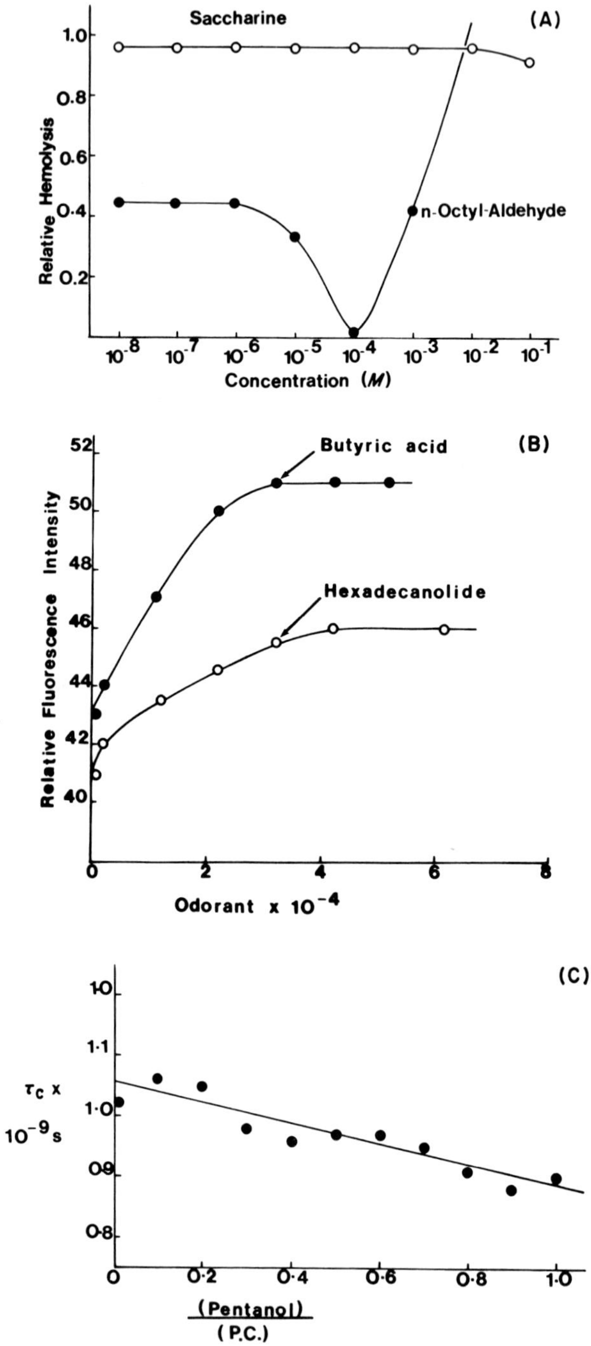
Saccharine
(A)
1.0
0.8
0.6
0.4
0.2
Relative Hemolysis
n-Octyl-Aldehyde
10^{-8} 10^{-7} 10^{-6} 10^{-5} 10^{-4} 10^{-3} 10^{-2} 10^{-1}
Concentration (*M*)
Butyric acid
(B)
52
50
48
46
44
42
40
Relative Fluorescence Intensity
Hexadecanolide
0 2 4 6 8
Odorant x 10^{-4}
(C)
1·0
1·1
τ_c x
1·0
10^{-9}s
0·9
0·8
0 0·2 0·4 0·6 0·8 1·0
(Pentanol)
(P.C.)

responds (Cherry *et al.*, 1970). Odorants also increase the surface pressure of phospholipid monolayers (Koyama and Kurihara, 1972; see also this volume, Chapter 13). This effect can also be correlated with the threshold of the odorants. Again, further investigation awaits isolation of ciliary membrane phospholipids.

The equations derived by Davies (1970) and used as a basis of the correlation between odorant threshold and observed membrane effects do not specifically support the hypothesis that phospholipids are the binding sites for odorants, since the parameters used apply equally well to the adsorption of an odorant to a hydrophobic binding site on a membrane protein.

B. Studies on Receptor Proteins

The receptor molecules of many ligands—including hormones, neurotransmitters, and antigens—are proteins. There is a strong possibility of similar receptors occurring in the olfactory system, so that certain classes of odorants will bind to receptor proteins in the cell membrane. Electron micrographs of the cilia of the primary neurons show the presence of numerous granules on the membrane surface (Menco *et al.*, 1976; Kerjaschki and Horandner, 1976). These are suggested as possible receptor proteins.

The isolation of receptor proteins from the olfactory system is more difficult than with hormone and drug receptors. The problem is one of recognition. One criterion for recognition of a drug or hormone receptor is that the affinity of binding of agonists or antagonists *in vitro* can be correlated with the biological activity of the system *in vivo*.

Another drawback is that odor–structure studies have been carried out chiefly with humans. However, biochemical work on the olfactory system is conveniently carried out using experimental animals. Thus most workers in the field tend to assume, in the absence of direct experimental evidence, that odor–structure relationships are analogous between humans and experimental animals. Despite these drawbacks, Price and others have made attempts to isolate proteins from the olfactory epithelium, which bind odorants (see Chapter 4).

C. Investigations with Membranes and Homogenates

Several attempts have been made to fractionate the olfactory epithelium with the aim of isolating pure plasma membrane fragments of the neuronal

Fig. 2. (A) An example of erythrocyte stabilization by an odorant. Experimental conditions as described by Seeman and Weinstein (1966). (B) Perturbation induced in an olfactory plasma membrane fraction by odorants, using the fluorescent probe ANS. (C) Pertubation induced in phosphatidylcholine vesicles by the odorant pentanol, detected by a substituted fatty acid nitroxide probe (Dodd *et al.*, 1970).

cells. Koch and Norring (1968) used differential- and sucrose-density gradient centrifugation techniques to separate subcellular components from rabbit brain and olfactory tissue homogenates. The enzyme markers cytochrome oxidase and Na^+–K^+-ATPase were used to identify and quantitate membranes from various cell fractions. Electron micrograph evidence indicated that a certain subfraction contained cilia and cell bodies. However, subsequent work (Koyama *et al.*, 1971; Menco *et al.*, 1974; Menevse, 1977; Menco, 1977) failed to achieve clear separations of neuronal plasma membranes using these techniques. In the absence of pure plasma membrane preparations, efforts have been made to seek, in homogenates of olfactory epithelium, proteins that interact with odorants.

Ash (1968) reported a decrease in absorbance at 267 nm when a homogenate of olfactory mucosa was mixed with the odorants linalool or linalyl isobutyrate. This effect was ascribed to formation of an odorant–olfactory homogenate complex, leading to a conformational change, which triggered depolarization of the olfactory cell membrane. However, it has been shown that this change in absorbance at 267 nm occurs in other tissues as well as olfactory tissue and is due to oxidation of ascorbic acid.

Fesenko *et al.* (1979) reported that a membrane fraction of rat and frog olfactory epithelium homogenates contains structures capable of binding camphor with an affinity constant of $6.7 \times 10^8 M$. These were absent in other tissues studied—tongue, lung, liver, and brain—and it was suggested that they belong to a class of receptor molecules for camphoraceous molecules. It was found that binding was abolished by trypsin and pronase, that sulfhydryl reagents decreased the amount of binding, and that a pH of 7.4 produced maximum binding. These results indicated that the binding component was protein in nature, and isoelectric focusing indicated a MW of 120,000. However, as with the other binding studies, no evidence has yet been presented that this binding component is indeed an olfactory receptor.

Gennings *et al.* (1977; see also this volume, Chapter 1) have investigated binding of the swine pheromone 5α-androst-16-en-3-one to sow olfactory tissue homogenates. Their results provide tentative evidence for a receptor to 5α-androst-16-en-3-one present only in olfactory tissue, with saturable specific binding being found. An affinity constant of 8.3×10^8 M was measured. Subsequent work has shown some irreproducibility of these results but this does not necessarily negate the results reported (D. Gower, personal communication; Persaud, 1980; see also this volume, Chapter 1).

In our laboratory binding of the urinous odorant 5α-[16,17-^{3}H]-androstan-3-one to homogenates of sheep olfactory epithelium was investigated. This odorant is closely related to the 5α-androst-16-en-3-one used by Gennings *et al.* (1977). It was not known, when we commenced this work, whether the sheep perceives 5α-androstan-3-one as an odorant.

Though this compound is both extremely hydrophobic and relatively nonvolatile, and therefore difficult to work with in the vapor phase, small but significant EOGs were obtained from preparations of sheep olfactory mucosa artificially maintained with oxygenated Ringers solution (Fig. 3A) (Squirrell, 1978). This suggested that sheep have the capacity to perceive this compound as an odorant. Binding studies using crude homogenates proved inconclusive and irreproducible, due mainly to the extreme hydrophobicity of the ligand. Binding studies, using an 11,000 *g* supernatant fraction of olfactory tissue homogenate with dextran-coated charcoal to remove "free" ligand, produced the results shown in Fig. 3B. Evidence for the presence of specific, saturable binding was obtained, but the amount of nonspecific bind-

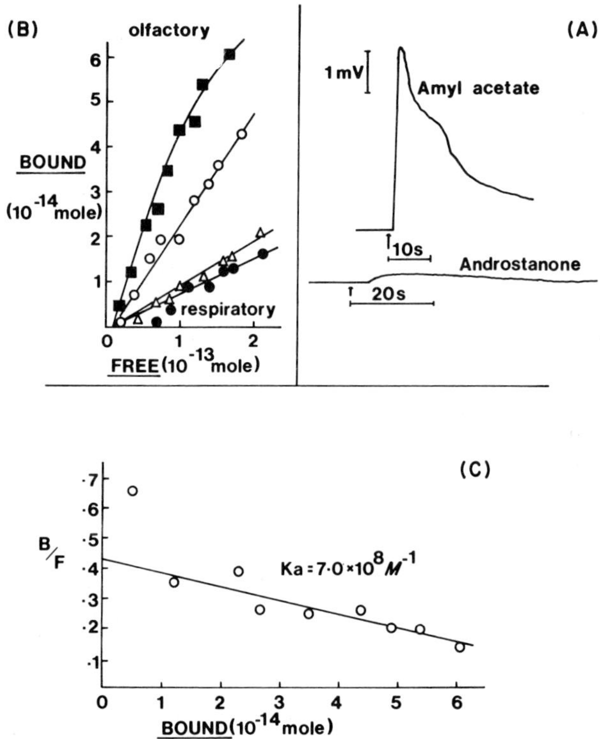

Fig. 3. Binding of 5α-[16,17-^{3}H]androstan-3-one to sheep olfactory epithelium. (A) EOG from intact sheep olfactory epithelium in response to stimulation by androstanone and amyl acetate. (B) Binding of androstanone to the supernatant fractions of sheep olfactory and respiratory epithelia, following centrifugation at 12,000 *g* for 15 min. The nonspecific binding (○, ●) was determined by adding excess unlabeled androstanone to the tissue and labeled steroid. The free and bound ligands were separated using the standard charcoal method. (C) Scatchard plot of binding data shown in Fig. 3B.

ing observed was large. This means that any estimate of an affinity constant is subject to large error. A Scatchard plot of the results (Fig. 3C) indicates an K_A of 7.0×10^8 *M*. This plot is representative of seven other plots obtained from four separate tissue preparations (Persaud *et al.*, 1980a). 5α-Androstan-3-one binding is inhibited by the following ligands in the rank order shown:

1. 5α-androst-16-en-3-one (urinous)
2. *cis*-2-methyl-2-(4′-*tert*-butylcyclohexyl)pentan-4-one (urinous)
3. 5α-androst-16-en-3α-ol (sandalwood, musk)
4. 5α-androst-16-en-3β-ol (nearly odorless)

It is not yet clear whether the observed binding is due to the presence of an olfactory receptor protein or a steroid hormone-binding protein.

A variety of spectroscopic probes detect perturbations in the crude olfactory plasma membrane when the membranes are titrated with high concentrations of odorant (Dodd, 1971) (Fig. 2B). In addition, a model system of phosphatidylcholine vesicles with a fatty acid nitroxide probe showed perturbation due to the odorant pentanol (Fig. 2C) (Dodd *et al.*, 1970). Similar changes are found with postsynaptic membrane preparations. Spectroscopic methods will be informative when pure membrane preparations are available. Biochemical work on olfactory tissue homogenates is still in its infancy. Future work will need to correlate results obtained from subcellular fractions with that of intact olfactory epithelium.

D. Isolation of Cells

Standard methods of isolation of viable cells from a variety of tissues, such as kidney, brain and pancreas, have been in use for years. These methods have been tried on the olfactory system. These isolation techniques attempt to break intercellular connections by a combination of mechanical, chemical and enzymatic methods; free cells are separated by differential centrifugation or differential adsorption to a surface. The basic difficulties involved in isolating intact viable olfactory neurons are (a) the olfactory neuron has to be severed on the distal side of the axon, so the plasma membrane has to reseal this break without loss of cell viability, (b) the cilia, which form a network intertwining with cilia from other neurons and with microvilli, have to be disentangled without rupturing them, and (c) heterogeneity of the olfactory epithelium.

The first attempts to isolate olfactory neurones were made by Ash *et al.* (1966) before any specific biochemical markers for these cells were known. Homogenization of rabbit olfactory epithelium and low speed centrifugation yielded a cell suspension that contained primary neurons, sustentacular, and other cells. Attempts to isolate and release the cells from the tissue by use of proteases and also by shaking the tissue with glass beads led to cell destruction.

Cell suspensions from rat olfactory epithelium have recently been obtained and comprehensively characterized (Hirsch and Margolis, this volume, Chapter 15). In the intervening period, Margolis discovered a number of specific biochemical markers for the olfactory tissue, which greatly aid the identification of cell types. The key steps involved were (a) treatment of the tissue with a collagenase–hyaluronidase mixture containing trypsin inhibitor, (b) consecutive washes with EDTA, Ca^{2+} and Mg^{2+}, and (c) centrifugation of the suspended tissue through a 4% BSA solution.

A partial enrichment of the cell suspension with neurons was obtained. The importance of the isolation of pure neuronal plasma membranes has been stressed elsewhere (see Section II,A). With further development of isolation techniques, it may be possible to use cultured cells for further biochemical research on the primary neurons of the olfactory system.

E. Experiments with Intact Olfactory Epithelium

Ligand binding studies on the intact olfactory epithelium have been limited. Mozell and Jagodowicz (1974) have shown that differential sorption behavior is found when comparing polar and nonpolar odorants, whereas Gennings *et al.* (1977) have shown that steroid pheromones are adsorbed to pig olfactory epithelium. Most of the experiments at this level have consisted of treating the tissue with reagents, frequently in solution, and monitoring the EOG response to odorants.

1. Involvement of Cyclic Nucleotides in the EOG Response

The involvement of cyclic nucleotides in an olfactory receptor process can be established if a set of experimental criteria, originally formulated by Sutherland, are fulfilled (Menevse *et al.*, 1977a). Cyclic nucleotides are involved as the second messenger system in receptors of several cells, including neurons (Greengard, 1976). Experiments have been performed using the EOG response of the tissue; the results show that the intracellular levels of cyclic nucleotides affect the magnitude of the EOG signal (Minor and Sakina, 1973; Menevse, 1977). An example is shown in Fig. 4. The effects are complex. For example, the potent phosphodiesterase inhibitor, SQ 20,009, has opposite effects on the epithelium from frog and sheep, respectively. The sheep tissue does not respond to the inhibitor or to dibutyryl cAMP (Fig. 4), but does respond to an increased level of cGMP with a transient increase in the amplitude of the EOG (Squirrell, 1978).

By only using EOG responses, it is not possible to eliminate the possibility

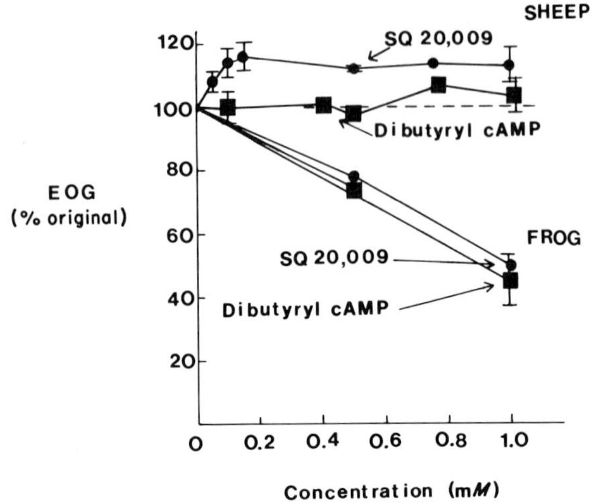

Fig. 4. Peak amplitudes of EOGs to pentyl acetate from various preparations of olfactory epithelium after 2.5-min exposure.

that the effects observed are due to modulation of secretory activity in the supporting cells rather than alteration of the cyclic nucleotide-dependent enzyme system in the terminal swelling of the olfactory neurons. An investigation of the problem with more sophisticated electrophysiological methods may resolve this issue. The inability to penetrate the primary olfactory neurons with microelectrodes means that straightforward pharmacological experiments, which can be carried out on other cells with excitable membranes such as protozoa (Doughty and Dodd, 1978), can not be carried out on the mammalian olfactory system. The results to date are at least consistent with the possibility that olfactory transduction may involve a classical cyclic nucleotide second messenger system.

2. *Chemical Modification Approach to Identification of Olfactory Receptors*

One of the established methods for identifying binding sites on proteins that interact with small molecules involves irreversible labeling of the site with a reactive chemical. A large number of ligand sites in a diverse array of biopolymers and membranes have been investigated (Katzenellenbogen, 1977). For example, marked selectivity of labeling can be obtained, enabling a regulatory site on an allosteric enzyme to be labeled with no modification of the catalytic site.

Chemical modification methods permit the design of incisive experiments on identification of olfactory receptors. For example, we can imagine that olfactory coding in the sensory membranes of olfactory neurons depends on the presence of sets of receptor proteins. This model fits with general concepts of receptors (Greaves, 1976) and is a readily testable hypothesis. Structure–activity relationships from human olfactory studies suggest that there may be receptor proteins, at least for the following classes of odorants: urinous, musky, floral, fruity, minty, and camphoraceous (Beets, 1978). The existence of such postulated receptor proteins could be verified if it were possible to selectively block the activity of each of these sites in an olfactory epithelium while retaining the activity of the other sites. Blocking would be achieved by covalently labeling one or more amino acid side chains at the odorant binding site. The interpretation of such experiments would depend on the binding properties of the proteins. If well-defined odors bind specifically to a single class of receptor protein with a high affinity, it should be possible to inactivate this class of proteins without affecting the activity of the other classes of olfactory proteins. However, if certain types of odorants bind to several classes of receptor proteins, chemical modification may lead to differential inhibition of the activity to various classes of odorants. For this case, in contrast with the previous one, large-scale mapping experiments with many odorants would be required to identify the selectivity of the proteins. If the binding sites for odorants are the phospholipid phase of the sensory membranes, it should still be possible to carry out chemical modification studies though interpretation of the experiments would be much more difficult than for specific binding sites in proteins.

There are three distinct types of chemical modification methods. The application of each type to identification of olfactory receptor proteins in vertebrates is discussed below. It is useful to bear in mind the following features, which are common to all three types of modification.

1. The odorant site is covalently labeled, leading to abolition of the characteristic response from that type of receptor protein. The activity would be regained only by biosynthesis of new protein.

2. The labeling may be prevented by adsorbing the characteristic odorant and congeners to the site. Odorants from structurally unrelated families would not prevent the labeling.

3. When the specificity of the labeling has been established, it is possible to label the site with a radioactive version of the labeling reagent. The protein is thus tagged with a radioactive marker, and this facilitates the isolation of the receptor protein.

A dilemma occurs in planning chemical modification experiments on the olfactory mucosa. Most reagents have been developed for investigation of proteins and membranes in aqueous solution. Hence, most of the reagents are nonvolatile and water soluble. However, the normal stimulus for the terrestrial vertebrate olfactory epithelium approaches the tissue as a dilute vapor, which does not disturb the mucus layer overlying the sensory cells. Application of a reagent in solution to the mucosa may disturb the mucus and alter the ionic composition or some other feature. Stringent control experiments must be performed to ensure that application of a particular reagent in solution is not introducing artifacts.

a. Group-Specific Reagents. These reagents manifest specificity only in the type of chemical reaction they undergo. For example, a variety of "thiol reagents" react only with cysteine sidechains; various alkylating reagents react with several nucleophilic amino acid side chains, including serine, cysteine, and lysine, and tetranitromethane specifically labels the phenolic sidechain in tyrosine. Since most proteins contain several amino acids of one type, specific labeling of one of the set of residues will depend on the labeled residue occurring in a unique solvent microenvironment, which enhances the reactivity of the sidechain. The specific labeling of the active site serine in the "serine proteases" is a well-known example of this phenomenon.

For this class of reagent to work, it is necessary that the odorant binding site have one of the several classes of polar amino acid sidechains. This type of sidechain is typically one of the catalytic functional groups at the active sites of enzymes where they participate directly in catalysis. Since there is no reason to suppose that odorants are chemically modified at their binding sites, but rather that they simply diffuse away after desorption, there is no special requirement for such a polar sidechain at an odorant binding site. On the contrary, because many odorants for terrestrial vertebrates are hydrophobic molecules, we might expect a preponderance of apolar amino acid sidechain residues at the binding site, forming a relatively hydrophobic site. A polar sidechain with an appropriate function group may be present as a hydrogen bonding partner for the polar "osmophoric group" of an odorant.

The EOG responses to a wide variety of odorants obtained from the frog olfactory epithelium are inhibited by various group specific reagents (Menevse, 1977). The inhibition by thiol reagents has been investigated in detail. The classical thiol reagent, *N*-ethyl maleimide, when applied in solution to the olfactory mucosa irreversibly decreases the EOG response to the fruity odorant ethyl butyrate (Getchell and Gesteland, 1972). Inhibition is also found when the tissue is treated using the reagent in the vapor phase (Menevse, 1977). Protection experiments using solutions of the odorant showed that the inhibition could be prevented. There are two features of this pioneering study, which complicate interpretation. First, the reagent can permeate membranes; thus the observed inhibition could be due to reaction with an intracellular protein involved in ion conductance rather than with a binding portein on the external face of the sensory membrane. Second, high concentrations (10^{-2} M) of the odorant were required for protection. At this concentration, a fraction of the lipophilic odorant may have permeated the plasma membrane.

The thiol groups involved in the olfactory response were further investigated using a nonpermeable thiol reagent, mersalyl (Menevse *et al.*, 1978), and the fruity odorant pentyl acetate. Protection by solutions of this odorant was observed, and interestingly, the response to some odors was protected to a greater extent than to others. This differential response can provide evidence for the existence of classes of receptor proteins. The concentration of protecting reagent (1 mM), though lower than that used by Getchell and Gesteland (1972), was sufficiently high to possibly induce membrane stabilization and related effects that resulted from interaction of the odorant with the phospholipid regions of the membranes.

Many investigations can be expected on this topic in the future. One potentially fruitful class of chemicals are bifunctional cross-linking reagents. In a study on an interesting model system for olfactory neurons, the ciliated single cell of Paramecium, the use of a cross-linking reagent altered the chemotactic behavior of the cell towards chemical stimuli (Doughty and Dodd, 1978).

b. Affinity Labeling. This method is a logical extension of that described above. The reagents use the same types of chemically reactive groups, but the selectivity of labeling is enhanced by incorporating the reactive moiety into a molecule having the general structure and conformation as the odorant of interest. Thus the affinity label will preferentially bind to the site of interest and label the appropriate site residue. This more sophisticated chemical modification has been strikingly successful with several types of ligand-binding sites (Jakoby and Wilchek, 1977).

Receptor identification using affinity odorants has been tried in the frog (Menevse *et al.*, 1977b; Squirrell, 1978) and in the rat (Persaud *et al.*, 1980b). The general strategy makes use of the assumptions outlined in the previous section and are illustrated with one type of chemically reactive group. Using the bromoacetate group as an analogue for the propionate group, a series of potential affinity odorants have been synthesized. By

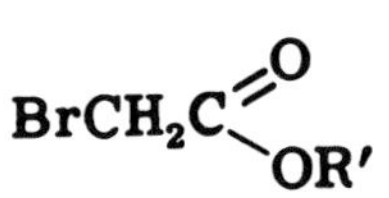

varying the size and shape of the R′ group, labels directed at the following classes of odorant sites have been prepared: fruity, minty, floral, and woody. Since the approach is a general one, it should be feasible to synthesise affinity odorants for many other classes of sites.

A comprehensive vapor-phase affinity-labeling study of the fruity site in the frog olfactory epithelium has been performed. A typical finding is shown in Fig. 5. The affinity label ethyl bromoacetate gives a good EOG response in the frog. It preferentially inhibits its own response. The rank order of inhibition obtained to various odors is interesting (Table III). The response of the fruity odorants is affected to a greater extent than the minty and camphoraceous odorants. This differential response provides a method that may be used to identify the types of receptor sites.

The responses of the rat olfactory epithelium to these affinity reagents, as studied in an *in vitro* preparation, differs from the frog preparation. Whether this indicates a difference in the types of olfactory receptor proteins in the two animals or whether it is an artifact of the *in vitro* preparation is not yet clear. During the course of the work with the rat preparation, it has been observed that the tissue can be exposed to some odorants for long periods (>10 min), without significant adaptation, whereas rapid deterioration of the response is obtained after a comparable exposure to other odorants.

c. Photoaffinity Labeling. A disadvantage of both of the previous methods is that the chemically reactive group in the labeling reagent or odorant makes it a potent stimulus for the trigeminal system. Photoaffinity labeling overcomes this disadvantage by generating the reactive moiety photolytically at the binding site. The selectivity of the labeling depends both on the properties of the label as an analogue for the odorant of interest and also on the kinetics of the labeling. This is an extremely powerful method for specific

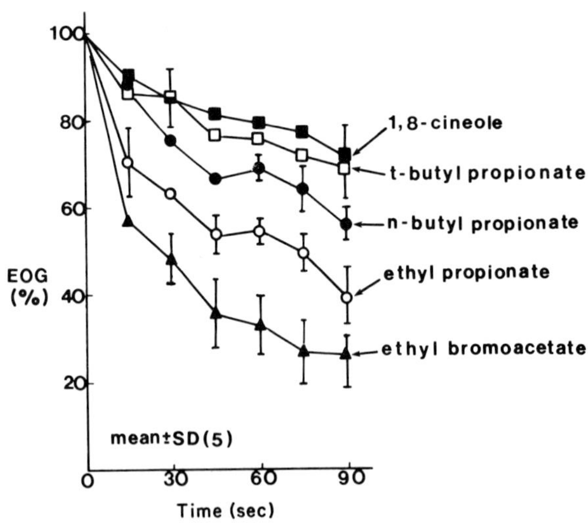

Fig. 5. Affinity labeling with ethyl bromoacetate. Alkyl group of esters were varied. EOGs were obtained in the frog as described by Menevse *et al.* (1977b).

TABLE III Differential Labeling with the Affinity Odorant Ethyl Bromoacetate[a]

Odorant	Odor	Inhibition of EOG between 60–90-sec labeling (%) (mean ± SD) (3–5)
1,8-Cineole	Camphoraceous	23.1 ± 8.7
tert-Butyl propionate	Minty	27.4 ± 8.0
Ethyl pivalate	Minty	34.8 ± 11.6
Ethyl propionate	Fruity	49.7 ± 3.9
Ethyl acetate	Fruity	51.1 ± 10.5
Ethyl bromoacetate	Fruity and irritant	72.3 ± 6.5

[a] From Squirrell (1978).

labeling of receptor sites and can be illustrated by the studies on the chemoreceptors of the bacterium *Pseudomonas aeruginosa*. This cell responds chemotactically to tyrosine, and the tyrosine response is abolished by photolabeling the cells using the tyrosine analogue *p*-azidophenylalanine (Barber *et al.*, 1979). Photoaffinity labeling of the frog olfactory epithelium has been obtained using a range of aromatic azides. All of these molecules are chemically unreactive in the dark and give good EOG responses. Irradiation of the olfactory mucosa with UV light does not itself affect the response

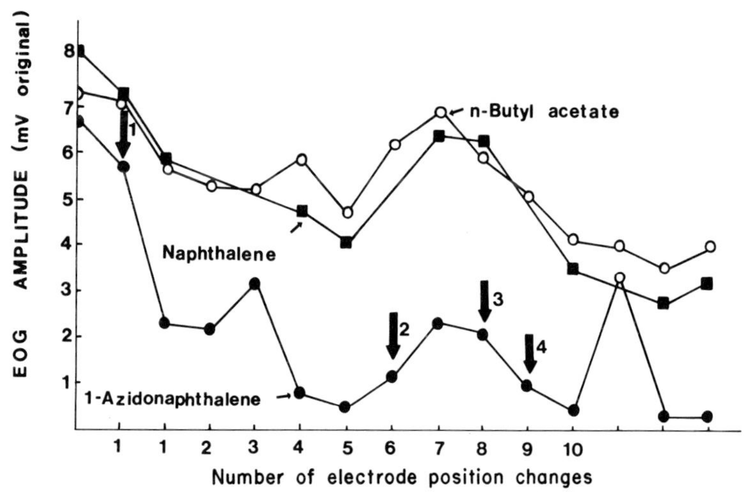

Fig. 6. The effect of light on the EOG responses elicited by 1-azidonaphthalene in different parts of the frog olfactory mucosa. Vertical arrows indicate illumination of the tissue at 326 nm. Conditions as described by Menevse *et al.* (1977b).

of the tissue, but when illumination occurs during application of a puff of odorant vapor, it leads to abolition of the response to the photoaffinity odorant. These labels give a differential inhibition of the response (Fig. 6) to various odorants as was found for the affinity odorants (Fig. 5). However the results are complex and many more types of photoaffinity odorants must be studied before the labeling pattern can be interpreted simply.

IV. PROSPECTS FOR FUTURE STUDIES

In the context of the experimental approaches outlined in this chapter, the following lines of investigation would be rewarding.

1. A systematic search for a vertebrate with large primary neurons, which would permit direct intracellular recording. This would simplify studies on the intact tissue along the lines outlined above. If the animal had a distinct behavioral response to odors and could be easily bred, further advantages would accrue; it would be possible to use the powerful genetic approach to biochemical problems. For studies at the molecular level, olfaction needs the olfactory equivalent of both the squid giant fibers and the electroplax preparation of electric eels.
2. Isolation of a pure preparation of olfactory cilia using the wide range of techniques, which are now available for characterization of the tissue.
3. Many more studies on specific labeling of olfactory receptors using both the antibody approach, the chemical modification approach and other methods. The results from studies using a large number of odorant families may be necessary before the pattern of results can be easily interpreted. Many more investigators in the field are required.
4. Since it will be more difficult to approach the quality coding mechanism in humans using biochemical methods, this problem can be approached by systematically synthesizing and purifying the members of the odorant families, measuring their olfactory properties and correlating these properties with molecular structure using quantitative methods. Curiously, few systematic studies of this type have been performed. A pointer in this direction are the elegant odor–structure studies of Ohloff (Ohloff and Giersch, 1980).

ACKNOWLEDGMENT

We thank the Science Research Council and the Medical Research Council for financial support, and Mr. P. Wood for assisting with the manuscript.

REFERENCES

Amoore, J. (1970). "Molecular Basis of Odor." Thomas, Springfield, Illinois.

Arctander, S. (1969). "Perfume and Flavour Chemicals." Arctander, Montclair, New Jersey.

Ash, K. O. (1968). Chemical sensing: An approach to biological molecular mechanisms using difference spectroscopy. *Science* **162**, 452–454.

Ash, K. O., Bransford, J. E., and Koch, R. B. (1966). Studies on dispersion of rabbit olfactory cells. *J. Cell Biol.* **29**, 554–562.

Bannister, L. H. (1974). Possible functions of mucus at the gustatory and olfactory surfaces. *In* "Transduction Mechanisms in Chemoreception" (T. M. Poynder, ed.), pp. 39–46. IRL, London.

Barber, R., Primrose, S., and Dodd, G. H. (1979). Photoaffinity labelling of a chemoreceptor protein in *Pseudomonas*. *FEBS Lett.* **105**, 43–46.

Beets, M. G. J. (1978). "Structure-Activity Relationships in Human Chemoreception." Applied Science Publ., London.

Beidler, L. M., ed. (1971). "Handbook of Sensory Physiology, Vol. IV/1, Chemical Senses." Springer-Verlag, Berlin and New York.

Benz, G., ed. (1976). "Structure-Activity Relationships in Chemoreception." IRL, London.

Boelens, H. (1974). Relationship between the chemical structure of compounds and their olfactive properties. *Cosmet. Perfum.* **89**, 1–7.

Bostock, H. (1974). Diffusion and the frog EOG. *In* "Transduction Mechanisms in Chemoreception" (T. M. Poynder, ed.), pp. 27–36. IRL, London.

Browning, J. L., Nelson, D. L., and Hansma, H. G. (1976). Ca influx across the excitable membrane of behavioural mutants of paramecium. *Nature (London)* **259**, 491–494.

Cagan, R. H., and Zeiger, W. N. (1978). Biochemical studies of olfaction: Binding specificity of radioactively labelled stimuli to an isolated olfactory preparation from rainbow trout (*Salmo gairdneri*). *Proc. Natl. Acad. Sci. U.S.A.* **75**, 4679–4683.

Chapman, D., ed. (1968). "Biological Membranes," Vol. 1. Academic Press, New York.

Chapman, D., and Wallach, D. F. H., eds. (1973). "Biological Membranes," Vol. 2. Academic Press, New York.

Cherry, R. J., Dodd, G. H., and Chapman, D. (1970). Small molecule-lipid membrane interactions and the puncturing theory of olfaction. *Biochim. Biophys. Acta* **211**, 409–416.

Cuatrecasas, P., and Hollenberg, M. D. (1976). Membrane receptors and hormone action. *Adv. Protein Chem.* **30**, 252–450.

Davies, J. T. (1970). Recent developments in the "Penetration and Puncturing" theory of odour. *In* "Taste and Smell in Vertebrates" (G. Wolstenholme and J. Knight eds.), pp. 265–291. Churchill, London.

De Reuck, A. V. S., and Knight, J. eds. (1968). "Hearing Mechanisms in Vertebrates" Ciba Foundation Symposium. Churchill, London.

Deutsch, S. (1967). "Models of the Nervous System." Wiley, New York.

Dodd, G. H. (1971). Studies on olfactory receptor mechanisms. *Biochem. J.* **123**, 31–32.

Dodd, G. H. (1974). Structure and function of chemoreceptor membranes. *In* "Transduction Mechanisms in Chemoreception" (T. M. Poynder, ed.), pp. 103–113. IRL, London.

Dodd, G. H. (1976). Structure-activity relationships in chemoreception. A preview. *In* "Structure-Activity Relationships in Chemoreception" (G. Benz, ed.), pp. 1–9. IRL, London.

Dodd, G. H. (1978). Biochemical aspects of coding and transduction in the primary olfactory neurones. *Drug Res.* **28**, 2362.

Dodd, G. H., Barratt, M. D., and Rayner, L. (1970). Spin probes for binding site polarity. *F.E.B.S. Lett.* **8**, 286–288.

Dodd, G. H., Menevse, A., and Poynder, T. M. (1977). An allosteric-membrane-enzyme hypothesis for olfaction. *In* "Olfaction and Taste VI" (J. Le Magnen and P. MacLeod, eds.), p. 59. IRL, London.

Doughty, M. J., and Dodd, G. H. (1978). Chemical modification of the excitable membrane in *Paramecium aurelia:* effect of a cross-linking reagent. *Comp. Biochem. Physiol. C* **59c**, 21–32.

Eldefrawi, M. E., and Eldefrawi, A. T. (1977). Acetylcholine receptors. *In* "Receptors and Recognition," Series A (P. Cuatrecasas and M. Greaves, eds.), Vol. 4, pp. 197–258. Chapman & Hall, London.

Fesenko, E. E., Novoselov, V. I., and Krapivinskaya, L. D. (1979). Molecular mechanisms of olfactory receptors. IV. Some biochemical characteristics of the camphor receptor from rat olfactory epithelium. *Biochim. Biophys. Acta* **587**, 424–433.

Gennings, J. N., Gower, D. B., and Bannister, L. H. (1977). Studies on the receptors to 5α-androst-16-en-3-one and 5α-androst-16-en-3α-ol in sow nasal mucosa. *Biochim. Biophys. Acta* **496**, 547–556.

Getchell, M. L., and Gesteland, R. C. (1972). The chemistry of olfactory reception: Stimulus-specific protection from sulfhydryl reagent inhibition. *Proc. Natl. Acad. Sci. U.S.A.* **69**, 1494–1498.

Greaves, M. F. (1976). Cell surface receptors: A biological perspective. *In* "Receptors and Recognition," Series A (P. Cuatrecasas and M. F. Greaves eds.), Vol. 1, pp. 3–32. Chapman & Hall, London.

Greengard, P. (1976). Possible role for cyclic nucleotides and phosphorylated membrane proteins in postsynaptic actions of neurotransmitters. *Nature (London)* **260**, 101–108.

Hagins, W. A. (1979). Excitation in vertebrate photoreceptors. *In* "The Neurosciences: Fourth Study Program" (F. O. Schmitt and F. G. Worden, eds.), pp. 183–192. MIT Press, Cambridge, Massachusetts.

Jakoby, W. B., and Wilchek, M., eds. (1977). "Affinity Labelling", Methods in Enzymology, Vol. XIVI, Academic Press, New York.

Katzenellenbogen, J. A. (1977). Affinity labelling as a technique in determining hormone mechanisms. *In* "Biochemical Actions of Hormones" (G. Litwack, ed.), Vol. 4, pp. 2–77. Academic Press, New York.

Kerjaschki, D., and Horandner, H. (1976). The development of mouse olfactory vesicles and their cell contacts: A freeze-etching study. *J. Ultrastruct. Res.* **54**, 420–444.

Koch, R. B., and Norring, N. L. (1968). Fractionation of olfactory tissue homogenates. Isolation of a concentrated plasma membrane fraction. *J. Neurochem.* **16**, 145–157.

Koyama, N., and Kurihara, K. (1972). Effect of odorants on lipid monolayers from bovine olfactory epithelium. *Nature (London)* **236**, 402–404.

Koyama, N., Kazuko, S., and Kurihara, K. (1971). Isolation and some properties of plasma membranes from bovine olfactory epithelium. *Biochim. Biophys. Acta* **241**, 42–48.

Menco, B. (1977). A qualitative and quantitative investigation of olfactory and nasal mucosal surfaces of cow and sheep based on various ultrastructural and biochemical methods. Ph.D. Thesis, Wageningen, Wageningen, Netherlands.

Menco, B., Menevse, A., and Dodd, G. H. (1974). Studies on olfactory cilia. Abstract 1st ECRO Congress. *Chemoreception Abstr.* **2**, 23.

Menco, B. P., Dodd, G. H., Davey, M., and Bannister, L. H. (1976). Presence of membrane particles in freeze-etched bovine olfactory cilia. *Nature (London)* **263**, 597–599.

Menco, B. P., Leunissen, J. L., Bannister, L. H., and Dodd, G. H. (1978). Bovine olfactory and nasal respiratory epithelium surfaces. *Cell Tissue Res.* **193**, 503–524.

Menevse, A. (1977). Biochemical studies on olfaction. Ph.D. Thesis, Univ. of Warwick, Warwickshire, England.

Menevse, A., Dodd, G. H., and Poynder, T. M. (1977a). Evidence for the specific involvement of cyclic AMP in the olfactory transduction mechanism. *Biochem. Biophys. Res. Commun.* **77**, 671–677.

Menevse, A., Dodd, G. H., Poynder, T. M., and Squirrell, D. (1977b). A chemical modification approach to the olfactory code. *Biochem. Soc. Trans.* **5**, 91–194.

Menevse, A., Dodd, G. H., and Poynder, T. M. (1978). A chemical modification approach to the olfactory code: studies with a thiol-specific reagent. *Biochem. J.* **176**, 845–854.

Michell, R. H., Jafferti, S. S., and Jones, L. M. (1976). Receptor occupancy dose-response curve suggests that phosphatidylinositol breakdown may be intrinsic to the mechanism of the muscarinic cholinergic receptor. *FEBS Lett.* **69**, 1–5.

Minor, A. V., and Sakina, N. L. (1973). The role of cAMP in olfactory reception. (In Russ.) *Neirofiziologia* **5**, 414–422.

Moncrieff, R. W. (1951). "The Chemical Senses." Leonard Hill Ltd., London.

Moulton, D. G. (1974). Dynamics of cell populations in the olfactory epithelium. *Ann. N.Y. Acad. Sci.* **237**, 52–61.

Mozell, M., and Jagodowicz, M. (1974). Mechanisms underlying the analysis of odorant quality at the level of the olfactory mucosa. I. Spatiotemporal sorption patterns. *Ann. N.Y. Acad. Sci.* **237**, 76–91.

Mustaparta, H. (1971). Spatial distribution of receptor responses to stimulation with different odours. *Acta Physiol. Scand.* **82**, 154–166.

Ohloff, G., and Flament, I. (1979). The role of heteroatomic substances in the aroma compounds of foodstuffs. *Prog. Chem. Org. Nat. Prod.* **36**, 231–284.

Ohloff, G., and Giersch, W. (1980). 8. Stereochemistry-activity relationships in olfaction. Odorants containing a proton donor / proton acceptor unit. *Helv. Chim. Acta* **63**, 76–94.

Ohloff, G., and Thomas, A. K., eds. (1971). "Gustation and Olfaction." Academic Press, New York.

Persaud, K. (1980). Olfactory discrimination mechanisms. PhD. Thesis, Univ. of Warwick, Warwickshire, England.

Persaud, K., Pelosi, P., and Dodd, G. H. (1980a). Binding of 5α-androstan-3-one to sheep olfactory mucosa. *Abstr., Int. Symp. of Olfaction Taste, 7th*, p. 101.

Persaud, K., Wood, P., and Dodd, G. H. (1980b). An approach to affinity labelling of receptor sites in the rat olfactory epithelium. *Abstr., Int. Symp. Olfaction Taste 7th*, p. 98.

Poynder, T. M., ed. (1974). "Transduction Mechanisms in Chemoreception." IRL, London.

Seeman, P., and Weinstein, J. (1966). Erythrocyte membrane stabilisation by tranquilizers and antihistamines. *Biochem. Pharmacol.* **15**, 1737–1752.

Squirrell, D. (1978). A study of olfactory mechanisms. Ph.D. Thesis, Univ. of Warwick, Warwickshire, England.

Uttal, W. R. (1973). "The Psychobiology of Sensory Coding." Harper, New York.

PART III

Discussion

Gesteland: Why do Drs. Kurihara and Brand believe that the calcium ionophores opened by taste stimuli are located in the basal synaptic region of the cell rather than on the microvilli where the stimulus presumably arrives? I am surprised at that assumption.

Kurihara: We propose that the electrical current flows from the microvilli to the synapse which depolarizes the synapse. The calcium channel usually is voltage dependent and is opened by the depolarization.

Gesteland: Why can't a calcium channel open on the microvilli with voltage-dependent turnoff as is found in invertebrate systems and all other calcium systems?

Kurihara: Microvillous membranes are not permeable to calcium ions. When we change the calcium concentration in the stimulating solution, we cannot observe any appreciable effect on the response.

Brand: One of the interesting things about calcium inside the cell is that it is subjected to extremely fine control. Attempts to inject calcium into a cell and measure its diffusion through the cell demonstrate that it is immediately sequestered, primarily by the endoplasmic reticulum and the mitochondria. The reason calcium is shown entering at the base is simply because it would never reach the synapse from the microvilli. Calcium entering at the microvilli would be buffered by regulating systems that exist along the length of the taste receptor cell.

DeSimone: A teleological explanation would be that the fluid of the oral cavity is subject to arbitrary and capricious change. It is not a reliable source of any ion, so that taste cells cannot rely upon ions in the oral cavity to depolarize.

Gesteland: You are not looking for a source of ions. You are looking for a conductive change in current flow for which it is a reliable source. That is how all calcium-mediated processes operate that have been studied.

DeSimone: What's the source? The saliva?

Gesteland: Of course. Extracellular calcium is always high compared with intracellular calcium.

DeSimone: But you can adapt the tongue to solutions without calcium and still get a response.

Senseman: With respect to calcium entering the base of the taste cell, there are clearly precedents for specialized, localized calcium channels at the synaptic region. I am surprised that DeSimone, as a physical chemist, did not present a more cogent argument against calcium entering at the apical end! Knowing the mobility of calcium in free solution one can calculate the transit time for calcium to pass through a 20–30 μm length cell. It would take far longer than the response latency data suggest that it could. The available evidence is clearly in favor of calcium entering the cell near the synapse rather than at the apical end.

Brand: The differential effects that Dr. Kurihara and his colleagues have demonstrated for the trivalent, divalent, and the monovalent ions on the receptor surface suggest that multivalent ions are titrating the surface to a degree greater than the monovalent ions. Might this be due to the trivalent and divalent ions simply binding to the receptors whereas the monovalent ions are not?

DeSimone: It is very likely.

Kurihara: In order to explain the dependence of the surface potential on valence of cations according to Gouy-Chapman theory, it is not necessary to postulate the binding of cations to the membrane, although the binding of cations actually seems to occur even when monovalent cations are applied to the membrane.

DeSimone: In the phospholipid monolayer model systems that we have examined, the di- and trivalent ions apparently do something in addition to electrostatically screening the surface. They seem to bind, although the word is ill-defined. For sodium and other small monovalent cations, however, all of the effects on surface pressure and surface potential that we observe with monolayers can be explained as simple electrostatic screening without assuming mass action or stoichiometric relationships between the surface and the ion. This is less true for divalent and trivalent ions. For ferric ion, you can't account for the results at all in terms of screening; it is obviously a strong binder and its effects are not easily reversed.

Margolis: It is my understanding, from Dr. Kurihara's data, that perfusion of the lingual artery with cyclic nucleotides alters the taste cell response to various compounds. You interpret this as due to cyclic nucleotide effects within the cell. Is there any evidence that these cyclic nucleotides do get inside the taste cells?

Kurihara: There is no evidence. It is very difficult to determine where the cyclic nucleotides modulate. The mechanism proposed is based on the observation that adenylate cyclase and phosphodiesterase are located at the microvilli membrane. This localization was examined histochemically by Nomura.

Margolis: Are they inside the cell or outside the cell?

Kurihara: In general those enzymes are located inside.

Price: In Kurihara's transduction mechanism, neutralizing charge on the apical cell surface apparently opens voltage-sensitive control gates at the basal end of the cell. How do you get from the apical end to the basal end? Do you envision pressure waves, a physicochemical process that involves nothing but the lipid, or specific receptor sites for activators and inhibitors?

Kurihara: I think the surface potential change, as well as the intramembrane diffusion potential, creates a current which opens the voltage-sensitive gates at the basal end.

Price: How do you get a current from the surface potential?

Kurihara: We don't want to deny that the microvilli membrane has channels through which an electrical current flows. The surface potential change brings about a change in the total membrane potential and then the emf at the microvilli membrane is decreased, which produces a circulating current.

Price: So a current is created at the apical end? Aren't gates opening at the basal end of the cell to allow calcium to get back?

Kurihara: Since the Ca^{2+} concentration in the intercellular medium is higher than that in the taste cell, the gate opening at the basal end allows Ca^{2+} influx into taste cells.

Gesteland: Is an almost open channel not a short circuit?

Kurihara: Short-circuiting will occur, but a current from the microvilli to the basal end will also flow. Just how much depends on the resistance of the membrane. If the resistance is not small, the current effectively stimulates the basal end.

Gesteland: What are the current carriers?

Kurihara: I don't know. Ca^{2+} channels at the basal end can be opened by a very weak current, and then small amounts of ions are enough; but which ions I don't know.

Cagan: The accessibility of the cyclic nucleotides into the cell are of concern, particularly since we attempted, several years ago, to examine perturbations of cAMP labeling in bovine taste papillae in response to taste stimuli. The small perturbations could not be ascribed to cAMP as a second messenger. Why don't you measure cyclic nucleotide levels and changes?

Kurihara: We are doing that.

Mozell: From DeSimone's analysis it would appear that the odorant molecules must diffuse to a point below the main body of the cilia before producing a response. This suggests that the receptor sites are either not in the cilia at all or are only in the proximal parts of the cilia.

DeSimone: The model would suggest that if there are receptors in the distal cilia they are probably not the first functional layer of receptors, and that the stimulus must travel some distance below the gas/mucus interface before transduction can occur. A conclusion from the model is that the proximal cilia or the knob contain the functional receptors.

Rhein: The cilia may not stick up towards the air–mucus interface. Micrographs of cilia often show them laying flat. They could therefore be further down in the mucus.

Cancalon: I am doing experiments using scanning electron microscopy and electrophysiology. With 0.03% Triton, I can completely remove the cilia except for a very short stump, leaving only the knob. If you examine it by scanning EM you see the knob with about 1/10 of the cilia left. One hour following the Triton treatment, we did not notice any EOG response, but 50–60% of the activity was back within 2 hr and the cilia are not regrowing before 3–4 days.

Cagan: How many cells do you examine to conclude that you completely removed the cilia? That claim was made in an abstract many years ago and continues to be cited as fact. Bronshtein and Minor showed that following removal of cilia the EOG responses decline.

Cancalon: I have looked at the entire olfactory area of a large number of olfactory lamellae in the catfish and in each case find that only a few cilia remained. The respiratory epithelium is not damaged at all and the respiratory cilia appear intact.

Gesteland: We are doing similar experiments with frogs. Following deciliation with 0.03% Triton, we examine the live mucosa with Nomarski microscopy. We find no cilia, no EOGs, and no single cell responses. The cells show a burst of about 20 spikes and then are quiet for about 2 min and then burst and then are quiet. We cannot modulate that with odorants. Within 2 hr the cilia start regrowing, within 4 hr the resting activity is returning to its irregular slow rate, and the EOG is back to one third amplitude. After 12 hr we have normal, responding single cells and normal EOGs. In 16 hr you cannot distinguish this mucosa from the mucosa before deciliation. The site of the stimulus action is the proximal part of the cilia where there are 9 + 2 filaments and axonemes with dynein arms. The mechanism is opening calcium gates by the stimulus and it is all blocked by anything that removes calcium or binds to calcium-binding sites. I believe that the experiments are convincing. We cannot find any way around these observations.

Price: When Bronshtein and Minor showed EOGs essentially back to normal, the cilia had regenerated to about 25% of the control length.

Gesteland: Yes, 30 microns of cilia are needed to get an EOG in frogs.

Getchell: Several of our experiments are in complete concurrence with Gesteland's, although at this time we cannot ascribe any of the voltage responses to calcium gates or calcium spikes. From our calculations the sites of the molecular receptors appear to be on the proximal part of the cilia or perhaps the knob. Using denervation to remove mature receptor cells, the EOG perfectly paralleled the presence, the loss, and the recovery of cilia, but over a longer time course than by applying Triton to the surface of the epithelium.

Karnovsky: It would be wise to refine the terminology regarding "affinity labels." The studies of Dr. Dodd with the halogenated esters are interesting, but they are really competition studies. What evidence is there that they are affinity labeling studies? It is absolutely essential to make certain that the substance is attached covalently to a protein or a lipid before speaking of it as "affinity labeling." I do not think there is the slightest evidence that those halogens, which are quite stable in the alkyl groups, will spontaneously be eliminated as the hydrochloride or hydrobromide and they just will attach to the protein or lipid.

Dodd: They are very reactive. They are classical alkylating reagents that are used for labeling, for example, protease active sites, by reacting with a carbonyl group. They will react with nucleophiles.

Karnovsky: What is the evidence that you really have labeled something?

Dodd: We examined EOGs as a function of time and used congeners, such as the series of esters. The EOG declines, and if the preparation remains for up to 4 hr, that is an irreversible change. Because I have not used a labeled reagent I cannot specifically answer the question. I can, however, quote prior art because they have been used widely for labeling a variety of sites.

Karnovsky: But haven't they been somewhat more complex substances than these rather simple ones that you've used? Shouldn't one start from basic chemistry and first show that something is actually labeled before speaking of "affinity labeling"?

Dodd: The anion of iodoacetate is a classical alkylating agent. It is a group-specific agent which will, for example, label histidines of the active site of RNAase. That demonstrates its mechanism reacting with the nucleophile at the active site. That is the strategy on which we based our approach.

Karnovsky: It also reacts with thiol groups. I suggest that we not use the biology to define the chemistry, but use the chemistry to define the biology. A simple experiment would be to use ^{14}C-labeled reagent and demonstrate that under your conditions you obtain a labeled protein fraction.

Dodd: That is our aspiration, but our immediate ambition is to do a wider range of studies to find something that gives the largest effects possible before we begin to use the labeled version.

Kurihara: A hydrophobic compound such as quinine will sometimes inhibit responses to other stimuli. This may also be true with inhibitors of phosphodiesterase, which are very hydrophobic. The suppressive effect of the inhibitors may be due to their hydrophobic properties. It should also be noted that dibutyryl cGMP is hydrophobic.

Dodd: Cyclic nucleotide effects are observed in several systems, such as insects, and cyclic nucleotides show effects on the olfactory epithelium. Whether this suggests a classical cyclase cascade second messenger system in transduction is an open question.

Kurihara: If you increase the concentration of dibutyryl cGMP, does that decrease the transient effect in the sheep?

Dodd: In the sheep preparation, the dibutyryl cGMP has a temporal effect, whereas in the frog the response to a given concentration of cAMP is constant with time.

Gower: Did you tritiate the androstenone to make androstanone for your studies?

Dodd: Yes. We tritiate across the double bond.

Gower: We confirm the "stickiness" of androstanone and androstenone and, for this reason, we had to completely change the odor delivery system for our attempted EOG studies with the sow olfactory mucosa. Even with Teflon tubing, androstenone is adsorbed. An all-silanized system ensured delivery of the ketone but the EOG was very small indeed and we doubt its validity. Dr. Dodd showed a particular form of specific binding of androstanone with the sheep olfactory mucosa.

Dodd: In examining some of our earlier binding data, we discovered that there were effects in several preparations, but they were on the limit. Binding studies with other androgens give a smaller amount of nonspecific binding. There is clearly an effect when the cold ligands were added; the binding decreased in four separate sheep preparations. Whether that is sufficient to indicate specific binding which is physiologically significant, I don't know at the present time.

Cagan: How do you view the androstanone receptor localization and molecular characteristics of the receptor? Isn't it in the soluble fraction?

Gower: Yes, it is in the soluble fraction as are most steroid receptors. As far as I know, no one has shown physiological effects of androstanone on the sheep. Are there reductases for androstanone in the sheep?

Dodd: Yes. The reductases described in the pig are all present in the sheep, in the soluble fraction after a 12,000 g spin. We don't know if they are steroid reductases which simply happen to reduce androstanone. They may not be specific.

Quinn: Dr. Hirsch, is there saturation of binding to the isolated olfactory epithelial cells with muscarinic acid? Also, is there any comparison between muscarinic acid binding to the epithelial cells as opposed to other areas of the brain?

Hirsch: I have done saturation kinetics on every ligand illustrated and determined the apparent K_D and B_{max} for each. With the exception of α-adrenergic sites, of which the epithelium has more than the brain, the other cases show fewer numbers of receptors in the epithelium (by about fivefold) than in the brain. The affinities are generally somewhat lower as well. For example, in the brain the affinity for QNB is 0.2–0.5 nM, while in the epithelium it is 5–6 nM. Although the affinity was lower in the epithelium and the number of sites was smaller, the pharmacologic specificity of epithelial receptors was the same as the corresponding brain sites except for the benzodiazepine.

Brand: In the allosteric model, how is the parameter L determined to be 9?

Mooser: L was determined from preliminary data using NaCl as the stimulus. NaCl was chosen because its equation contains the smallest number of variable parameters (dissociation constants for sodium ion binding to the active and inactive states and L). The calculated value for L was then kept constant for the remaining data. Subsequently, I did multiple nonlinear regressions to determine if the initial value for L provided the best fit when all of the stimuli and all of the parameters were considered together. This resulted in a relatively small change in L and is the final value reported.

Cagan: Although you chose the rat system because of the plateau, isn't that an aberrant situation with a long-lasting plateau of response rather than adaptation?

Mooser: Yes it is. I do not think it would be useful if it adapted. This experimental characteristic is mandatory when assuming a steady state in the system. Without the assumption, the equations become unreasonably complex. Even though the rat and other animals receive critical sensory information in pre and post steady state portions of the electrophysiologic response, the basic system can still be consistent with a two-state mechanism. The fact that the experimental support for the model is based on an uncommon, but exploitable, characteristic of the rat response does not invalidate its fundamental importance. Nonsteady state portions of the response are probably superimposed with other events such as adaptation.

Cagan: Isn't there something very unusual about the system that may not involve a receptor mechanism, but rather involve synaptic events that are unusual in the rat? In addition, in terms of the inhibition data being at the cell membrane, isn't that an article of faith rather than a demonstrable fact based on nerve recordings? There is multiple innervation by fibers, synapses, and branching; there is considerable modulation.

Mooser: Synaptic events or other events not associated with the receptor cell membrane cannot account for the data. The reasoning behind this is as follows. Under steady state conditions, all intermediate steps between stimulus-induced activation and recording the neural response remain at a constant level. In addition, changes in stimulus concentration are the only experimental variables. This makes it possible, after satisfying a few assumptions, to directly correlate changes in magnitude of the neural response with receptor-cell membrane events. The assumptions are: (1) the stimulus acts at the level of the receptor cell membrane, and (2) stimulus-induced events are linearly related to the magnitude of the neural response. The first assumption is generally accepted, but the second requires experimental support. For this reason, we spent considerable time on the inhibition studies before proceeding with other experiments. Irreversible protein modification reagents, like HNB-dmS, inhibit the NaCl-stimulated response at a rate consistent with a classical psuedo first-order decay. Based on our studies with N-substituted maleimides, HNB-dmS would not be expected to cross the receptor cell membrane. If nonlinear modulation of the response were to occur between the points of membrane activation (or inhibition) and summated neural activity, the inhibition would not follow a first-order decay unless the higher order events were multiple and counteracting to give an overall relationship which has the appearance of a first-order reaction. I find this possibility unlikely and concluded that the steady state magnitude of summated neural activity over the range observed during HNB-dmS inhibition directly reflects receptor-cell membrane alterations. This relationship held for inhibition of the response at a number of NaCl concentrations and provided the necessary confidence to analyze stimulus concentration–response data in terms of specific models.

Cagan: The fact that you can fit the data to a model does not prove it. Caution is needed in extrapolating back to the membrane, even with the fit of the model. There may be other models that fit that do not involve allosteric factors in the membrane.

Mooser: That's a good point. This type of analysis never proves a mechanism; it eliminates potential mechanisms, but never proves one. I started with as simple a case as possible without trying to devise all of the possible models, which are infinite, and looked for consistency. In no way can this be the only model that fits, but I am not sure there is a simpler model which is consistent with the data.

Cagan: Will it be possible to examine aspects of your model by intracellular recordings?

Mooser: Obtaining the data would be very difficult.

Dodd: If we use William of Ockham's razor on the point, assuming that the events are ligand phenomena, can they be described by a model simpler than a two-state one?

Mooser: No. Somewhere there has to be change in the system; whether it's a change in physicochemical events or alterations in conformation is debatable. However, changes in physicochemical properties are no more simple than applying ligand binding energy to displace the resting state of an equilibrium.

DeSimone: On the other hand, if we agree that there is charge on the membrane and that it operates in a concentration range in excess of 0.1 mM for monovalent salts, then physicochemical events, such as changes in surface potential and changes in surface pressure, will occur and could conceivably relate to taste responses. We don't have to assume it because we know charged interfaces respond to salts with altered potential and pressure. Therefore a model based on those events is probably more desirable than one which supposes a more complicated state of affairs, in accord with Ockham's razor.

Cancalon: The S-100 protein used to be known as a good marker for glial cells. Now it has been found in neurons. And Graziadei assures me that it is not in sustentacular cells. Could you comment on this problem?

Margolis: There is accumulating evidence that the S-100 protein is not a single protein but a family of related proteins. Differences may depend on the isolation techniques, the antiserum, etc. There is no doubt it is largely glial in nature, but it is also true that there are uncertainties about its distribution now that didn't exist previously. I recall that Graziadei has seen S-100 staining in the olfactory epithelium but I don't recall which cell types stained. He used antiserum supplied by Dr. Claire Zomzely-Neurath.

Wysocki: A basic assumption has been made that the olfactory epithelium in the pig mediates the behavioral responses to androsterol and androstenone. As a cautionary note, the vomeronasal epithelium might play a role in the behavioral responses. Perhaps EOGs and binding studies using the vomeronasal epithelium might give different results.

Gower: We measured EOG responses of the vomeronasal organ and had no more success than with the olfactory tissue.

Wysocki: It is difficult in general to stimulate and get responses from the vomeronasal epithelium.

Dodd: There is no doubt for humans that it is a powerful odorant with a low olfactory threshold, and that it acts for humans only on the olfactory system.

Price: The small EOG might mean that there are few cells that respond to it. It does not necessarily have anything to do with its molecular weight or sticking to Teflon.

PART IV

Neurotransmitters in Taste and Olfaction

17

Neurotransmitter Biochemistry of the Mammalian Olfactory Bulb

FRANK L. MARGOLIS

We are all more influenced by smell than we know. (Hercule Poirot)
"Murder in Retrospect," Agatha Christie

I. INTRODUCTION

The role of an organism's nervous system may be considered to be threefold in nature. The first of these can be viewed as being responsible for monitoring changes in the internal and external milieu and is, therefore, essentially a *detection* mode. Next, this gathered information must be subjected to *evaluation*, so that it can be ordered with regard to relative importance. The final aspect relates to the *generation of responses*, which will

Biochemistry of Taste and Olfaction

ISBN 0-12-154450-8

permit the organism to cope with the changes detected. This is clearly a dynamic iterative process in which all three aspects are proceeding simultaneously. One of our ultimate goals is to comprehend the molecular basis of organismic behavior. To do this, we must have detailed information about the mechanisms of intercellular signaling and information processing. In the olfactory system, this can be reduced to a preliminary set of questions, some of which are addressed by this chapter.

What are the molecules used by the various cells for information transfer? Specifically, what is the neurotransmitter at the first synapse? In the course of olfactory chemoreceptor cell turnover, what are the biochemical responses of the secondary neurons in the pathway to continual change of their input sources? Are there specific molecules regulating this cellular turnover, and what are they? Is there target selectivity of the newly invading olfactory axons? What molecular basis permits them to select the proper partners in the bulb? Is there biochemical information transfer from the afferent olfactory axon to the target cell in the olfactory bulb? Does molecular information move in a retrograde direction from the cells in the bulb to the neurons in the epithelium, and, if so, what role does it play in the overall economy of this system?

In order to approach some of these questions, my laboratory has been studying various aspects of the biochemistry of the mammalian olfactory system. This chapter deals primarily, but not exclusively, with an overview of the current status of transmitter biochemistry in the olfactory bulb. The bibliographic coverage is not intended to be inclusive but rather to give the interested reader access to the relevant literature.

II. ANATOMICAL ORGANIZATION

Because the anatomical organization has played an important role in determining research strategies, it is desirable to briefly outline the anatomy of this region of the nervous system. The olfactory and vomeronasal systems comprise two parallel sensory pathways. In each, the primary sense cells reside in the periphery, and their axons project to the main or accessory olfactory bulb, respectively, within the cranium, where they synapse with the dendrites of the major output neurons, as well as with those of certain interneurons in complex synaptic structures called *glomeruli.* The second-order output neurons project to various loci in the limbic system. In addition to the primary afferent input to the bulb, there are centrifugal fiber inputs from deeper in the CNS, many of which derive from sites to which bulbar output neurons project, thus forming a series of circuit loops. Within the main olfactory bulb are two types of output neurons, the mitral and tufted cells. A final subclass of cells includes several types of intrinsic or local circuit neurons. These consist of peri- or juxtaglomerular cells, a large population of granule cells, and a series of morphologically distinct populations of short axon neurons (Schneider and Macrides, 1978).

The olfactory bulb like the retina, to which it bears an interesting resemblance, is a highly laminated structure (Fig. 1) in which various neuron cell bodies and areas of synaptic interaction are distributed in characteristic arrays (Broadwell, 1977; Shepherd, 1979).

In both the olfactory and vomeronasal systems, the primary sense cells are apparently unique among neurons in that they are continually being shed and replaced from progenitor cells in the chemosensory neuroepithelium (Graziadei and Monti Graziadei, 1978, 1979; Barber and Raisman, 1978a,b).

Finally, the gross anatomical organization permits independent access to the cell bodies of afferent neurons, to their site of synapse in the bulb, to the secondary output neuron axons, and to their terminal synaptic projection sites. The types of lesions utilized for these purposes have been both surgical and chemical in nature. Thus, the olfactory bulb has been disconnected from

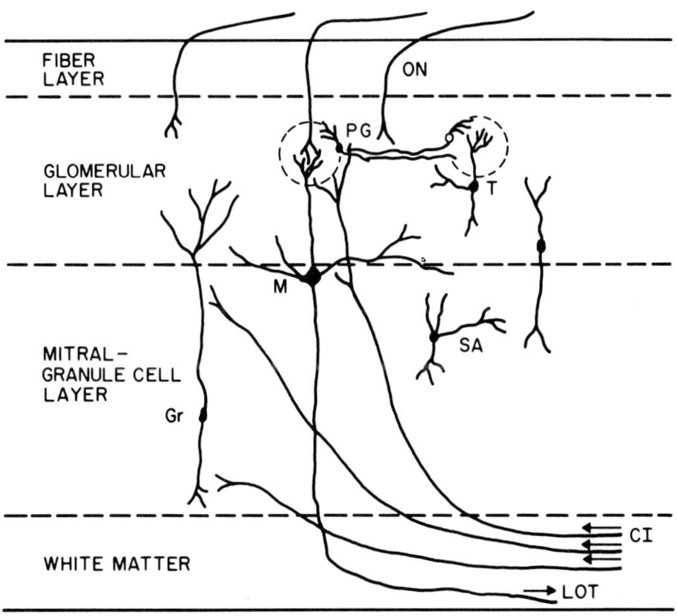

Fig. 1. Schematic diagram depicting the anatomy of the mammalian olfactory bulb and boundaries used in the microdissection of the dog olfactory bulb and the cells obtained from each layer. Abbreviations: CI, centrifugal input; Gr, granule cell; LOT, lateral olfactory tract; M, mitral cell; ON, olfactory nerve; PG, periglomerular cell; SA, short-axon cell; and T, tufted cell. (From Nadi *et al.*, 1980.)

the olfactory epithelium by surgical axotomy of the olfactory nerve (Takagi, 1971; Harding *et al.*, 1977; Graziadei and Monti Graziadei, 1978; Cancalon and Elam, 1979; Simmons and Getchell, 1979; Rochel and Margolis, 1980) to evaluate the influence of chemoreceptor neuron degeneration following regeneration and reinnervation of the bulb. In addition to the use of surgical olfactory nerve section, a variety of chemical agents have played a role in equivalent studies. Intranasal irrigation with zinc sulfate, if used under certain well-defined conditions, can cause extensive destruction of the olfactory neuroepithelium and subsequent synaptic terminal degeneration in the bulb with little or no regeneration (Margolis *et al.*, 1974; Singh *et al.*, 1976; Harding *et al.*, 1978). Other authors using alternate conditions have achieved less permanent results (see Alberts, 1974, for detailed review; see also Matulionis, 1976; Slotnick and Gutman, 1977). Other agents such as detergents, antimitotics, or alkylating agents have also been shown to generate chronic or transient neuroepithelial and bulbar synaptic degeneration depending on the compound and mode of administration (Margolis and Grillo, 1978; Nadi and Margolis, 1978; Harding and Wright, 1979; Nadi *et al.*, 1981; Rochel and Margolis, 1980). One additional approach to achieve this end has been nostril cauterization (Meisami, 1976). The bulb can be isolated entirely from the rest of the brain surgically to evaluate the influence on centrifugal fibers (Ross *et al.*, 1979), or the lateral olfactory tract can be selectively lesioned to study the effect on deeper CNS regions (Broadwell, 1977; Wenk *et al.*, 1977) or on mitral cell function in the bulb (Macrides *et al.*, 1976).

The nature of the interplay between the two bulbs can be studied following removal of one of them (Hirsch and Margolis, 1980; Rochel and Margolis, 1980). Clearly, selective lesion of fiber tracts or nuclei, whose cells project to the bulb can help to evaluate the nature of specific centrifugal inputs to the bulb (Broadwell, 1977; Wenk *et al.*, 1977; Macrides *et al.*, 1981). Finally, the use of toxins and pharmacological agents such as kainic acid, 6OHDA, and reserpine, to selectively destroy or inactivate selected pathways also can be used. These approaches permit one to selectively manipulate one portion of this system and to evaluate the biochemical responses of other portions in a manner not always feasible in other CNS pathways.

III. TRANSMITTER BIOCHEMISTRY

A. Introduction

Evaluation of which neuroactive compounds are present in the bulb and their distributions among the various fiber and cell types has become something of a minor "cottage industry" of late. The approaches used have included autoradiographic and immunohistochemical techniques, as well as direct chemical measurements of the compounds themselves, their binding sites and their enzymes of synthesis and degradation. But demonstration of the presence of a compound in the tissue is only a first step; more detailed microanatomical localization and ultimately demonstration of function are required. To this end, many authors have coupled biochemical measurements with selective lesioning techniques. This has been a very productive approach but one dependent on a critical assumption. It assumes that the disappearance of a compound or enzyme activity following a lesion implies the prior presence of that component solely within the degenerating fiber, cell, or terminal. Although this assumption is generally valid, it is not universally true. In a later section, a demonstration of postlesion transsynaptic biochemical alterations in the olfactory bulb is presented, which will illustrate some of the potential pitfalls awaiting the unwary.

B. Acetylcholine

All the cholinergic activity in the mammalian bulb derives from centrifugal fibers (Chapter 20). The key sources of this input seem to originate caudal to the lateral preoptic area, as demonstrated by lesions of the diagonal band from the studies of Wenk *et al.* (1977) and Youngs *et al.* (1979). The distribution of this input across the bulb has also been evaluated by consideration of the laminar distribution of muscarinic binding sites in the dog olfactory bulb (Nadi *et al.*, 1980). In that study, muscarinic binding showed nearly uniform distribution in all layers as though the cholinergic input exerts a general modulatory influence throughout the bulb. Observation of α-bungarotoxin binding in the glomerular neuropil has been interpreted to indicate that mitral cell dendrites may be cholinergic (Hunt and Schmidt, 1977), but the absence of immunohistochemical staining for choline acetyltransferase within olfactory glomeruli (Kan *et al.*, 1980) argues against this postulate.

C. Serotonin

Serotonin has been measured in the olfactory bulb (Haubrich and Denzer, 1973). Based on autoradiography, fluorescence immunohistochemistry, and cytochemistry, it seems to be present only in fibers and not in cell bodies (Dahlström *et al.*, 1965; Halász *et al.*, 1977, 1978). The origin of these fibers is presumably neurons with cell bodies in the raphé which are known to project to the olfactory bulb (Broadwell and Jacobowitz, 1976). These fibers terminate in the glomerular region (Moore *et al.*, 1978) and are thought to exert their influence via interneurons (Halász *et al.*, 1977, 1978).

D. Catecholamines

The presence of the catecholamines, norepinephrine and dopamine, in the olfactory bulb has been demonstrated chemically (Fallon and Moore, 1978; Palkovits *et al.*, 1979; Jaffé and Cuello, 1980; Nadi *et al.*, 1981). Details of anatomical localization for the amines themselves and for their synthetic enzymes derive from the work of several investigators (Dahlström *et al.*, 1965; Broadwell, 1977; Halász *et al.*, 1977, 1978; Fallon and Moore, 1978; Youngs *et al.*, 1979; Nadi *et al.*, 1980).

Norepinephrine is present in fibers but not in cell bodies in the bulb, and it has been demonstrated that most if not all of the norepinephrine content can be eliminated by lesions of tracts originating from the locus coeruleus (Broadwell, 1977; Youngs *et al.*, 1979). Quantitation of the distribution of this amine in various lamina of the dog olfactory bulb indicates that norepinephrine is distributed throughout the bulb (Table I) with a tendency to higher concentration in deeper rather than in more superficial layers of the dog (Nadi *et al.*, 1981), but with an apparent peak in the external plexiform layer of the rat (Jaffé and Cuello, 1980).

Definite assignment of all the norepinephrine exclusively to neuronal terminals seems difficult since norepinephrine innervation of the vascular bed must also be considered. The binding sites for α- and β-aminergic ligands (clonidine and dihydroalprenolol, respectively) (Nadi *et al.*, 1980) show a differential distribution across the layers, which does not coincide with the distribution of norepinephrine (Table I). Epinephrine cannot be a complicating factor in these attempts to generate a coherent interpretation of the biochemical observations, as none could be demonstrated to be present in the bulb (Nadi *et al.*, 1981), nor could phenylethanolamine-*N*-methyltransferase be demonstrated immunohistochemically (Halász *et al.*, 1977).

In contrast with the norepinephrine innervation of the bulb, which has been referred to as the extrinsic catecholamine innervation (Fallon and Moore, 1978), dopamine has been called the intrinsic catecholamine innervation of the bulb. Dopamine exhibits a very different bulbar distribution from that described above for norepinephrine (Broadwell and Jacobowitz, 1976; Fallon and Moore, 1978). There is essentially no extrinsic dopaminergic innervation of the bulb (Fallon and Moore, 1978), and lesions that significantly reduce norepinephrine content of the bulb and other forebrain areas have virtually no effect on dopamine levels (Fallon and Moore, 1978; Macrides *et al.*, 1981), suggesting that the intrinsic juxtaglomerular dopaminergic neurons contain essentially all the dopamine of the bulb. Additional studies (Halász *et al.*, 1978; Priestley *et al.*, 1979) have demonstrated a population of cells located in the periglomerular position that take up [^{3}H]dopamine, stain for tyrosine hydroxylase and dopa decarboxylase, but

TABLE I Distribution of Catecholamine Related Parameters in Dog Olfactory Bulb[a]

Layer	Norepinephrine (pg/mg protein)	Dopamine (pg/mg protein)	Tyrosine hydroxylase (pmole/mg protein/hr)	Total bound ligand (%)		
				Dihydroalprenolol	Spiroperidol	Clonidine
Fiber	332	444	89	25	10	30
Glomerular	314	1168	264	10	15	35
Mitral–granule	558	680	150	28	55	10
White matter	597	269	23	37	20	25

[a] For details, see Nadi *et al.*, (1980, 1981).

not for dopamine-β-hydroxylase, and are presumptive dopaminergic cells. There is some question as to whether these are a subpopulation of periglomerular cells or are superficial tufted cells, but there is no doubt as to their location. When the laminar distribution of chemically measured dopamine was evaluated in the bulb, it showed a very different distribution from that for norepinephrine (Table I). Dopamine was primarily localized in the glomerular layer in both the rat and the dog (Jaffé and Cuello, 1980; Nadi *et al.*, 1981), consistent with the morphological studies noted above. More striking was the observation that the activity of the first enzyme in the biosynthetic pathway, tyrosine hydroxylase (Nadi *et al.*, 1981) (Table I), exhibited this exact pattern and did not coincide with the norepinephrine distribution. This suggested that the dopaminergic neurons are more dependent upon local synthesis than are the noradrenergic fibers, which either have a much longer transmitter half life or are more dependent upon reuptake processes than are the dopaminergic cells. Since dopamine appears to be primarily associated with periglomerular cells, it seemed logical to evaluate the laminar distribution of dopamine binding sites in the bulb. Curiously, spiroperidol binding (Table I) was highest not in the glomerular layer but in the mitral–granule layer, which was where we observed a peak of tyrosine content (Nadi *et al.*, 1980). These observations are difficult to reconcile with the other data presented above, and they indicate that the pattern of catecholamine ligand binding will not yield to a simplistic interpretation and that more information is required.

Since these two catecholamines differ in laminar distribution within the bulb and in cell type of origin, it seemed probable that they are subject to different regulatory influences. One major influence on the dopamine-containing cells would seem to be the continual synaptic remodeling occurring at the level of their dendrites due to afferent olfactory cell turnover. Thus, this seemed to be a system in which one could evaluate the influence of changes in afferent input on second-order neuron function.

In pursuit of this thought and in consideration of whether the continual synaptic remodeling occurring in this system due to olfactory turnover might modulate transmitter function, we began to evaluate catecholamine responses to deafferentation. Since we could select treatments that would cause acute or chronic deafferentation, it was clearly of interest to compare these conditions. Dopamine levels and tyrosine hydroxylase activity in the olfactory bulb were critically dependent upon intact afferent input (Table II). Conditions that caused a fall in the level of the specific olfactory marker protein (for reviews, see Margolis, 1975, 1977, 1980a)—which is present only in the chemoreceptor neuron axons and their synaptic terminals—and a decline in the level of the dipeptide carnosine (see below) also caused declines in the quantity of dopamine and in tyrosine hydroxylase activity (Table

II). At the same time norepinephrine levels seem to undergo a transient rise. The magnitude of the effect on dopamine and tyrosine hydroxylase is very large and is reversible only where there is evidence that reinnervation of the bulb is occurring as exemplified here by the reversal after Triton X-100 treatment. In the absence of other information, one could contend that these data argue for the presence of dopamine in the primary afferents. This argument is untenable in the face of all of the histochemical and immunohistochemical data cited above, as well as the laminar analyses performed chemically. The most rational explanation seems to be that the state of afferent innervation does in fact regulate biochemical expression of the juxtaglomerular dopaminergic neurons. Other examples of transsynaptic regulation of catecholamine metabolism are known (for review, see Thoenen and Otten, 1977), as is evidence for environmental determination of transmitter expression by neurons in tissue culture (Patterson, 1978).

In retina, which exhibits organizational analogies to olfactory bulb, regulation of dopamine synthesis by stimulation of afferent input is also seen (Iuvone *et al.*, 1978), although in retina the response time course is much more rapid than observed in bulb. Since there are many periglomerular cells in the bulb that are "GABAergic" and which are presumably subject to the same influences after deafferentation as are the dopaminergic cells, it seemed possible that they too would show a transsynaptic response. However, this was not the case (F. L. Margolis, unpublished), indicating the specificity of this phenomenon. Since the periglomerular GABAergic cells are but a small fraction of the total population of bulbar GABAergic cells, it is possible that a change in their biochemistry might be small against a large background. To test this properly would require either additional laminar microchemistry or immunocytochemical evaluation. Although the mechanism of this trans-synaptic regulation of dopamine synthesis in the bulb is not yet understood, it seems amenable to evaluation. Possible explanations could be that direct synaptic contact is required or that the transport of some molecule from the chemoreceptor neuron terminals to the juxtaglomerular cell dendrites stimulates expression of tyrosine hydroxylase activity and dopamine synthesis. Two obvious candidates for this role are the olfactory marker protein and the dipeptide carnosine. In this regard, it is of interest to comment on the unusual response of the membrane associated carnosine binding site to deafferentation. Unlike many other transmitter systems the carnosine binding site, which we presume to be postsynaptic, declines after deafferentation as though the presence of carnosine were required for its maintenance (Hirsch *et al.*, 1978). However, the site is not gone but is only in a cryptic state because its presence can be demonstrated on incubation with high concentrations of ligand (Hirsch and Margolis, 1979). Possibly, the switch from a cryptic to a noncryptic state is somehow involved in regulation of tyrosine

TABLE II **Influence of Chemical Deafferentation on Biochemical Parameters of Mouse Olfactory Bulb**[a]

Agent	Days	Percent of control activity				
		Dopamine	Norepinephrine	Tyrosine hydroxylase	Carnosine	Olfactory marker protein
$ZnSO_4$ (0.17 *M*)	4	42	147	58	13	55
	10	25	141	30	5	30
	20	15	211	—	12	10
	50	30	133	6	19	15
Triton X-100 (0.5%)	4	51	95	75	19	65
	10	48	158	—	22	30
	15	—	—	18	—	—
	20	32	134	—	39	60
	50	99	89	100	87	90

[a] Data modified from Nadi *et al.* (1981) and Margolis and Grillo (unpublished).

hydroxylase expression in periglomerular cells. In this regard, it is of interest to note that carnosine is reported to influence periglomerular cell neurophysiology (MacLeod and Straughan. 1979).

Since the cell specific olfactory marker protein is expressed by the chemoreceptor neurons virtually concomitant with their formation of synapses in the bulb (Farbman and Margolis, 1980), it is possible that this specific protein may play a trophic role in transsynaptic regulation of dopamine synthesis in this neural system.

A more circuitous mechanism might involve afferent synaptic activity indirectly by regulating the level of some circulating hormone, which could in turn feed back on the periglomerular cells and subsequently regulate their genetic expression. In the absence of data, these remain only hypotheses in need of study.

E. Amino Acids

Of the various amino acids studied for their roles as transmitters in the olfactory bulb, perhaps the most intensive study has been devoted to GABA. Biochemical evidence from studies of high affinity uptake (Margolis *et al.*, 1974), autoradiography (Halász *et al.*, 1979; J. M. Palacios and M. J. Kuhar, personal communication), laminar analysis of GABA levels, and muscimol binding (Nadi *et al.*, 1980) and immunocytochemistry (Ribak *et al.*, 1977) supports the role of GABA as the transmitter in granule cells and in most periglomerular cells (see Chapter 18).

The major output neuron of the olfactory bulb is the mitral cell, which has three classes of sites of synaptic interaction. Two are by way of dendrodendritic synapses onto periglomerular and granule cells, while the third is more classical in nature. This last site is the synaptic input to the olfactory cortex by way of mitral cell axons, which constitute the lateral olfactory tract (LOT). It is assumed, but not known, that the same transmitter is used at all three sites. Biochemical evidence that the transmitter used by these cells at the synaptic site in the olfactory cortex might be an acidic amino acid derives from the observation that surgical removal of the olfactory bulb with subsequent degeneration of the LOT is accompanied by preferential losses of aspartate (− 30%) and glutamate (− 20%) from the olfactory cortex (Harvey *et al.*, 1975). More recently, Collins (1979a) has shown that 4–10 days after unilateral bulbectomy there is a selective decline of aspartate content in rat olfactory cortex at the depth where many of the LOT fibers are thought to terminate. Although there is some indication of glutamate decline as well, these data suggest that at least some of the LOT fiber terminals in the olfactory cortex utilize aspartate as their synaptic transmitter. This is further supported by the observation of Ca^{2+}-dependent aspartate release from iso-

TABLE III Distribution of Amino Acid Related Parameters in Dog Olfactory Bulb[a]

Layer	Compound present (μmole/g wet weight)				Total bound ligand (%)		
	Taurine	Carnosine	GABA	Glutamate	Muscimol	Kainic acid	Carnosine
Fiber	8.9	1.3	1.5	9.6	5	5	20
Glomerular	7.5	1.0	1.7	11.2	30	15	40
Mitral–granule	7.9	0.3	7.0	8.0	50	55	20
White matter	8.5	(0)	4.5	5.2	20	25	20

[a] For details, see Nadi *et al.* (1980).

lated olfactory cortex when the LOT is electrically stimulated (Collins, 1979b). Whether this amino acid is also the transmitter at the dendrodendritic synapses in the bulb is not known. Isolation of dendrodendritic synaptosomes from bovine olfactory bulbs was reported (Kornguth *et al.*, 1979), and Quinn and Cagan (this volume, Chapter 18) show biochemically the presence of glutamate decarboxylase and of GABA binding sites. Binding sites for kainic acid, which may or may not label glutamate-binding sites, show a maximum in the mitral–granule layer of the dog olfactory bulb (Nadi *et al.*, 1980). However, there is no indication of a discrete distribution of aspartate or glutamate to support a transmitter role for these compounds (Nadi *et al.*, 1980). This is not surprising since the bulk of the bulbar content of these amino acids probably subserves a metabolic role.

The distribution of other amino acids in various layers or cell types in the olfactory bulb has been studied by several techniques (Austin *et al.*, 1978; Halász *et al.*, 1979; Ross *et al.*, 1979; Nadi *et al.*, 1980) (Table III). To date there is little biochemical evidence for a significant transmitter role in the bulb for any amino acid other than those discussed above (i.e., GABA, glutamate, and aspartate).

F. Peptides

A major thrust in recent years has been the characterization of peptide distribution in various neural tissues. The olfactory bulb has not been immune to this activity, although much less is known than desired. A brief summary of the peptides reported to be present in the olfactory bulb demonstrates their diversity (Table IV). Much of this information derives from immunochemical approaches. In view of the well-known, potential problems with cross reactivity of even the best antisera, it is essential, if at all possible, to confirm the presence of the peptide by unambiguous chemical means. An example of the kind of problem that can arise was the recent report of a high

TABLE IV Peptide Content of Olfactory Bulb

Peptide	Species	Content	Reference
Met-enkephalin	Rat	1.3 ng/mg protein	Yang *et al.* (1978)
Met-enkephalin	Mouse	493 pmole/g	R. V. Lewis and F. L. Margolis, (unpublished)
Cholecystokinin	Pig	271 pmole/g	Rehfeld (1978)
Gastrin	Pig	<0.1 pmole/g	Rehfeld (1978)
Substance P	Rat	27 ng/g	Brownstein (1977)
Somatostatin	Rat	20 ng/g	Brownstein (1977)
Insulin	Rat	60 ng/g	Havrankova *et al.* (1978)
Luteinizing hormone–releasing hormone (LHRH)		Present	Hoffman *et al.* (1979)
Vasoactive intestinal polypeptide (VIP)	Rat	6.7 pmole/g	Fahrenkrug *et al.* (1978)

molecular weight protein with β-endorphin and ACTH immunoreactivity in human placenta (Julliard *et al.*, 1980). Ultimately, it was shown that this protein is in fact the heavy chain of IgG, of which a 17 amino acid fragment gave 0.2% immune cross reactivity on a molar basis. Clearly, depending upon the relative proportions of the authentic and unknown materials, one can be fooled by reliance on immunological techniques exclusively.

With a few exceptions, the detailed anatomical localization of these peptides is not known, and therefore consideration of their functions in the bulb are premature. Although commonly thought to act as neurotransmitters or neuromodulators, peptides may play other roles as well with regard to neuronal organization, glial activity, or vascular function.

The endogenous level of the opioid peptide met-enkephalin, monitored by a high performance liquid chromatographic procedure, is unaffected by peripheral deafferentation (R. V. Lewis and F. L. Margolis, unpublished), indicating it to be absent from primary afferents. Opiate binding activity was quite high in the fiber layer of the olfactory bulb (Nadi *et al.*, 1981), but total opiate binding did not alter in response to deafferentation (Hirsch *et al.*, 1978). Thus, the interrelationships of these observations remain unclear. Of perhaps greater interest is the observation that unilateral bulbectomy results in a rapid, selective decline in opiate binding in the contralateral bulb (Hirsch and Margolis, 1980) to about 50% of control in 2 weeks. This suggests that degenerating fibers from the contralateral anterior olfactory nucleus receive opioid peptide input presynaptically. These observations predict that one or more populations of bulbar interneurons would be the source of the opioid peptides. Preliminary immunohistochemical data are in support of this (N. Brecha, personal communication; F. Macrides, personal communication).

Immunohistochemical evidence in support of the presence of LHRH containing fibers in the bulb (Hoffman *et al.*, 1979) and of insulin and insulin binding sites (Havrankova *et al.*, 1978), as well as of several other gastrointestinal and neuroendocrine peptides, are consistent with the role of the olfactory bulb in modulating a wide variety of activities related to limbic system function (Cain, 1975; Macrides, 1976; Cartas-Heredia *et al.*, 1978; Pager, 1978; Robinzon *et al.*, 1979). If the various immunochemical studies are chemically validated, this implies that we can anticipate significant and exciting future advances in the role of these compounds.

One peptide system in the olfactory bulb about which a great deal has been learned is that of the dipeptide carnosine, β-alanyl-L-histidine. The presence of high levels of this dipeptide in olfactory tissue was reported independently by my laboratory and that of Amos Neidle within a few hours of each other at the 1974 Meeting of the American Society for Neurochemistry and later in more formal form (Margolis, 1974; Neidle and Kandera, 1974). As is evident in Table V, other laboratories have confirmed and extended this observation in a variety of vertebrate species. In contrast with the homologous peptide homocarnosine (GABA-L-histidine), which does not exhibit any striking variation in CNS regional distribution, carnosine is very highly concentrated in the olfactory bulb and olfactory epithelium. In this region of the brain, carnosine tends to be present at millimolar concentrations, which is at least one order of magnitude higher than its content anywhere else in the brain. The related peptide anserine also is present in the brain of some species (Fisher *et al.*, 1977), although we have never seen it in mice. Carnosine and anserine are well known as constituents of muscle (Crush, 1970), and homocarnosine was recently reported to occur in myenteric plexus (Jessen *et al.*, 1979).

Because the primary afferent transmitter of the olfactory chemoreceptor neurons at their synapses in the bulb was unknown and as none of the compounds discussed here gave any evidence of fulfilling that role and as carnosine exhibited a highly localized regional distribution, we studied the possibility that this dipeptide might be the primary afferent transmitter of the olfactory pathway. Our approach has been wide ranging and has included studies of the peptide distribution and synthesis, axonal transport, and purification and characterization of enzymes involved in synthesis and degradation, as well as evaluation of a membrane associated binding site.

One of our key observations was that intranasal irrigation of the mouse with zinc sulfate solutions, which are known to cause destruction of the olfactory neurons in the nasal mucosa, also caused degeneration of their axons and synaptic terminals in the olfactory bulbs. When monitored biochemically, the level of carnosine in the bulb fell but the level of a wide variety of other amino acids and enzymes remained the same (see

TABLE V Distribution of Carnosine and Homocarnosine in Olfactory Bulbs and Whole Brain of Various Species[a]

Species	Olfactory bulb (μmole/mg tissue)		Whole brain (μmole/mg tissue)		Reference
	Carnosine	Homocarnosine	Carnosine	Homocarnosine	
Mouse	1.8	0.07	0.04	0.08	Neidle and Kandera (1974)
Mouse	0.9–1.9	0.04–0.12	—	0.08	Ferriero and Margolis (1975)
Hamster	1.2	—	—	—	Margolis (unpublished)
Pig	0.5	—	—	—	Margolis (unpublished)
Dog	1.2	—	—	—	Margolis (unpublished)
Rabbit	0.5	—	—	—	Margolis (unpublished)
Gecko	3.9	—	0.16	—	Margolis (unpublished)
Rat	2.7	—	0.03	0.12	Quinn and Fisher (1977)

[a] A more detailed summary has been published (Margolis, 1980).

Tables II and VI). This strongly suggested that the dipeptide was in fact localized in the chemoreceptor neurons and their bulbar terminals. Further support for this derived from studies of the distribution of amino compounds in dissected layers of the olfactory bulb. It is evident that carnosine is highly localized to the fiber and glomerular layers, in contrast with the distribution of several other compounds (Table III). As noted above, deafferentation has little or no effect on most other compounds or enzymes in the bulb, arguing against their presence in the olfactory neuron terminals; the histochemical studies on various substances by other laboratories are consistent with this postulate (also see Section III, C).

In order to evaluate whether carnosine is synthesized locally or obtained by an uptake mechanism, several approaches were used. Neidle and Kandera (1974) showed that olfactory bulb *in vitro* could convert histidine into carnosine. Subsequently, we demonstrated (Margolis and Grillo, 1977) that intranasal irrigation with [^{14}C]β-alanine or [^{14}C]histidine resulted in synthesis of [^{14}C]carnosine, some of which was transported to the olfactory bulb where is was localized primarily in the fiber and glomerular layers (Nadi *et al.*, 1980). This phenomenon is dependent upon an intact olfactory nerve since axotomy blocks this process (Harding and Wright, 1979) and causes almost total loss of carnosine synthetase activity (Harding *et al.*, 1977). However, because of the unique regenerative properties of this neural system,

TABLE VI Effect of Deafferentation on Biochemical Parameters of Mouse Olfactory Bulb[a]

Parameter	% of Control	Reference
Carnosine	5	Margolis (1974)
GABA	90	Margolis (1974)
Glutamic acid	90	Margolis (1974)
Met-enkephalin	93	R. V. Lewis and F. L. Margolis (unpublished)
cAMP	100	Margolis *et al.* (1974)
GABA uptake	125	Margolis *et al.* (1974)
Choline uptake	110	Margolis *et al.* (1974)
β-Alanine transaminase	76	Harding and Margolis (1976)
Choline acetyltransferase	90	Harding and Margolis (1976)
Carnosine synthetase	6	Harding and Margolis (1976)
Olfactory marker protein	18	Harding and Margolis (1976)
Carnosinase	97	F. L. Margolis and M. Grillo (unpublished)
Monoamine oxidase	95	F. L. Margolis (unpublished)
Etorphine binding	100	Hirsch *et al.* (1978)
GABA binding	100	Hirsch *et al.* (1978)
Carnosine binding	10	Hirsch *et al.* (1978)

[a] Mice were killed 3–4 weeks after intranasal irrigation with 100 μl $ZnSO_4$ (0.17 *M*).

this is a reversible phenomenon and the ability to synthesize and transport carnosine progressively returns to normal within a few weeks after surgery (Harding *et al.*, 1977; Harding and Wright, 1979). Intranasal lavage with a variety of cauterizing, alkylating, or antimitotic agents or detergents can cause acute or chronic blockade of carnosine synthesis and transport and associated anosmia (Table VII). These and other related approaches have enabled us to determine that the turnover of olfactory carnosine is on the order of 10–20 hr, while that in muscle is on the order of 10 days (Neidle and Margolis, 1976; Margolis and Grillo, 1977). Thus, carnosine can be synthesized and degraded locally within the olfactory pathway and in the bulb, and it is located within those layers containing the chemoreceptor axons and their synaptic terminals. Direct confirmation of this by autoradiographic analysis would, at the electron microscopic level, permit determination of whether carnosine occurs in vesicles.

The enzymes responsible for metabolism of carnosine in the olfactory pathway have been partially characterized. Carnosine synthetase has been purified 600-fold from mouse olfactory bulb. It requires ATP and has quite high specificity for histidine and β-alanine as substrates and is clearly different from the chicken muscle enzyme (Horinishi *et al.*, 1978). It is a cytosolic enzyme (Ng and Marshall, 1978; Harding and O'Fallon, 1979), and only a very small portion remains particulate-associated after osmotic lysis.

TABLE VII Effect of Various Drugs Administered Intranasally on Mouse Olfactory Bulb[a]

	[1-^{14}C]β-ala → Carnosine (cpm/mg bulb)		Bulb weight (mg)		
Drug	1 day	1 month	1 day	1 month	Percentage anosmic (1 month)
Colchicine, 10 m*M*	22	212	26	14	100
Vinblastine Sulfate, 10 m*M*	18	1	27	14	100
DMS Br, 50 m*M*	12	130	25	22	25
Triton X-100, 0.5%	28	252	26	24	0
NEM, 5 m*M*	29	276	26	24	0
$ZnSO_4$, 0.17 *M*	5	15	28	16	70
Saline, 0.15 *M*	282	301	26	30	0

[a] Compounds were administered intranasally 100 μl/mouse. One μCi[1-^{14}C]β-alanine was administered intranasally 18 hr before death; [^{14}C]carnosine was isolated from the bulb by ion exchange chromatography. Anosmia is operationally defined as the inability to locate a buried Noyes pellet in 300 sec after overnight starvation. NEM, *N*-ethylmaleimide, DMS Br, dimethyl-(2-hydroxy-5-nitrobenzyl)sulfonium bromide. (F. L. Margolis, M. Grillo, and J. S. Margolis, unpublished.)

When evaluated for distribution between neuronal and glial enriched fractions, the activity is primarily associated with the neuronal enriched fractions (Ng *et al.*, 1977). An interesting observation is that although the K_m for β-alanine for the purified enzyme is 5–10 mM, the tissue concentration of β-alanine is only 0.1–0.2 mM. Thus, either compartmentation of β-alanine within the tissues raises its concentration in the vicinity of the enzyme or the enzyme is always subsaturated with regard to β-alanine. If the latter is so, then elevation of the tissue β-alanine concentration should cause a large increase in carnosine content. In fact, chronic but not acute elevation of β-alanine (Table VIII) does result in elevated tissue carnosine levels, although not as much as expected, demonstrating that the system is undersaturated with β-alanine and has excess synthetic capacity. Perhaps plasma β-alanine plays a role in regulating carnosine synthesis.

Carnosine degradation also seems complex. There are two carnosinase hydrolyzing activities in the mouse olfactory pathway. One, primarily in the epithelium, is immunologically and enzymologically identical to the carnosinase we have purified from mouse kidney (Margolis *et al.*, 1979). This enzyme is almost totally cytosolic (Margolis *et al.*, 1979), is a metalloenzyme, and has very high substrate specificity for carnosine among a variety of histidine-containing dipeptides (Brown *et al.*, 1977; Margolis *et al.*, 1979). The second enzyme, which has not been purified, seems to be labile and is almost inactive in the absence of added Mn^{2+}. This enzyme is present in both bulb and epithelium, as well as in other tissues, and is only partially cytosolic. About one-third of this activity in neural tissue is tightly membrane-associated (unpublished observations). In several reports, the portion of carnosinase activity associated with the crude particulate fraction from different species has varied considerably (Ng *et al.*, 1977; Ng and Marshall, 1978; Harding and O'Fallon, 1979; Margolis *et al.*, 1979). Since in the mouse there are two enzymes with different properties, some of these discrepancies may also relate to the assay conditions used rather than to differences in enzyme distribution. Although the microanatomical location of the particulate carnosinase activity is not known, its existence is at least consistent with an extracellular degradative mechanism that would be required if carnosine functions as a neuroaffector in this pathway. The lack of response of carnosinase activity to selective chemical and surgical manipulations, which influence carnosine synthetase activity (Margolis and Grillo, 1978, and unpublished), argues that the latter, but not the former, is associated with the chemoreceptor neurons and their synaptic terminals.

For a putative neuroaffector to influence cellular function directly, it must either enter the cell or interact with a binding site–receptor complex on the membrane surface. Since we could not demonstrate uptake of carnosine into synaptosomal fractions (A. Keller and F. L. Margolis, unpublished), it

TABLE VIII Effect of Acute and Chronic β-Alanine Administration to Mice[a]

Acute administration: Hours after β-Alanine	β-Alanine (nmole/mg) Bulb	Epithelium	Plasma	Carnosine (nmole/mg) Bulb	Epithelium
0	0.15	0.23	N.D.[b]	1.92 ± 0.24	1.83 ± 0.60
2	1.00	10.40	—	—	—
6	0.66	6.07	0.80	1.79 ± 0.52	2.33 ± 0.69
27	0.19	0.61	N.D.	2.54 ± 0.62	3.06 ± 0.50

Chronic administration	nmole Carnosine/mg tissue: Saline	AOAA	β-Alanine	β-Alanine + AOAA
Bulb	1.8 ± 0.3	2.2 ± 0.2	3.8 ± 1.5	5.0 ± 1.1
Epithelium	3.9 ± 1.5	3.4 ± 0.4	6.5 ± 2.6	12.7 ± 2.9

[a] β-Alanine: 2 g/kg ip; Aminooxyacetic acid (AOAA): 15 mg/kg ip at 15 min before β-alanine. β-Alanine data are average of two assays of 2–3 mice/assay. Carnosine data are mean ± SD for five individual mice. For chronic administration, drug dosage as above, but repeated twice daily for 4 days, killed 16 hr after last dose. Data are mean ± SD for six individual mice.

[b] ND, below detection limit.

seemed most promising to look for a membrane-associated binding site. Initial studies using proton magnetic resonance (^{1}HMR) spectrometry as the probe suffered from lack of sensitivity (Brown *et al.*, 1977). Subsequent studies with [^{3}H]carnosine of high specific activity were more encouraging (Hirsch *et al.*, 1978; Hirsch and Margolis, 1979). A binding site was present in higher concentration in bulb than elsewhere in brain and was of high affinity, stereospecific, saturable, and selective. Evaluation of its anatomical distribution in the bulb (Nadi *et al.*, 1980) indicated that it was predominantly located in the glomerular layer, unlike any of the other binding sites studied (Tables I and III). This is exactly what would be expected for a compound active as the primary afferent transmitter in the bulb.

Thus, based on occurrence, localization of synthetic and degradative enzyme activities and membrane associated binding sites, this dipeptide is currently the best candidate for the primary afferent transmitter of the olfactory chemoreceptor neurons at their synapses with mitral and periglomerular cell dendrites in the olfactory bulb. Clearly, a definitive neurophysiological study of its activity is the evidence still required to test this postulate.

Preliminary neurophysiological studies of the role of carnosine in the olfactory bulb have recently appeared. Carnosine was applied iontophoretically to the glomerular layer or to the external plexiform layer of the olfactory bulb of a gecko (Tonosaki and Shibuya, 1979) while the olfactory epithelium was being stimulated with amyl acetate. Recordings were made of mitral cell responses. Carnosine applied to the glomerular layer blocked the excitatory response to amyl acetate, but when applied to the external plexiform layer, it augmented both the frequencies of response and of spontaneous activity. Application of carnosine to the glomerular layer, which was exhibiting a suppression response to amyl acetate, caused further suppression, but it still augmented the response when applied to the external plexiform layer. In a similar study in rat olfactory bulb (MacLeod and Straughan, 1979), i.e., recording from mitral cells subsequent to iontophoretic application of carnosine, the authors observed both excitatory and inhibitory effects of carnosine, which were dependent on the pH. "The present results present poor support for carnosine as an excitatory transmitter in the bulb...it is of interest, however, to note that in our experiments carnosine was a particularly effective inhibitor of periglomerular cells." (MacLeod and Straughan, 1979). Therefore, in both of these preliminary reports, carnosine influences mitral cell activity, possibly by affecting periglomerular cells. Recently the effect of carnosine was studied on the averaged evoked potentials (AEP) recorded from the main olfactory bulb of the rabbit in response to LOT stimulation. "Carnosine caused an increase in the frequency, a decrease in the rate of onset and a decrease in the decay rate of the AEP's." (Gonzalez and Freeman, 1979). More extensive studies are required in order to fully understand the role played in this pathway by the dipeptide carnosine. Because the neuronal circuitry within the bulb is very complex, it is highly encouraging that carnosine has neurophysiological activity in the appropriate location within the olfactory bulb.

IV. SUMMARY AND FUTURE

Transmitter biochemistry of the olfactory bulb is no more complex than it has ever been. What has changed is knowledge of the subject. The expansion of our knowledge in the last 10 years has served to dramatize how ignorant we were, and still are, of the intimate molecular interactions among the various cell types in this region of the central nervous system. Of the seven questions asked in the Introduction (Section I) to this chapter, partial answers have been given only to the first two or three. Clearly, much remains to be done as has been indicated throughout this chapter.

The use of selective lesion techniques has been a very powerful tool here as in other neural systems to help identify the cellular location of various putative transmitters. This approach is not without potential dangers in interpretation, as is evident in our observations of transsynaptic regulation of dopamine synthesis. It should also be clearly kept in mind that other biochemical changes can occur subsequent to or concomitant with neural lesions. Thus, major changes in cyclic nucleotide phosphodiesterase isozyme patterns (Margolis, 1977), in ornithine decarboxylase activity (Rochel and Margolis, 1980), in ATPase activity (Meisami and Manoochehri, 1977), as well as in adenyl cyclase activity (N. S. Nadi and F. L. Margolis, unpublished), have been observed after lesions of the afferent input to the olfactory bulbs.

The rate at which new endogenous transmitter-like compounds are being reported in neural tissue assures us that the biochemistry of intercellular information transfer in the bulb, as elsewhere, will grow increasingly complex. Especially for the peptides, our progress is hampered by a lack of knowledge of their specific functions and by the absence of specific antagonists. In addition, there is no reason to assume *a priori* that a given peptide will exhibit the same activity at all sites (Barker and Smith, 1980). For many other systems, the ability to selectively regulate synthesis, degradation, release. and response has led to a greater increase in comprehension.

Especially for carnosine, we need the means for selective alteration of the synthetic and degradative enzymes to be able to monitor the influence of alterations in levels and turnover of this peptide on information transfer, transsynaptic regulation, and ultimate behavioral responses. Carnosine and some of its cogeners are selective metal ion chelators (Brown and Antholine, 1979, 1980). It is not known whether this property plays a role in its function. In what way, if any, are carnosine and the olfactory marker protein coupled in olfactory neuron function? Is the specific protein involved in regulation of dopamine expression by the periglomerular neurons? Does it have trophic activity?

More specifically, in the olfactory pathway progress in studying these questions will depend on the continuing development of appropriate *in vitro* models. Organotypic culture of olfactory bulb (Corrigal *et al.*, 1976) and epithelium (Farbman, 1977) and long-term maintenance of disaggregated cell suspensions of both epithelium and bulb will be essential to permit evaluation of various mechanistic aspects of the cellular interactions under controlled conditions.

Finally, transmitter biochemistry of the olfactory bulb is but one aspect of intercellular signaling in the primary olfactory system. Therefore, it is essential that the approaches to the study of these problems be multidisciplinary not parochial, utilizing as many techniques as possible to gain the broadest insight possible into this fascinating and complex area.

ACKNOWLEDGMENT

I wish to gratefully acknowledge the cooperation, tolerance, stimulation, and exuberance of my co-workers over the past several years, whose diligence has made these studies possible—not the least of whom is my secretary, Jill Cluesmann.

REFERENCES

Alberts, J. R. (1974). Producing and interpreting experimental olfactory deficits. *Physiol. Behav.* **12**, 657–670.

Austin, L., Récasens, M., Mathur, R. L., and Mandel, P. (1978). The distribution of taurine, cysteinyl sulphinilic acid decarboxylase and cysteinyl sulphinilic acid α-ketoglutarate transaminase in the rat olfactory bulb and olfactory nucleus. *Neurosci. Lett.* **9**, 59–63.

Barber, P. C., and Raisman, G. (1978a). Cell division in the vomeronasal organ of the adult mouse. *Brain Res.* **141**, 57–66.

Barber, P. C., and Raisman, G. (1978b). Replacement of receptor neurons after section of the vomeronasal nerves in the adult mouse. *Brain Res.* **147**, 297–313.

Barker, J. L., and Smith, T. G., Jr. (1980). "Role of Peptides in Neuronal Function." Dekker, New York.

Broadwell, R. D. (1977). Neurotransmitter pathways in the olfactory system. *In* "Society for Neuroscience Symposia III" (J. A. Ferrendelli, ed.), pp. 131–166. Soc. Neurosci., Bethesda, Maryland.

Broadwell, R. D., and Jacobowitz, D. M. (1976). Olfactory relationships of the telencephalon and diencephalon in the rabbit. III. The ipsilateral centrifugal fibers to the olfactory bulbar and retrobulbar formations. *J. Comp. Neurol.* **170**, 321–346.

Brown, C. E., and Antholine, W. E. (1979). Multiple forms of the colbalt(II)-carnosine complex. *Biochem. Biophys. Res. Commun.* **88**, 529–536.

Brown, C. E., and Antholine, W. E. (1980). Evidence that carnosine and anserine may participate in Wilson's disease. *Biochem. Biophys. Res. Commun.* **92**, 470–477.

Brown, C. E., Margolis, F. L., Williams, T. H., Pitcher, R. G., and Elgar, G. (1977). Carnosine in olfaction. Proton magnetic resonance spectral evidence for tissue-specific carnosine binding sites. *Neurochem. Res.* **2**, 555–579.

Brownstein, M. J. (1977). Biologically active peptides in the mammalian central nervous system. *In* "Peptides in Neurobiology" (A. Gainer, ed.), pp. 145–169. Plenum, New York.

Cain, D. (1975). Effects of insulin injection on responses of olfactory bulb and amygdala single units to odors. *Brain Res.* **99**, 69–83.

Cancalon, P., and Elam, J. S. (1979). Rates of growth and composition of axonally transported proteins in regenerating olfactory nerve. *Soc. Neurosci. Abstr.* **5**, 676.

Cartas-Heredia, L., Guevara-Aguilar, R., and Aguilar-Baturoni, H. U. (1978). Oestrogenic influences on the electrical activity of the olfactory pathway. *Brain Res. Bull.* **3**, 623–630.

Collins, G. G. S. (1979a). Effect of chronic bulbectomy on the depth distribution of amino acid transmitter candidates in rat olfactory cortex. *Brain Res.* **171**, 552–555.

Collins, G. G. S. (1979b). Evidence of a neurotransmitter role for aspartate and γ-aminobutyric acid in the rat olfactory cortex. *J. Physiol. (London)* **291**, 51–60.

Corrigall, W. A., Crain, S. M., and Bornstein, M. B. (1976). Electrophysiological studies of fetal mouse olfactory bulb explants during development of synaptic functions in culture. *J. Neurobiol.* **7**, 521–53.

Crush, K. G. (1970). Carnosine and related substances in animal tissues. *Comp. Biochem. Physiol.* **34**, 3–30.

Dahlström, A., Fuxe, K., Olsson, L., and Ungerstedt, U. (1965). On the distribution and possible function of monoamine nerve terminals in olfactory bulb of the rabbit. *Life Sci.* **4**, 2071–2074.

Fahrenkrug, J., Galbo, H., Holst, J. J., and Schaffalitzky de Muckadell, O. B. (1978). Influence of the autonomic nervous system on the release of the vasoactive intestinal polypeptide from the porcine gastrointestinal tract. *J. Physiol. (London)* **280**, 405–422.

Fallon, J. H., and Moore, R. Y. (1978). Catecholamine innervation of the basal forebrain. III. Olfactory bulb, anterior olfactory nuclei, olfactory tubercle and piriform cortex. *J. Comp. Neurol.* **180**, 533–544.

Farbman, A. I. (1977). Differentiation of olfactory receptor cells in organ culture. *Anat. Rec.* **189**, 187–200.

Farbman, A. I., and Margolis, F. L. (1980). Olfactory marker protein during ontogeny: Immunohistochemical localization. *Dev. Biol.* **74**, 205–215.

Ferriero, D., and Margolis, F. L. (1975). Denervation in the primary olfactory pathway of mice. II. Effects on carnosine and other amine compounds. *Brain Res.* **86**, 75–86.

Fisher, D. E., Amend, J. F., and Strumeyer, D. V. (1977). Anserine and carnosine in chicks (*Gallus gallus*) rat pups (*Rattus rattus*) and ducklings (*Anas platyrhynchos*): Comparative ontogenetic observations. *Comp. Biochem. Physiol. B* **56**, 367–370.

Gonzalez, M. T., and Freeman, W. J. (1979). Effects of carnosine on rabbit olfactory bulb evoked potentials. *Fed. Proc. Fed. Am. Soc. Exp. Biol.* **38**, 900.

Graziadei, P. P. C., and Monti Graziadei, G. A. (1978). The olfactory system: A model for the study of neurogenesis and axon regeneration in mammals. *In* "Neuronal Plasticity" (C. W. Cotman, ed.), pp. 131–153. Raven, New York.

Graziadei, P. P. C., and Monti Graziadei, G. A. (1979). Neurogenesis and neuron regeneration in the olfactory system of mammals. I. Morphological aspects of differentiation and structural organization of the olfactory sensory neurons. *J. Neurocytol.* **8**, 1–18.

Halász, N., Ljungdahl, Å., Hökfelt, T., Johansson, O., Goldstein, M., Park, D., and Biberfeld, P. (1977). Transmitter histochemistry of the rat olfactory bulb. I. Immunohistochemical localization of monoamine synthesizing enzymes, support for intrabulbar, periglomerular dopamine neurons. *Brain Res.* **126**, 455–474.

Halász, N., Ljungdahl, Å., and Hökfelt, T. (1978). Transmitter histochemistry of the rat olfactory bulb. II. Fluorescence histochemical, autoradiographic and electron microscopic localization of monoamines. *Brain Res.* **154**, 253–271.

Halász, N., Ljungdahl, Å., and Hökfelt, T. (1979). Transmitter histochemistry of the rat olfactory bulb. III. Autoradiographic localization of [^{3}H]GABA. *Brain Res.* **167**, 221–240.

Harding, J. W., and Margolis, F. L. (1976). Denervation in the primary olfactory pathway of mice. III. Effect on enzymes of carnosine in metabolism. *Brain Res.* **110**, 351–360.

Harding, J. W., and O'Fallon, J. V. (1979). The subcellular distribution of carnosine, carnosine synthetase, and carnosinase in mouse olfactory tissues. *Brain Res.* **173**, 99–109.

Harding, J. W., and Wright, J. W. (1978). Reversible effects of olfactory nerve section on behavior and biochemistry in mice. *Brain Res. Bull.* **4**, 17–22.

Harding, J. W., and Wright, J. W. (1979). Effects of intranasal irrigation with mitotic inhibitors on olfactory behavior and biochemistry in mice. *Brain Res.* **168**, 31–41.

Harding, J. W., Graziadei, P. P. C., Monti Graziadei, G. A., and Margolis, F. L. (1977). Denervation in the primary olfactory pathway of mice. IV. Biochemical and morphological evidence for neuronal replacement following nerve section. *Brain Res.* **132**, 11–28.

Harding, J. W., Getchell, T. V., and Margolis, F. L. (1978). Denervation of the primary olfactory pathway in mice. V. Long-term effect of intranasal $ZnSO_4$ irrigation on behavior, biochemistry and morphology. *Brain Res.* **140**, 271–285.

Harvey, J. A., Scholfield, C. N., Graham, L. T., Jr., and Aprison, M. H. (1975). Putative transmitters in denervated olfactory cortex. *J. Neurochem.* **24**, 445–449.

Haubrich, D. R., and Denzer, J. S. (1973). Simultaneous extraction and fluorometric measurement of brain serotonin, catecholamines, 5-hydroxyindoleacetic acid and homovanillic acid. *Anal. Biochem.* **55**, 306–312.

Havrankova, J., Schmechel, D., Roth, J., and Brownstein, M. (1978). Identification of insulin in rat brain. *Proc. Natl. Acad. Sci. U.S.A.* **75**, 5737–5741.

Hirsch, J. D., and Margolis, F. L. (1979). L-[^{3}H]Carnosine binding in the olfactory bulb. II. Biochemical and biological studies. *Brain Res.* **174**, 81–94.

Hirsch, J. D., and Margolis, F. L. (1980). Influence of unilateral olfactory bulbectomy on opiate and other binding sites in the contralateral bulb. *Brain Res.* **199**, 39–47.

Hirsch, J. D., Grillo, M., and Margolis, F. L. (1978). Ligand binding studies in the mouse olfactory bulb: Identification and characterization of a L-[^{3}H]carnosine binding site. *Brain Res.* **158**, 407–422.

Hoffman, G. E., Davis, B. J., and Macrides, F. (1979). LHRH perikarya send axons to the olfactory bulb in the hamster. *Soc. Neurosci, Abstr.* **5**, 528.

Horinishi, H., Grillo, M., and Margolis, F. L. (1978). Purification and characterization of carnosine synthetase from mouse olfactory bulbs. *J. Neurochem.* **31**, 909–914.

Hunt, S., and Schmidt, J. (1977). Are mitral cells cholinergic? *In* "Society for Neuroscience Symposia III" (J. A. Ferrendelli, ed.), pp. 204–218. Soc. Neurosci., Bethesda, Maryland.

Iuvone, P. M., Galli, C. L., Garrison-Gund, C. K., and Neff, N. H. (1978). Light stimulates tyrosine hydroxylase activity and dopamine synthesis in retinal amacrine neurons. *Science* **202**, 901–902.

Jaffé, E. H., and Cuello, A. C. (1980). The distribution of catecholamines, glutamate decarboxylase and choline acetyltransferase in layers of the rat olfactory bulb. *Brain Res.* **186**, 232–237.

Jessen, K. R., Mirsky, R., Dennison, M. E., and Burnstock, G. (1979). GABA may be a neurotransmitter in the vertebrate peripheral nervous system. *Nature (London)* **281**, 71–74.

Julliard, J. H., Shibasaki, T., Ling, N., and Guillemin, R. (1980). High-molecular-weight immunoreactive β-endorphin in extracts of human placenta is a fragment of Immunoglobulin G. *Science* **208**, 183–185.

Kan, K.-S., Giolli, R. A., and Chao, L.-P. (1980). Immunohistochemical localization of choline acetyltransferase in forebrain. *Trans. Am. Soc. Neurochem.* **11**, 203.

Kornguth, S. E., Scott, G., and Jaramillo, R. (1979). Ultrastructural and biochemical comparisons of dendrodendritic synaptosomes from bovine superior colliculi and olfactory bulbs with axodendritic and axosomatic synaptosomes. *J. Neurochem.* **33**, 875–882.

MacLeod, N. K., and Straughan, D. W. (1979). Responses of olfactory bulb neurons to the dipeptide carnosine. *Exp. Brain Res.* **34**, 183–188.

Macrides, F. (1976). Olfactory influences on neuroendocrine function in mammals. *In* "Mammalian Olfaction, Reproductive Processes, and Behavior" (R. L. Doty, ed.), pp. 29–65. Academic Press, New York.

Macrides, F., Firl, A. C., Jr., Schneider, S. P., Bartke, A., and Stein, D. G. (1976). Effects of one-stage or serial transecting of the lateral olfactory tracts on behavior and plasma testosterone levels in male hamsters. *Brain Res.* **109**, 97–109.

Macrides, F., Davis, B. J., Youngs, W. M., Nadi, N. S., and Margolis, F. L. (1981). Cholinergic and catecholaminergic afferents to the olfactory bulb in the hamster: A neuroanatomical, biochemical and histochemical investigation. *J. Comp. Neurol.* (in press).

Margolis, F. L. (1974). Carnosine in the primary olfactory pathway. *Science* **184**, 909–911.

Margolis, F. L. (1975). Biochemical markers of the primary olfactory pathway: A model neural system. *Adv. Neurochem.* **1**, 193–246.

Margolis, F. L. (1977). Biochemical studies of the primary olfactory pathway. *In* "Society for Neuroscience Symposia III" (J. A. Ferrendelli, ed.), pp. 167–188. Soc. Neurosci., Bethesda, Maryland.

Margolis, F. L. (1980a). A marker protein for the olfactory chemoreceptor neuron. *In* "Proteins of the Nervous System" (R. A. Bradshaw and D. Schneider, eds.). Raven, New York. pp. 59–84.

Margolis, F. L. (1980b). Carnosine: An olfactory neuropeptide. *In* "Role of Peptides in Neuronal Function" (J. L. Barker and T. G. Smith, Jr., eds.). Dekker, New York. pp. 545–572.

Margolis, F. L., and Grillo, M. (1977). Axoplasmic transport of carnosine (β-alanyl-L-histidine) in the mouse olfactory pathway. *Neurochem. Res.* **2**, 507–519.

Margolis, F. L., and Grillo, M. (1978). An approach to chemical nerve section in the mouse olfactory pathway. *Soc. Neurosci. Abstr.* **4**, 254.

Margolis, F. L., Roberts, N., Ferriero, D., and Feldman, J. (1974). Denervation in the primary olfactory pathway of mice: Biochemical and morphological effects. *Brain Res.* **81**, 469–483.

Margolis, F. L., Grillo, M., Brown, C. E., Williams, T. H., Pitcher, R. G., and Elgar, G. J. (1979). Enzymatic and immunological evidence for two forms of carnosinase in the mouse. *Biochim. Biophys. Acta* **570**, 311–323.

Matulionis, D. H. (1976). Light and electron microscopic study of the degeneration and early regeneration of olfactory epithelium in the mouse. *Am. J. Anat.* **145**, 79–100.

Meisami, E. (1976). Effects of olfactory deprivation on postnatal growth of the rat olfactory bulb utilizing a new method for production of neonatal unilateral anosmia. *Brain Res.* **107**, 437–444.

Meisami, E., and Manoochehri, S. (1977). Effects of early bilateral chemical destruction of olfactory receptors on postnatal growth, Mg-ATPase and Na-K-ATPase activity of olfactory and non-olfactory structures of the rat brain. *Brain Res.* **128**, 170–175.

Moore, R. Y., Halaris, A. E., and Jones, B. E. (1978). Serotonin neurons of the midbrain raphé: Ascending projections. *J. Comp. Neurol.* **180**, 417–438.

Nadi, N. S., and Margolis, F. L. (1978). A simple method for the elimination of amine contaminants in buffers for single-column amino acid analysis in physiological samples at picomole levels. *Anal. Biochem.* **91**, 180–185.

Nadi, N. S., Hirsch, J. D., and Margolis, F. L. (1980). Laminar distribution of putative neurotransmitter amino acids and ligand binding sites in the dog olfactory bulb. *J. Neurochem.* **34**, 138–146.

Nadi, N. S., Head, R., Grillo, M., Hempstead, J., Grannot-Reisfeld, N., and Margolis, F. L. (1981). Chemical deafferentation of the olfactory bulb: Plasticity of the levels of tyrosine hydroxylase, dopamine and norepinephrine. *Brain Res.* **211** (in press).

Neidle, A., and Kandera, J. (1974). Carnosine: An olfactory bulb peptide. *Brain Res.* **80**, 359–364.

Neidle, A., and Margolis, F. L. (1976). Carnosine turnover in olfactory bulb and other tissues of the mouse. *Trans. Am. Soc. Neurochem.* **7**, 117.

Ng, R. H., and Marshall, F. D. (1978). Regional and subcellular distribution of homocarnosine-carnosine synthetase in the central nervous system. *J. Neurochem.* **30**, 187–190.

Ng, R. H., Marshall, F. D., Henn, F. A., and Sellström, A. (1977). Metabolism of carnosine and homocarnosine in subcellular fractions and neuronal and glial cell-enriched fractions of rabbit brain. *J. Neurochem.* **28**, 449–452.

Pager, J. (1978). Ascending olfactory information and centrifugal influences contributing to a nutritional modulation of the rat mitral cell responses. *Brain Res.* **140**, 251–269.

Palkovits, M., Záborszky, L., Brownstein, M. J., Fekete, M. I. K., Herman, J. P., and Kanyicska, B. (1979). Distribution of norepinephrine and dopamine in cerebral cortical areas of the rat. *Brain Res. Bull.* **4**, 593–601.

Patterson, P. H. (1978). Environmental determination of autonomic neurotransmitter functions. *Annu. Rev. Neurosci.* **1**, 1–17.

Priestley, J. V., Kelly, J. S., and Cuello, A. C. (1979). Uptake of [^{3}H]dopamine in periglomerular cells of the rat olfactory bulb: An autoradiographic study. *Brain Res.* **165**, 149–155.

Quinn, M. R., and Fisher, H. (1977). Effect of dietary histidine deprivation in two rat strains on hemoglobin and tissue concentrations of histidine-containing dipeptides. *J. Nutr.* **107**, 2044–2054.

Rehfeld, J. F. (1978). Localisation of gastrins to neuro- and adeno-hypophysis. *Nature (London)* **271**, 771–773.

Ribak, C. E., Vaughn, J. E., Saito, K., Barber, R., and Roberts, E. (1977). Glutamate decarboxylase localization in neurons of the olfactory bulb. *Brain Res.* **126**, 1–18.

Robinzon, B., Katz, Y., and Rogers, J. G., Jr. (1979). The involvement of the olfactory bulbs in the regulation of gonadal and thyroidal activities of male red-winged blackbirds, exposed to short-day light regime. *Brain Res. Bull.* **4**, 339–345.

Rochel, S., and Margolis, F. L. (1980). The response of ornithine decarboxylase during neuronal degeneration and regeneration in olfactory epithelium. *J. Neurochem.* **35**, 850–860.

Ross, C. D., Godfrey, D. A., and Carter, J. A. (1979). Effect of lesions on amino acid distribution in rat olfactory bulb. *Soc. Neurosci. Abstr.* **5**, 132.

Schneider, S. P., and Macrides, F. (1978). Laminar distributions of interneurons in the main olfactory bulb of the adult hamster. *Brain Res. Bull.* **3**, 73–82.

Shepherd, G. M. (1979). "The Synaptic Organization of the Brain," 2nd ed. Oxford Univ. Press, London and New York.

Simmons, P. A., and Getchell, T. V. (1979). Neurogenesis in the olfactory epithelium. *Soc. Neurosci. Abstr.* **5**, 134.

Singh, P. J., Tucker, A. M., and Hofer, M. A. (1976). Effects of nasal $ZnSO_4$ irrigation and olfactory bulbectomy on rat pups. *Physiol. Behav.* **17**, 373–382.

Slotnick, B. M., and Gutman, L. A. (1977). Evaluation of Intranasal zinc sulfate treatment on olfactory discrimination in rats. *J. Comp. Physiol. Psychol.* **91**, 942–950.

Takagi, S. F. (1971). Degeneration and regeneration of the olfactory epithelium. *In* "Handbook of Sensory Physiology" (L. M. Beidler, ed.), Vol. IV/1, pp. 76–94. Springer-Verlag, Berlin and New York.

Thoenen, H., and Otten, J. (1977). Molecular events in transsynaptic regulation of the synthesis of macromolecules. *In* "Essays in Neurochemistry and Neuropharmacology" (M. B. H. Yarding, W. Lovenberg, D. F. Sharman, and J. K. Lagnado, eds.), Vol. 1, pp. 73–101. Wiley, New York.

Tonosaki, K., and Shibuya, F. T. (1979). Action of some drugs on gecko olfactory bulb mitral cell responses to odor stimulation. *Brain Res.* **167**, 180–184.

Wenk, H., Meyer, U., and Bigl, V. (1977). Centrifugal cholinergic connections in the olfactory system of rats. *Neuroscience* **2**, 797–800.

Yang, H.-Y., Hong, J. S., Fratta, W., and Costa, E. (1978). Rat brain enkephalins: Distribution and biosynthesis. *Adv. Biochem. Psychopharmacol.* **18**, 140–159.

Youngs, W. M., Nadi, N. S., Davis, B. J., Margolis, F. L., and Macrides, F. (1979). Evidence for a cholinergic projection to the olfactory bulb from the magnocellular preoptic area. *Soc. Neurosci. Abstr.* **5**, 135.

18

Neurochemical Studies of the γ-Aminobutryic Acid System in the Olfactory Bulb

MICHAEL R. QUINN AND ROBERT H. CAGAN

I. INTRODUCTION

Olfactory sensory information is important for mediating many behaviors such as suckling by neonatal humans (MacFarlane, 1975) and rats (Singh *et al.*, 1976; Teicher and Blass, 1977), locating food sources by macroosmotic animals (Garcia *et al.*, 1974), home stream migration by salmonid fish (Hasler *et al.*, 1978), and social interactions by a multitude of adult animal species (Doty, 1976). The neural and biochemical mechanisms underlying integration of olfactory information are only beginning to become understood.

The neuroanatomy of the olfactory bulb is reasonably well established.

Biochemistry of Taste and Olfaction
ISBN 0-12-154450-8

The first synapse in the olfactory pathway occurs in the olfactory bulb of the central nervous system. Of the laminated layers of the olfactory bulb (Fig. 1; see also this volume Chapters 17 and 20), the outermost layer is composed exclusively of unmyelinated axons (Gasser, 1956), which originate from the olfactory bipolar receptor neurons in the olfactory mucosa (see this volume, Chapter 15) of the nasal cavity (Shepherd, 1972). The olfactory receptor axons course centrally through the cribiform plate and terminate in a layer of the olfactory bulb composed of discrete spherical regions of neuropil, called glomeruli. The glomeruli are easily discerned in histological sections of the olfactory bulb using light microscopy (Fig. 1) and are structurally similar to those observed in the cerebellum and thalamus (Shepherd, 1974).

Within the glomeruli, the olfactory receptor axons ramify and synapse

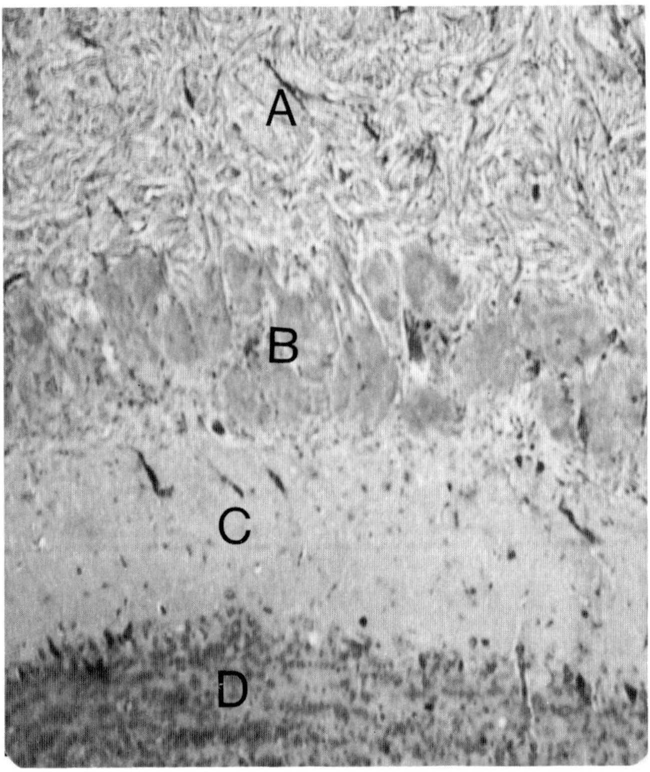

Fig. 1. Photomicrograph of a longitudinal histological section of a rat olfactory bulb stained with a modified lithium hematoxylin stain. It shows the olfactory nerve layer (A), the glomerular layer (B), the external plexiform layer (C), and the mitral cell body layer (D) (280 ×). (Taken from Robinzon and Quinn, 1978.)

onto the apical region of the primary dendrites of the mitral cells (Shepherd, 1972). The axons of the mitral cells leave the olfactory bulb via the lateral olfactory tract and terminate in higher centers of the brain. It has been estimated that for every 1000 olfactory receptor axons entering the olfactory bulb of the rabbit, only one mitral cell axon exits (Allison and Warwick, 1949). The axodendritic synapses between the receptor cell axons and the mitral cell dendrites represent the first stage of integration of olfactory sensory information.

Significant modulation and focusing of sensory information occurs at the dendrodendritic synapses between the granule cell dendrites and the secondary dendrites of the mitral cells. The dendrodendritic synapses are suggested to be critical for information processing (Schmitt *et al.*, 1976) and are thought to be an important component of the coding system for olfactory information within the olfactory bulb. It is estimated that about 90% of the synapses that occur in the third laminar layer of the olfactory bulb, the external plexiform layer (Fig. 1), are dendrodendritic synaptic contacts (Reese and Shepherd, 1972). Within the external plexiform layer, the secondary dendrites of the mitral cells form reciprocal dendrodendritic synapses with the dendritic gemmules of the granule cells (Shepherd, 1972). Only 19% of the synapses in the glomerular layer are dendrodendritic (White, 1973), and they occur between the periglomerular cells and the mitral cells.

The physiological function of the synaptic interactions between granule cell and mitral cell dendrites is analogous to a filter. The mitral cell primary dendrites are excited by afferent input from the axon terminals of olfactory receptor cells. The secondary dendrite of the mitral cell excites the granule cell, which in turn inhibits the mitral cell through a negative feedback loop (Fig. 2). The negative feedback is thought to occur either by reciprocal dendrodendritic contacts or by serial dendrodentritic synapses. Excitation of granule cells also creates field potentials that inhibit neighboring mitral cells (Shepherd, 1974). This type of "horizontal communication" within the olfactory bulb acts as a filtering mechanism to tune incoming olfactory sensory information.

Centrifugal fibers arising from higher centers in the central nervous system (i.e., the anterior olfactory nucleus, the anterior commissure, and the diagonal limb of the horizontal band) apparently exert their influence on the olfactory bulb through the granule cells (Schneider and Macrides, 1978; Dahlström *et al.*, 1965; Felix and McLennen, 1971). The centrifugal fibers either synapse directly onto granule cells and thereby indirectly cause inhibition of mitral cells, or synapse onto short-axon cells, which may inhibit the granule cells and thereby cause less inhibition (disinhibition) of the mitral cells (McLennan, 1971; Schneider and Macrides, 1978). Hence the granule cells are important in processing olfactory information as well as in coordinat-

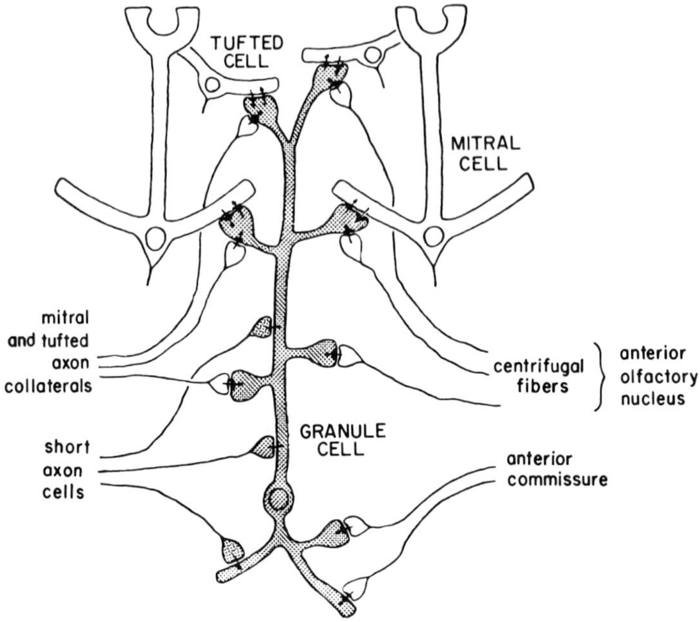

Fig. 2. Schematic representation of the synaptic connections in the olfactory bulb involving the granule cells. Shaded areas represent inhibitory terminals, and clear areas represent excitatory terminals. The arrows indicate the directional flow of information across the synapse. (Taken with slight modifications from Shepherd, 1972, with permission.)

ing sensory processing with other information (nutritional state, hunger, behavior, hormonal) relayed by centrifugal fibers from higher centers in the central nervous system.

II. GABA IN THE OLFACTORY BULB

A relatively high concentration of γ-aminobutyric acid (GABA) in the olfactory bulb, compared with other brain regions, was noted by Popov *et al.* (1967) and confirmed by several others (Baxter, 1970; Neidle and Kandera, 1974; Margolis, 1974; Quinn, unpublished). Osborne *et al.* (1976) reported that a synaptosomal fraction prepared from rat olfactory bulbs releases GABA upon electrical stimulation. The synaptosomes also selectively release aspartate and glutamate, amino acids postulated to have a transmitter role in the olfactory bulb, possibly in the mitral cells (Osborne *et al.*, 1976; Harvey *et al.*, 1975). Bergez *et al.* (1978) measured [^{14}C]GABA and [^{3}H]taurine release from olfactory bulb slices. These investigators observed that both GABA and taurine are released from rat olfactory bulb slices by Ca^{2+}-

dependent K^+-stimulation, similar to that observed for cerebral cortex slices (Davies *et al.*, 1975; Mulder and Snyder, 1974; Srinivasan *et al.*, 1969), whole retina (Mandel and Pasantes-Morales, 1976), and synaptosomes from rat cerebral cortex (Sieghart and Heckl, 1976).

Pharmacologic evidence using electrophysiological techniques indicate that GABA is a neurotransmitter in granule cells, causing inhibition of mitral cells (Nicoll, 1970, 1971; Felix and McLennan, 1971; McLennan, 1971). It was noted, while recording extracellularly from mitral cells in the olfactory bulb, that bicuculline and picrotoxin block the inhibition of mitral cells induced by electrophoretic application of GABA (Felix and McLennan, 1971; Nicoll, 1971; McLennan, 1971). In addition, the excitatory amino acids glutamate, aspartate, DL-homocysteate, and also norepinephrine were found to inhibit the mitral cells. This inhibitory action on the mitral cells is reversed and becomes excitatory upon treatment with the GABA antagonists, picrotoxin and bicuculline (Nicoll, 1971). The inhibitory action of these amino acids and of norepinephrine is thought to occur through the granule cells. When the granule cell is excited, release of GABA at the dendrodendritic synapses causes inhibition of the mitral cells. If the effects of GABA on the mitral cells are blocked by an antagonist, then the inhibitory influence of the granule cells is prevented. With the granule cell component of the dendrodendritic synapse blocked, the excitatory amino acids and norepinephrine excite the mitral cell directly (Nicoll, 1971; Felix and McLennan, 1971).

Uptake of [^{3}H]GABA into the various layers of the olfactory bulb was measured autoradiographically (Halász *et al.*, 1979). Although [^{3}H]GABA was strongly taken up by glial cells throughout all layers of the olfactory bulb, the primary neuronal uptake was observed in dendrites of granule cells and of periglomerular cells. Some short-axon cells and axon terminals of neurons that were speculated to originate from parts of the central nervous system outside of the olfactory bulbs were also reported to take up [^{3}H]GABA (Halász *et al.*, 1979). It was suggested by these authors that GABA may be responsible for the reported inhibitory nature of the short-axon cells (Shepherd, 1972), but the origins within the central nervous system of the labeled axon terminals were not determined.

III. GLUTAMIC ACID DECARBOXYLASE

Synthesis of GABA from L-glutamate is catalyzed by glutamic acid decarboxylase (GAD; L-glutamate 1-carboxy-lyase; EC 4.1.1.15), which is the rate-limiting enzyme (Roberts and Kuriyama, 1968). The activity of GAD determines steady-state levels of GABA in the central nervous system

(Roberts and Kuriyama, 1968). GAD requires pyridoxal-5′-phosphate as a co-factor. During states of vitamin B_6 deficiency, *in vivo* GAD activity in the olfactory bulb decreases to about 50% of that of the control (Quinn and Chan, 1979), similar to the effects on GAD in other regions of the brain.

GAD is unevenly distributed among the laminated layers of the olfactory bulb, with the highest activity and GABA content in the external plexiform layer, the granule cell layer, and the glomerular layer (Graham, 1973). Although these results were essentially confirmed by Austin *et al.* (1979), some discrepancy was reported concerning the level of GAD activity in the mitral cell body layer. Graham (1973) reported low GAD activity, whereas Austin *et al.* (1979) reported that the mitral cell body layer contains the highest GAD activity of all the layers in the olfactory bulb. It is noted that different assay methods for measuring GAD activity were used and that the activity of the enzyme was about fourfold higher in all olfactory bulb layers as measured by Graham (1973).

In one report (Austin *et al.*, 1979), the internal plexiform layer was not separated from the mitral cell layer, and it was suggested that the internal plexiform layer may be relatively high in GAD activity. It was stated (Austin *et al.*, 1979) that perhaps Graham (1973) dissected the internal plexiform layer away from the mitral cell layer and that this may account for some of the discrepancies in the distribution of GAD. It does not seem likely that the internal plexiform layer contains relatively high levels of GAD activity because this layer contains few cell bodies and is primarily a narrow zone of neuronal passage (Shepherd, 1972). GAD is distributed throughout the granule cell cytoplasm (Ribak *et al.*, 1977), and the granule cell dendrites that pass through the internal plexiform layer would be expected to contain a relatively small proportion of the GAD activity. Therefore, it is not likely that the methods of dissection can account for the discrepancy between the two studies.

The distribution of GAD within neuronal cells of the olfactory bulb has been visualized immunocytochemically. Antisera were prepared against GAD purified from a synaptosomal fraction of whole brains of mice (Ribak *et al.*, 1977). With these antisera, immunocytochemical techniques allowed visualization of the intracellular distribution of GAD in the olfactory bulb with light and electron microscopy. Ribak *et al.* (1977) demonstrated that GAD-positive reaction product is present in the dendritic gemmules and cell bodies of the granule cells. The dendrites and gemmules of a portion of the periglomerular cells were also observed to contain GAD-positive reaction product. There appears to be two populations of periglomerular cells, one that utilizes GABA as an inhibitory neurotransmitter and another that utilizes dopamine as an excitatory neurotransmitter (Lichtensteiger, 1966; Hökfelt *et al.*, 1975; Halász *et al.*, 1977, 1978; Ribak *et al.*, 1977).

IV. DENDRODENDRITIC SYNAPTOSOMES

Synaptosomes are vesiculated nerve endings that form when brain tissue is homogenized under isosmotic or slightly hyperosmotic conditions (Gray and Whittaker, 1962). Synaptosomal preparations have proven to be excellent model systems to study central nervous system neurotransmitters (Cotman and Matthews, 1971). When synaptosomes form, the postsynaptic membrane may be retained as an adhesion to the presynaptic synaptosomal membrane along with the intrasynaptic material (Jones, 1975). Synaptosomes serve as an *in vitro* preparation of nerve terminals because they maintain their biochemical integrity; they contain vesicles, neurotransmitters, and the enzymes that catalyze synthesis of the transmitters, neuronal mitochondria, and cytoplasm. They retain morphological characteristics of the synaptic complex; they possess metabolic properties such as respiratory patterns and active transport systems; and the intact functional synaptosomal membranes contain receptor sites for relevant biologically active compounds. They maintain an electrical resting potential and respond to depolarizing stimuli similar to intact neurons (DeBelleroche and Bradford, 1973).

Although synaptosomes are useful for studying synapses without the presence of other neuronal and nonneuronal components, one disadvantage is that they are generally derived from a heterogeneous population of synaptic contacts. Kornguth *et al.* (1976) reported the isolation and morphological characterization of a synaptosomal fraction derived from the dendrodendritic synapses of swine olfactory bulb. In contrast with axodendritic and axosomatic synaptosomes, which are prepared from a crude mitochondrial fraction (10,000 *g*), the dendrodendritic synaptosomes of the olfactory bulb (Kornguth *et al.*, 1976; Quinn and Cagan, 1979, 1980) and of the superior colliculi (Kornguth *et al.*, 1979) are sedimented at a low gravitational force (1000 *g*). The dendrodendritic synaptosomal fraction is prepared from the top buffy region of the crude nuclear pellet (1000 *g*).

About the same time as the report by Kornguth *et al.* (1976) appeared, Osborne *et al.* (1976) reported the presence of a high concentration of GABA, when expressed as a percentage of the total amino acids measured, in a "heavy" synaptosomal fraction prepared from the entire crude nuclear pellet (P_1; 1000 *g*) isolated from rat olfactory bulbs. The GABA contained in the synaptosomes was retained during incubation, possibly representing a high-affinity uptake system, and was released upon electrical stimulation (Osborne *et al.*, 1976). They suggested that the "heavy" synaptosomes may have been derived from dendrodendritic synapses. Our evidence (see below) supports their suggestion.

Based on the available information, we hypothesized that if GABA is the granule cell neurotransmitter that mediates inhibitory influences on mitral cells, then a substantial portion of olfactory bulb GAD should be present in the dendrodendritic synaptosomal fraction (Quinn and Cagan, 1979). Because the postsynaptic membrane, and often the intact postsynaptic synaptosome, adheres to the presynaptic synaptosome (Kornguth *et al.*, 1976), a relatively large amount of the GABA postsynaptic receptor is also postulated to be present in the dendrodendritic synaptosomal fraction.

V. GLUTAMIC ACID DECARBOXYLASE IN DENDRODENDRITIC SYNAPTOSOMES

To test the above hypotheses, a dendrodendritic synaptosomal fraction was prepared as follows. Male Charles River CD rats (8–10 weeks old) were decapitated, and the olfactory bulbs were quickly dissected and placed into ice-cold sucrose (0.32 *M*) containing potassium phosphate buffer (0.4 m*M*, pH 7.0), 1 m*M* $MgCl_2$, 0.2 m*M* pyridoxal-5′-phosphate, and 1 m*M* aminoethylisothiouronium bromide (Kornguth *et al.*, 1976; Quinn and Cagan, 1980). The rat olfactory bulbs (1.0 g wet weight) were homogenized and

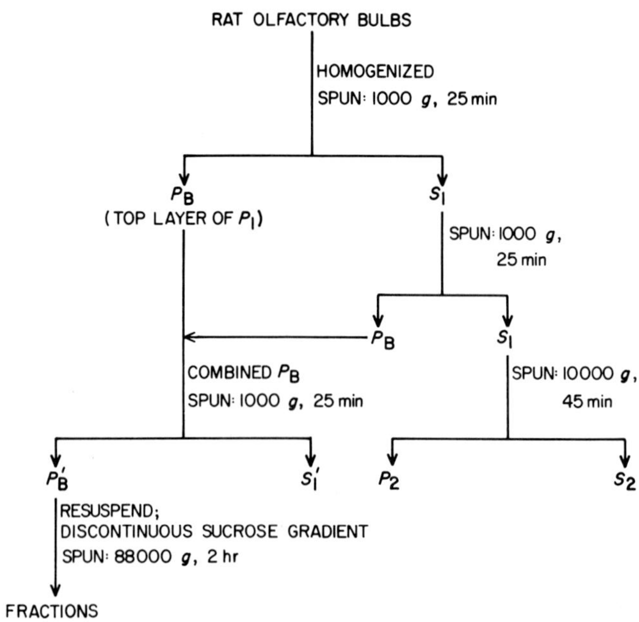

Fig. 3. Schematic representation of the procedure used to obtain a crude fraction of dendrodendritic synaptosomes, modified from that described by Kornguth *et al.* (1976).

fractionated (Quinn and Cagan, 1980) using a modification of the procedure described by Kornguth *et al.* (1976). The fractionation is shown schematically in Fig. 3.

The dendrodendritic synaptosomes (DDS) are obtained in the top buffy region (P_B) of the crude nuclear pellet (P_1, 1000 g). The activity of GAD was assayed in the various fractions obtained from rat olfactory bulbs (Fig. 4) by measuring $^{14}CO_2$ evolution from L-[1-^{14}C]glutamic acid (Wilson *et al.*, 1972; Buhler, 1962; Quinn and Cagan, 1980). Following centrifugation of the homogenate, 31–43% of the homogenate GAD activity was present in P_B, 6–13% in P_1, 6–8% in P_2, and the remainder of the recovered enzyme activity was soluble (S_1' and S_2; Table I). It is not surprising that an appreciable portion of the recovered GAD activity was present in the supernatant fraction because GAD is distributed throughout the granule cell (Ribak *et al.*, 1977) and would be liberated from the cell soma and dendritic shafts during homogenization.

Synaptosomes originating from axosomatic and axodendritic synaptic contacts are known to sediment with the crude mitochondrial pellet (P_2; 10,000 g). In our studies only 6–8% of the GAD activity in the olfactory bulb homogenate was obtained in P_2 (Table I). Ultimately, the GAD activity recovered after differential centrifugation was approximately equally distributed between the crude dendrodendritic synaptosomal fraction, P_B', and the supernatant, S_2. Additional centrifugation of S_2 at 27,000 *g* did not result in any

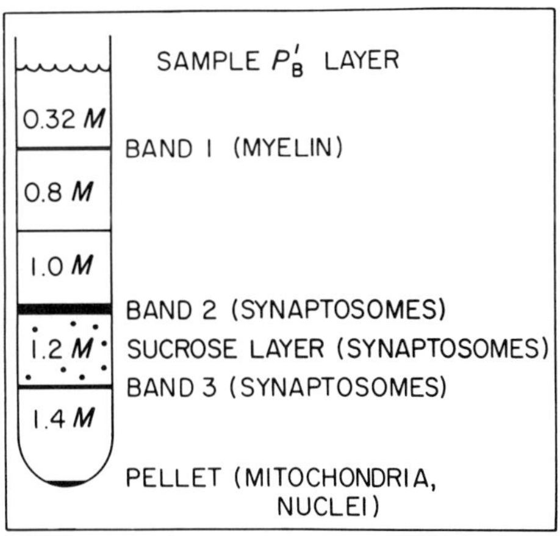

Fig. 4. Schematic representation of the distribution of subcellular fractions of rat olfactory bulbs. Fraction P_B' was centrifuged in a discontinuous sucrose density gradient containing 1 m*M* aminoethylisothiouronium bromide and 0.2 m*M* pyridoxal-5′-phosphate in a swinging-bucket rotor at 88,000 *g* for 2 hr. The resulting fractions are described on the right of the diagram. (Taken from Quinn and Cagan, 1980, with permission.)

TABLE I Distribution of GAD Activity in Fractions of Rat Olfactory Bulb Following Differential Centrifugation[a]

	Specific activity (pmole/min-mg protein)		Total activity (nmole/min)	
Fraction	Experiment 1	Experiment 2	Experiment 1	Experiment 2
H	1130 ± 61	1251 ± 47	72.2 (100)	127.4 (100)
P_B	1132 ± 106	1065 ± 105	31.0 (43)	39.4 (31)
P_1[b]	1007 ± 70	948 ± 76	9.4 (13)	7.6 (6)
S_1	1152 ± 43	918 ± 75	20.3 (28)	33.9 (27)
P_B'[c]	994 ± 40	1220 ± 98	18.2 (25)	28.2 (22)
S_1'	727 ± 3	754 ± 167	5.5 (8)	6.7 (5)
P_2	794 ± 146	1162 ± 59	4.4 (6)	9.9 (8)
S_2	1213 ± 28	851 ± 79	14.9 (21)	25.5 (20)

[a] Each value is the mean ± SD of 3–4 replicates. The total activity is calculated from the mean specific activity. In parentheses are shown the percentages of the homogenate activity that were recovered in the respective fractions. For explanation of fraction symbols, see text and Fig. 3.

[b] P_1 in the table includes only the lower regions of the crude nuclear pellet.

[c] This is the fraction placed onto the discontinuous sucrose gradient (See Table II). Taken with modification from Quinn and Cagan (1980), with permission.

further appreciable sedimentation of GAD. We suspect that the small amount of GAD recovered in P_2 may have been derived from axosomatic or axodendritic synapses involving GABAergic neurons, possibly short-axon cells or axon terminals of GABAergic neurons orginating extrinsically (Halász *et al.*, 1979), although the latter is less likely. The olfactory receptor cells of the garfish and pike contain substantial quantities of GAD activity and GABA (Roskoski *et al.*, 1974), but it is doubtful that the GAD recovered in rat olfactory bulb P_2 was derived from axodendritic synaptic contacts between the olfactory receptor cell axons and the primary dendrites of the mitral cells. The olfactory receptor cells of the rat do not appear to utilize GABA as a neurotransmitter. The olfactory receptor cell axon terminals do not take up appreciable quantities of [^{3}H]GABA (Halász *et al.*, 1979). When the olfactory receptor cells are caused to degenerate by axotomy or $ZnSO_4$ irrigation of the nasal cavity, there is no appreciable change in GAD or GABA content or in the high-affinity uptake system for GABA in the denervated olfactory bulb (Margolis *et al.*, 1974). If GABA were present in the olfactory receptor cell axon terminals, a decrease of GABA and GAD activity would be expected in the denervated olfactory bulb as is observed for carnosine and carnosine synthetase (Margolis *et al.*, 1974; Harding and Margolis, 1976; see also this volume, Chapter 17). MacLeod (1978) iontophoretically applied several compounds to mitral cells while recording electrophysiologically. He observed inhibition of the mitral cells when GABA was applied to their

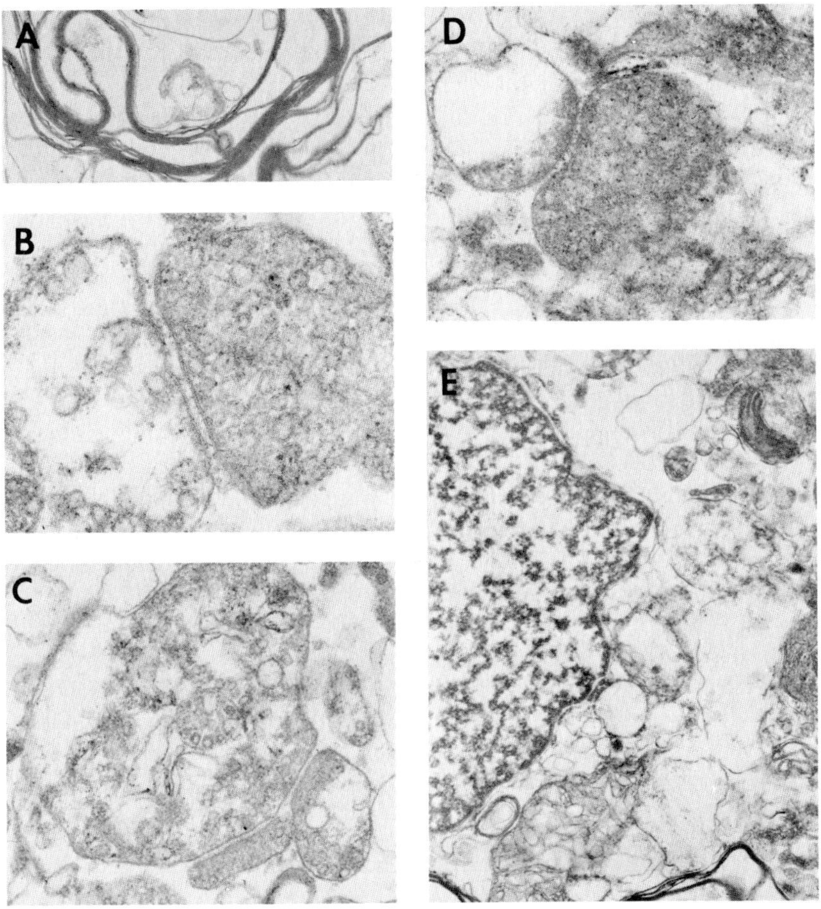

Fig. 5. Electron micrographs of subcellular fractions of rat olfactory bulbs. (A) Band 1 was recovered from the gradient at the 0.32/0.8 *M* sucrose interface; it contains myelin (× 14,000). (B) Band 2 was recovered from the gradient at the 1.0/1.2 *M* sucrose interface; it shows typical DDS with the characteristic asymmetrical postsynaptic thickening (× 49,000). (C) Band 2 showing a DDS triplex (× 24,000). (D) Band 3 was recovered from the gradient at the 1.2/1.4 *M* sucrose interface; it shows typical synaptosomes (× 31,000). (E) Pellet recovered from the bottom of the sucrose gradient, showing nuclei, mitochondria, and membranes of unknown origin (× 14,000). Electron micrographs courtesy of Dr. Paula Orkand. (Taken from Quinn and Cagan, 1980, with permission.)

primary dendrites. Mitral cells are known to be excited by the receptor cell axons and to transmit excitatory impulses to higher centers in the brain (see this volume, Chapter 17); therefore GABA apparently is not a neurotransmitter candidate in mammalian olfactory receptor neurons.

It was verified that the GAD present in P_B' was of DDS origin by further purifying fraction P_B' in a discontinuous sucrose density gradient according to the method described by Kornguth *et al.* (1976), except that 0.2 m*M* pyridoxal-5′-phosphate and 1 m*M* aminoethylisothiouronium bromide were added to each sucrose layer (Quinn and Cagan, 1980). It was reported that their addition to purified GAD stabilizes the enzyme activity for several months (Wu, 1976). Experiments performed in our laboratory demonstrated that the GAD enzyme activity of a supernatant fraction prepared from rat olfactory bulb homogenate was stabilized for at least one month at −65°C by addition of these compounds.

Three major bands were obtained following centrifugation of P_B' in the sucrose gradient (Fig. 4). Band 1 (Fig. 5) was composed of myelin and coiled tubular structures and contained a negligible amount of GAD (Table II). Band 2 contained many DDS; when considered with the particles suspended in the 1.2 *M* sucrose layer, this region accounted for 78–86% of the GAD activity placed onto the gradient (Table II). Band 3 contained 14–22% of the GAD, and a few DDS were observed.

In contrast with the reports of Kornguth *et al.* (1976, 1979), the DDS in the present experiments banded at the interface of the 1.0/1.2 *M* sucrose layers rather than at the 1.2/1.4 *M* interface. Recent experiments in our laboratory demonstrate that the addition of pyridoxal-5′-phosphate and

TABLE II Distribution of GAD Activity in Discontinuous Sucrose Gradient[a]

Fraction	Specific activity (pmole/min-mg protein) Experiment 1	Specific activity (pmole/min-mg protein) Experiment 2	Total activity (nmole/min) Experiment 1	Total activity (nmole/min) Experiment 2
Band 1	672 ± 114	None[b]	0.12 (1)	None[b]
Band 2	1186 ± 51	1551 ± 205	5.22 (60)	2.95 (43)
1.2 *M* Sucrose layer	1113 ± 145	1437 ± 68	2.24 (26)	2.44 (35)
Band 3	926 ± 116	1265 ± 59	1.22 (14)	1.52 (22)
Pellet	969 ± 24	None[b]	1.16 (13)	None[b]
Recovery			(114)	(100)

[a] Each value is the mean ± SD of 3–4 replicates. The total activity is calculated from the mean specific activity. In parentheses are shown the percentages of the total activity placed on the gradient that were recovered in the respective fractions.

[b] The indicated fraction was not observed in this experiment. (Taken from Quinn and Cagan, 1980, with permission.)

aminoethylisothiouronium bromide to the sucrose layers in the gradient cause the DDS to band at the lighter buoyant density. When these compounds were omitted from the homogenizing buffer (sucrose A) and the gradient layers, the DDS banded at the 1.2/1.4 *M* sucrose interface as reported by Kornguth *et al.* (1976, 1979).

VI. GABA BINDING TO MEMBRANES FROM DENDRODENDRITIC SYNAPTOSOMES

The highest specific GABA binding activity in the CNS, after the cerebellum, is in the olfactory bulb and the cerebral cortex (Beaumont *et al.*, 1978; Williams and Risley, 1979), yet no detailed study of GABA binding to synaptosomal membranes from olfactory bulbs has been made. Using [^{3}H]GABA as a ligand, sodium-dependent and sodium-independent binding to neuronal membrane preparations are classified as representing high-affinity uptake sites (presynaptic or glial) and postsynaptic receptors, respectively (Enna and Snyder, 1975, 1977). The sodium-dependent binding can be decreased by freeze–thawing the membranes, by treatment with Triton X-100, and by performing the binding assay in a Na^+-free medium (Enna and Synder, 1975, 1977; Wong and Horng, 1977; Horng and Wong, 1979).

Na^+-Independent GABA binding sites in the postsynaptic membrane differ from other putative neurotransmitter sites because they are not solubilized by treatment with low concentrations of Triton X-100. Incubation of crude synaptic membranes from various neuronal sources with 0.01–0.5% (v/v) Triton X-100 for 30 min at 37°C is reported to solubilize Na^+-dependent binding sites, to increase the Na^+-independent specific GABA binding activity by tenfold, to cause a tenfold decrease in the K_D, and to discriminate the GABA postsynaptic receptor sites as two distinct classes: high-affinity and low-affinity sites. Recently, Toffano *et al.* (1978) demonstrated that an endogenous protein inhibitor (modulin) was solubilized from synaptic membranes by Triton X-100 treatment. The inhibitor noncompetitively blocks the high-affinity sites for GABA. It was suggested by Toffano *et al.* (1978) that the endogenous protein may act *in vivo* by allosterically modulating the affinity of the postsynaptic receptor for GABA.

Based on the evidence discussed in this chapter, the DDS are postulated to use GABA as a neurotransmitter in the negative feedback loop between dendrites of the granule cells and secondary dendrites of the mitral cells in the olfactory bulb. This leads us to predict that the mitral cell synaptic membrane component of the DDS contains GABA postsynaptic receptors. Fraction P_B', which contains DDS, was prepared from rat olfactory bulbs as described above. Fraction P_B' was lysed in deionized water, and the crude synaptic membranes (CSM) were sedimented by centrifugation (48,000 *g*, 15 min) and suspended in 50 m*M* Tris-citrate buffer (pH 7.1). The CSM were either assayed for binding activity (fresh) or frozen (−25°C) for at least 18 hr, thawed, and assayed. To subject the CSM to Triton-treatment, the freeze–thawed CSM were incubated (37°C, 30 min) in 50 m*M* Tris-citrate buffer (pH 7.1) containing 0.05% (v/v) Triton X-100. Following incubation, the Triton-treated CSM were centrifuged (48,000 *g*, 15 min), washed 3 times to remove the detergent, and resuspended in buffer.

Binding of [^{3}H]GABA to CSM was measured by incubating CSM (100–300 μg protein) with [^{3}H]GABA in 50 m*M* Tris-citrate buffer (pH 7.1) for 10 min at 0°C. The free ligand was separated from that bound to the CSM by filtration through a Millipore filter (Krueger and Cagan, 1976; see also this volume, Chapters 3, 10, and 15). Specific binding of [^{3}H]GABA is defined as the total counts bound minus the counts bound in the presence of a large excess (8 m*M*) of unlabeled GABA. After filtering the incubation medium, the CSM-ligand complex remaining on the Millipore filter was rinsed once with 10 ml of ice-cold Tris–citrate buffer. The filter was dissolved in scintillation fluid and counted in a Packard Tri-Carb liquid scintillation counter at a counting efficiency for tritium of 36%.

The influence of 100 m*M* NaC1 on specific binding of [^{3}H]GABA to fresh, frozen–thawed, and Triton-treated CSM is shown in Table III. Fresh CSM bind approximately 12-fold more [^{3}H]GABA in the presence of 100 m*M* NaC1 and the frozen–thawed CSM show an increase of about threefold. The relatively small amount of Na^+-dependent binding to the CSM that remains after freeze–thawing appears to be completely abolished by the Triton-treatment, as reported for CSM from other GABAergic regions of the central nervous system (Horng and Wong, 1979; Toffano *et al.*, 1978; Enna and Snyder, 1977). Freeze–thawing the CSM increases the specific Na^+-inde-

TABLE III [^{3}H]GABA Binding to Crude Synaptic Membranes[a]

Treatment	Specific binding (fmole/mg protein) Control	+100 m*M* NaCl
Fresh	17 ± 0.1	198 ± 6
Frozen–thawed	25 ± 3	69 ± 2
Triton-treated	720 ± 7	653 ± 47

[a] Crude synaptic membranes were prepared from fraction P_B' and treated as described in the text. Binding was measured using 8 n*M* [^{3}H]GABA. The values are corrected for nonspecific binding and represent the mean ± SEM of three replicate samples.

pendent [^{3}H]GABA binding by 1.5-fold over that of the fresh preparation, and treatment with Triton X-100 causes a further increase of approximately 30-fold. It does not appear likely that the influence of Triton-treatment on GABA binding in the presence of Na^+ can be explained by an "unmasking" of the Na^+-dependent sites. This is because muscimol, a potent GABA agonist, which does not bind to Na^+-dependent GABA uptake sites, has its binding also stimulated by Triton-treatment (Quinn, unpublished; Beaumont *et al.*, 1978). Also, 3-aminopropanesulfonic acid does not displace GABA bound to fresh rat brain CSM in the presence of 100 mM NaC1, but it does inhibit Na^+-independent GABA binding to fresh and to Triton-treated CSM (Enna and Snyder, 1977). Its ability to displace GABA bound to Triton-treated CSM, but not affect Na^+-dependent binding to fresh CSM, argues against an "unmasking" hypothesis.

To evaluate the subcellular distribution of the Na^+-independent GABA binding sites a homogenate of rat olfactory bulbs was prepared and fractionated by differential centrifugation as described above. Specific Na^+-independent [^{3}H]GABA binding activity was determined with Triton-treated membrane preparations. The highest binding activity occurs with the CSM prepared from fraction P_B' (Table IV); this fraction also accounts for the highest proportion of the total binding activity recovered from the homogenate. When the DDS in P_B' were further purified in a discontinuous sucrose gradient as described by Kornguth *et al.* (1976), essentially all of the [^{3}H]GABA binding activity was recovered in the synaptic membranes of the DDS. These results demonstrate that the GABA postsynaptic receptors in the olfactory bulb are predominantly associated with the dendrodendritic synapses.

The specific Na^+-independent [^{3}H]GABA binding to Triton-treated CSM

TABLE IV Distribution of [^{3}H]GABA Binding Activity Following Differential Centrifugation[a]

Fraction	Specific binding (fmole/mg protein)	Total binding (pmole)
H	458 ± 81	13.1
P_1[b]	286 ± 20	2.3
P_B'	629 ± 19	6.3
P_2	256 ± 2	0.9

[a] Crude synaptic membranes were prepared from each fraction and treated with Triton X-100. Binding was measured using 8 nM [^{3}H]GABA. The values are corrected for nonspecific binding and represent the mean ± SD of duplicate samples. For explanation of fraction symbols, see text and Fig. 3.

[b] Lower regions of crude nuclear pellet.

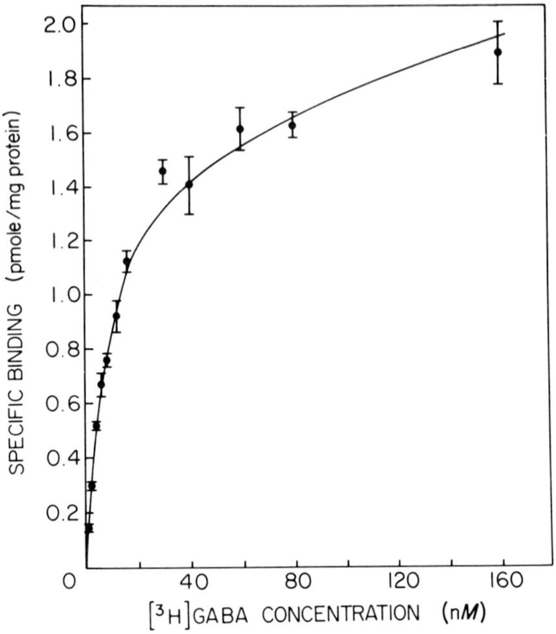

Fig. 6. Binding of [^{3}H] GABA. Frozen-thawed crude synaptic membranes from fraction P_B' were treated with Triton X-100. Binding was measured at several concentrations of [^{3}H]GABA and corrected for nonspecific binding by assaying samples in parallel with a large excess of unlabeled GABA. Each point represents the mean ± SEM of 2–4 samples compiled over three experiments.

from P_B' was also demonstrated to be saturable (Fig. 6) and linear with protein concentration up to about 500 μg CSM protein/2.2 ml. Scatchard analysis of the saturation data reveals a high-affinity binding site ($K_D = 8.0 \times 10^{-9}\,M$; B_{max} = 1.5 pmole/mg protein) and a second site of lower affinity ($K_D = 2.4 \times 10^{-8}\,M$; B_{max} = 2.1 pmole/mg protein). The parameters of GABA binding as well as the distribution of GAD and Na^+-independent GABA binding provide neurochemical evidence to support the hypothesis that GABA is a neurotransmitter at the dendrodendritic synapses in the olfactory bulb.

VII. FUTURE PROSPECTS

Additional research should enable characterization of the GABA system at the dendrodendritic synapses in the external plexiform layer of the olfactory bulb. The synaptosomal fraction and the CSM preparation from P_B' described in this chapter should enable studies of the effects of pharmacological agents on inhibition of GABA synthesis, on high-affinity GABA uptake, and on inhibition of GABA postsynaptic receptors. Ultimately, such pharmacological agents could be used *in vivo* to delineate more clearly the physiological role of the dendrodendritic synapses in the olfactory bulb. To obtain meaningful results, it would be necessary to localize the injection of substances in the external plexiform layer of the olfactory bulb. We would predict that aberrations in olfactory ability would become apparent, especially in the ability of the animal to discriminate concentrations of an odor over its dynamic range of responsiveness.

Another area of interest is the influence of vitamin B_6 nutriture on the synaptic GABA system. It was demonstrated that during vitamin-B_6 deficiency the activity of glutamic acid decarboxylase in the olfactory bulb is decreased to about 50% of that of the control (Quinn and Chan, 1979), but it is not known whether the deficiency affects GAD activity in the dendritic terminals of the granule cell or in other metabolic compartments. It also is possible that the affinity (K_D) or maximal binding (B_{max}) of the GABA postsynaptic receptor may change to compensate for the decrease in GABA, thereby preventing deficits in olfactory ability. Other nutritional states, such as perinatal protein–calorie malnutrition, are also of interest because the olfactory bulb undergoes extensive postnatal synaptogenesis (Hinds and Hinds, 1976) and suckling behavior in the neonate is mediated by olfaction.

The CSM prepared from the DDS may be useful for producing monoclonal antibodies (see Chapter 6) directed against the GABA postsynaptic receptor and is currently under investigation in our laboratory. This would allow visualization of the distribution of GABA receptors in histological sections of the olfactory bulb and perhaps of other GABAergic regions in the central nervous system. Monoclonal antibodies against the GABA receptor would also expedite isolation and characterization of the GABA receptor macromolecules.

ACKNOWLEDGMENT

The authors are grateful for the excellent technical assistance of Barbara M. Small. This work was supported in part by United States Public Health Service Research Grant No. AM18932 from National Institute of Arthritis, Metabolism, and Digestive Diseases and National Institutes of Health Research Service Postdoctoral Award No. 1F32NS-05958 to M. R. Quinn from National Institute of Neurological and Communicative Disorders and Stroke.

REFERENCES

Allison, A. C., and Warwick, R. T. T. (1949). Quantitative observations on the olfactory system of the rabbit. *Brain* **72**, 186–197.

Austin, L., Recasens, M., and Mandel, P. (1979). GABA in the olfactory bulb and olfactory nucleus of the rat: The distribution of gamma-aminobutyric acid, glutamic acid decarboxylase, GABA transaminase and succinate semialdehyde dehydrogenase. *J. Neurochem.* **32**, 1473–1477.

Baxter, C. F. (1970). The nature of γ-aminobutyric acid. *In* "Handbook of Neurochemistry" (A. Lajtha, ed.), Vol. 3, pp. 289–353. Plenum, New York.

Beaumont, K., Chilton, W. S., Yamamura, H. I., and Enna, S. J. (1978). Muscimol binding in rat brain: Association with synaptic GABA receptors. *Brain Res.* **148**, 153–162.

Bergez, F., Urban, P. F., and Mandel, P. (1978). Potassium-evoked release of [^{14}C]GABA and [^{3}H]taurine from rat olfactory bulb slices. *Neurosci. Lett.* **8**, 137–142.

Buhler, D. R. (1962). A simple scintillation counting technique for assaying $^{14}CO_2$ in a Warburg flask. *Anal. Biochem.* **4**, 413–417.

Cotman, C. W., and Matthews, D. A. (1971). Synaptic plasma membranes from rat brain synaptosomes: Isolation and partial characterization. *Biochim. Biophys. Acta* **249**, 380–394.

Dahlström, A., Fuxe, K., Olson, L., and Ungerstedt, U. (1965). On the distribution and possible function of monoamine nerve terminals in the olfactory bulb of the rabbit. *Life Sci.* **4**, 2071–2074.

Davies, L. P., Johnston, G. A. R., and Stephanson, A. L. (1975). Postnatal changes in the potassium-stimulated, calcium-dependent release of radioactive GABA and glycine from slices of rat central nervous tissue. *J. Neurochem.* **25**, 387–392.

DeBelleroche, J. S., and Bradford, H. F. (1973). The synaptosome: An isolated, working, neuronal compartment. *Prog. Neurobiol.* **1**, 277–298.

Doty, R. L., ed. (1976). "Mammalian Olfaction, Reproductive Processes, and Behavior." Academic Press, New York.

Enna, S. J., and Snyder, S. H. (1975). Properties of γ-aminobutyric acid (GABA) receptor binding in rat brain synaptic membrane fractions. *Brain Res.* **100**, 81–97.

Enna, S. J., and Snyder, S. H. (1977). Influences of ions, enzymes, and detergents on γ-aminobutyric acid-receptor binding in synaptic membranes of rat brain. *Mol. Pharmacol.* **13**, 442–453.

Felix, D., and McLennan, H. (1971). The effect of bicuculline on the inhibition of mitral cells of the olfactory bulb. *Brain Res.* **25**, 661–664.

Garcia, J., Hankins, W. G., and Rusiniak, K. W. (1974). Behavioral regulation of the milieu interne in man and rat. *Science* **185**, 824–831.

Gasser, H. S. (1956). Olfactory nerve fibers. *J. Gen. Physiol.* **39**, 473–496.

Graham, L. T., Jr. (1973). Distribution of glutamic acid decarboxylase activity and GABA content in the olfactory bulb. *Life Sci.* **12**, Part I, 443–447.

Gray, E. G., and Whittaker, V. P. (1962). The isolation of nerve endings from brain: An electron-microscopic study of cell fragments derived by homogenization and centrifugation. *J. Anat.* **96**, 79–88.

Halász, N., Ljungdahl, Å., Hökfelt, T., Johansson, O., Goldstein, M., Park, D., and Biberfeld, P. (1977). Transmitter histochemistry of the rat olfactory bulb. I. Immunohistochemical localization of monoamine synthesizing enzymes. Support for intrabulbar, periglomerular dopamine neurons. *Brain Res.* **126**, 455–474.

Halász, N., Ljungdahl, Å., and Hökfelt, T. (1978). Transmitter histochemistry of the rat olfactory bulb. II. Fluorescence histochemical, autoradiographic and electron microscopic localization of monoamines. *Brain Res.* **154**, 253–271.

Halász, N., Ljungdahl, Å., and Hökfelt, T. (1979). Transmitter histochemistry of the rat olfactory bulb. III. Autoradiographic localization of [^{3}H]GABA. *Brain Res.* **167**, 221–240.

Harding, J., and Margolis, F. L. (1976). Denervation in the primary olfactory pathway of mice. III. Effect on enzymes of carnosine metabolism. *Brain Res.* **110**, 351–360.

Harvey, J. A., Scholfield, C. N., Graham, L. T., Jr., and Aprison, M. H. (1975). Putative transmitters in denervated olfactory cortex. *J. Neurochem.* **24**, 445–449.

Hasler, A. D., Scholz, A. T., and Horrall, R. M. (1978). Olfactory imprinting and homing in salmon. *Am. Sci.* **66**, 347–355.

Hinds, J. W., and Hinds, P. L. (1976). Synapse formation in the mouse olfactory bulb. I. Quantitative studies. *J. Comp. Neurol.* **169**, 15–40.

Hökfelt, T., Halász, N., Ljungdahl, Å., Johansson, O., Goldstein, M., and Park, D. (1975). Histochemical support for a dopaminergic mechanism in the dendrites of certain periglomerular cells in the rat olfactory bulb. *Neurosci. Lett.* **1**, 85–90.

Horng, J. S., and Wong, D. T. (1979). γ-Aminobutyric acid receptors in cerebellar membranes of rat brain after a treatment with Triton X-100. *J. Neurochem.* **32**, 1379–1386.

Jones, D. G. (1975). "Synapses and Synaptosomes. Morphological Aspects," pp. 135–173. Wiley, New York.

Kornguth, S., Juhl, U., Knobeloch, L., Sunderland, E., Johnson, T., and Scott, G. (1976). Isolation of dendrodendritic synapses from swine olfactory bulbs. *Brain Res.* **105**, 423–435.

Kornguth, S. E., Scott, G., and Jaramillo, R. (1979). Ultrastructural and biochemical comparisons of dendrodendritic synaptosomes from bovine superior colliculi and olfactory bulbs with axodendritic and axosomatic synaptosomes. *J. Neurochem.* **33**, 875–882.

Krueger, J. M., and Cagan, R. H. (1976). Biochemical studies of taste sensation. Binding of L-[^{3}H]alanine to a sedimentable fraction from catfish barbel epithelium. *J. Biol. Chem.* **251**, 88–97.

Lichtensteiger, W. (1966). Uptake of norepinephrine in periglomerular cells of the olfactory bulb of the mouse. *Nature (London)* **210**, 955–956.

MacFarlane, A. (1975). Olfaction in the development of social preferences in the human neonate. *Parent-Infant Interact., Ciba Found. Symp.* **33**, 103–113.

McLennan, H. (1971). The pharmacology of inhibition of mitral cells in the olfactory bulb. *Brain Res.* **29**, 177–184.

MacLeod, N. (1978). Is carnosine (β-ala-L-his) a neurotransmitter in the primary olfactory pathway? *In* "Iontophoresis and Transmitter Mechanisms in the Mammalian Central Nervous System" (R. W. Ryall and J. S. Kelly, eds.), pp. 117–119. Elsevier/North-Holland, Amsterdam.

Mandel, P., and Pasantes-Morales, H. (1976). Taurine: A putative neurotransmitter. *Adv. Biochem. Psychopharmacol.* **15**, 141–151.

Margolis, F. L. (1974). Carnosine in the primary olfactory pathway. *Science* **184**, 909–911.

Margolis, F. L., Roberts, N., Ferriero, D., and Feldman, J. (1974). Denervation in the primary olfactory pathway of mice: Biochemical and morphological effects. *Brain Res.* **81**, 469–483.

Mulder, A. H., and Snyder, S. H. (1974). Potassium-induced release of amino acids from cerebral cortex and spinal cord slices of the rat. *Brain Res.* **76**, 297–308.

Neidle, A., and Kandera, J. (1974). Carnosine—An olfactory bulb peptide. *Brain Res.* **80**, 359–364.

Nicoll, R. A. (1970). GABA and dendrodendritic inhibition in the olfactory bulb. *Pharmacologist* **12**, 236.

Nicoll, R. A. (1971). Pharmacological evidence for GABA as the transmitter in granule cell inhibition in the olfactory bulb. *Brain Res.* **35**, 137–149.

Osborne, R. H., Duce, I. R., and Keen, P. (1976). Amino acids in "light" and "heavy" synaptosome fractions from rat olfactory lobes and their release by electrical stimulation. *J. Neurochem.* **27**, 1483–1488.

Popov, N., Pohle, W., Rösler, V., and Matthies, H. (1967). Regionale verteilung von γ-aminobuttersäure, glutaminsäure, asparaginsäure, dopamin, noradrenalin und serotonin im rattinhirn. *Acta Biol. Med. Ger.* **18**, 695–702.

Quinn, M. R., and Cagan, R. H. (1979). Glutamate decarboxylase in dendrodendritic synaptosomes from rat olfactory bulbs. *Fed. Proc., Fed. Am. Soc. Exp. Biol. (Abstr.)* **38**, 516.

Quinn, M. R., and Cagan, R. H. (1980). Subcellular distribution of glutamate decarboxylase in rat olfactory bulb: High content in dendrodendritic synaptosomes. *J. Neurochem.* **35**, 583–590.

Quinn, M. R., and Chan, M. M. (1979). Effect of vitamin B-6 deficiency on glutamic acid decarboxylase activity in rat olfactory bulb and brain. *J. Nutr.* **109**, 1694–1702.

Reese, T. S., and Shepherd, G. M. (1972). Dendro-dendritic synapses in the central nervous system. *In* "Structure and Function of Synapses" (G. D. Pappas and D. P. Purpura, eds.), pp. 121–136. Raven, New York.

Ribak, C. E., Vaughn, J. E., Saito, K., Barber, R., and Roberts, E. (1977). Glutamate decarboxylase localization in neurons of the olfactory bulb. *Brain Res.* **126**, 1–18.

Roberts, E., and Kuriyama, K. (1968). Biochemical–physiological correlations in studies of the γ-aminobutyric acid system. *Brain Res.* **8**, 1–35.

Robinzon, B., and Quinn, M. R. (1978). Simple method for staining unmyelinated axons in rat olfactory bulb. *Brain Res. Bull.* **3**, 563–566.

Roskoski, R., Jr., Ryan, L. D., and Diecke, F. J. P. (1974). γ-Aminobutyric acid synthesized in the olfactory nerve. *Nature (London)* **251**, 526–529.

Schmitt, F. O., Dev, P., and Smith, B. H. (1976). Electrotonic processing of information by brain cells. *Science* **193**, 114–120.

Schneider, S. P., and Macrides, F. (1978). Laminar distributions of interneurons in the main olfactory bulb of the adult hamster. *Brain Res. Bull.* **3**, 73–82.

Shepherd, G. M. (1972). Synaptic organization of the mammalian olfactory bulb. *Physiol. Rev.* **52**, 864–917.

Shepherd, G. M. (1974). "The Synaptic Organization of the Brain: An Introduction," pp. 111–144. Oxford Univ. Press, London and New York.

Sieghart, W., and Heckl, K. (1976). Postassium-evoked release of taurine from synaptosomal fractions of rat cerebral cortex. *Brain Res.* **116**, 538–543.

Singh, P. J., Tucker, A. M., and Hofer, M. A. (1976). Effects of nasal $ZnSO_4$ irrigation and olfactory bulbectomy on rat pups. *Physiol. Behav.* **17**, 373–382.

Srinivasan, V., Neal, M. J., and Mitchell, J. F. (1969). The effect of electrical stimulation and high potassium concentrations on the efflux of [^{3}H]γ-aminobutyric acid from brain slices. *J. Neurochem.* **16**, 1235–1244.

Teicher, M. H., and Blass, E. M. (1977). First suckling response of the newborn albino rat: The roles of olfaction and amniotic fluid. *Science* **198**, 635–636.

Toffano, G., Guidotti, A., and Costa, E. (1978). Purification of an endogenous protein inhibitor of the high affinity binding of γ-aminobutyric acid to synaptic membranes of rat brain. *Proc. Natl. Acad. Sci. U.S.A.* **75**, 4024–4028.

White, E. L. (1973). Synaptic organization of the mammalian olfactory glomerulus: New findings including an intraspecific variation. *Brain Res.* **60**, 299–313.

Williams, M., and Risley, E. A. (1979). Characterization of the binding of [^{3}H]muscimol, a potent γ-aminobutyric acid agonist, to rat brain synaptosomal membranes using a filtration assay. *J. Neurochem.* **32**, 713–718.

Wilson, S. H., Schrier, B. K., Farber, J. L., Thompson, E. J., Rosenberg, R. N., Blume, A. J., and Nirenberg, M. W. (1972). Markers for gene expression in cultured cells from the nervous system. *J. Biol. Chem.* **247**, 3159–3169.

Wong, D. T., and Horng, J. S. (1977). Na^+-Independent binding of GABA to the Triton X-100 treated synaptic membranes from cerebellum of rat brain. *Life Sci.* **20**, 445–452.

Wu, J.-Y. (1976). Purification, characterization, and kinetic studies of GAD and GABA-T from mouse brain. *In* "GABA in Nervous System Function" (E. Roberts, T. N. Chase, and D. B. Tower, eds.), pp. 7–55. Raven, New York.

19

Neurochemistry of the Olfactory Tubercle

NEIL R. KRIEGER

I. INTRODUCTION

The rat olfactory tubercle is a thin sheet of cortex less than 1 mm in thickness that extends over the ventral surface of much of the forebrain. It consists of three well-defined histological laminae: plexiform, pyramidal, and polymorphic. It derives major inputs from the mitral and tufted cells of the olfactory bulb, and from the dopaminergic cell bodies of the midbrain A_{10} region (Fig. 1).

Biochemistry of Taste and Olfaction

ISBN 0-12-154450-8

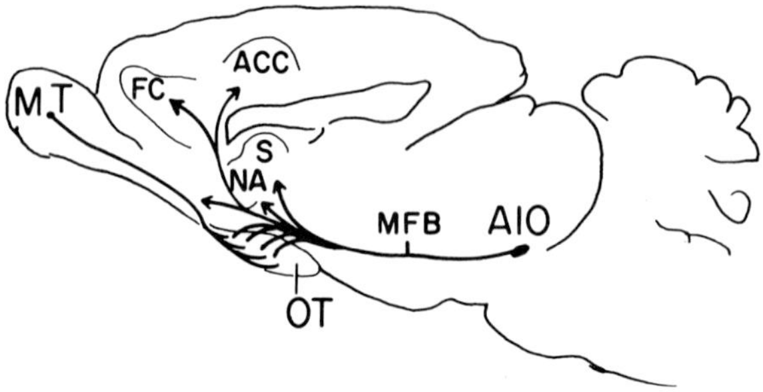

Fig. 1. Schematic sagittal representation of bulbar and dopaminergic (Björklund and Lindvall, 1978) projections to the olfactory tubercle. M, mitral cells; T, tufted cells; FC, frontal cortex; ACC, anterior cingulate cortex; NA, nucleus accumbens; S, septum; MFB, median forebrain bundle; OT, olfactory tubercle.

Dopamine, GABA, and acetylcholine concentrations within the tubercle are among the highest in the brain. The laminar organization, the well-separated major cell types, and the high levels of transmitter-related substances make this region attractive for neurochemical studies. The presence of identified inputs from the bulb and midbrain mean that the tubercle can function as a stage for the study of the chemistry and physiology of those regions as well. We have been interested in localizing neurochemicals to anatomically identified cell types and in establishing neurochemical correlates for neuronal function. We have carried out localization studies for the enzymes and metabolites associated with dopaminergic, GABAergic, and cholinergic transmission, and we have measured changes in cyclic AMP content in the separate laminae of the olfactory tubercle.

Changes in cyclic AMP levels may afford a neurochemical correlate of physiological function. The dopaminergic inputs from the midbrain to the tubercle are suitable for the study of the pharmacology of neuroleptic agents. Similarly, input from the bulb could afford a monitoring post for sensory activity in the olfactory tubercle.

II. ANATOMY AND BIOCHEMISTRY

A. Anatomy

The olfactory tubercle is located posterior to the anterior olfactory nucleus, lateral to the septal area, and medial to the lateral olfactory tract (Fig. 2). The tubercle, because of its connections with other parts of the brain, is considered part of the limbic system. Thus, it has inputs not only from the olfactory bulb (Shepherd, 1979; Anderson and Westrum, 1972) and primary

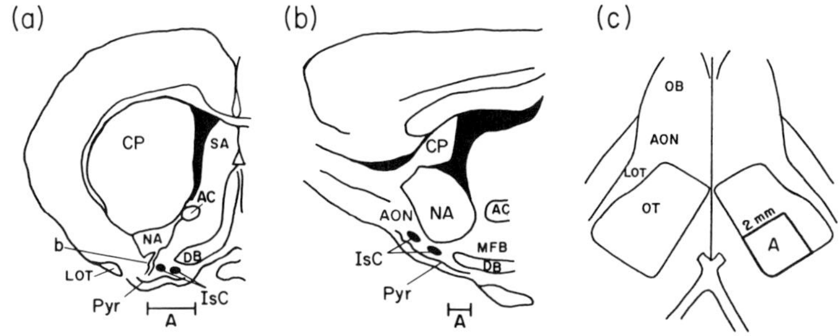

Fig. 2. Schematic drawings of the rat forebrain. (a) coronal (Hedreen and Chalmers, 1972); (b) sagittal (König and Klippel, 1963); (c) ventral. The letter A designates the pedestal area within the olfactory tubercle that was sectioned for biochemical experiments. In panels (a) and (b), the bar indicates the extent of intersection with those coronal and sagittal sections. Pyr, pyramidal layer of the olfactory tubercle; IsC, islands of Calleja; b, cell bridge; MFB, medial forebrain bundle; DB, diagonal band; NA, nucleus accumbens; AC, anterior commissure; CP, caudate putamen; OB, olfactory bulb; AON, anterior olfactory nucleus; LOT, lateral olfactory tract; SA, septal area; OT, olfactory tubercle.

olfactory cortex (Valverde-Garcia, 1965; Price, 1974) but also from the amygdala (Valverde-Garcia, 1965; De Olmos, 1972), the prefrontal cortex (Leonard, 1969), and the temporal cortex (Van Hoesen *et al.*, 1976). Limbic structures are also the targets of its output, for example, the nuclei gemini of the hypothalamus (Scott and Chafin, 1975) and the mediodorsal nucleus of the thalamus (Heimer, 1972).

There are three well-defined histological laminae within the rat olfactory tubercle (Fig. 3): a superficial plexiform layer (1), a pyramidal cell layer (II), and a deep polymorphic layer (III) [See Heimer (1978) for a thorough review of the anatomy of the olfactory tubercle]. In the caudal tubercle (A in Fig. 2C), the tissue layers are planar and parallel, as illustrated in Fig. 3. The plexiform layer contains a few neuronal cell bodies. The pyramidal layer consists of a dense sheet of small and medium-sized pyramidal cell bodies (10–15 μm diameter). The deepest layer (III) consists of scattered medium-sized pyramidal cells, granule cells (5–10 μm diameter), known as islands of Calleja, and polymorphic cells (up to 40 μm in diameter). These cells have been described from Golgi impregnations (Calleja, 1893; Beccari, 1910; Cajal, 1911, 1955; Price, 1973; Heimer, 1978) and, to a limited extent, from observations with the electron microscope (Anderson and Westrum, 1972; Hosoya and Hirata, 1974; Hosoya, 1973). Electron micrographs of these three cell types are shown in Figs. 4–6. The pedestal location for the neurochemical studies described below was at A in the caudal tubercle (Fig. 2). Sections cut tangentially at A contain tissue primarily from one histological layer.

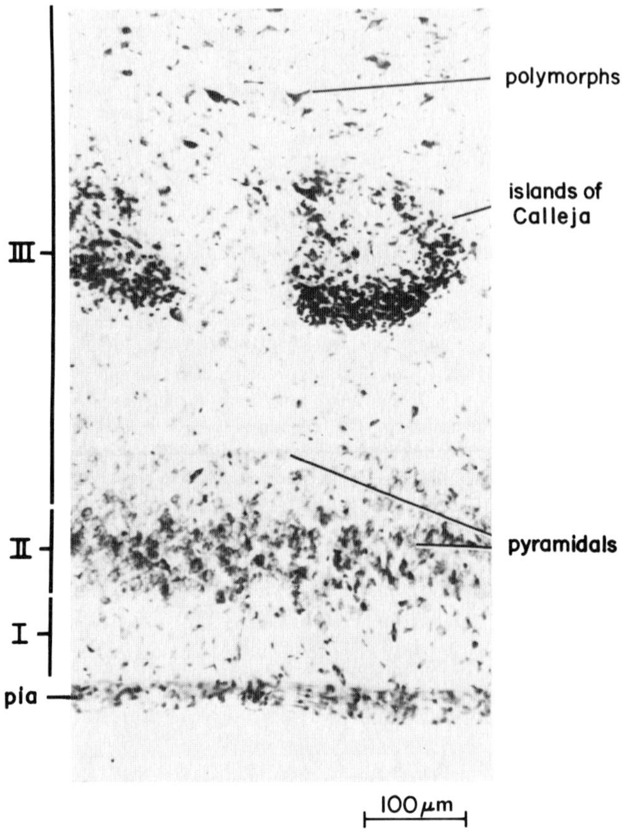

Fig. 3. Coronal section showing the layers of the olfactory tubercle at the position of pedestal A (see Fig. 2). The stain is formol–thionine. (Note that the dimensions of the perfused and fixed tissue are somewhat different from those of the frozen tissue.) The plexiform, pyramidal, and polymorphic layers are labeled I, II, and III, respectively.

Within each histological layer, there is considerable heterogeneity. In addition to their intrinsic cell types, each contains glial elements and a rich variety of inputs from other regions. Consider the plexiform layer. Its processes include apical dendrites of the pyramidal cells (Price, 1974; Heimer, 1978) and extrinsic axons and terminals from the olfactory bulb (Anderson and Westrum, 1972), from the olfactory cortex (Price, 1974), and from

---

Fig. 4. Pyramidal cell of rat olfactory tubercle. Apical dendrite (AD) points ventrally toward pia. Basal dendrite (BD) and axon initial segment (IS) point dorsally toward islands of Calleja. Oval, noninvaginated nucleus (N) is typical of these neurons. Scale = 2 μm.

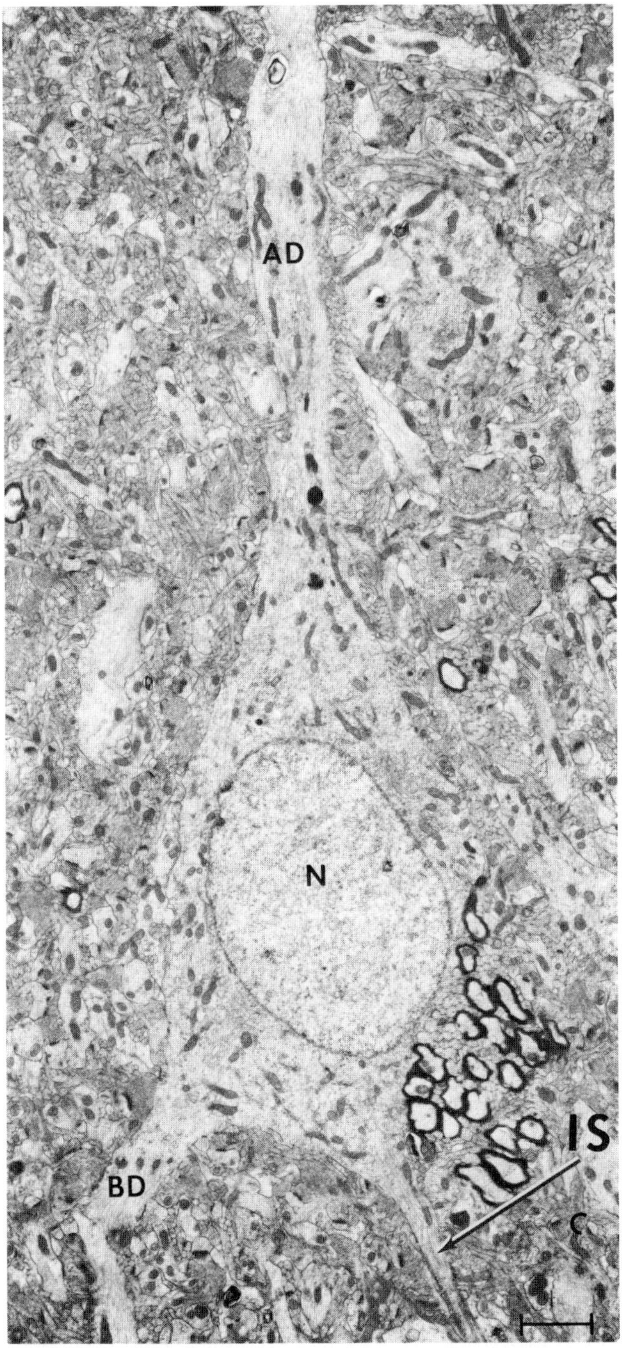
AD
N
IS
BD
C

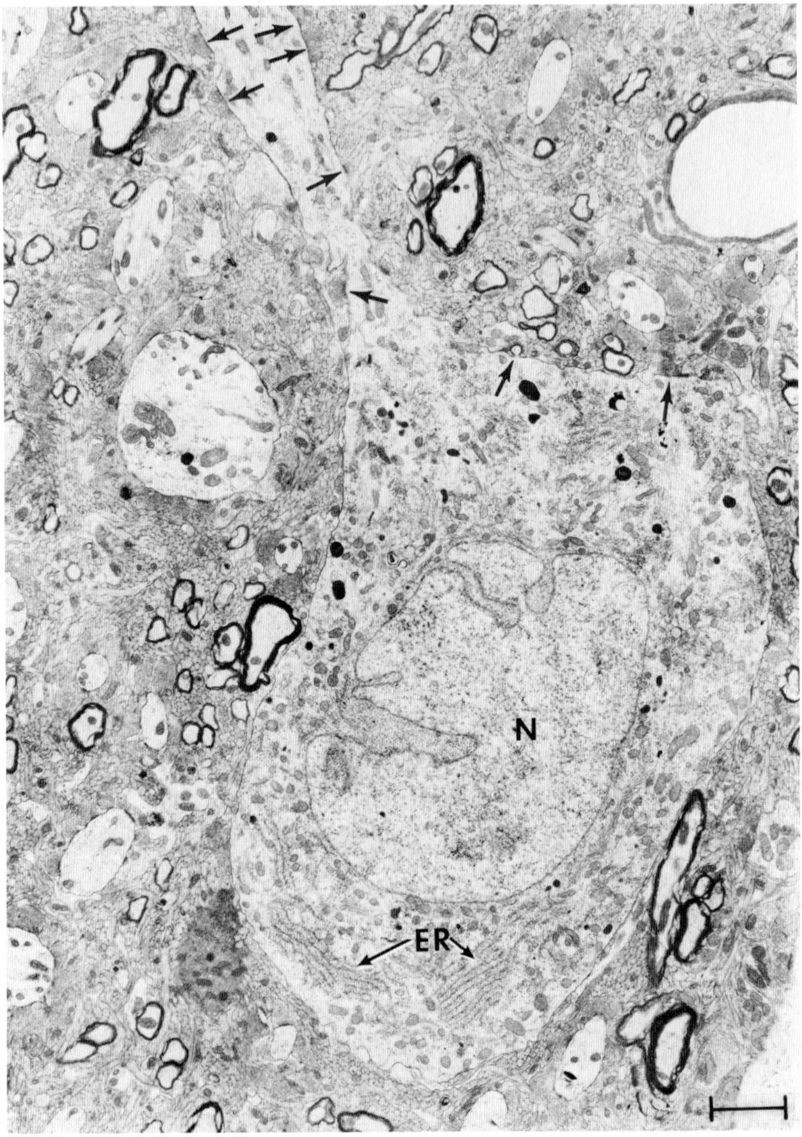

Fig. 5. Polymorphic cell (rat). Note typical invaginated nucleus (N), stacks of rough endoplasmic reticulum (ER), and synaptic contacts (arrows) on soma and dendrite. Scale = 2 μm.

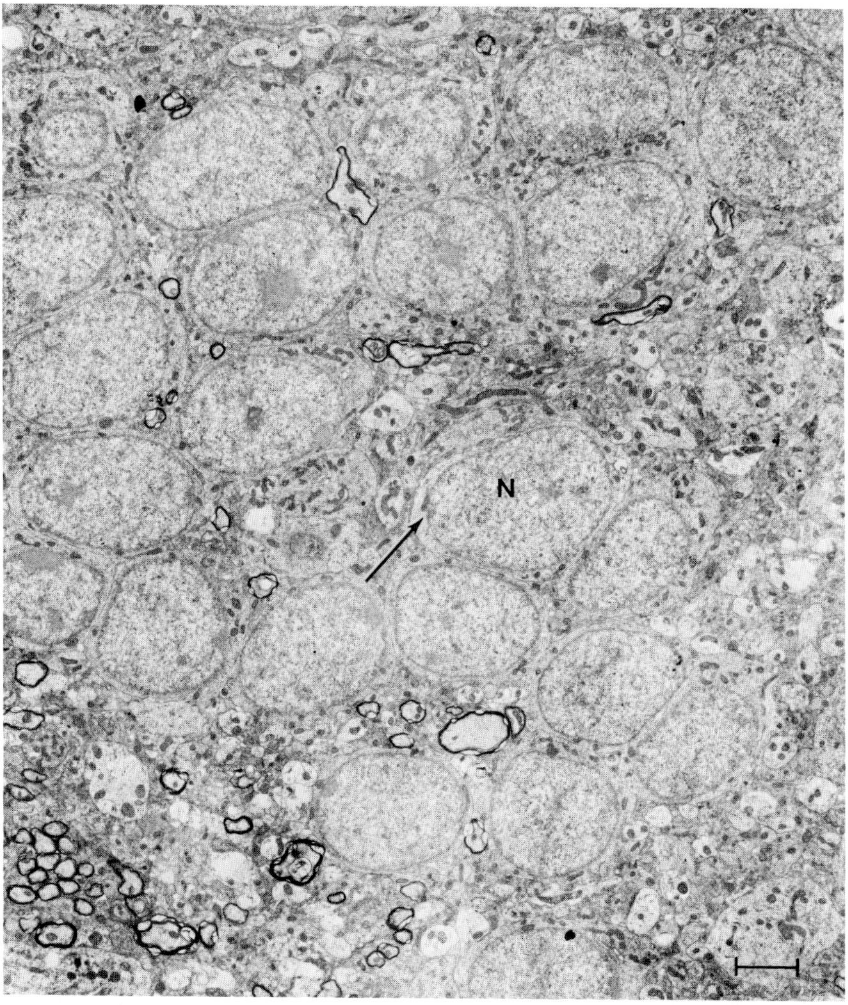

Fig. 6. Island of Calleja (rat). Tiny neurons devoid of synaptic contacts are closely packed with their plasma membranes in apposition. Nucleus (N) fills each cell almost to the exclusion of cytoplasm (arrow). Scale = 2 μm.

dopaminergic cells in the brain stem (Ungerstedt, 1971). There are also occasional islands of Calleja and axons and dendrites ascending from deeper layers.

Still, the laminar organization, the well-separated and relatively few cell types, and the high neurotransmitter concentrations of the caudal tubercle have provided excellent conditions for neurochemical localization studies.

B. Neurochemical Localization

The laminar distributions for three of the enzymes associated with the neurotransmitter systems for dopamine, GABA, and acetylcholine are shown in Fig. 7. Activities for dopamine-sensitive adenylate cyclase (DSAC), glutamic acid decarboxylase (GAD), and choline acetyltransferase (CAT) are plotted as functions of depth in the tubercle (Krieger, 1980; Krieger and Heller, 1979; C. R. Gordon and N. R. Krieger, unpublished, 1980).

For these studies, male Sprague-Dawley rats, weighing approximately 300 g, were killed by decapitation. The hemisected brain was blocked and rapidly frozen in powdered Dry Ice while resting on its ventral surface on a

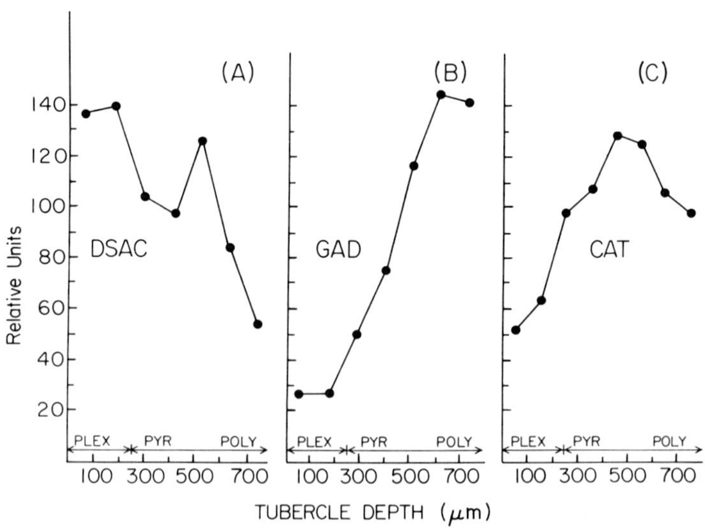

Fig. 7. Enzymatic activities in the olfactory tubercle as functions of depth. (A) Dopamine-sensitive adenylate cyclase, DSAC (Krieger, 1980). (B) Glutamic acid decarboxylase, GAD (Krieger and Heller, 1979). (C) Choline acetyltransferase, CAT (Gordon and Krieger, unpublished 1980). The plotted values are means from three or more independent experiments. The depths of the midpoints of the assayed sections are plotted on the abscissa. The labels PLEX, PYR, and POLY indicate the approximate locations of the plexiform, pyramidal, and polymorphic layers, respectively.

teflon plate. In a cryostat at −15°C, the tissue surrounding the olfactory tubercle was pared away, leaving a pedestal of approximately 2 mm on each side (Fig. 2c). Consecutive tangential sections were cut. Every seventh section was stained with toluidine blue and examined under the microscope to monitor the depth and orientation of section. The six intervening sections (40 μg protein) were pooled for measurement of enzyme activity.

Homogenizations, incubations, and assays were carried out as previously described (Krieger *et al.*, 1977; Krieger and Heller, 1979). For the assay of dopamine-sensitive adenylate cyclase, assays were according to the methods of Kebabian *et al.* (1972). Cyclic AMP values were determined by the procedure of Brown *et al.* (1971). Glutamic acid decarboxylase was assayed according to the method of Albers and Brady (1959) with modifications similar to those described by Tappaz *et al.* (1976). Choline acetyltransferase was assayed according to the method of Schrier and Shuster (1967) and protein by the method of Lowry *et al.* (1951).

The distributions of the three enzymatic activities (Fig. 7) are sufficiently different from one another as to clearly indicate separate localizations. These distributions are consistent with neuronal localizations to the pyramidal cells for dopamine-sensitive adenylate cyclase, the granule or polymorph cells for glutamic acid decarboxylase, and the granule cells for choline acetyltransferase. Additional experiments are needed to rule out localizations to glial cells or to processes entering from outside the tubercle. For dopamine-sensitive adenylate cyclase these additional experiments are described in Section IV. Similar observations have been reported for glutamic acid decarboxylase (Okada *et al.*, 1977) and for choline acetyltransferase (Godfrey *et al.*, 1980; see also this volume, Chapter 20).

Table I assembles values from the literature for the three enzymatic activities shown in Fig. 7. The values observed in the tubercle for all three of these activities are quite high compared with values observed in other brain regions (see references for extensive comparisons). The range of absolute values reflects differences in conditions used in the preparation and assay of tissue, in addition to biological variability.

C. Metabolites

The chemical composition of the plexiform layer is in part derived from mitral and tufted neurons (Fig. 1). Chemical study of this layer can give information about these populations of bulb neurons. The distributions reported for the amino acids GABA, glutamate, and glycine in the tubercle (Krieger and Heller, 1979) and for cyclic AMP (see Section II) can be viewed in this context.

Recently we carried out studies on the tripeptide TRH (thyrotropin releas-

TABLE I Enzyme Activities for Rat Olfactory Tubercle

Enzyme	Units	Olfactory tubercle	Caudate putamen	Hippocampus	Reference
Glutamic acid decarboxylase	$\frac{\text{pmole}}{\text{h} \times \mu\text{g protein}}$	526	270	409	Tappaz *et al.* (1976)
	$\frac{\mu\text{mole}}{\text{h} \times \text{g wet wt}}$	27.5			Gilad and Reis (1979)
	$\frac{\mu\text{mole}}{\text{h} \times \text{g dry wt}}$	250			Fonnum *et al.* (1977)
	$\frac{\text{nmole}}{\text{h} \times \text{mg protein}}$	100			Krieger and Heller (1979)
Choline acetyltransferase	$\frac{\text{pmole}}{\text{h} \times \mu\text{g protein}}$	290	261	55	Hoover *et al.* (1978)
	$\frac{\text{pmole}}{\text{h} \times \mu\text{g protein}}$	60		19.5	Palkovitz *et al.* (1974)
	$\frac{\mu\text{mole}}{\text{h} \times \text{wet wt}}$	10.5			Gilad and Reis (1979)
	$\frac{\mu\text{mole}}{\text{h} \times \text{g dry wt}}$	84			Fonnum *et al.* (1977)
	$\frac{\text{nmole}}{\text{h} \times \text{mg protein}}$	65			Gordon and Krieger (unpublished 1980)
Dopamine-sensitive adenylate cyclase (100 μM dopamine)	$\frac{\text{pmole}}{\text{min} \times \text{mg wet wt}}$	9	18		Miller *et al.* (1974)
	$\frac{\text{pmole}}{\text{min} \times \text{mg wet wt}}$	4.2	24		Clement-Cormier *et al.* (1974)
	$\frac{\text{pmole}}{\text{min} \times \text{mg protein}}$	100			Krieger (1980)

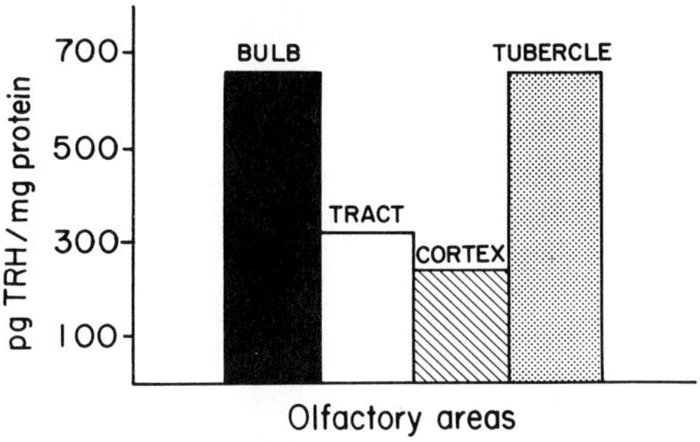

Fig. 8. TRH levels (immunoreactivity) in the rat olfactory areas: bulb, tract, cortex, and tubercle. (A. Winokur and N. R. Krieger, unpublished, 1980.)

ing hormone; pyroGlu-His-Pro NH_2 (Kreider *et al.*, 1981; Winokur and Krieger, unpublished 1980). Figure 8 shows the distribution for TRH in the olfactory areas of the rat brain. Assays were carried out by the radioimmunoassay method of Bassiri and Utiger (1972). We have shown that immunoreactivity corresponds with TRH concentration in the olfactory bulb (Kreider *et al.*, 1980). TRH levels in the tubercle are high compared with those in the olfactory tract and adjacent olfactory cortex, and they are comparable with the values in the bulb. Further experiments will establish whether this suggestive pattern reflects a genuine connection between tubercle and bulb TRH.

III. PHYSIOLOGY

Little is known about the function of the olfactory tubercle. Its limbic and olfactory connections are suggestive, but their significance remains elusive. A role in motor and stereotypic behavior has been proposed and partially explored (Pijnenburg *et al.*, 1976; Costall *et al.*, 1975, 1977; Tassin *et al.*, 1978). We have used the chemistry of the tubercle as an approach to study its physiology and as an approach to the physiology of the related bulb and midbrain regions.

A. Cyclic AMP Levels

Since the layers of the olfactory tubercle are rich in adenylate cyclase and dopamine-sensitive adenylate cyclase, measurement of cyclic AMP levels within these layers might constitute a basis for correlating neurochemical change with neuronal function. Both sensory input from the bulb and dopaminergic input from the midbrain might be expected to dictate or modulate changes in these levels.

We have carried out experiments that demonstrate the feasibility of such an approach, and we have measured increases in cyclic AMP in the various layers of the olfactory tubercle. We used the rapid synthesis of cyclic AMP that follows decapitation (Breckinridge, 1964) as a simple means to increase cyclic AMP levels (Schmidt *et al.*, 1972). In Fig. 9, the plotted values show cyclic AMP tissue levels as a function of time after decapitation. The level at zero time was obtained from assays carried out after killing by microwave irradiation (1300 W; 1.3 sec/100 g). This procedure inactivates the enzymes that synthesize and degrade cyclic AMP within a few seconds, thus blocking any rise in cyclic AMP levels (Schmidt *et al.*, 1972).

The curve shows that at the earliest measurable time point after decapitation (1 min) the increase in cyclic AMP was already maximal. The lesser values seen at later times probably reflect declining intracellular ATP stores. In the interval from 1–5 min after decapitation, cyclic AMP levels in the tubercle are essentially constant.

Figure 10 compares the cyclic AMP levels at successive depths of the olfactory tubercle at 3 min after decapitation (hatched bar) with the levels seen immediately following microwave killing (bar). Frozen sections (16 μm)

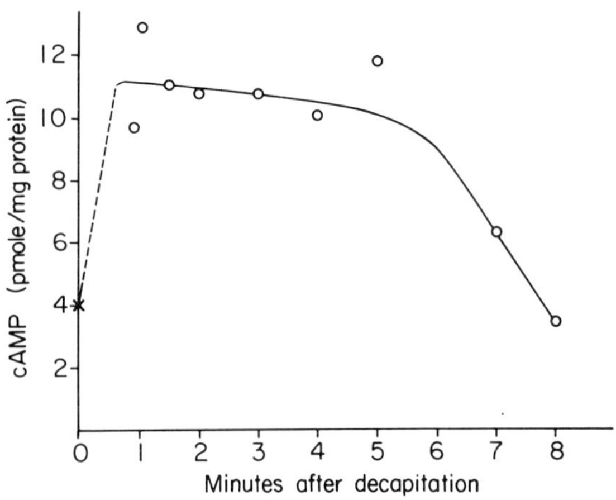

Fig. 9. Cyclic AMP levels in the olfactory tubercle as a function of time after decapitation. (O—O) Cyclic AMP levels after decapitation. The brain was rapidly removed, hemisected, and frozen on dry ice to stop enzymatic reactions. The frozen tubercle was excised and then boiled for 5 min (2 m*M* trismaleate; 1 m*M* isobutylmethylxanthine pH 7.4) to denature the enzymes. The tissue was homogenized and assayed for cyclic AMP (Brown *et al.*, 1971) and protein (Lowy *et al.*, 1951). Each point is the mean of two determinations from one tubercle. (X) cAMP levels after microwave irradiation. The plotted value is the mean of determinations from eight tubercles.

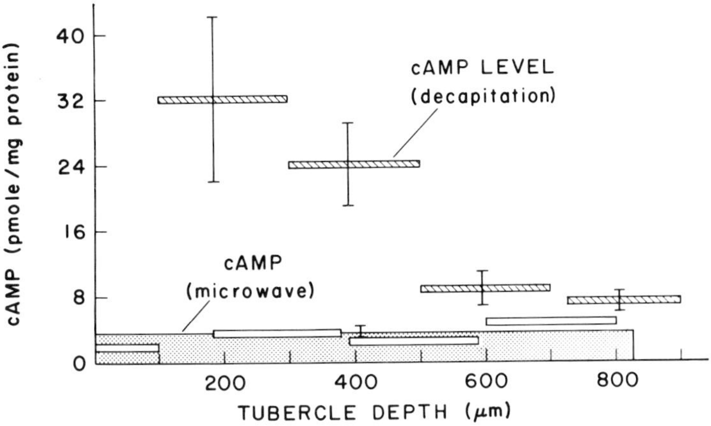

Fig. 10. Cyclic AMP levels as a function of depth in the olfactory tubercle. Hatched bar, cAMP levels after decapitation. Values are from four independent experiments and are means ± SEM of duplicate determinations. Unfilled bar, cAMP levels after microwave irradiation. Values are from a single tubercle. Dotted bar, cAMP levels after microwave irradiation. The indicated level is the mean ± SEM of eight separately assayed whole tubercles.

were cut as described in Section II, pooled into groups of 12, homogenized, and assayed for cyclic AMP and protein. The postdecapitation increases in cyclic AMP in the laminae of the olfactory tubercle are substantial in the plexiform and pyramidal layers, whereas in the polymorphic layer there is almost no change. This pattern parallels the distribution of the pyramidal cells (see the discussion far Fig. 3). We suggest that the increases reflect rising cyclic AMP concentrations within the pyramidal cell bodies and their dendrites.

B. Sprouting: A Model for Plasticity in the Central Nervous System

Dopamine fibers sprout in the olfactory tubercle following bulbectomy. Moore (1974) documented this sprouting by observations of formaldehyde-induced fluorescence. Reis and co-workers (Pickel *et al.*, 1977; Gilad and Reis, 1979a), using the PAP method of Sternberger (1974) to visualize tyrosine hydroxylase at the light microscope level, observed an increased density of dopamine terminals in the plexiform and pyramidal layers 7 days after bulbectomy. Activities of tyrosine hydroxylase, glutamic acid decarboxylase, and choline acetyltransferase increased by up to 30% in the 2 weeks following bulbectomy (Gilad and Reis, 1979b). Still needed is knowledge about the interconnections among the cell types of the tubercle and localizations for glutamic acid decarboxylase and choline acetyltransferase to identified cell types. Then changes in activity of these neurochemical markers can provide useful indices of function.

C. Steroid Targets in the Olfactory Tubercle

A possible role for the tubercle in expression of sexual behavior is suggested by the work of Pfaff and Keiner (1973). They mapped uptake sites for exogenous [^{3}H]estradiol using autoradiography of frozen sections and light microscopy. Their data show marked concentration of [^{3}H]estradiol by the granule cells of the olfactory tubercle. The laminar distributions of glutamic acid decarboxylase and choline acetyltransferase (Fig. 7) show that both activities are high at the depths at which the granule cells are situated. These enzymes may provide neurochemical markers for these neurons and thus facilitate the study of steroid responsitivity in the tubercle.

IV. PHARMACOLOGY

One of the strongest of the "dopamine hypotheses" of schizophrenia is the hypothesis that antipsychotic drugs produce their therapeutic effects by blocking dopamine receptors (Meltzer and Stahl, 1976). This possibility has contributed to the recent sustained interest in the study of these receptors. Because the olfactory tubercle is rich in both presynaptic autoreceptors for dopamine, (Aghajanian and Bunney, 1974; Carlsson, 1975) and the postsynaptic enzyme receptor, dopamine-sensitive adenylate cyclase (Clement-Cormier *et al.*, 1974; Miller *et al.*, 1974), and because of its unusually favorable cytoarchitecture the tubercle is a region of the mammalian brain tailored for the study of these receptors.

Kebabian and Calne (1979) distinguished, on biochemical and pharmacological grounds, between dopamine receptors of types I and II. Type I receptors are linked to the enzymatic activity of adenylate cyclase and provide the focus for discussion in this section. Type II receptors are not linked to adenylate cyclase, and may further be distinguished from type I by their 10 times greater responsivity to apomorphine (Carlsson, 1975; Carlsson *et al.*, 1976). This category includes the presynaptic autoreceptor.

Pyramidal Neurons in the Olfactory Tubercle: Possible Targets for the Actions of Antipsychotic Drugs

Although the dopaminergic fibers of the CNS have been described in elegant detail (Ungerstedt, 1971; Fuxe, 1965a,b; Björklund and Lindvall, 1978), the cells that receive these inputs have still to be identified. These cells have dopamine receptors. Dopamine-sensitive adenylate cyclase (DSAC) is a dopamine receptor (Kebabian *et al.*, 1972; Kebabian and Calne, 1979) and can therefore serve as a marker for these cells.

In Section II, we showed that the distribution of DSAC corresponds closely with the distribution of the pyramidal cells in the olfactory tubercle. Fig. 11A presents this distribution in more detail. Adenylate cyclase activity is plotted as a function of depth in the tubercle, both for the assay condition in the absence of dopamine and for that in the presence of dopamine. The activity in the absence of dopamine is distributed evenly over the various layers. When assayed in the presence of dopamine, the specific activity is 140 pmole/min/mg protein in the outer plexiform (0–250 μm) and 50 pmole/min/mg protein in the deepest part of the polymorphic layer (700–800 μm). Expressed in terms of percent activation by dopamine over the same interval, the activation declines from 140 to 50%.

Figure 11B illustrates the effect of the dopamine antagonist trifluoperazine (a phenothiazine) on the activation of the enzyme by dopamine. In the presence of 2 μM trifluoperazine, the activation at all depths is blocked by 70–90%. Control experiments on sections of the tubercle showed that trifluoperazine did not affect adenylate cyclase activity in the absence of added

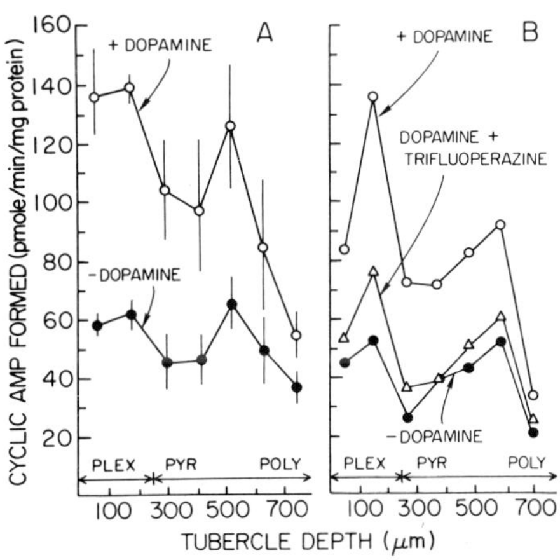

Fig. 11. Adenylate cyclase activity as a function of depth in the olfactory tubercle. (A) The plotted activities are means of specific activities ± SEM calculated at each depth from three independent experiments. In each experiment, each homogenate was assayed in triplicate, both in the absence (●—●) and presence of 100 μM dopamine (○—○). (B) Data are from a single tubercle. At each depth and for each condition homogenates were assayed in triplicate. The SEM is less than 10%. ●—●, no dopamine; ○—○, 100 μM dopamine; △—△, 100 μM dopamine plus 2 μM trifluoperazine. The depths of the midpoints of the assayed sections are plotted on the abscissa. The labels: PLEX, PYR, and POLY indicate the approximate locations of the plexiform, pyramidal, and polymorphic layers, respectively.

dopamine. The failure of propranolol to block the stimulation by dopamine (Krieger *et al.*, 1977) indicates that the dopamine is not acting through a β-receptor. Thus, in this *in vitro* system, dopamine-sensitive adenylate cyclase is a target for this antipsychotic drug. The K_i of $10^{-8}M$ calculated from these experimental data is consistent with the value reported by Clement-Cormier *et al.* (1974).

To prove a neuronal as opposed to a glial or axonal localization for dopamine-sensitive adenylate cyclase, we have used selective chemical lesions with kainic acid and 6-OH dopamine (Church *et al.*, 1979). Kainic acid has been shown to cause a severe loss of neural perikarya, while leaving neural processes of passage and glial cells intact (McGeer *et al.*, 1976; Schwarcz and Coyle, 1977). 6-OH Dopamine destroys those neurons that have high-affinity uptake systems for dopamine (Ungerstedt and Arbuthnott, 1970). Three days after injecting these agents into the olfactory tubercle, homogenates of the olfactory tubercle were assayed for DSAC (Kebabian *et al.*, 1972). The kainic acid lesion reduced the DSAC activity by 70–90% compared with values from sham lesioned or unlesioned controls. Homogenates of 6-OH dopamine-treated tubercles did not differ in DSAC activity from untreated controls. The lesions were assessed by light and fluorescence microscopy. Sections from kainic acid-treated tubercles showed extensive neuronal losses with increased numbers of glial cells. Sections from 6-OH dopamine-treated tubercles appeared normal under the light microscope. Examination of 6-OH dopamine-treated tissue by glyoxylic acid-induced histofluorescence established the loss of dopaminergic terminals. The marked decrease of DSAC accompanying the selective loss of neurons (kainic acid treatment) but not accompanying the loss of dopaminergic terminals (6-OH dopamine treatment) suggests that this enzyme occurs in the neurons and not in the glia or in the dopaminergic terminals of the region.

The hypothesis that pyramidal cells contain DSAC is consistent with the observed distribution of this activity. Their cell bodies are found in both the dense pyramidal layer and the deeper polymorphic layer (Fig. 3), and their dendrites are known from Golgi studies to be arborized both superficially and centrally (Calleja, 1893; Price, 1974; Heimer, 1978). This distribution of dendrites could account for the levels of dopamine-sensitive adenylate cyclase activity seen in the laminae of the tubercle (Fig. 11). Kuhar and coworkers (1978; Klemm *et al.*, 1979) have reported similar findings based on an autoradiographic approach.

These results and interpretations are consistent with evidence concerning the localization of the dopaminergic input that originates in the ventral tegmentum. The data of Fuxe (1965a,b) show a rich fluorescence for the dopamine axon terminals in the plexiform and pyramidal layers with decreases in the deeper layers. Similarly, stains for degenerating axon terminals after lesions of the ventral tegmentum in the brainstem show the greatest localization of silver grains in the plexiform and pyramidal cell body layers (Hedreen and Chalmers, 1972; Maler *et al.*, 1973). The islands of Calleja tend not to fluoresce for dopamine (Fuxe, 1965a) or to show the presence of degenerating terminals (Hedreen and Chalmers, 1972).

If we consider the likely possibility that the same pyramidal population discussed above is also the target for axons from the olfactory bulb, as has been established for the pyramidal cell population of the prepyriform cortex (Shepherd, 1979), their importance to the synaptic organization of the olfactory tubercle becomes apparent, and they emerge as specific candidates for the neurochemical approach to the physiology of the bulb and midbrain advanced in Section III.

V. PATHOLOGY

The anterior perforated substance is the region of human brain that corresponds to the olfactory tubercle of lower mammals (Calleja, 1893; Beccari, 1911; Cajal, 1911, 1955; Crosby and Humphrey, 1941; Humphrey, 1967; Takimoto *et al.*, 1962; Olson *et al.*, 1973a,b; Björklund and Lindvall, 1978). The perforated appearance of its surface is the result of numerous entering blood vessels. It is a rhomboid-shaped region bounded by the olfactory trigone and the optic tract (Fig. 12a). The diagonal band of Broca forms its immediate posterior border. The slight elevation of its rostral portion is said

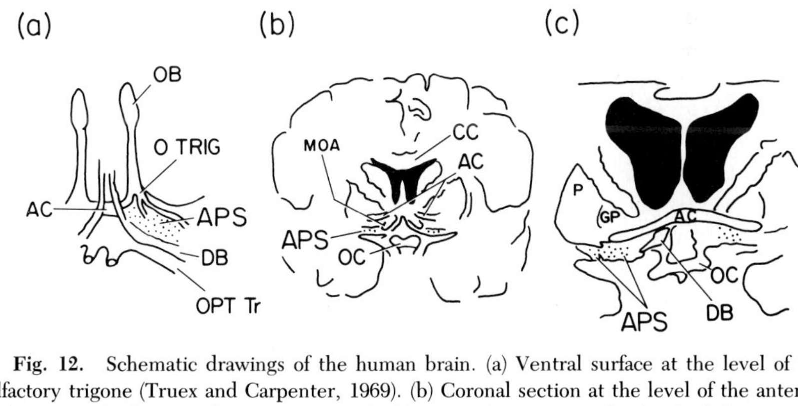

Fig. 12. Schematic drawings of the human brain. (a) Ventral surface at the level of the olfactory trigone (Truex and Carpenter, 1969). (b) Coronal section at the level of the anterior commissure (Roberts and Haneway, 1971). (c) Coronal section at the level of the anterior commissure (De Armond *et al.*, 1974). AC, anterior commissure; OB, olfactory bulb; O TRIG, olfactory trigone; DB, diagonal band; OPT Tr, optic tract; MOA, medial olfactory area; CC, corpus collossum; OC, optic chiasm; P, putamen; GP, globus palidus; APS, anterior perforated substance.

to be a rudiment of the tubercle (Truex and Carpenter, 1969). In coronal sections (Figs. 12b,c), it lies ventral to the anterior commissure.

The plexiform, pyramidal, and polymorphic layers of this region of the human brain were first described by Beccari (1911) and Cajal (1911). The organization of the domains, layers, and cell types in the human anterior perforated substance are not as well-defined as in the rat olfactory tubercle. Nevertheless, the presence of the same neurochemicals and transmitter-related enzyme systems that have been described for the rat, including those for dopamine, GABA, and acetylcholine, compel the attention of the neurochemist.

The neurochemistry of the anterior perforated substance is of particular consequence to neuropathology for the comparisons it affords with the caudate nucleus and the putamen. These regions constitute anatomically and chemically related nuclei that are of established importance in the pathology of such illnesses as Huntington's and Parkinson's diseases (Klawans, 1973). Heimer (1978) has drawn this comparison for the monkey and for the rat with careful detail at the anatomical level. Our neurochemical findings (Krieger, 1980) and those of others (Tables I and II) support this comparison.

The work of Hornykiewicz (Ehringer and Hornykiewicz, 1960), McGeer and McGeer (1976), Bird and Iversen (1974), and others has demonstrated specific chemical deficits in brain autopsy tissue from victims of various diseases compared with autopsy tissue from normals. Most striking have been the deficits of dopamine observed in the caudate, putamen, and substantia nigra after deaths due to Parkinson's Disease (Ehringer and Hornykiewicz, 1960; Bernheimer *et al.*, 1973) and of GABA observed in the caudate and putamen following deaths related to Huntington's Disease (Bird and Iversen, 1974; McGeer and McGeer, 1976).

Table II assembles values from the literature for enzyme activities and metabolite levels that have been assayed in homogenates prepared from the anterior perforated substance. The table compares autopsy tissue from normal deaths and diseased states. It demonstrates that for normal brain these enzyme activities and metabolite levels (e.g., those related to dopamine, GABA, and acetylcholine metabolism) are high in the anterior perforated substance compared with elsewhere in the brain. It also documents one neurochemical difference between pathological and normal tissue, (i.e, for homovanillic acid).

All the neurochemicals listed in the table are extremely vulnerable to the catabolic processes accompanying tissue death. Losses from degradation of the enzymes by proteolysis are of prime concern. For example, in the study of McGeer and McGeer (1974), the activities of tyrosine hydroxylase, choline acetyltransferase, and glutamic acid decarboxylase were all substantially decreased, 52, 35, and 40%, respectively in the first 10 postmortem hr. The

TABLE II Neurochemical Data for Human Autopsy Tissue[a]

Neurochemical	Units	Anterior perforated substance			Caudate			Hippocampus			Reference[c]
		Normal	PD	HD	Normal	PD	HD	Normal	PD	HD	
Serotonin	μg/g wet weight	0.86	—	—	0.35	—	—	0.10			Farley *et al.* (1977)
		93%			0.65			30%			Mackay *et al.* (1978b)
Norepinephrine	μg/g wet weight	0.37	—	—	0.12	—	—	0.024			Farley and Hornykiewicz (1977)
		0.5			0			0			Mackay *et al.* (1978b)
Dopamine	μg/g wet weight	0.61	0.5	—	4.50	1.5	—	0.030			Farley *et al.* (1977)
		29%	—	—	0.42	—	—	0%			Mackay *et al.* (1978b)
Homovanillic acid	μg/g wet weight	4.43	1.5	—	6.08	2.3	—	—			Price *et al.* (1978)
		116%	—	—	1.74	—	—	9%			Mackay *et al.* (1978b)
Glutamic acid decarboxylase	μmole/h × 100 mg protein	124%	—	—	4.51	2.98	1.74	66%	[b]	[b]	McGeer and McGeer (1976)
	μmole/h × g wet weight	70%			0.71	—	—	50%	—	—	Mackay *et al.* (1978a)
Choline acetyltransferase	μmole/h × 100 mg protein	44%	—	—	10.71	12.0	4.08	8.1%	[b]	[b]	McGeer and McGeer (1976)
	μmole-h × g wet weight	92%			4.18	—	—	18%	—	—	Mackay *et al.* (1978a)

(*continued*)

TABLE II (*Continued*)

Neurochemical	Units	Anterior perforated substance			Caudate			Hippocampus			Reference[c]
		Normal	PD	HD	Normal	PD	HD	Normal	PD	HD	
Acetylcholine esterase	mmole/h × 100mg protein	61%	—	—	1.54	[b]	[b]	14%	[b]	[b]	McGeer and McGeer (1976)
	mmole/h × g wet weight	52%	—	—	1.74	—	—	24%	—	—	Mackay *et al.* (1978a)
Tyrosine hydroxylase	nmole/h × 100mg protein	85%	—	—	33.6	9.2	35.5	8.2	3.2	5.4	McGeer and McGeer (1976)
	nmole/h × wet weight	5%	—	—	90.7	—	—	6%	—	—	Mackay *et al.* (1978a)
Dopa decarboxylase	nmole/h × 100mg protein	70%	—	—	153	—	—	—	—	—	McGeer and McGeer (1976)
	nmole/h × g wet weight	9%	—	—	682	—	—	0.7%	—	—	Mackay *et al.* (1978a)

[a] Values that appear in the table as a percentage are to be compared with the corresponding value for the normal caudate. —, Not determined; PD, Parkinson's Disease; HD, Huntington's Disease.

[b] Observed value was not statistically different from the value for the normal autopsy tissue from the region. —, Not determined.

[c] In the studies of Farley *et al.* (1977) and Price *et al.* (1978), the anterior perforated substance was assayed as part of the larger anatomical region: lateral olfactory area.

freezing and thawing attendant upon storage at low temperatures is another serious source of inactivation. Although these technical problems have received considerable attention, it seems likley, considering their nature, that they will not be solved quickly.

The separated territories of the pyramidal, granule, and polymorphic neurons of the anterior perforated substance permit the neurochemical study of each with minimal interference from the others. The heterogeneous cell types of the caudate and the putamen (Kemp and Powell, 1971; Fox *et al.*, 1972; Pasik *et al.*, 1976) have precluded such an approach. The promise from such studies over the long term is the correlation with specified disease states of specific neurochemical differences in morphologically identified neurons.

ACKNOWLEDGMENT

The author thanks Mr. John Megill, Dr. Allen Church, Mrs. Harriet Abriss, and Dr. Solomon Erulkar for their help and their suggestions.

REFERENCES

Aghajanian, G. K., and Bunney, G. S. (1974). *In* "Frontiers in Neurology and Neuroscience Research" (P. Seeman and G. M. Brown, eds.), pp. 4–11. Univ. of Toronto Press, Toronto.

Albers, R. W., and Brady, R. O. (1959). The distribution of glutamate decarboxylase in the nervous system of the Rhesus monkey. *J. Biol. Chem.* **234**, 926–928.

Anderson, C. A., and Westrum, L. E. (1972). An electron microscopic study of the normal synaptic relationships and early degenerative changes in the rat olfactory tubercle. *Z. Zellforsch. Mikrosk. Anat.* **127**, 462–482.

Bassiri, R. M., and Utiger, R. D. (1972). The preparation and specificity of antibody to thyrotropin releasing hormone. *Endocrinology* **90**, 722–727.

Beccari, N. (1910). Il lobo paraolfattorio nei mammiferi. *Arch. Ital. Anat. Embriol.* **9**, 173–220.

Beccari, N. (1911). La sostanza perforata anteriore e i suoi rapporti col rinencefalon el cervello dell'uomo. *Arch. Ital. Anat. Embriol.* **10**, 261–328.

Bernheimer, H., Birkmayer, W., Hornykiewicz, O., Jellinger, K., and Seitelberger, F. (1973). Brain dopamine and the syndromes of Parkinson and Huntington. *J. Neurol. Sci.* **20**, 415–455.

Bird, E. D., and Iversen, L. L. (1974). Huntington's Chorea post mortem measurement of glutamic acid decarboxylase, choline acetyltransferase and dopamine in basal ganglia. *Brain* **97**, 457–472.

Björklund, A., and Lindvall, O. (1978). The meso-telencephalic dopamine neuron system. *In* "Limbic Mechanisms" (K. E. Livingston and O. Hornykiewicz, eds.), pp. 307–331. Plenum, New York.

Breckinridge, B. McL. (1964). The measurement of cyclic adenylate in tissues. *Proc. Natl. Acad. Sci. U.S.A.* **52**, 1580–1586.

Brown, B. L., Albano, J. D. M., Ekins, R. P., and Sgherzi, A. M. (1971). A simple and sensitive saturation assay method for the measurement of adenosine 3′, 5′-cyclic monophosphate. *Biochem. J.* **121**, 561–562.

Cajal, S. R. (1977). "Histologie du Systeme Nerveux," Vol. 11. Maloine, Paris.

Cajal, S. R. (1955). "Studies on the Cerebral Cortex" (transl. by L. M. Kraft), pp. 70–75. Lloyd-Luke, London.

Calleja, C. (1893). "La Region Olfactoria Del Cerebro." Madrid.

Carlsson, A. (1975). *In* "Chemical Tools in Catecholamine Research" (O. Almgren, Carlsson, and J. Engle, eds.), Vol. 7, pp. 219–225. North-Holland Publ., Amsterdam.

Carlsson, A., Kehn, W., and Lindquist, M. (1976). *In* "Advances in Parkinsonism, Biochemistry, Physiology, Treatment" (W. Birkmayer and O. Hornykiewicz, eds.). Editions Roche, Basel.

Church, A. C., Bunney, B. S., and Krieger, N. R. (1979). Localization of dopamine-sensitive adenylate cyclase in the rat olfactory tubercle. *Neurosci. Abstr.* **5**, 586.

Clement-Cormier, Y. C., Kebabian, J. W., Petzold, G. L., and Greengard, P. (1974). Dopamine-sensitive adenylate cyclase in mammalian brain: a possible site of action of antipsychotic drugs. *Proc. Natl. Acad. Sci. U.S.A.* **71**, 1113–1117.

Costall, B., Naylor, R. J., and Neumeyer, J. L. (1975). Differences in the nature of the stereotyped behavior induced by apomorphine derivatives in the rat and in their actions in extrapyramidal and mesolimbic brain areas. *Eur. J. Pharmacol.* **31**, 1–16.

Costall, C., Marsden, D., Naylor, R. J., and Pycock, C. (1977). Stereotyped behavior patterns and hyperactivity induced by amphetamine and apomorphine after discrete 6-hydroxydopamine lesions of extrapyramidal and mesolimbic nuclei. *Brain Res.* **123**, 89–111.

Crosby, E. C., and Humphrey, T. (1941). Studies of the vertebrate telencephalon. *J. Comp. Neurol.* **74**, 309–352.

De Armond, S. J., Fusco, M. M., and Dewey, M. M. (1974). "Structure of the Human Brain," p. 100. Oxford Univ. Press, London and New York.

De Olmos, J. S. (1972). The amygdaloid projection field in the rat as studied with the cupric-silver method. *In* "The Neurobiology of the Amygdala" (B. E. Eleftherion, ed.), pp. 145–204. Plenum, New York.

Ehringer, H., and Hornykiewicz, O. (1960). Verteilung von noradrenalin und dopamin (3-hydroxytyramin) in gehirn des menschen and ihr verhatten bei erkrankungen des extrapyramidalen systems. *Klin. Wochenschr.* **38**, 1236–1239.

Farley, I. J., and Hornykiewicz, O. (1977). Noradrenaline distribution in subcortical areas of the human brain. *Brain Res.* **126**, 53–62.

Farley, I. J., Price, K. S., and Hornykiewicz, O. (1977). Dopamine in the limbic regions of the human brain: Normal and abnormal. *Adv. Biochem. Psychopharmacol.* **16**, 57–64.

Fonnum, F., Walaas, I., and Iversen, E. (1977). Localization of GABAergic, cholinergic and aminergic structures in the mesolimbic system. *J. Neurochem.* **29**, 221–230.

Fox, C. A., Andrade, A. N., Schwyn, R. C., and Rafols, J. A. (1972). The aspiny neurons and the glia in the primate striatum: a Golgi and electron microscopic Study. *J. Hirnforsch.* **13**, 341–362.

Fuxe, K. (1965a). The distribution of monoamine terminals in the central nervous system. *Acta Physiol. Scand., Suppl.* **247**, 37–85.

Fuxe, K. (1965b). Evidence for the existence of monoamine neurons in the central nervous system. *Z. Zellforsch. Mikrosk. Anat.* **65**, 573–596.

Gilad, G. M., and Reis, J. D. (1979a). Collateral sprouting in central mesolimbic dopamine neurons: biochemical and immunocytochemical evidence of changes in the activity and distribution of tyrosine hydroxylase in terminal fields and in cell bodies of A_{10} neurons. *Brain Res.* **160**, 17–36.

Gilad, G. M., and Reis, J. D. (1979b). Transneuronal effects of olfactory bulb removal on choline acetyltransferase and glutamic acid decarboxylase activities in the olfactory tubercle. *Brain Res.* **178**, 185–190.

Godfrey, D. A., Ross, C. D., Herrmann, A. D., and Matschinsky, F. M. (1980). Distribution and derivation of cholinergic elements in the rat olfactory bulb. *Neuroscience* **5**, 273–292.

Hedreen, J. C., and Chalmers, J. P. (1972). Neuronal degeneration in rat brain induced by 6-hydroxydopamine: a histological and biochemical study. *Brain Res.* **47**, 1–36.

Heimer, L. (1972). The Olfactory connections of the diencephalon: an experimental light-and-electron microscopic study with special emphasis on the problem of terminal degeneration. *Brain Behav. Evol.* **6**, 484–523.

Heimer, L. (1978). The olfactory cortex and the ventral striatum. *In* "Limbic Mechanisms" (K. E. Livingston and O. Hornykiewicz, eds.), pp. 95–187. Plenum, New York.

Hoover, D. B., Muth, E. A., and Jacobowitz, D. M. (1978). A mapping of the distribution of acetylcholine, choline acetyltransferase and acetylcholinesterase in discrete areas of rat brain. *Brain Res.* **153**, 295–306.

Hosoya, Y. (1973). Electron microscopic observations of the granule cells (Calleja's island) in the olfactory tubercle of rats. *Brain Res.* **54**, 330–334.

Hosoya, Y., and Hirata, Y. (1974). The fine structure of the "dwarf-cell cap" of the olfactory tubercle in the rat's brain. *Arch. Histol. Jpn.* **36**, 407–423.

Humphrey, T. (1967). The development of the human tuberculum olfactorium during the first three months of embryonic life. *J. Hirnforsch.* **9**, 437–469.

Kebabian, J. W., and Calne, D. B. (1979). Multiple receptors for dopamine. *Nature (London)* **277**, 93–96.

Kebabian, J. W., Petzold, G. L., and Greengard, P. (1972). Dopamine-sensitive adenylate cyclase in caudate nucleus of rat brain and its similarity to the "dopamine receptor." *Proc. Natl. Acad. Sci. U.S.A.* **69**, 2145–2149.

Kemp, J. M., and Powell, T. P. (1971). The structure of the caudate nucleus of the cat. *Philos. Trans. R. Soc. London, Ser. B* **262**, 383–401.

Klawans, H. L., Jr. (1973). "The Pharmacology of Extrapyramidal Movement Disorders." Karger, Basel.

Klemm, N., Murrin, L. C., and Kuhar, M. J. (1979). Neuroleptic and dopamine receptors: autoradiographic localization of (^{3}H) spiperone in rat brain. *Brain Res.* **169**, 1–9.

König, J. F. R., and Klippel, R. A. (1963). "The Rat Brain. A Stereotaxic Atlas of the Forebrain and Lower Parts of the Brain Stem." Williams & Wilkins, Baltimore, Maryland.

Kreider, M. R., Winokur, A., and Krieger, N. R. (1981). *Brain Res.* (in press).

Krieger, N. R. (1980). Localization of dopamine-sensitive adenylate cyclase within the rat olfactory tubercle. *Brain Res.* **183**, 383–391.

Krieger, N. R., and Heller, J. S. (1979). Localization of glutamic acid decarboxylase within laminae of the rat olfactory tubercle. *J. Neurochem.* **33**, 299–302.

Krieger, N. R., Kauer, J. S., Shepherd, G. M., and Greengard, P. (1977). Dopamine-sensitive adenylate cyclase within laminae of the olfactory tubercle. *Brain Res.* **131**, 303–312.

Kuhar, M. J., Murrin, L. C., Malout, A. T., and Klemmn, N. (1978). Dopamine receptor binding *in vivo:* the feasibility of autoradiographic studies. *Life Sci.* **22**, 203–210.

Leonard, C. M. (1969). The prefrontal cortex of the rat. I. Cortical projections of the mediodorsal nucleus. II. Efferent connections. *Brain Res.* **12**, 321–343.

Lowry, O. H., Rosebrough, N. J., Farr, A. L., and Randall, R. J. (1951). Protein measurement with the Folin phenol reagent. *J. Biol. Chem.* **193**, 265–275.

McGeer, E. G., Innanen, V. T., and McGeer, P. L. (1976). Evidence on the cellular localization of adenyl cyclase in the neostriatum. *Brain Res.* **118**, 356–358.

McGeer, P. L., and McGeer, E. G. (1976). Enzymes associated with the metabolism of catecholamines, acetylcholine, and GABA in human controls and patients with Parkinson's Disease and Huntington's Chorea. *J. Neurochem.* **26**, 65–76.

Mackay, A. V. P., Davis, P., Dewar, A. T., and Yates, C. M. (1978a). Regional distribution of enzymes associated with neurotransmission by monoamines, acetylcholine, and GABA in the human brain. *J. Neurochem.* **30**, 827–839.

Mackay, A. V. P., Yates, C. M., Wright, A., Hamilton, P., and David, P. (1978b). Regional distribution of monoamines and their metabolites in the human brain. *J. Neurochem.* **30**, 841–848.

Maler, L., Fibiger, H. C., and McGeer, P. L. (1973). Demonstration of the nigrostriatal projection by silver staining after nigral injections of 6-hydroxy-dopamine. *Exp. Neurol.* **40**, 505–515.

Meltzer, H. Y., and Stahl, S. M. (1976). The dopamine hypothesis of schizophrenia: a review. *Schizophr. Bull.* **2**, 19–76.

Miller, R. J., Horn, A. S., and Iversen, L. L. (1974). The action of neuroleptic drugs on dopamine-stimulated adenosine cyclic 3′,5′-monophosphate production in rat neostriatum and limbic forebrain. *Mol. Pharmacol.* **10**, 759–766.

Moore, R. Y. (1974). *In* "Dynamics of Regeneration and Growth in Neurons" (K. Fuxe, K. Olson, and Y. Zotterman, eds.), p. 381. Pergamon, New York.

Okada, Y., Hosoya, Y., and Kurosawa, F. (1977). High concentration of γ-aminobutyric acid (GABA) and glutamate decarboxylase (GAD) activity in the regions of medial forebrain bundle, diagonal band of Broca, and the third layer of the olfactory tubercle. *Proc. Jpn. Acad., Ser. B.* **53**, 236–240.

Olson, L., Boreus, O., and Seiger, A. (1973a). Histochemical demonstration and mapping of 5-hydroxytryptamine-and catecholamine-containing neuron systems in the human fetal brain. *Z. Anat. Entwicklungsgesch.* **139**, 259–282.

Olson, L., Nystrom, B., and Seiger, A. (1973b). Monoamine fluorescence histochemistry of human post mortem brain. *Brain Res.* **63**, 231–247.

Palkovitz, M., Saavedra, J., Kobayashi, R. M., and Brownstein, M. (1974). Choline acetyltransferase content of limbic nuclei of the rat. *Brain Res.* **79**, 443–450.

Pasik, P., Pasik, T., and DiFiglia, M. (1976). Quantitative aspects of neuronal organization in the neostriatum of the Macaque monkey. *In* "The Basal Ganglion" (M. D. Yahr, ed.), pp. 57–90. Raven, New York.

Pfaff, D., and Keiner, M. (1973). Atlas of estradiol-concentrating cells in the central nervous system of the female rat. *J. Comp. Neurol.* **151**, 121–158.

Pickel, V. M., Joh, T. H., and Reis, D. J. (1977). Regional and ultrastructural localization of tyrosine hydroxylase by immunocytochemistry in dopaminergic neurons of the mesolimbic and nigroneostriatal systems. *Adv. Biochem. Psychopharmacol.* **16**, 321–329.

Pijnenburg, A. J., Honig, W. M., Van der Heydn, J. A., and Van Rossum, J. M. (1976). Effects of chemical stimulation of the mesolimbic dopamine system upon locomotor activity. *Eur. J. Pharmacol.* **35**, 45–58.

Price, J. L. (1973). An autoradiographic study of complementary laminar patterns of termination of afferent fibers of the olfactory cortex. *J. Comp. Neurol.* **150**, 87–108.

Price, K. S., Farley, I. J., and Hornykiewicz, O. (1978). Neurochemistry of Parkinson's Disease: relation between striatal and limbic dopamine. *Adv. Biochem. Psychopharmacol.* **19**, 293–300.

Roberts, M., and Hanaway, J. (1971). "Atlas of the Human Brain," p. 9. Lea & Febiger, Philadelphia, Pennsylvania.

Schmidt, M. J., Schmidt, D. E., and Robison, G. A. (1972). Cyclic AMP in the rat brain: microwave irradiation as a means of tissue fixation. *Adv. Cyclic Nucleotide Res.* **1**, 425–434.

Schrier, B. K., and Shuster, L. (1967). A simplified radiochemical assay for choline acetyltransferase. *J. Neurochem.* **14**, 977–985.

Schwarcz, R., and Coyle, J. T. (1977). Striatal lesions with kainic acid: neurochemical characteristics. *Brain Res.* **127**, 235–249.

Scott, J. W., and Chafin, B. R. (1975). Origin of olfactory projections to lateral hypothalamus and nuclei gemini of the rat. *Brain Res.* **88**, 64–68.

Shepherd, G. M. (1979). "The Synaptic Organization of the Brain." Oxford Univ. Press, London and New York.

Sternberger, L. A. (1974). "Immunocytochemistry." Prentice-Hall, Englewood Cliffs, New Jersey.

Takimoto, T., Sakamoto, E., and Mitsui, Y. (1962). A cytoarchitectural study of the anterior perforated substance and its neighboring structures in man. *Tokushima J. Exp. Med.* **9**, 8–23.

Tappaz, M. L., Brownstein, M. J., and Palkovits, M. (1976). Distribution of glutamate decarboxylase in discrete brain nuclei. *Brain Res.* **108**, 371–379.

Tassin, J., Stinus, L., Simon, H., Blanc, G., Thierry, A., Moal, M., Cardo, B., and Glowinski, J. (1978). Relationship between the locomotor hyperactivity induced by A_{10} lesions and the destruction of the fronto-cortical dopaminergic innervation in the rat. *Brain Res.* **141**, 267–281.

Truex, R. C., and Carpenter, M. B. (1969). "Human Neuroanatomy," 6th ed., pp. 519–522. Williams & Wilkins, Baltimore, Maryland.

Ungerstedt, U. (1971). Sterotaxic mapping of the monoamine pathways in the rat brain. *Acta Physiol. Scand., Suppl.* **367**, 1–48.

Ungerstedt, U., and Arbuthnott, G. W. (1970). Quantitative recording of rotational behavior in rats after 6-hydroxy-dopamine lesions of the nigrostriatal dopamine system. *Brain Res.* **24**, 485–493.

Valverde-Garcia, F. (1965). "Studies on the Piriform Lobe." Harvard Univ. Press, Cambridge, Massachusetts.

Van Hoesen, G. W., Meuslam, M. M., and Haaxma, R. (1976). Temporal cortical projections to the olfactory tubercle in the rhesus monkey. *Brain Res.* **109**, 375–381.

20

Quantitative Histochemistry of Gustatory and Olfactory Cholinergic Pathways

FRANZ M. MATSCHINSKY, DONALD A. GODFREY, C. DAVID ROSS, AND WILLIAM T. NORFLEET

I. INTRODUCTION

Cholinergic mechanisms seem to be involved in the early processing of signals in sensory systems (Godfrey *et al.*, 1980). In this chapter results of quantitative histochemical measurements of choline acetyltransferase and acetylcholinesterase in the three well-known types of taste papillae of the rat tongue and in the laminar layers of the olfactory bulb of the rat are reported. In the case of the olfactory system, the effect of surgical lesions behind the bulb on the enzyme distributions was also investigated.

Biochemistry of Taste and Olfaction

ISBN 0-12-154450-8

II. QUANTITATIVE HISTOCHEMICAL METHODOLOGY

The quantitative histochemical methodology has been previously described in much detail (Lowry and Passonneau, 1972; Godfrey and Matschinsky, 1976; Matschinsky *et al.*, 1968). Of particular importance in the present study was the application of a new approach to three-dimensional mapping of quantitative histochemical measurements (Godfrey and Matschinsky, 1976). This procedure for localizing quantitative histochemical measurements makes possible a permanent, objective record of the location of each sample and makes the investigation of complex CNS structures feasible.

The tissue is fixed by freezing in Freon-12 cooled to −150°C with liquid nitrogen. Tissue blocks are cut in a cryostat at −20°C; the section thickness is between 12–20 μm. Sections are either freeze-dried or mounted on glass slides, to be stained with thionine or for acetylcholinesterase. Dissection of identified samples from freeze-dried tissue was done with the microtools developed and described by Lowry. A drawing attachment on a dissecting microscope (Wild or Zeiss) permitted precise mapping of the location of all samples as previously described. Samples were weighed on quartz fiber microbalances with sample weights ranging between 25–500 ng.

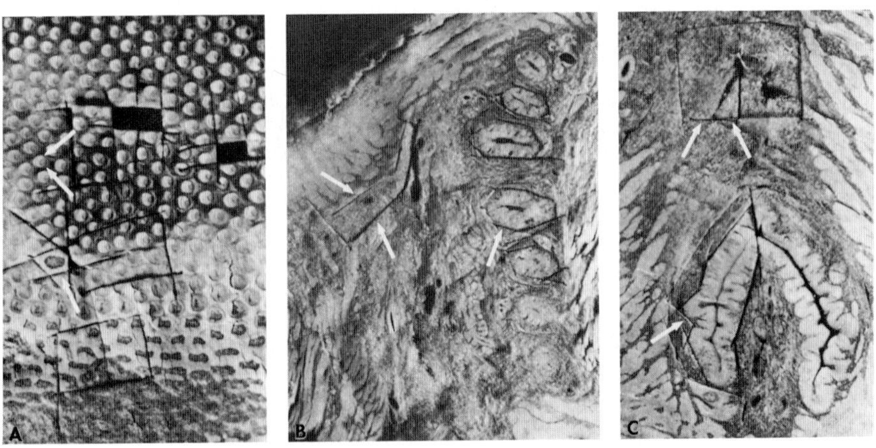

Fig. 1. Unstained freeze dried sections of rat tongue taste bud material. (A) Cross section of a fungiform papilla still in place but separated from the surrounding epithelium by several knife cuts (single arrow). Cross sections of numerous filiform papillae can be recognized (double arrow). (B) Section showing five foliate papillae, still in place but separated from the surrounding epithelium by several knife cuts (single arrow). Reference tissue (two arrows). (C) Section of circumvallate papilla with dissecting marks (single arrow). The outline of the control samples is also clearly visible (two arrows).

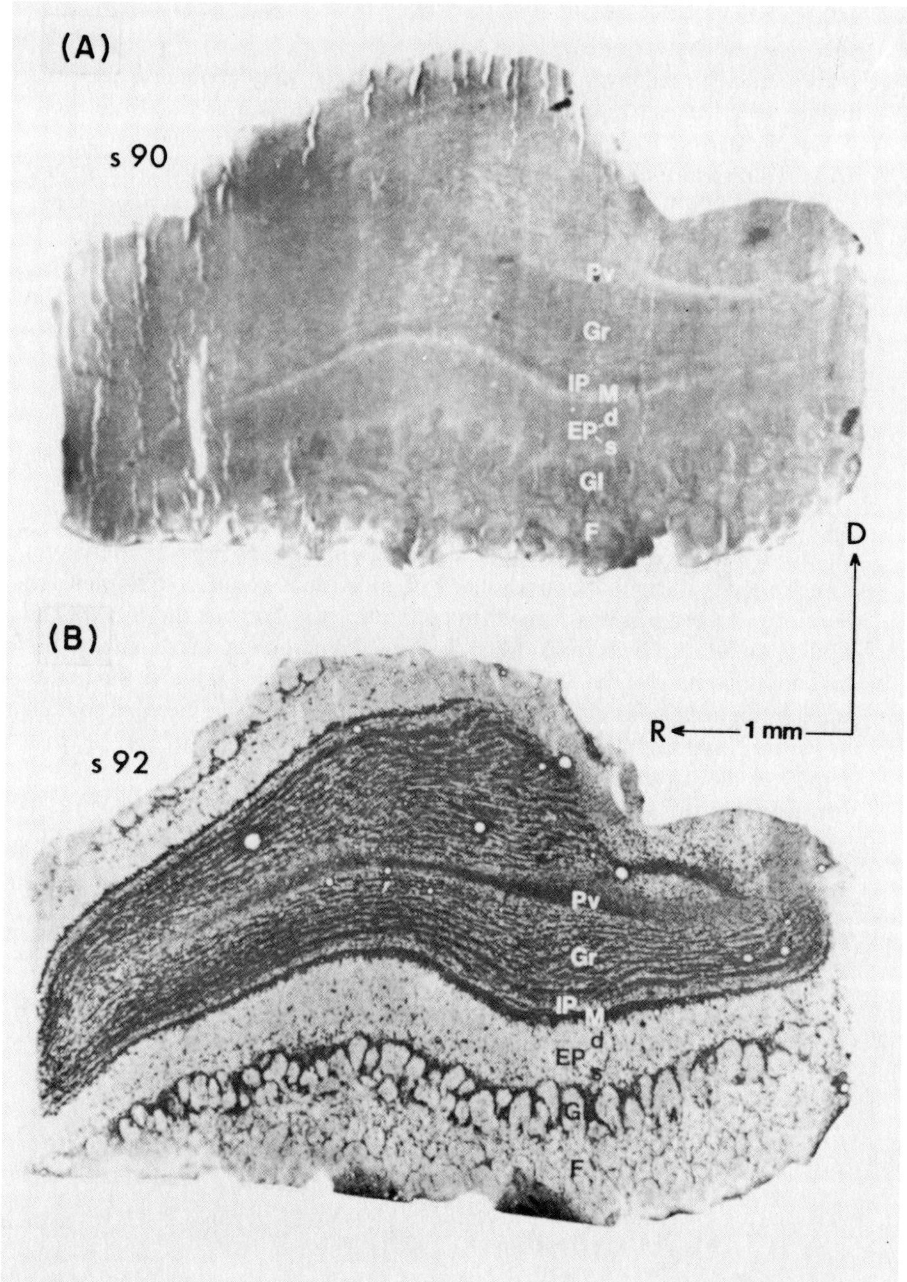

Fig. 2. Comparison of unstained freeze dried (s 90, part A) and thionin stained frozen (s 92, part B) sections through the rat olfactory bulb. The sections are cut in a sagittal plane and are nominally 20-μm thick. Abbreviations of layers are: F, olfactory nerve fiber layer; Gl, glomerular layer; EP, external plexiform layer (subscripts s and d refer to superficial and deep portions of the layer); M, mitral cell body layer; IP, internal plexiform layer; Gr, granular layer; PV, periventricular layer. (Modified from Godfrey *et al.*, 1980.)

The assay procedures used for quantitating acetylcholinesterase and choline acetyltransferase are both radiometric methods modified from those of McCaman and Hunt (1965; Ross and McDougal, 1976; Godfrey *et al.*, 1977). To measure choline acetyltransferase, tissue samples were incubated with [1−^{14}C]acetyl–CoA and choline. The [^{14}C]acetylcholine produced was extracted into 3-heptanone containing tetraphenylboron. To measure acetylcholinesterase activity, samples were incubated with [1 −^{14}C]acetylcholine iodide. The acetate produced was extracted at low pH into ethyl acetate. The amount of ^{14}C labeled product was quantified by liquid scintillation counting.

A. Sampling of Taste Papillae and Olfactory Bulb Structures

The three types of taste papillae were easily recognized in freeze dried sections of the rat tongue (Figs. 1A–1C). In sections placed tangentially to the surface of the tongue, the fungiform papillae present as cross sections and dissecting results in horizontal slices of individual papillae. Samples are therefore heterogeneous, derived from the tip, the body or the base of the papilla. An attempt was made to collect equal amounts of material from the tip, body, and base of the buds. In samples of the circumvallate or the foliate papillae the taste buds are sectioned longitudinally and the samples contain the entire length of the cells. For comparison, tongue epithelium in close vicinity of the papillae was dissected.

In freeze dried sections of the olfactory bulb the various layers were in general also easily recognized (Fig. 2). Boundaries and structures not readily

TABLE I Enzymes Involved in Cholinergic Innervation of Taste Papillae of Rat Tongue

Papillae	Source of Tissue	Choline acetyl transferase (μmole) kg dry wt/min	Acetylcholinesterase (mmole) kg dry wt/min
Circumvallate	Papillae with buds (n = 7)	22.3 ± 3.7	10.5 ± 1.3
	Papillae without buds (n = 5)	1.0 ± 0.4	2.1 ± 0.3
	Nearby tissue (n = 2)	14.3, 22.2	6.2, 11.2
	Distant tissue (n = 1 or 2)	1.3, 8.8	1.0, —
Foliate	Papillae with buds (n = 6)	40.8 ± 6.2	22.4 ± 1.3
	Papillae without buds (n = 5)	−0.2 ± 1.2	3.5 ± 0.3
	Nearby tissue (n = 1 or 2)	7.6, —	14.6, 15.8
	Distant tissue (n = 1 or 2)	1.1, —	6.8, 1.3
Fungiform	Papillae with buds (n = 2)	0.0, 3.9	1.9, 2.7
	Nearby epithelium (n = 2)	0.6, 1.2	0.5, 0.3

discerned could be identified by reference to the nearby thionine-stained sections.

B. Enzyme Distribution in Taste Papillae of the Rat Tongue

Activities of the two enzymes involved in acetylcholine synthesis and degradation were detected in the circumvallate and foliate papillae but were absent from the fungiform papillae and from epithelial samples free of taste buds (Table I). The activities found in the taste bud material are of the same order of magnitude as reported for the cholinergic systems of Islets of

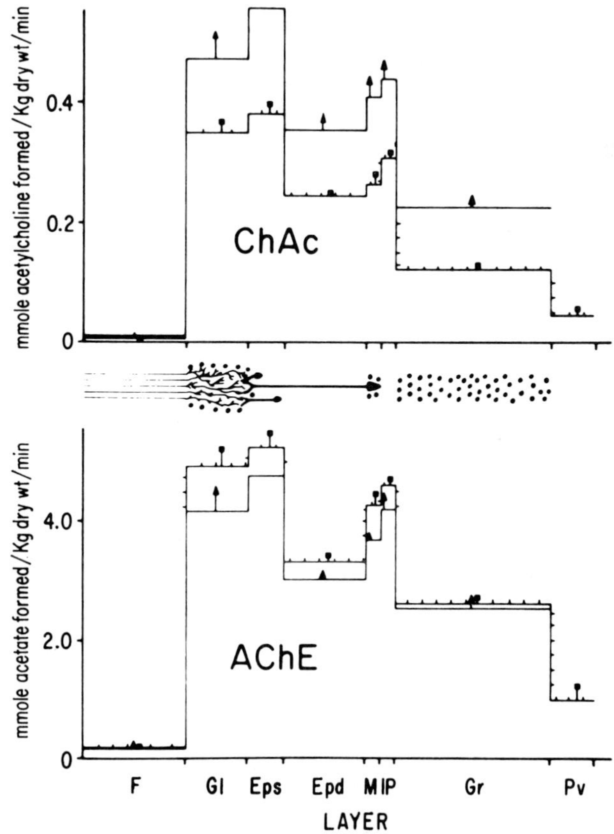

Fig. 3. Distribution of choline acetyltransferase (ChAc) and acetylcholinesterase (AChE) activities across the layers of the olfactory bulbs of two control rats. The diagram between the plots for the two enzymes represents a schematic view of the olfactory bulb structure including, from left to right, olfactory nerve fibers, small periglomular cells, tufted cells, mitral cells, and granule cells. For abbreviations see legend to Fig. 2. (Modified from Godfrey *et al.*, 1980.)

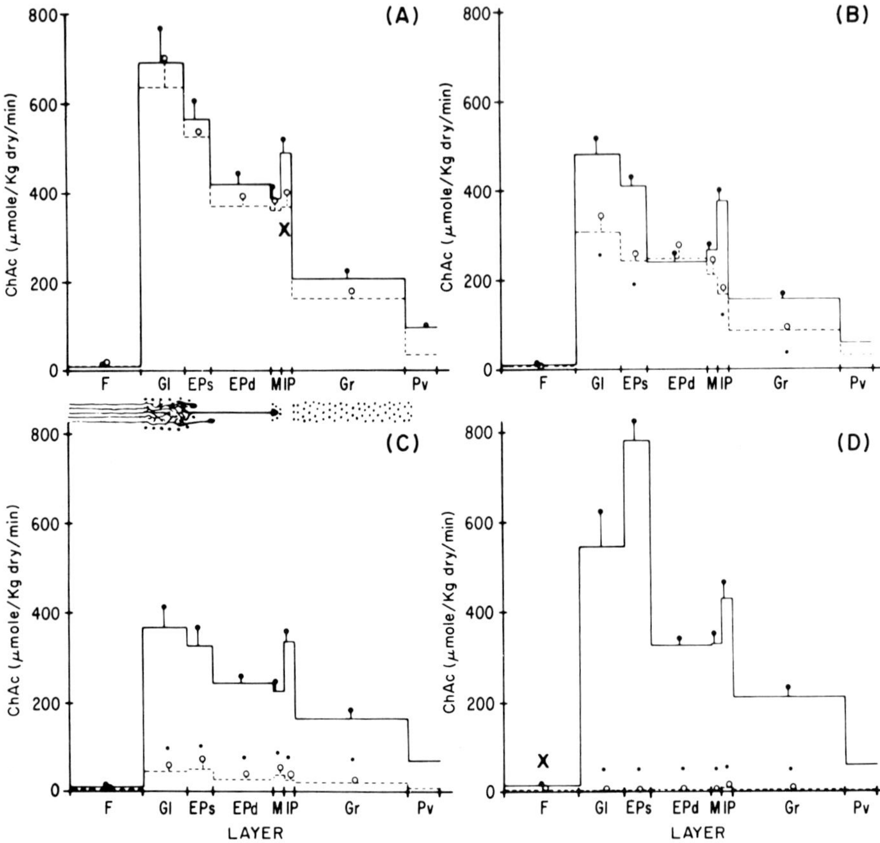

Fig. 4. Distribution of choline acetyltransferase activities (left) and of acetylcholinesterase activities (right, p. 449) across the layers of the olfactory bulbs of four rats (A–D), in which surgical lesions were placed behind the olfactory bulb. Solid and broken lines represent the mean values of the control and the lesioned rats, respectively. Filled and open circles represent the corresponding standard errors of the mean. Asterisks indicate differences between control and lesioned side significant at $p < 0.005$, crosses indicate differences significant at $p < 0.05$.

The locations of the lesions are briefly described in the text. Abbreviations, see legend to Fig. 2. (Modified from Godfrey *et al.*, 1980.)

Langerhans (Godfrey and Matschinsky, 1975). This comparison may help to view these data in the proper perspective. Acetylcholine, released from the vagus nerve, is physiologically important and regulates insulin release in the cephalic phase of food intake (Berthaud *et al.*, 1980). With this reference in mind, cholinergic innervation of taste buds of a magnitude similar to that observed in islet tissue could therefore be physiologically very important.

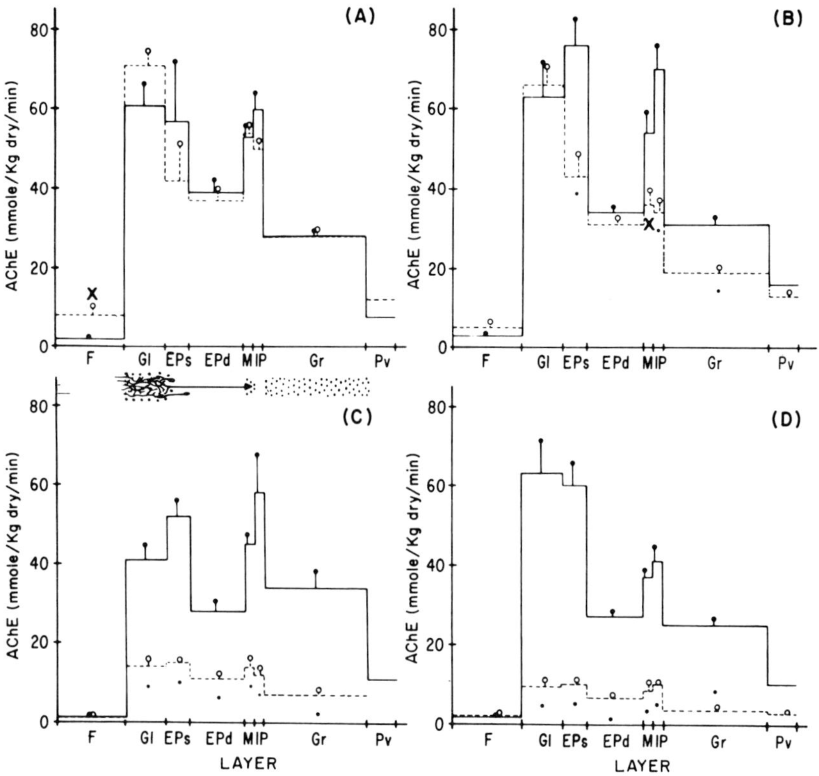

Fig. 4. *(continued)*

Noteworthy is the apparent lack of cholinergic innervation of fungiform papillae in contrast to the circumvallate and foliate papillae. The two groups are innervated by two different nerves, which may explain the difference.

C. Enzyme Distribution in the Rat Olfactory Bulb

The profiles of distribution of acetylcholinesterase and of choline acetyltransferase in the layers of the rat olfactory bulb show extreme differences of the lowest and the highest activities (Fig. 3). They are lowest in the fiber layers and highest in the external plexiform layer. Patterns of activities of both enzymes run parallel. The data are highly reproducible between individual animals. These quantitative profiles of distribution are in general agreement with the observations made on sections stained for acetylcholinesterase (not shown).

D. Effect of Surgical Lesioning on Distribution of Acetylcholinesterase and Choline Acetyltransferase of Olfactory Bulb Layers in the Rat

Surgical lesions were placed behind the bulb of the right eye in four rats and the enzyme profiles of the lesioned bulb compared with the profiles of the left control bulb (Figs. 4,5). Tissue was taken from the animal one week

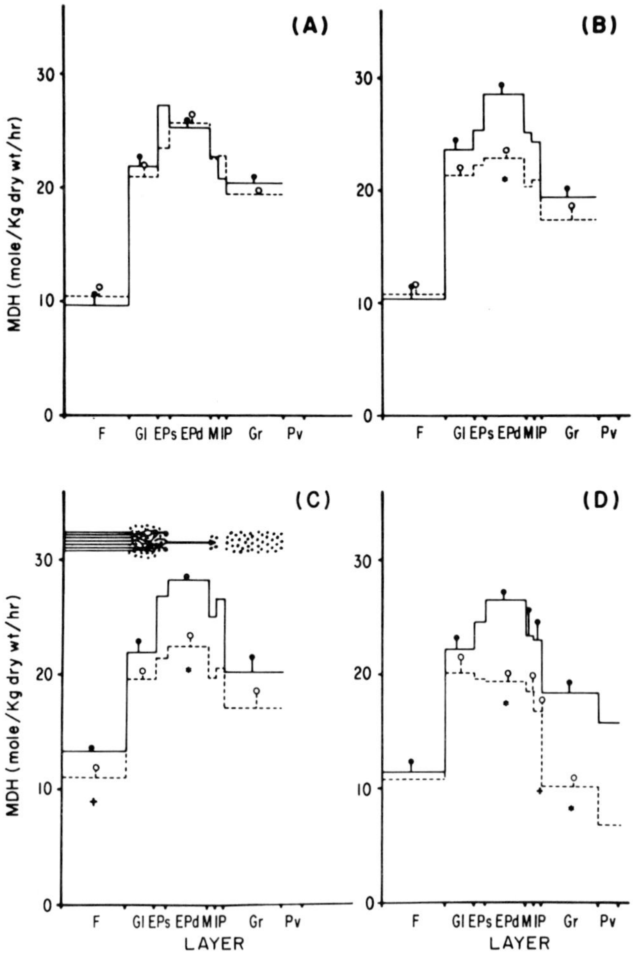

Fig. 5. Distribution of malic dehydrogenase activities across the layers of the olfactory bulbs of four rats, in which surgical lesions are placed behind the olfactory bulb. For additional information see legends to Figs. 2 and 4 and the text. (Modified from Godfrey *et al.*, 1980.)

after placing the lesions. The extent and precise locations of the lesions were ascertained histologically. In animals A and B, the lesions were aimed to cut the lateral olfactory tract alone. The actual lesions as determined histologically included not only the entire lateral olfactory tract but also tissue deep to the tract. The extent of this deep lesion was much smaller in rat A than in rat B. In rats C and D, the cut was aimed through the olfactory peduncle to a depth sufficient to cut all pathways entering the right olfactory bulb from caudal locations. The lesion was essentially complete in rat D as verified histologically. The results of the lesioning experiments are clear cut. Only the deep lesions affect the cholinergic innervations of the bulbar layers (Fig. 4).

Malic dehydrogenase was measured in bulbar layers as an unspecific marker. The observed changes were small compared with the effect that lesioning had on the profiles of cholinergic enzymes (Fig. 5).

The data suggest that all cholinergic elements within the olfactory bulb represent cholinergic centrifugal fibers and their terminals. The existence of centrifugal cholinergic innervation of the olfactory bulb is consistent with a general scheme of centrifugal cholinergic pathways to the peripheral relays and/or receptors of the sensory system. Similarly, the present data on gustatory structures favor such a concept.

III. SUMMARY

Quantitative histochemical mapping procedures have been employed to study the distribution of choline acetyltransferase and acetylcholinesterase activities in rat olfactory and gustatory pathways. The following structures were examined: light microscopic layers of the olfactory bulb, the circumvallate, foliate, and fungiform papillae of the tongue and appropriate reference tissues for structures of both pathways. In studies of the olfactory system, surgical lesions were placed behind the bulb on one side, and the effects of this lesion on enzyme patterns were studied. From the results several conclusions are suggested: (1) The olfactory receptor cells and mitral cells are not cholinergic; (2) essentially all cholinergic elements in the olfactory bulb derive from some location caudal to the bulb, with the bulk of the centrifugal cholinergic fibers to the bulb traveling in regions deeper than the lateral olfactory tract; (3) circumvallate and foliate papillae and submucosal tissue adjacent to these structures contain both enzymes of the cholinergic system in contrast to fungiform papillae in which these enzymes seem to be absent; (4) these cholinergic innervations of the olfactory and gustatory systems may be part of efferent neural projections. The physiological role of the cholinergic pathways described is unknown.

ACKNOWLEDGMENT

This work was sponsored by National Institute of Arthritis, Metabolism and Digestive Diseases Research Grant AM 22119 and The National Institute of Neurological and Communicative Disorders and Stroke Research Grant NS 14843.

REFERENCES

Berthaud, H. R., Trimble, E. R., Siegel, E. G., Bereiter, D. A., and Jeanreneaud, B. (1980). Cephalic phase insulin secretion in normal and pancreatic islet transplanted rats. *Am. J. Physiol.* **238**, E336–E340.

Godfrey, D. A., and Matschinsky, F. M. (1975). Enzymes of the cholinergic system in Islets of Langerhans. *J. Histochem. Cytochem.* **23**, 645–651.

Godfrey, D. A., and Matschinsky, F. M. (1976). Approach to three dimensional mapping of quantitative histochemical measurements applied to studies of the cochlear nucleus. *J. Histochem. Cytochem.* **24**, 697–712.

Godfrey, D. A., Williams, A. D., and Matschinsky, F. M. (1977). Quantitative histochemical mapping of enzymes of the cholinergic system in cat cochlear nucleus. *J. Histochem. Cytochem.* **25**, 397–416.

Godfrey, D. A., Ross, C. D., Hermann, A. D., and Matschinsky, F. M. (1980). Distribution and derivation of cholinergic elements in the rat olfactory blub. *Neuroscience* **5**, 273–292.

Lowry, O. H., and Passonneau, J. V. (1972). "A Flexible System of Enzymatic Analysis," pp. 221–260, Academic Press, New York.

McCaman, R. E., and Hunt, J. M. (1965). Microdetermination of choline acetylase in nervous tissue. *J. Neurochem.* **12**, 253–259.

Matschinsky, F. M., Lowry, O. H., and Passonneau, J. V. (1968). Quantitative histochemical analysis of glycolytic intermediates and cofactors with an oil well technique. *J. Histochem. Cytochem.* **16**, 29–39.

Ross, C. D., and McDougal, D. B., Jr. (1976). The distribution of choline acetyltransferase activity in vertebrate retina. *J. Neurochem.* **26**, 521–526.

PART IV

Discussion

Fisher: What is the current status of the neurotransmitters responsible for olfaction?

Margolis: With regard to the primary afferent transmitter, the best candidate is the dipeptide carnosine. Some positive and some negative physiological data have been reported but it still awaits a definitive neurophysiological demonstration.

Quinn: Negative electrophysiological evidence with carnosine has been reported. How definitive it is, I'm not certain. Carnosine is the only valid candidate at the moment.

Matschinsky: There are glutamate and aspartate. In lesioning studies we examined both the olfactory bulb and the pyriform cortex. Glutamate and aspartate levels in the different layers of the olfactory bulb were not affected, but in the pyriform cortex, for example in the fiber layers, both glutamate and aspartate fell by 80%. You could make a case for these as neurotransmitters.

Margolis: But they would be mitral cell transmitters from bulb to tubercle, not from epithelium to bulb.

DeSimone: Using immunohistochemistry, many of the gastrointestinal polypeptides have been identified in different systems outside of the GI tract. Are these substances present in the olfactory and taste systems? Because of their known relation to chemoreception in the GI tract they might be prime candidates for neurotransmitters in the olfactory or gustatory systems.

Margolis: There is good evidence that almost every peptide found anywhere is found in the olfactory bulb as well. For example, substance P, cholecystokinin, insulin, and somatostatin. With the exception of one chemical study on opiate peptides demonstrating the presence of met-enkephalin in mouse bulb (R. V. Lewis and F. L. Margolis, unpublished), all of the evidence for other peptides derives from cross-reaction with antisera. Extreme caution is needed when attributing antisera cross-reactivity to the chemical presence of the peptides. With that cautionary note, virtually anything that is elsewhere in the brain is also in the bulb.

Cagan: What are the future prospects of the work on the cholinergic innervation of the taste system?

Matschinsky: There are two necessary experiments. The first is a lesioning study with the glossopharyngeal nerve. The second is the physiological question of whether interference with acetylcholine release or its action has any effect on biophysical parameters of taste perception.

Cagan: The absence of enzyme activity in fungiform papillae was striking. Because only one taste bud is present in each fungiform papilla in the rat, could it have been missed?

Matschinsky: No. We considered possible cholinergic fibers at the base of the fungiform papilla. In the fungiform, we made an effort to section all the way through the tongue and collect samples beneath the taste bud. I doubt that we missed fibers from adjacent sections or from around the fungiform papilla. We did only two animals because it was so laborious.

Margolis: There were declines in cholinesterase activity after separation of the bulb from the balance of the brain. What is the evidence that it is pre- or postsynaptic?

Matschinsky: We have no hard information on the localization of the biochemical in substructures.

Murphy: Is there any indication of receptor density, as for example dopamine receptors versus β-receptors after denervation?

Hirsch: Regarding the effects of zinc sulfate deafferentation on these binding sites in the bulb, my data indicate that at short times after denervation, within 30 days, the only receptor in the bulb modified is the carnosine binding site. Other binding sites for α-receptors, β-receptors, GABA-receptors, opiate receptors, etc., are essentially unaffected. Other types of lesions, such as unilateral bulbectomy, could be possible for observing effects in the other bulb over long time periods. With the zinc sulfate technique there is essentially no change over 180 days.

Cancalon: Did the zinc sulfate destroy both the olfactory neurons and the basal nerve cells?

Margolis: After long times, possibly 5–10% remain. There is well over 90% destruction of everything, virtually down to the bone.

Wysocki: The return of the system is impressive. Have you studied the behavior of the animal and the histology of the epithelium?

Margolis: We did rather crude behavioral studies in determining the ability of an animal to find a food pellet buried underneath shavings and they were in close agreement with the biochemical findings. We have not done histology except after zinc sulfate treatment.

Matschinsky: Regarding the activation of adenylate cyclase in the tubercle and the change of cyclic AMP levels in ischemia, what time is required?

Krieger: The delay time illustrated was 3 min, but it can be observed after 1 min. It indicates the presence of the synthetic capacity. The problem is when you want to be rid of it, to make observations on the effect of incoming dopamine fibers. Microwave is needed to eliminate it.

Dodd: In the synaptosomal preparation, what happens to the other enzymes of glutamate metabolism? Could glutamate dehydrogenase be liberated and convert glutamate to α-ketoglutarate which can then be decarboxylated?

Quinn: Aminooxyacetic acid (AOAA) would not inhibit in that case. Also, the stoichiometry of the GABA:CO_2 indicates that this does not occur.

Karnovsky: Is aminooxyacetate an obligatory component in assaying glutamate dehydrogenase? What is the stoichiometry? And what is the stoichiometry in the presence of AOAA? Also, I'm impressed that Triton treatment enhanced binding of GABA by 40-fold.

Quinn: The stoichiometry of CO_2:GABA was 0.76:1.0 mole/mole. Therefore $^{14}CO_2$ production may slightly underestimate the activity of glutamate decarboxylase. AOAA is used routinely to demonstrate inhibition of glutamate decarboxylase. Upon adding AOAA, essentially 90% of the activity is abolished and the stoichiometry is not as close to unity.

Cagan: It is interesting that this stoichiometry has a long history in reports of glutamate decarboxylase in other tissues. No one seems to have an adequate explanation.

Quinn: The effect of Triton X-100 is a puzzling phenomenon. The GABA receptor is unusual in being a synaptic receptor that is not solubilized by Triton X-100. Costa and Toffano, using cerebellum, have dialyzed the supernatant following treatment with Triton X-100. What remained in the dialysis bag, when added to the binding assay, decreased [^{3}H]GABA binding. Its approximate molecular weight was 15,000, and they suggested that it may allosterically modulate the interaction of GABA with the GABA receptor site.

Hirsch: As far as I know, the GABA system is unique in responding to Triton in that particular way. The technique was originally developed by Enna and Snyder to remove Na^+-

dependent binding of GABA. Others have tried to determine what is in the Triton supernatant. The material referred to by Dr. Quinn apparently regulates both GABA receptors and the benzodiazepine receptor. The benzodiazepine drugs competitively compete with this protein for regulation of the GABA receptor. I have done the Triton treatment many times with the GABA system and routinely find 20- to 40-fold stimulation of specific GABA or muscimol binding.

PART V

Analogous Chemoreceptors

21

Chemical Sensing by Bacteria

NICHOLAS F. PAONI, ANN M. MADERIS, AND D. E. KOSHLAND, JR.

I. INTRODUCTION

Bacterial chemotaxis has emerged as a potential model for understanding the role of the cell in sensory transduction and sensory processing. It is perhaps amazing that a simple bacterium has a sensing system at all, yet the processing patterns of the bacterium and of complex species all seem to fit a general pattern as outlined in Fig. 1. An entire organism has various receptors for the stimuli that it wishes to receive, such as light, sound, taste, and olfaction. These receptors generate a signal that is initially processed through a specialized system, i.e., the visual, the auditory, the gustatory, and the olfactory, and there is usually a central response, which integrates this signal with others to produce a final behavioral response. If one examines an individual cell, the same type of processing occurs, except that the multiplicity of stimuli are reduced. Some cells in fact can respond only to a single stimulus, but others respond to many.

Biochemistry of Taste and Olfaction

ISBN 0-12-154450-8

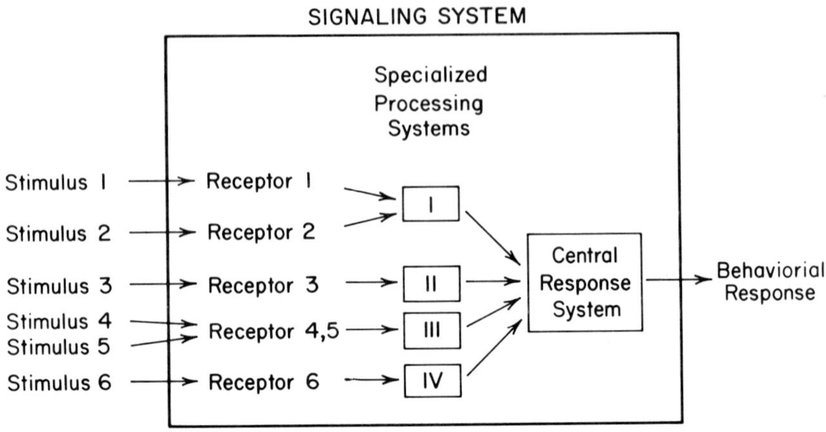

Fig. 1. Signal processing in a single cell. External stimuli activate receptors, which convey information to specialized processing parts of the system and then to a central response integrative network. The output is the behavioral response of tumbling in the bacterium and an electrical signal or a neurotransmitter release in the case of a neuron.

Viewed in this general framework, a bacterial system has a number of advantages for experimental study. It is a single cell that can be grown in large numbers, making biochemical analysis easier since large amounts of an individual protein can be produced. Second, the single cell is the entire organism and hence its behavioral responses can be studied *in vivo* under natural conditions without the necessity of large tissue culture arrangements and/or the imposition of an artificial environment. Third, the use of recombinant DNA technology has made genetic manipulations of bacteria extremely advantageous. In this chapter we will discuss some aspects of the bacterial system and their implications for the more complex systems that form the focus of this volume.

II. BRIEF DESCRIPTION OF THE BACTERIAL SYSTEM

The chemotactic response, originally discovered by Pfeffer in 1884 and Englemann in 1881, allows the bacterium to swim toward chemicals and environments optimal for its survival and away from environments that are dangerous or deleterious. It is a highly selected system that has optimized many features for the benefit of the bacterium. In essence it provides for the bacterium a feedback system in relation to the environment equivalent to the pain–pleasure feedback system of higher species.

The behavioral response of bacteria, which allows them to migrate up or down gradients, is achieved by the control of tumbling frequency (Berg and

Brown, 1972; Macnab and Koshland, 1972). When bacteria are progressing in a direction that they "consider" favorable, they suppress tumbling and continue swimming in approximately straight lines. When they are swimming in a direction that their sensory system tells them is unfavorable, they generate tumbling and head in a new direction.

The new direction after a tumble is not oriented specifically to be more favorable. Each tumble causes a random reorientation (Berg, 1971). Hence, unlike some species, it is not the direction of orientation of the tumble that results in the migration, rather it is the frequency of tumbling. For the bacterium heading in the wrong direction, a tumble can have two possible outcomes. It can by chance continue to swim in the wrong direction, in which case it will not move very far before it tumbles again. If as a result of a tumble it reorients and starts swimming in a favorable direction, tumbling will be suppressed, and the bacterium will swim a longer than normal distance. Thus the manipulation of the tumbling frequency is sufficient to bias the random walk of the bacterium to move in a favorable direction.

Not only is the movement efficient in energy terms, but it is efficient in regard to simplicity of behavioral response. Tumbling or not tumbling is the behavioral output of the processing system. Thus, to a first approximation, the output of the cell is an ON–OFF phenomenon, such as the electrical discharge of a neuron or the secretion of a neurotransmitter.

The gram-negative bacteria, *Escherichia coli* and *Salmonella typhimurium*, are the organisms on which the bulk of the chemotactic studies (Koshland, 1979b; Macnab, 1979; Springer *et al.*, 1979) have been made, although chemotaxis has been observed in gram-positive organisms and many other species (Doetsch and Hageage, 1968; Weibull, 1960; Macnab, 1979). *E. coli* and *S. typhimurium* are very similar to each other and are each about 1-μm wide and 2-3-μm long. The receptors, which detect the signals from the external environment, are located either in the inner membrane or in the periplasmic space just outside of the inner membrane. About 30 receptors already have been identified that can detect signals from the external environment and deliver them to the processing system. This processing system requires approximately ten gene products specifically devoted to the transduction and interpretation of the chemotactic response. A wide variety of other components provide a support system either in the form of the energy needed to drive the flagellar apparatus or the level of S-adenosylmethionine involved in adaptation. These support systems have a direct analogy with the energy, oxygen, structural components, etc., needed for the operation of the brain, and they can similarly be assumed to be a constant background in the analysis by the specific sensory system. At times, of course, the interface between the two systems must be taken into account. But in general we can emphasize the properties of the ten proteins that can be considered to be "the brains" of the bacterium and the 30 receptors that represents its "tongue and nose."

III. THE RECEPTORS

The receptors are designed for a wide range of compounds, of which the nutrients are mostly carbohydrates and amino acids (Adler, 1975; Koshland, 1979a). Receptors are also apparently present for repellents that are largely noxious chemicals, although some, such as acetate, may be an indicator of crowded conditions.

Each receptor is given a name identified with its main activator, although few of the receptors are specific for a single compound. For example, in the case of the galactose receptor, galactose and glucose bind strongly (K_d's of approximately 10^{-7}), but a variety of other sugars such as arabinose bind at higher concentrations ($\sim 10^{-3}$) and would be detected by the organisms in the absence of the more strongly binding sugars (Anraku, 1968; Boos *et al.*, 1972; Zukin *et al.*, 1977). Similarly the receptor for serine has been found to bind cysteine, alanine, and glycine but at lower affinities than serine (Adler, 1966). The ribose receptor binds ribose with a dissociation constant (K_d) of approximately 10^{-7}, and the only other sugar that elicits an effect from this receptor is allose, which binds 1000 times less strongly (Aksamit and Koshland, 1974). Since allose is not a prevalent sugar found naturally in high concentrations, it is probable that this receptor is physiologically significant for ribose only. Thus the receptors appear to be designed specifically to provide response for a limited group of compounds important to the metabolism and survival of the organism.

Evidence for the existence of the periplasmic receptors has been obtained by competition, purification, and genetic studies in the case of the galactose (Anraku, 1968), ribose (Aksamit and Koshland, 1974), and maltose receptors (Hazelbauer, 1975). A combination of behavioral studies and genetics identified the other receptors (Adler, 1969, 1975). A new technique made it possible to provide direct evidence for the aspartate and serine receptors (Clarke and Koshland, 1979). As discussed below, these experiments identify the aspartate and serine receptors as a direct part of, or intimately connected with, the tar and tsr proteins in the bacterial membrane. Some of these receptors are constitutive in the sense that they are present in every cell of the appropriate species; some are inducible, i.e., they can be formed if the bacterium is grown under the proper conditions (Adler, 1969; Fahnestock and Koshland, 1979; Koshland, 1979a). This introduces a level of plasticity into the system.

IV. THE PROTEINS OF THE PROCESSING SYSTEM

Bacterial techniques for the selection of mutants are highly developed and have been applied to these systems. A generally nonchemotactic mutant is indicated by the loss of the chemotactic response to all stimuli. These genes, together with the phenotypic properties, are shown in Table I. They fall into two categories. Nine genes are identified with the central processing of all stimuli (the first nine listed) (Adler, 1969; Warrick *et al.*, 1977). The remaining three genes listed (the *tsr*, *tar* and *trg* genes) involve processing of several stimuli, but their products can be deleted without damage to other parts of the system (Ordal and Adler, 1974; Silverman and Simon, 1977a; Strange and Koshland, 1976). It is intriguing that even a monocellular species develops some compartmentalization of the sensory stimuli.

Although it cannot be said that the function of all the genes of Table I have been identified, the rather extensive searches for other chemotaxis genes that have been made indicate that if additional genes are to be identified with the system, they will be few in number (Silverman *et al.*, 1977; DeFranco *et al.*, 1979). Thus one receives a picture of a moderately complicated system but one sufficiently simple so that all of the components can be identified within a reasonable period of time. In fact, most of the gene products in the central processing system have already been identified and

TABLE I Properties of the Signal Processing Genes of *E. coli* and *S. typhimurium*[a]

Gene classification			
Salmonella typhimurium	*Escherichia coli*	Molecular weight of peptide product	No-gradient motility pattern
P	*A*	76,000	Smooth
Q	*Y*	8,000	Smooth
R	*X*	28,000	Smooth
S	—	—	Smooth
T	*Z*	24,000	Tumbly
U	*C*	—	Smooth, random tumbly
V	—	—	Smooth
W	*W*	12,000	Smooth
X	*B*	38,000	Tumbly
—	*tsr*(D)	60,000	Random
—	*tar*	60,000	Random
trg	*trg*	—	Random

[a] For more complete documentation of the original literature, see Koshland (1979b).

their molecular weights are shown in Table I (Silverman and Simon, 1976, 1977b).

V. THE REVERSIBLE METHYLATION SYSTEM

The clues in regard to the biochemical role of the processing proteins came from the finding that methionine is needed for chemotaxis (Adler and Dahl, 1967) and that methionine acts as the precursor for *S*-adenosylmethionine, which is the required chemotactic agent (Armstrong, 1972; Aswad and Koshland, 1975). The methionine requirement was established by growing auxotrophs, which could not make methionine, and showing that the bacteria could not respond to gradients in the absence of added nutrient methionine. The methionine requirement could be fulfilled over a very narrow concentration range. At 10^{-7} *M* methionine, essentially no chemotactic response was observed. At a nutrient level of 10^{-5} *M*, the methionine requirement was saturated. Chloramphenicol, a potent inhibitor of protein synthesis, has no effect on chemotaxis over a period of several hours so the methionine requirement was not to allow protein synthesis.

To discover the role of methionine, a temporal memory experiment was carried out on the methionine auxotroph with and without added methionine (Aswad and Koshland, 1974). With added methionine the response time was normal. Without added methionine, a much longer response time was observed. This experiment established two conclusions: (a) that methionine is required for sensing and is not merely a requirement for proper flagellar functioning and (b) that methionine is involved in the adaptation process, i.e., it affects the response time of the bacterium. In its absence the memory apparatus does not function normally.

Methionine is not itself an active compound and so it seemed logical that it was a precursor for another substrate. *S*-Adenosylmethionine seemed a logical choice (Armstrong, 1972; Aswad and Koshland, 1975). It is widely known as a substrate for methylation, but it was impossible to get *S*-adenosylmethionine to cross the membrane barrier. To prove that *S*-adenosylmethionine was indeed required, various inhibitors of SAM–synthetase were leaked into the cell, and it was shown that they altered the adaptation time of the bacteria, i.e., they produced the same behavior as methionine deprivation (Aswad and Koshland, 1975). This established that SAM is the substrate involved in the chemotactic response. Toluenized cells, which allowed SAM to pass through the cell barrier, and broken cell preparations were later used to confirm this conclusion (Springer and Koshland, 1977; Paoni and Koshland, 1979).

If SAM was acting as the methylating agent, what did it methylate? The next step was a highly intriguing one. Kort *et al.* (1975) found that *E. coli* membrane proteins in the 60,000-MW range were methylated in the chemotactic process. Subsequently, these proteins were shown to be the products of the *tar*, *tsr*, and *trg* genes (Silverman and Simon, 1977a; Springer *et al.*, 1977; Kondoh *et al.*, 1974). Similar proteins in *S. tymphimurium* were separated on SDS gels and are shown in Fig. 2. The level of methylation of the protein bands was shown to be influenced by the addition of attractants and repellents to the cells (Kort *et al.*, 1975; Paoni and Koshland, 1979). Moreover, the change in protein methylation occurs in the time span of the adaptation to the chemical stimuli.

In Fig. 3, the change in level of methylation on addition of the attractant, aspartate, is shown. Both *E. coli* and *Salmonella* show the same total increase in methylation level. Moreover, as shown in the figure, the fractional change is the same *in vivo* and in permeabilized cells.

If methylation occurs, could the central processing gene products be in-

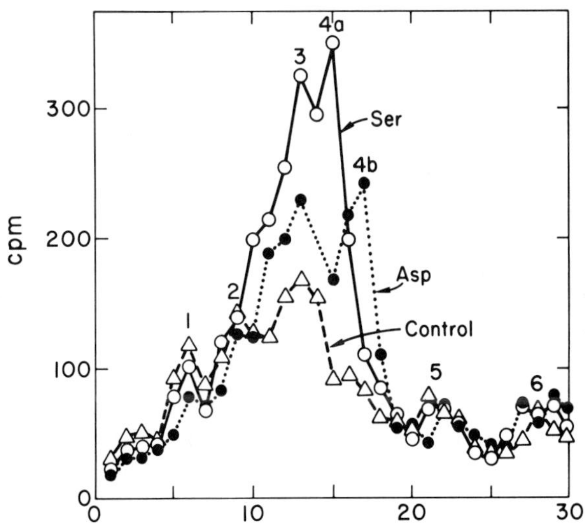

Fig. 2. Effect of attractants on banding pattern in *S. typhimurium* methylation. *S. typhimurium* strain ST2 cells were permeabilized by treatment with 0.14% toluene, 9.5 m*M* EDTA, and then radiolabeled with *S*-adenosyl-L-[methyl-^{3}H]methionine for 30 min at 30°. The cells were stimulated with 10 m*M* L-aspartate ●·····●, 10 m*M* L-serine ○——○, or buffer alone △---△ for 30 min before the methylation reaction was stopped by dilution of the cells into cold acetone. The samples were electrophoresed on 30-cm long sodium dodecyl sulfate slab gels, 7% in polyacrylamide as described by Laemmli (1970). The (50,000–65,000 MW) region of the dried gel was cut into 1-mm slices, and the radioactivity in each slice was determined.

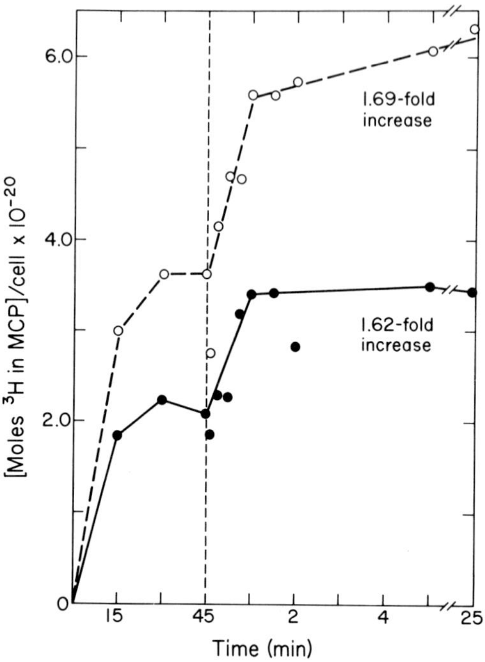

Fig. 3. Comparison of permeabilized cell (○) and *in vivo* (●) methylation. The intact cells were incubated with L-[methyl-^{3}H]methionine in the *in vivo* assay; the permeabilized cells were incubated with *S*-adenosyl-L-[methyl-^{3}H]methionine in the *in vitro* assay for 45 min. L-Aspartate was added to a final concentration of 1 m*M*, and samples were taken at the times indicated. For the purpose of this experiment the radioactivity in all the MCP bands was summed.

volved in this function? To check on such a possibility broken cell preparations of the wild type and various mutants were made. One of the mutants had been shown to give a low level of methylation so it was a likely candidate for a mutant lacking the enzyme that catalyzes methylation. Membranes from mutants were mixed with cytoplasm from wild type and membranes from wild type were mixed with cytoplasm from mutants in all combinations. It was found (Table II) that the cytoplasm of *cheR* mutants when mixed with wild type membranes failed to give methylation of the 60,000 dalton membrane protein (Springer and Koshland, 1977). On the other hand, the cytoplasm of the wild-type mixed with the mutant membrane gave full methylation. Thus a deficiency in the mutant cytoplasm existed, and it was the catalyst for methylation, not the protein to be methylated. The gene product was identified as a methyltransferase of molecular weight 40,000 (Springer and Koshland, 1977). It may be composed of two peptide chains, the cheR gene product of 28,000 and a 12,000 MW peptide (DeFranco *et al.*, 1979).

TABLE II Methyltransferase Activity in Mutants of *Salmonella typhimurium*[a]

Gene locus in which mutant appears	Wild-type methylation (%)
Wild type	100
cheA	123
cheY	150
cheR	7
cheS	81
cheZ	71
cheC	118
cheV	120
cheW	157
cheB	49
tar	110
tsr	105

[a] Membranes of wild type mixed with cytoplasm of mutant.

Work on the properties of the methylated protein established that a glutamyl residue was methylated (Eq. 1) (Kleene *et al.*, 1977; Van der Werf and Koshland, 1977). This is interesting because the known mammalian protein methylases are nonspecific. The bacterial sensing enzyme is apparently specific for this glutamyl residue in that no aspartyl residues are modified, and glutamyl residues in a wide variety of other proteins are not affected by the enzyme. This is what one would expect of a regulatory protein.

$$\text{Protein—CH}_2\text{CH}_2\text{COOH} + \text{SAM} \rightarrow \text{Protein—CH}_2\text{CH}_2\text{COOCH}_3 + \text{SAH} \quad (1)$$

Having obtained a method for preparing radioactive methylated proteins a systematic search for a demethylating enzyme was pursued (Stock and Koshland, 1978). As shown in Table III, the mutants lacking the esterase were found to be the *cheB* mutants. An interesting further fact emerged in that the *cheZ* mutants of *Salmonella* showed no esterase activity whereas the corresponding *cheZ* mutants of *E. coli* did. This suggested some interaction between the cheZ and cheB gene products, and this was found to be the case. However, the catalytic activity resides in the cheB gene product; this gene codes for a peptide of MW 38,000.

The information obtained so far on the methylating system indicates that the three proteins (tar, tsr, and trg proteins) in the membrane receive signals from receptors in the periplasmic space or the membrane or are themselves receptors (Clarke and Koshland, 1979). The influence of these periplasmic or direct binding proteins induces a conformational change in the 60,000 dalton

TABLE III Methylesterase Activity in Mutants of *E. coli* and *S. typhimurium*

Gene locus in which mutant appears	Relative methylesterase activity (wild type = 100%)	
	E. coli	*S. typhimurium*
Wild-type	100	100
cheA	102	60
cheB	8	1
cheC	—	66
cheW	64	87
cheR	55	45
cheY	59	57
cheZ	92	1
cheS	—	125
cheV	—	74
tar	277	—
tsr	317	—

proteins to make them available as substrates for the methyl transferase enzyme. A glutamic group is methylated by S-adenosylmethionine and is catalyzed by the transferase enzyme located in the cytoplasm. The receptor appears not to have a direct influence on the enzyme since the receptor is located outside the inner membrane and the enzyme is located in the cytoplasm. The methyl group is later removed by an esterase enzyme also located in the cytoplasm. If the receptor spanned the membrane and activated cytoplasmic enzymes, the transferase and esterase would act on tsr, tar and trg proteins without specificity. Hence it seems likely that the change in conformation of the membrane protein causes the level of methylation to change. It is control of these processes that allows response and adaptation.

An intriguing aspect of this methylation process is that the tar, tsr and trg gene products have multiple methylation (DeFranco and Koshland, 1980). It has been shown that as many as four methyl groups are covalently attached to the 60,000 dalton proteins in the sensing system under some conditions of stimulation. In the absence of a gradient, one or two methyl groups appear to be present. Increases in repellent or decreases in attractant reduce the number of methyl groups, whereas increases in attractant and decreases in repellent do the opposite. This finding of multiple methylation in these signaling proteins is particularly intriguing because multiple phosphorylation has been observed in rhodopsin in the visual system (Sanderson, 1972), the ribosomal proteins (Thomas *et al.*, 1979), glycogen synthesis (Smith *et al.*, 1971), and the acetylcholine receptor (Vandlen *et al.*, 1979).

A multiple modification system has quite different characteristics than a monosubstituted system in which an active versus inactive protein can easily explain the two states. Multiple modification seems to suggest a gradual change in properties instead of a simple ON–OFF switch. It also suggests that the timing of rates of methylation and demethylation is important. A multiple phosphorylation or methylation system also offers the advantage that it can respond immediately to either repellents or attractants. If there is an intermediate level of methylation in the absence of gradients, then attractants can lead to increased methylation and repellents to decreased levels. The receptors apparently alter the conformation of the substrate tar, tsr, trg, etc., proteins to make them more or less susceptible to transferase or esterase action. Thus the membrane-bound signaling proteins become substrates for the cytoplasmic enzymes.

VI. THE RESPONSE REGULATOR MODEL

Macnab and Koshland (1972) showed that bacteria respond to gradients by a temporal sensing system in which the "memory" of the bacterium plays a significant role. The rationale of the experiments is quite simple. Bacteria subjected to sudden temporal increases in attractant concentration are observed to suppress their tumbling frequency. Bacteria subjected to sudden temporal decreases in attractant concentration increase their tumble frequency. In both cases the bacteria adapt, i.e., gradually return to their normal tumbling frequency. These are exactly the responses required to direct migration favorably for bacteria swimming up spatial gradients of attractants (suppression of tumbling) or down gradients of attractants (increase of tumbling). They swim at uniform velocities and sense spatial gradients by a temporal sensing mechanism. This sensing mechanism integrates the gradient information through a "memory" apparatus within the bacterial cell.

The mechanism proposed to explain such a bacterial memory is shown schematically in Fig. 4. It has elements of temporal mechanisms proposed for a wide variety of species. It is postulated that some molecule or parameter called the *response regulator* (X) controls the frequency of tumbling. Bacteria produce a response regulator and decompose it continuously so that X would be present at a steady state level in a no-gradient situation.

If the pool size fluctuated in a Poissonian manner relative to the critical value (signified X_{crit}), then the random walk behavior observed in the bacteria could be explained. Whenever X (used to indicate the instantaneous value of X) was above this value of X_{crit}, tumbling would be suppressed. Whenever it was below the value, tumbling would be generated, and hence

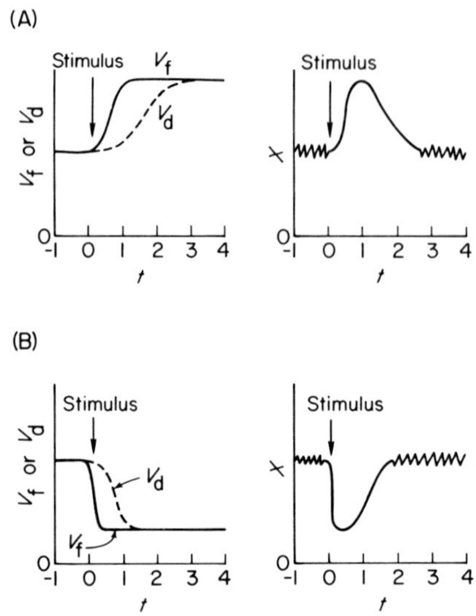

Fig. 4. Explanation of response and adaptation at molecular level. Assuming a response regulator is formed at a rate V_f and decomposed at a rate V_d then $V_f = V_d$ on average in absence of a gradient, but fluctuations in level of X, the response regulator, relative to threshold produces a random pattern. (A) When attractant is added or repellent removed or diluted, V_f and V_d change, but V_f responds more readily than V_d. As a result, a transient increase in response regulator occurs followed by adaptation. (B) An attractant decrease or repellent increase again causes a transient change (decrease in X in this case), which adapts back to the no-gradient situation in time.

the random variaton of the tumble regulator could be translated into the random behavior of the bacterium.

This model furthermore explains how increases in attractants or repellents could lead to the observed behavioral responses. Increases in attractants in this model cause increases in the rate of formation of the response regulator and also increases in the rate of decomposition. But the attractant effect on the formation constant (V_f) occurs more rapidly than the effect on the rate of decomposition (V_d) (Fig. 4). Hence the level of response regulator increases transiently. Such a hypothesis can readily be translated into molecular terms, as both rapid and slow conformational changes are known.

In each of the above cases, the model shows a return of the response regulator to the normal state after an appropriate interval of time. It is perhaps worth dwelling on this property for a moment since adaptation is extensive in biological species. There are two types of adaptation. One is the adaptation after a pulsed stimulus in which the stimulus is removed. Such phenomena are relatively easy to explain. The adaptation that is more difficult to explain is the one in which the new stimulus is maintained yet the species adapts. In bacteria, the increase in concentration of ribose leads to suppression of tumbling, but after this initial response the bacteria return to normal even though the concentration of ribose in the medium remains at the new higher level. Such adaptation to constant stimuli, both pleasurable and noxious, is observed in complex organisms and in individual cells. The principles shown in Fig. 4 can explain such adaptation in mathematical and chemical terms.

The mechanism shown in Fig. 4 is oversimplified. However, the more complex model, which details the role for individual receptors, is essentially a more detailed explanation of the fundamental percepts of Fig. 4. Figure 4 states two fundamental conclusions. First, the receptor-initiated stimulus event must occur at two levels in the system. If the receptor simply increased one process in the sensory transduction system, the bacteria would not adapt and return to the initial behavioral pattern. The receptor stimulus event therefore must act at two places in order for an eventual return of the response regulator to its initial level. Second, there must be a difference in the response times for the interaction of the receptor at these two locations. If the receptor signal initiated instantaneous responses or an equally slow response at the two points in the system, one might have internal compensation of all steps and therefore no sensory signal. At a step in another part of the system, a transient signal is recorded. The length and intensity of the signal will depend on the rate constants of the individual enzymatic steps. Hence, it is possible to design systems for whatever response time is needed, from extremely rapid for neuronal cells to extremely slow for hormonal cells. The bacteria lie somewhere in the middle and appear to have a similar type of sensory processing. The difference in the bacterial case appears to be that it carries out a methylation–demethylation reaction, whereas most higher mammalian cells carry out phosphorylation–dephosphorylation.

VII. SUMMARY AND CONCLUSIONS

Having described the bacterial chemotaxis system, it is perhaps appropriate to consider the generalities of such a system and compare them with other processing methods. In the bacterial cells a signal is received and transmitted across the bacterial membrane into the interior of the cell. There the information is processed to generate tumbling as a signal for a change and to suppress tumbling as a signal to continue on. It has been found that the processing system can integrate information from several stimuli, i.e., add or subtract information to come to an overall decision as to whether to continue swimming or to depart into a new direction. This information processing all occurs at the level of chemicals within the cytoplasm.

From what we know of neurons, it is apparent that the individual neuron proceeds much as the bacterium. It has receptors on its surface that transmit information across the membrane into the interior of the cell. A signal can generate an electrical impulse, but it also affects the cyclic nucleotide system, either cAMP or cGMP. The cyclic nucleotide system in turn causes phosphorylation and dephosphorylation of proteins in an analogy to the methylation and demethylation of the bacterial system. The behavioral output of the cell, the release of a neurotransmitter or of an electrical signal, is determined by these chemical events and the design of the cell. In some cases, the intracellular machinery is entirely designed toward maintaining the properties of the membrane to transmit the electrical signal, but in others, it takes a central role in processing information. In addition to the neurons identified with behavior, it is apparent that hormonal cells have an analogous processing system, and control of metabolism in fat cells also seems to follow such a model.

What is manifestly different in higher systems from the bacterial is the neural network. In such a network, an individual cell plays only a part in the overall processing machinery, so the wiring diagram is a key feature of the ultimate behavioral response. In the case of simpler systems, delocalized organization of cells provides the advantages of a multicellular response, but it may not be much more sophisticated than the processing of an individual cell. The knowledge already acquired in regard to the bacterial cell indicates that the individual neuron or the individual hormonal cell is capable of a great deal of information processing. The individual neurons of taste and olfactory systems are evidently far more complex than simple ON–OFF switches, and their chemistry will reveal a great deal of the manner in which sensory systems operate.

REFERENCES

Adler, J. (1966). Chemotaxis in bacteria. *Science* **153**, 708–716.

Adler, J. (1969). Chemoreceptors in bacteria. *Science* **166**, 1588–1597.

Adler, J. (1975). Chemotaxis in bacteria. *Annu. Rev. Biochem.* **44**, 341–356.

Adler, J., and Dahl, M. M. (1967). A method for measuring the motility of bacteria and for comparing random and non-random motility. *J. Gen. Microbiol.* **46**, 161–173.

Aksamit, R., and Koshland, D. E., Jr. (1974). Identification of the ribose binding protein as the receptor for ribose chemotaxis in *Salmonella typhimurium*. *Biochemistry* **13**, 4473–4478.

Anraku, Y. (1968). Transport of sugars and amino acids in bacteria. I. Purification and specificity of the galactose- and leucine-binding proteins. *J. Biol. Chem.* **243**, 3116–3122.

Armstrong, J. B. (1972). An S-adenosylmethionine requirement for chemotaxis in *Escherichia coli*. *Can. J. Microbiol.* **18**, 1695–1701.

Aswad, D., and Koshland, D. E., Jr. (1974). Role of methionine in bacterial chemotaxis. *J. Bacteriol.* **118**, 640–645.

Aswad, D., and Koshland, D. E., Jr. (1975). Evidence for an S-adenosylmethionine requirement in the chemotactic behavior of *Salmonella typhimurium*. *J. Mol. Biol.* **97**, 207–223.

Berg, H. C. (1971). How to track bacteria. *Rev. Sci. Instrum.* **42**, 868–871.

Berg, H. C., and Brown, D. A. (1972). Chemotaxis in *Escherichia coli* analyzed by three-dimensional tracking. *Nature (London)* **239**, 500–504.

Boos, W., Gordon, A. S., Hall, R. E., and Price, H. D. (1972). Transport properties of the galactose-binding protein of *Escherichia coli*. Substrate-induced conformational change. *J. Biol. Chem.* **247**, 917–924.

Clarke, S., and Koshland, D. E., Jr. (1979). Membrane receptors for aspartate and serine in bacterial chemotaxis. *J. Biol. Chem.* **254**, 9695–9702.

DeFranco, A. L., and Koshland, D. E., Jr. (1980). Multiple methylation in the processing of sensory signals during bacterial chemotaxis. *Proc. Natl. Acad. Sci. U.S.A.* (in press).

DeFranco, A. L., Parkinson, J. S., and Koshland, D. E., Jr. (1979). Functional homology of chemotaxis genes in *Escherichia coli* and *Salmonella typhimurium*. *J. Bacteriol.* **139**, 107–114.

Doetsch, R. N., and Hageage, G. J. (1968). Motility in procaryotic organisms: problems, points of view, and perspectives. *Biol. Rev. Cambridge Philos. Soc.* **43**, 317.

Fahnestock, M., and Koshland, D. E., Jr. (1979). Control of the receptor for galactose taxis in *Salmonella typhimurium*. *J. Bacteriol.* **137**, 758–763.

Hazelbauer, G. L. (1975). The maltose chemoreceptor of *Escherichia coli*. *J. Bacteriol.* **122**, 206–214.

Kleene, S. J., Toews, M. L., and Adler, J. (1977). Isolation of glutamic acid methyl ester from *E. coli* membrane protein imbibed in chemotaxis. *J. Biol. Chem.* **252**, 3214–3218.

Kondoh, H., Ball, C. B., and Adler, J. (1979). Identification of a methyl-accepting chemotaxis protein for the ribose and galactose chemoreceptors of *Escherichia coli*. *Proc. Natl. Acad. Sci. U.S.A.* **76**, 260–264.

Kort, E. N., Goy, M. F., Larsen, S. H., and Adler, J. (1975). Methylation of a membrane protein involved in bacterial chemotaxis. *Proc. Natl. Acad. Sci. U.S.A.* **72**, 3939–3942.

Koshland, D. E., Jr. (1977). A response regulator model in a simple sensory system. *Science* **196**, 1055–1063.

Koshland, D. E., Jr. (1979a). Bacterial chemotaxsis. *In* "The Bacteria" (J. R. Sokatch and L. N. Ornston, eds.), Vol. 7, pp. 111–166. Academic Press, New York.

Koshland, D. E., Jr. (1979b). A model regulatory system—bacterial chemotaxis. *Physiol. Rev.* **59**, 811–862.

Laemmli, U. K. (1970). Cleavage of structural proteins during the assembly of the head of bacteriophage T4. *Nature (London)* **227**, 680–685.

Macnab, R. M. (1979). Chemoresponsiveness in bacterial and unicellular eukaryotes. I. Bacterial chemotaxis. *In* "Encyclopedia of Plant Physiology, New Series, Physiology of Movements" (W. Haupt and M. E. Feinleib, eds.), Vol. 7, pp. 310–344. Springer-Verlag, Berlin and New York.

Macnab, R. M., and Koshland, D. E., Jr. (1972). The gradient-sensing mechanism in bacterial chemotaxis. *Proc. Natl. Acad. Sci. U.S.A.* **69**, 2509–2512.

Ordal, G. W., and Adler, J. (1974). Properties of mutants in galactose taxis and transport. *J. Bacteriol.* **117**, 517–526.

Paoni, N., and Koshland, D. E., Jr. (1979). Use of permeabilized cells for studies on the biochemistry of bacterial chemotaxis. *Proc. Natl. Acad. Sci. U.S.A.* **76**, 3693–3697.

Sanderson, K. E. (1972). Linkage map of *Salmonella typhimurium*. *Bacteriol. Rev.* **36**, 558–586.

Silverman, M., and Simon, M. (1976). Operon controlling motility and chemotaxis in *E. coli*. *Nature (London)* **264**, 577–579.

Silverman, M., and Simon, M. (1977a). Chemotaxis in *Escherichia coli:* Methylation of *che* gene products. *Proc. Natl. Acad. Sci. U.S.A.* **74,** 3317–3321.

Silverman, M., and Simon, M. (1977b). Identification of polypeptides necessary for chemotaxis in *Escherichia coli. J. Bacteriol.* **130,** 1317–1325.

Silverman, M., Matsumura, P., Hilmen, M., and Simon, M. (1977). Characterization of lambda *E. coli* hybrids carrying chemotaxis genes. *J. Bacteriol.* **130,** 877–887.

Smith, C. H., Brown, N. E., and Larner, J. (1971). Molecular characteristics of the totally dependent and independent forms of glycogen synthase of rabbit skeletal muscle. *Biochim. Biophys. Acta* **242,** 81–88.

Springer, M. S., Goy, M. F., and Adler, J. (1977). Sensory transduction in *Escherichia coli:* two complementary pathways of information processing that involve methylated proteins. *Proc. Natl. Acad. Sci. U.S.A.* **74,** 3312–3316.

Springer, M. S., Goy, M. F., and Adler, J. (1979). Protein methylation in behavioural control mechanisms and in signal transduction. *Nature (London)* **280,** 279–284.

Springer, W. R., and Koshland, D. E., Jr. (1977). Identification of a protein methyltransferase as the *cheR* gene product in the bacterial sensing system. *Proc. Natl. Acad. Sci. U.S.A.* **74,** 533–537.

Stock, J. B., and Koshland, D. E., Jr. (1978). A protein methylesterase involved in bacterial sensing. *Proc. Natl. Acad. Sci. U.S.A.* **75,** 3659–3663.

Strange, P. G., and Koshland, D. E., Jr. (1976). Receptor interactions in a signaling system: Competition between ribose receptor and galactose receptor in the chemotaxis response. *Proc. Natl. Acad. Sci. U.S.A.* **73,** 762–766.

Thomas, G., Siegmann, M., and Gordon, J. (1979). Multiple phosphorylation of ribosomal protein S6 during transition of quiescent 3T3 cells into early Gi, and cellular compartmentalization of the phosphate donor. *Proc. Natl. Acad. Sci. U.S.A.* **76,** 3952–3956.

Van der Werf, P., and Koshland, D. E., Jr. (1977). Identification of a γ-glutamyl methyl ester in a bacterial membrane protein involved in chemotaxis. *J. Biol. Chem.* **252,** 2793–2795.

Vandlen, R. L., Wu, W. C.-S., Eisenuch, J. C., and Raftery, M. M. (1979). Studies on the composition of purified torpedo Californium autyl choline receptor and its subunits. *Biochemistry* **18,** 1845–1854.

Warrick, H. M., Taylor, B. L., and Koshland, D. E., Jr. (1977). The chemotactic mechanism of *Salmonella typhimurium:* Preliminary mapping and characterization of mutants. *J. Bacteriol.* **130,** 223–231.

Weibull, C. (1960). Movement. *In* "The Bacteria" (I. C. Gunsalus and R. Y. Stanier, eds.), Vol. 1, p. 153. Academic Press, New York.

Zukin, R. S., Strange, P. G., Heavey, L. R., and Koshland, D. E., Jr. (1977). Properties of the galactose binding protein of *Salmonella typhimurium* and *Escherichia coli. Biochemistry* **16,** 381–386.

22

Biology and Physical Chemistry of Feeding Response of Hydra

HOWARD M. LENHOFF

Biochemistry of Taste and Olfaction

ISBN 0-12-154450-8

I. INTRODUCTION

The terms taste and olfaction, often grouped under the broad heading of chemoreception, may not refer to comparable phenomena among the lower metazoans. For example, a causal examination of the behavior of some aquatic invertebrates might suggest that the mechanisms by which they detect chemicals in potential food are analogous to those observed in taste and olfaction in higher terrestrial organisms. Yet on closer examination, these chemically evoked behavioral and physiological responses may show more similarity to hormonal and neurotransmitter systems of higher forms. Questions arising from categorizing chemosensory phenomena cannot be put off as simply a matter of semantics. The mere labeling of a phenomenon often determines how the problem will be investigated, and by whom, that is, what experimental approach will be taken, and whether the investigator will be a psychologist or a biochemist, an ethologist or a physiologist. In the case of the lower invertebrates, the demarcation between behavior and physiology becomes gossamer thin because many lower invertebrates lack endocrine systems, well-defined nervous systems, and structures usually associated with more highly organized animals.

Research on the control of feeding behavior of hydra and of some other cnidarians to exogenous chemicals points out some of these hazy areas of categorizing chemosensory phenomena. On the other hand, such research also gives insight into the evolution of receptors and sensory integration and into the mechanism of activation of receptors to specific chemicals. Although chemical activation of cellular receptors evolved in single cell organisms, it is in the lower metazoans that these chemical signals were first transduced from cell to cell to effect a coordinated organismal response. The cnidarians appear to offer a pivotal point in the evolution of chemical receptor systems into the more specialized ones of higher forms. The chemoreceptors of cnidarians always face the environment, i.e., the external fluids or the fluids in the gut, because these diploblastic animals are composed of basically two epithelial layers of cells separated by a thin acellular mesolamella. Thus, there exists the possibility that as tripoblastic animals with organ systems evolved, existing surface chemoreceptors of ancesteral diploblastic metazoans became internalized and took on new functions responding to circulating substances. Perhaps it is at this point in evolution that chemoreceptors and their corresponding effector systems started on the pathway toward specialization.

In our research on the feeding behavior of hydra and of some other cnidarians, we have focused primarily on the common feature of all chemosensory systems—the activation of a receptor by a specific molecule. This chapter gives an overview of five aspects of this research: (a) the biology of the feeding behavior in hydra as activated by reduced glutathione (γ-glutamylcysteinylglycine) (b) feeding activators in other cnidarians, (c) integration of feeding activators with other receptor–effector systems, (d) properties of the receptor, and (e) structure–activity relationships and conformational studies on glutathione analogs.

Our overall goal thus far has been to understand the biology of the animal and its feeding response sufficiently well to characterize properties of the receptor *in vivo* and determine the conformation and/or structure-activity relationships of the activator while it is at the receptor site. Such information, which is now available, will provide a firm basis of comparison when we reach the stage of characterizing the isolated receptor. On the other hand, in investigating chemoreceptor mechanisms in an animal as primitive as hydra, we have particularly avoided modeling our research on receptors after that done on other systems currently being investigated. Instead we exploit hydra's unique properties in order to find out what it can tell us. Research on hydra may not provide the whole picture, but we think it will provide some useful insights into basic properties of metazoan chemoreceptors, which may not be readily attainable from other systems.

II. BIOLOGY AND QUANTIFICATION OF GLUTATHIONE-ACTIVATED FEEDING BEHAVIOR

A hydra is shaped like a two-ply hollow tube, about 8 by 1 mm when extended, made up of both outer (ectodermal) and inner (endodermal) epithelial layers. At the posterior end of the tube is a basal disk with which the hydra usually attaches to a surface, and at the anterior end is a mouth surrounded by a ring of tentacles (Fig. 1). The tentacles are armed with many nematocytes, one of the seven cell types of hydra. The nematocytes contain nematocysts ("stinging capsules"), which start the feeding process by piercing the prey with long spear-like tubules.

Feeding in hydra, i.e., the capture and ingestion of food, consists of a number of separate steps: (a) A prey organism that accidentally blunders into an outstretched tentacle is captured, wounded, and poisoned through the action of the deadly nematocysts that line the tentacle; (b) following capture of the prey, the tentacles contract toward the mouth and the mouth opens; (c) on contact with the mouth, the food is ingested.

The second step (hereafter called the feeding response), that is, the contraction of the tentacle toward the mouth and the opening of the mouth, is under chemical control. It has long been known that food extracts elicit a

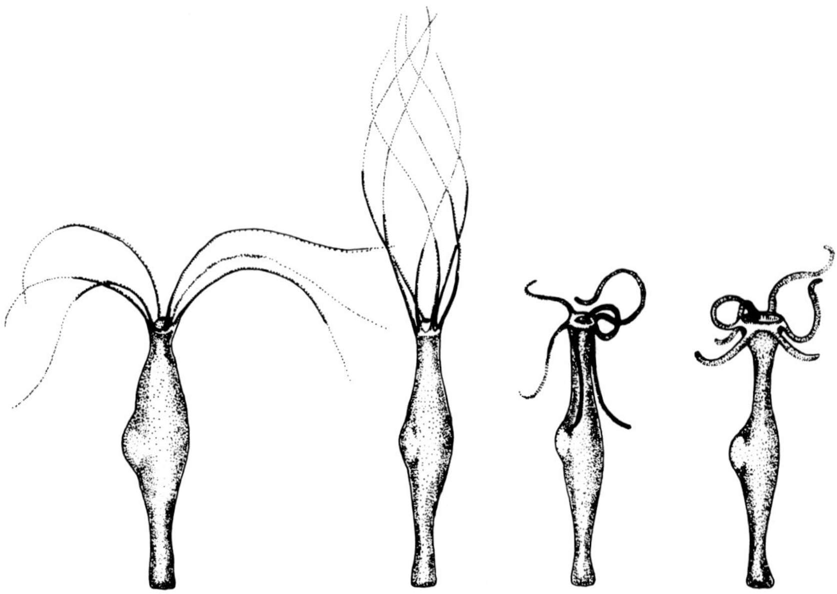

Fig. 1. Stages of the GSH-activated feeding response of hydra. See text. (Lenhoff, 1961a, reprinted with permission of *J. Gen. Physiol.*)

feeding response in hydra (Ewer, 1947). A landmark experiment was reported by Loomis (1955) who showed that the ubiquitous tripeptide reduced glutathione specifically activates the feeding response of *Hydra littoralis.*

In our laboratory, we have been investigating factors affecting the feeding response of hydra initiated by glutathione (GSH) to get insight into the general mechanisms by which cellular receptors are activated by a specific molecule. We found that although hydra did not offer some of the typical advantages for receptor research, such as easy accessibility to electrophysiological measurements or having readily identifiable chemoreceptors, it did offer other rather unique advantages. For one, the molecule GSH itself is sufficiently complex, though not too large, to allow making a significant number of analogs for determining the conformation(s) and structure–activity relationships of GSH necessary to activate the receptor.

A second advantage derives from the very nature of hydra itself: (a) It is of a simple tissue-level structure; (b) the receptor is on the surface of the outer epithelium; (c) the biological response can be readily quantified; (d) pure clones of the animals can be easily grown in the laboratory in kilogram quantities (wet weight); and (e) the fluid environment (medium) surrounding the receptor can be accurately controlled within a pH range of about 4–8 and over wide ranges of ion concentrations.

An extraordinary feature of investigating the glutathione response is that by merely recording visual observations of hydra's feeding behavior, it is possible to deduce many physicochemical parameters of both the receptor and the activator. Hence, special attention was given to development of this behavioral assay and to having animals that could respond in virtual synchrony and reliably to GSH.

A. Growth, Maintenance, and Selection of Hydra

To obtain repeatedly reliable and reproducible results, it was essential to have available a large number of hydra that were genetically alike, in the same stage of development and that had been grown in a rigorously controlled and defined environment. Methods are described for mass culture of hydra (Loomis and Lenhoff, 1956; Lenhoff and Brown, 1970) and for growth and maintenance of *Hydra littoralis* for experimentation on the feeding response (Lenhoff, 1961a, 1965). To ensure greater precision, we also selected the experimental animals from sparsely populated culture dishes because we have found that hydra from densely populated areas do not exhibit a maximal feeding response (Lenhoff, 1965). We also used animals of the same degree of pigmentation, as judged visually, because they seem to show less variation in their responses to GSH than do animals of differing pigmentation. In addition, before testing the animals, they are kept in a defined medium free of K^+ for at least 8 hr because K^+ ions are inhibitory (Lenhoff, 1965).

B. Description of the Feeding Response and Its Measurement

A hydra in the absence of glutathione is shown in Fig. 1A: the mouth is closed, and the tentacles are outstretched and relatively motionless. After addition of glutathione, the tentacles first begin to writhe and sweep inwards toward the central vertical axis of the animal (Fig. 1B). Next, the tentacles bend toward the mouth, and the mouth opens (Fig. 1C). Shown in this composite drawing (Fig. 1C) are the various positions that a tentacle takes before contracting. These movements, culminating in mouth opening, usually all take place within one-half a minute. Figure 1D shows how a hydra looks during the greater portion of the feeding reflex, its mouth open wide and the tentacles in various phases of contraction. Frequently, the tips of the tentacles are observed within the hydra's mouth, as shown in Fig. 1C and 1D.

To assure objectivity and to obtain quantitative data of the feeding response, we measured the time that elapsed between the moment the hydra were placed in the glutathione solution and the instant the mouths initially opened (initial time or and t_i) and finally closed (final time or t_f). The length of time that a mouth was open in response to glutathione was, therefore, $t_f - t_i$ (Lenhoff, 1961a). Each value for $t_f - t_i$ that we reported is the mean for a group of five animals. We have found that, using the aforementioned conditions for growing, maintaining, and selecting the hydra and for carrying out the experiments, each set of animals used per experiment usually responded to glutathione in near synchrony and gave $t_f - t_i$ values having little standard deviation. The initial time, t_i, probably represents the time it takes for glutathione to combine with the receptor and for all of the physiological events to occur in the interim between activation of the receptor and the opening of the mouth. Thus, $1/t_i$ would represent the rate at which these events take place.

The time that the mouth remains open, $t_f - t_i$, is interpreted to represent the time that it takes for some limiting substrate to be consumed. Alternatively, it could also indicate the time required for an inhibitor to be released and become effective. The $t_f - t_i$ can be shortened or lengthened by varying the temperature. Hence, some event(s) affecting the duration of the feeding response may be thermochemical (Lenhoff, 1961b).

III. OTHER FEEDING ACTIVATORS IN CNIDARIANS

Because only GSH activates the feeding behavior of all species of hydra that have been tested thus far, as well as behavioral responses in a tick (Galun and Kindler, 1965), a hagfish (Døving and Holmberg, 1974), and a snail (Kater and Rowell, 1973), there is no reason to a priori believe that something unique about this molecule makes it the only suitable activator of feeding among the cnidarians. A survey of over 25 cnidarians showed that (a) other compounds could serve as specific activators of the feeding behavior, and (b) in some cases, animals responded to two or more feeding activators in diverse, but specific, ways.

Animals tested from every class and most families of the Cnidaria elicit a feeding response to either one or a few small-molecule compounds (Table I). The molecules found most commonly to initiate a feeding response are the tripeptide GSH and the imino acid proline. In the Hydrozoa, the feeding response of each organism investigated was induced by a single specific compound (Lenhoff, 1974; Lenhoff and Heagy, 1977). Proline is especially prevalent as an activator among the athecate colonial marine hydroids. For example, all members of these groups thus far tested, i.e., *Cordylophora* (Fulton, 1963), *Pennaria* (Pardy and Lenhoff, 1968), *Tubularia* (Rushforth, 1976), and *Proboscidactyla* (Spencer, 1974), responded only to proline. All

TABLE I Chemical Activators of Feeding in the Cnidaria

Type of organism	Activator
I. Hydrozoans	
A. Hydroids	
1. Five species of hydra	GSH
2. Four species of colonial marine hydroids (without theca)	Proline
B. Siphonophores	
Two species (including Portuguese Man-of-War)	GSH
II. Anthozoans	
A. Sea anemones	
1. *Anthopleura*	GSH
2. *Boloceroides*	Valine
3. *Actinia*	Glutamate
4. *Haliplanella*	Leucine
5. *Calliactis*	GSH, proline
B. Colonial anemones	
1. *Palythoa*	Proline and/or GSH
2. *Zoanthus*	GSH
C. Corals	
1. Six species	Proline or GSH
2. One species	4 Amino acids
III. Scyphozoa	
One species (large jellyfish)	20 Amino acids, GSH, glyclglycine

hydras tested responded only to GSH (Loomis, 1955; Lenhoff, 1974), and the only other hydrozoans tested also responded to GSH (Lenhoff and Schneiderman, 1959; Mackie and Boag, 1963).

In the Anthozoa, we see three trends. In general, although more than one compound may elicit feeding behaviors, the animals exhibit varying ranges of specificity. For example, the most specificity is seen among the sea anemones: *Boloceroides* responds primarily to valine (Lindstedt *et al.*, 1968), *Anthopleura* to GSH (Lindstedt, 1971a), *Haliplanella* to leucine (Lindstedt, 1971b), *Actinia* to glutamic acid (Steiner, 1957), and *Calliactis* to GSH (Reimer, 1973). The specificity broadens with the colonial anemones; whereas *Zoanthus* responds primarily to GSH (Reimer, 1971a), *Palythoa* responds to relatively high concentrations of either GSH or proline or to low amounts of these two activators acting synergistically (Reimer, 1971b). Lastly, corals seem to respond best to proline alone or GSH alone, as well as to numerous other amino acids at relatively higher concentrations (Mariscal and Lenhoff, 1968; Mariscal, 1971; Lehman and Porter, 1973).

Chrysaora, the only example of the large jellyfishes tested, seems to respond to GSH and to a large number of amino acids (Loeb and Blanquet, 1973). More species need to be tested before making any generalizations about this group.

I am not necessarily convinced, however, that all the compounds reported in Table I are true activators of feeding. Likewise, I also feel that other feeding activators for those organisms listed may yet be detected. Lenhoff and Heagy (1977) posed three criteria for a compound to be accepted as a natural activator of a feeding behavior: (a) The compound should be active in low concentrations, i.e., 10^{-4} M or less; (b) sufficient analogs of the presumed activator should be tested to show that the receptor has some degree of specificity; and (c) an analog that is a nontoxic competitive inhibitor should reversibly inhibit the response activated by natural tissue extracts. Thus far in only one instance, that of hydra, have all these criteria been met. The next most thoroughly studied case is Fulton's (1963) work showing that the proline activation of feeding in *Cordylophora lacustris* could take place in less than 10^{-4} M and that the specificity resided in the imino region of a heterocyclic α-imino acid, which is neither substituted nor unsaturated in such a way as to affect the imino acid group.

Nonetheless, in addition to the cases where some anthozoans respond to either GSH or proline, there are two other unusual cases in which cnidarians respond to two substances. In both cases, each substance initiates a separate response. Yet in the presence of both substances, a very specific behavioral pattern is initiated.

Lindstedt (1971a) describes an unusual case in which two phases of the feeding response of the sea anemone *Anthopleura elegantissima* are controlled by different chemical activators. Asparagine controls the contraction and bending of tentacles which brings food to the mouth; reduced glutathione controls the ingestion of food once it contacts the mouth. A complete feeding response occurs only when both chemical activators are present.

An even more complex case occurs in hydra. In addition to the GSH receptor, there also exists a receptor to tyrosine on the surface of cells lining the hydra's gut. When hydra are activated by GSH present in the external environment and by tyrosine present in the gut, the animal exhibits a "neck response," i.e., a constriction of the upper one-third of the body tube. These neck constrictions apparently allow hydra to retain previously ingested food in the gut while swallowing newly captured prey. No other natural amino acid, including phenylalanine, could substitute for tyrosine. Analogs of tyrosine having either the α-amino or α-carboxyl blocked were inactive (Blanquet and Lenhoff, 1968).

From these experimental results we conclude that in addition to its external glutathione receptor, hydra has an enteroreceptor specific for tyrosine. The hydroxyl, the α-amino, and the α-carboxyl groups must all be present in order for the amino acid to be active (Blanquet and Lenhoff, 1968).

The existence in hydra of two chemoreceptor systems that must act in harmony represents, to our knowledge, the first report of two integrated, chemically mediated responses in the lower invertebrates. This system differs from the asparagine–glutathione system in the sea anemone *Anthopleura* (Lindstedt, 1971a) in which the molecules act in two sequential steps to activate feeding.

An Evolutionary Consideration

We have presented elsewhere (Lenhoff, 1974, 1975; Lenhoff *et al.*, 1976) arguments supporting the view that receptors in higher organisms, such as those involved in chemoreception and in neurotransmitter and hormone action, evolved from similar primitive general receptors, which originally functioned in pinocytosis in single cells and in activation of feeding behaviors in the lower metazoans. These arguments are based primarily upon research on the specificity of feeding activators of the cnidarians mentioned in Table I. The overall rationale of these arguments is that it would seem simpler for organisms during evolution to modify existing chemoreceptors to control new tasks rather than to develop completely new receptor–effector systems (Lenhoff, 1974, 1975). Such speculations are analogous to the generally accepted conservative view of the evolution of proteins.

IV. INTEGRATION OF GSH RECEPTOR–EFFECTOR SYSTEM WITH OTHER SENSORY SYSTEMS OF HYDRA

The interaction between the tyrosine and GSH receptor–effector systems represents only one of six effects of the GSH system, three of which involve interactions with other receptor–effector systems. These other actions include: (a) increasing the rates of tentacle-waving "concerts" (Rushforth and Hofman, 1972), (b) and (c) inhibition of two behavioral contraction responses of hydra, which are not part of the feeding response (Rushforth *et al.*, 1964), and (d) stimulated changes in bioelectric potentials related to the feeding response.

A. Tentacle Concerts

The flexing motions of the tentacles sweeping inward toward the central vertical axis above the mouth (Fig. 1A) are called *tentacle concerts.* These concerts occur spontaneously in hydra. Rushforth and Hofman (1972) found that reduced GSH in a concentration as low as 5×10^{-10} *M* gives rise to a statistically significant increase in the frequency of tentacle concerts in *Hydra littoralis* and that the frequency accelerates with increasing GSH concentration up to 5×10^{-9} *M*. At this higher GSH concentration, tentacle writhing commences. Tentacle concerts and tentacle writhing may be seen in Figs. 1B and 1C, respectively.

B. Glutathione Inhibition of Contraction Responses

Rushforth (1965) has conducted extensive and convincing experiments that show that the contractions of hydra induced either by light or by shaking, as well as the animals' spontaneous contractions, are inhibited while the animals are feeding on *Artemia salina*. Prompted by the experiments on the feeding response, Rushforth first tested *Artemia* extracts and then GSH; both inhibited the contractions. Using his quantitative procedures for measuring inhibition of the contraction response, he showed (Rushforth, 1965) that the "modes of action of glutathione are similar to those discovered . . . [in] studying the mouth opening response."

The mechanism by which GSH inhibits the contraction response induced by light or mechanical agitation is unknown. Possibly there is a direct means whereby GSH turns off contraction, or glutathione may inhibit the response as an indirect consequence of eliciting the contractile events involved in feeding behavior.

C. Electrophysiological Correlates of Glutathione-Activated Feeding Response

Extending his research on the contraction responses, Rushforth found both indirect and direct electrophysiological correlates of the GSH-activated feeding response (Rushforth, 1967a,b). Just as homogenates of *Artemia* or solutions of GSH inhibited the contraction response of hydra, so they inhibited the production of electrical potentials associated with either the spontaneous contractions of hydra or contractions induced by light. Furthermore, Rushforth found that the electrical potentials associated with the contraction of isolated tentacles of *Hydra pseudologactis* were not produced in the presence of 10^{-5} *M* reduced GSH.

More striking is Rushforth's discovery that, when reduced GSH inhibited the production of potentials associated with tentacle contraction, at the same time it directly initiated potentials associated with the GSH-induced asymmetric tentacle movements. As the tentacle adapted to GSH, the frequency of these potentials decreased, and the spontaneous tentacle contractions with their associated potentials were restored. Hence, these experiments

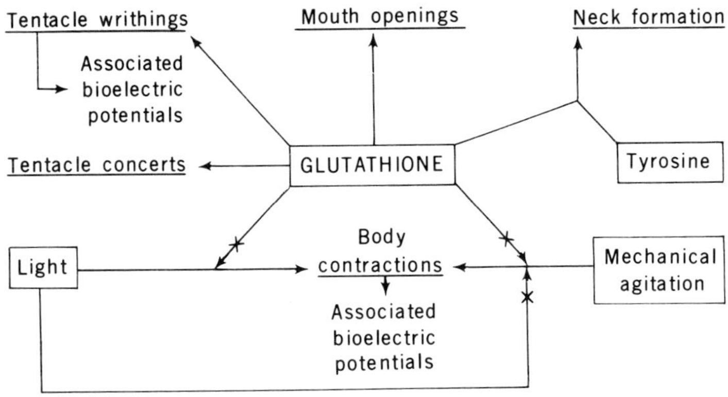

Fig. 2. Integration of receptor–effector systems of hydra. This diagram, which summarizes the various measurable effects stimulated by the action of glutathione on hydra, also emphasizes that the output of the glutathione receptor is linked with other receptor–effector systems of the animal. In addition, it points out an action of light in inhibiting the contraction response to mechanical agitation. (Lenhoff, 1968a, reprinted with permission of *Science*.)

not only present the first evidence of direct electrophysiological correlates of GSH action but also provide strong evidence for the presence of glutathione receptor sites on hydra tentacles.

D. Sensory Systems of Hydra

It is apparent from the above studies that the behavior of the "simple" hydra is not as straightforward as commonly thought. That is, hydra somersault to light, discharge nematocysts, and swallow food. The information summarized in Fig. 2 shows that hydra's receptor–effector systems interact in a number of ways. It is of interest to note that hydra possess the basic elements of three primary senses—those to chemical, light, and mechanical stimuli. Recent evidence shows that hydra lacking nerve cells (Campbell, 1976) neither contract nor respond to GSH (Heagy, Campbell, and Lenhoff, unpublished). Hence, it appears that although hydra have no central nervous system, the cells of the nerve net play a role in contraction and feeding behaviors.

V. PROPERTIES OF THE GSH RECEPTOR

A. Evidence of Specificity for GSH on the Outside Surface of the Animal

Earlier experiments of Loomis (1955) showed that GSH was the substance leaking from wounded prey that activated feeding in *Hydra littoralis*. To eliminate the possibility of activity caused by a contaminant in the GSH preparation, he showed that chemically synthesized GSH was also active. Later, while investigating the competitive action of GSH analogues, I confirmed that the GSH in the prey's fluids activate the response. Other early experiments (Lenhoff, 1961a,b) showed that the duration of the response depended upon the concentration of GSH applied to the hydra and that the response took place only with GSH in the solution. Furthermore, the GSH was neither consumed nor metabolized. These experiments suggested that, on the surface of the hydra's outer epithelium, there were receptor sites that become activated only in the presence of GSH.

A saturable receptor was indicated by analysis of plots of the duration of the feeding response against the concentration of GSH to which the hydra was exposed (Lenhoff, 1961a). Analysis of the resultant curve with concepts borrowed from enzymology suggests that there is a receptor that is saturated and gives a maximal response at GSH concentrations of 5×10^{-6} *M* and greater. A maximal response is considered analogous to the maximal velocity of an enzyme-catalyzed reaction; both occur during saturation of an active site.

B. Determination of the Apparent Dissociation Constant

The assumptions made in determining the apparent dissociation constant, K_A, between the activator (*A*) and the receptor (*R*) have been reported elsewhere (Lenhoff, 1965, 1968b). The effect of the activation is signified by ϵ and the maximum effect by ϵ_M. The equation derived

$$(A)/\epsilon = 1/\epsilon_M\,(A) + K_A/\epsilon_M \tag{1}$$

is analogous to the second form of the Lineweaver–Burk plot, the equation developed by Beidler (1954) for mammalian taste chemoreception, which is a form of the Langmuir adsorption isotherm. This equation is useful in analyzing chemoreception phenomena because it minimizes deviations in individual animal responses that occur at very low levels of activator.

Previous data (Lenhoff, 1969) have shown that this equation can be used to interpret the plot of $(A)/\epsilon$ against (A); we obtained straight lines at most glutathione concentrations (Fig. 3). From such plots, we can determine, for example, at pH 7, an apparent K_A of 10^{-6} *M*. Such a low K_A is meaningful from at least three viewpoints: (a) The low constant indicates a high affinity of the receptor for glutathione; (b) concentrations around 10^{-6} *M* are well within the physiological range expected under natural conditions of feeding; and (c) this constant provides a means of characterizing the receptor, that is, the glutathione receptor of *H. littoralis* may be said to have an apparent dissociation constant of 10^{-6} *M* under the given conditions. The constant is

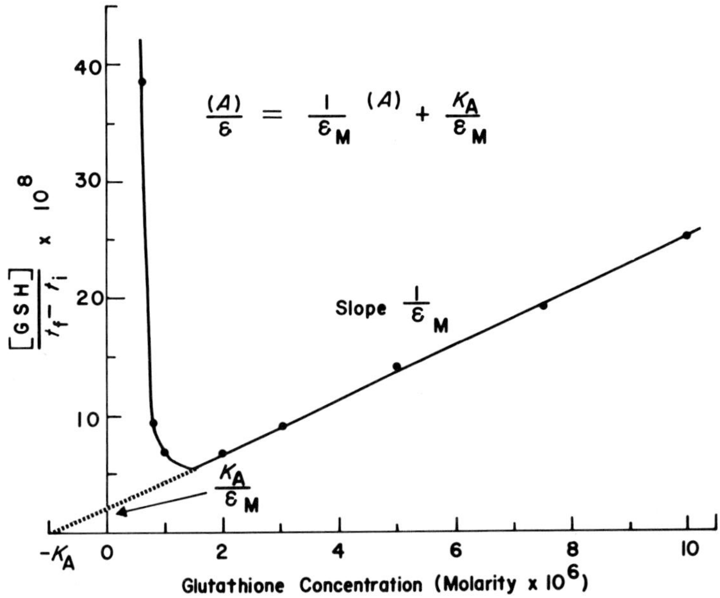

Fig. 3. Plot for determining constants of the combination of glutathione with its receptor. See text for details. (Lenhoff, 1965, reprinted with permission of *Am. Zool.*)

characteristic of the receptor and remains nearly the same whatever the nutritional state of the hydra (Lenhoff, 1961a,b). Similarly, experiments in which the buffer anion is varied alter the maximal response but not the dissociation constant (Lenhoff, 1969).

C. *In Vivo* Determination of pH Profile of Receptor

Changes in K_A with pH can be used to determine the pK's of the ionizable groups on glutathione or at the receptor site which are involved in the combination with glutathione. The pK measurements were made by means analogous to enzymology in determining the pK's of ionizable groups at the active sites of enzymes. For our purposes, we needed an equilibrium equation similar to Dixon's (1953; Dixon and Webb, 1964) for enzymes, which would take into account the influence of pH on the dissociation constant. This modified equation (Lenhoff, 1969) assumed that if the activator, receptor site, or activator–receptor complex ionizes, then each component in the expression for equilibrium (A, R, AR) equals its concentration multiplied by a term that is a function of pH. For example, if the activator ionized, then the total concentration of the activator, A_t, would be A times the pH function of A, or f_a (pH). The logarithimic form of the equation is:

$$pK_A = pK_A^0 + \log f_{ar}\,(\text{pH}) - \log f_r\,(\text{pH}) - \log f_a\,(\text{pH}) \tag{2}$$

Here pK_A refers to the negative logarithm of the dissociation constant of *AR*, whereas pK_A^0 is the same constant if none of the components has ionic groups; if no component ionizes, then pK_A and pK_A^0 are equal. The derivation of this equation is explained elsewhere (Lenhoff, 1968b).

The foregoing equation indicates that a plot of pK_A against pH will consist of a series of straight lines joined by short curved parts, and holds true for the glutathione–hydra system (Fig. 4). The results (Lenhoff, 1969) followed almost exactly the predictions from the modified Dixon equations. The following interpretations were made (Lenhoff, 1969): (a) Ionizable groups of the receptor site participate in binding glutathione because significant variations in pK_A occurred with change in pH. (b) The concave downward inflections at pH's 4.6, 4.8, 6.5, and 7.6 represent p*K*'s of ionizable groups at the receptor site. These p*K*'s probably do not represent ionizable groups of glutathione, which have p*K*'s either below pH 4 (2.1 and 3.5) or above pH 8 (8.7 and 9.6) (Wieland, 1954). If the receptor site is protein, then the p*K*'s may represent two β-carboxyls of peptide aspartic acid (or γ-carboxyls of peptide glutamic acid), an imidazole group, and a terminal α-amino group, respectively. (c) The horizontal lines indicate pH values that do not affect the combination of glutathione with the receptor site. (d) The quenching of the charges (Dixon and Webb, 1964) at about pH 4 and 8 indicate that receptor-site groups having p*K*'s of 4.6 and 7.6 may be associated with complementary charged groups of glutathione.

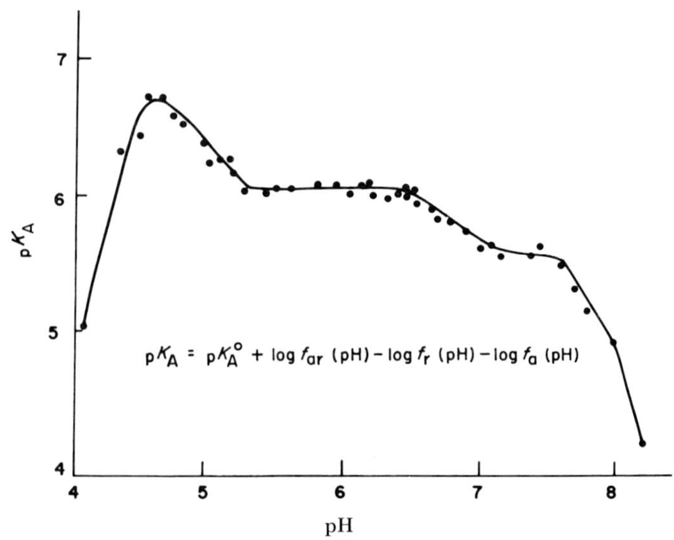

Fig. 4. Effect of pH on the dissociation constant, K_A, between glutathione and its receptor. (Lenhoff, 1965, reprinted with permission of *Am. Zool.*)

D. Other Factors Influencing the Response

The feeding response of hydra is also influenced by a number of external factors, such as the ionic composition of the medium (Table II) and certain proteases (Lenhoff and Bovaird, 1960). Environmental cations, for example, affect the response activated by GSH in many ways. Without calcium ions, hydra do not respond to reduced GSH (Lenhoff and Bovaird, 1959). The calcium requirement is pH-dependent (Lenhoff, unpublished), and a concentration of about 10^{-4} *M* was necessary for a maximal response. Strontium was the only ion that could substitute for calcium, and even it was much less effective (Lenhoff and Bovaird, 1959). The chelating agent ethylenediaminetetraacetic acid (EDTA) also inhibited feeding. This inhibition was completely reversed only by calcium ions and to some degree by strontium ions (Lenhoff and Bovaird, 1959). Magnesium ions were not required; in high concentrations they inhibited the responses by competing with calcium ions (Lenhoff, 1961b). Sodium likewise competed with calcium, but less effectively than did magnesium (Lenhoff and Bovaird, 1959).

Potassium ions inhibit the feeding response (Lenhoff, 1965), but unlike magnesium and sodium ions, they did not act by competing with calcium ions. Concentrations of K^+ as low as 10^{-4} *M* could lower the response to glutathione significantly, and this inhibition could be reversed by placing the animals in a potassium-free medium for a few hours (Lenhoff, unpublished). Since K^+ plays an important role in bioelectric potential, it may be that these ions act by affecting the cell membrane potential of hydra. More recent work using *H. attenuata* shows that the presence of Na^+ in the environment is an absolute requirement for the glutathione-mediated feeding response (Asbill, 1975).

Temperature effects on the feeding response are complex (Lenhoff, 1961b), but they indicate a rate-limiting thermochemical step in the effector system with an activation energy of about 13,000–14,000 calories (Lenhoff,

TABLE II Effects of Cations[a]

Cation	Required (C_{50})[b] (M)	Inhibits (I_{50})[c] (M)
Na^+	10^{-4}	—
K^+	—	10^{-5}
Ca^{2+}	10^{-4}	—
Mg^{2+}	—	10^{-3}

[a] Data summarized from unpublished experiments.

[b] C_{50} refers to concentration needed for maximal responses in the presence of 10^{-5} *M* GSH.

[c] I_{50} refers to concentration inhibiting the response given by hydra in the presence of 10^{-5} *M* GSH and 10^{-4} *M* Ca^{2+}.

unpublished). This step is thought to involve consumption of some regenerable substance, such as ATP.

A number of nontripeptides can activate a feeding response in some coelenterates in the absence of added GSH (Lenhoff and Bovaird, 1960; Lenhoff and Zwisler, 1963). Of special interest are the proteolytic enzymes papain, ficin, and trypsin (Lenhoff and Bovaird, 1960) because it is now recognized that numerous control mechanisms can be activated by proteases (Reich *et al.*, 1976).

E. Cyclic Nucleotides

As with higher organisms, cyclic nucleotides are thought to play in lower forms a similar function as second messengers. The first report that cAMP functions in such a fashion in cnidarians is that of Gentleman and Mansour (1974) for the sea anemone *Anthopleura elegantissima*. They showed that, on addition to GSH, the concentration of cAMP in the oral disk and pharynx tissues increased. Preliminary work with hydra shows that addition of GSH causes rapid changes in the levels of both cGMP and cAMP (Cobb *et al.*, 1980).

VI. STRUCTURE–ACTIVITY AND CONFORMATIONAL RELATIONSHIPS OF GSH AND ITS ANALOGS TO THE RECEPTOR

A. Earlier Work on Active Structure of Glutathione

To determine the structure of GSH present at the receptor site, it was necessary to quantify the ability of different structural analogs of GSH either to activate or inhibit a response. The aforementioned bioassay experiments were used to determine the effectiveness (K_A and ϵ_M) of synthetic agonists (Lenhoff, 1961b). The relative ability of analogs that bind but do not activate, i.e., antagonists, was determined by measuring their ability to inhibit competitively the activity of GSH in eliciting the feeding response (Lenhoff and Bovaird, 1961).

Data from these earlier investigations established that: (a) the thiol is not required for activation, because ophthalmic acid (γ-Glu-Abu-Gly), norophthalmic acid (γ-Glu-Ala-Gly), and *S*-methylglutathione (*S*-Me-GSH) also activated the response (Cliffe and Waley, 1958; Lenhoff and Bovaird, 1961); (b) activation of the response requires the intact tripeptide backbone of glutathione, because the just-mentioned analogs activated feeding, whereas amino acids, dipeptides, and a number of tripeptide analogs with large and charged substituents at the sulfhydryl group did not activate (Lenhoff, 1961b; Lenhoff and Bovaird, 1961); (c) the receptor has a high affinity for the glutamyl part of the tripeptide because glutamic acid and glutamine were the only amino acids to show competitive inhibition (Lenhoff and Bovaird, 1961), and the tripeptide asparthione (β-Asp-Cys-Gly) did not initiate the response (Loomis, 1955; Lenhoff and Bovaird, 1961); and (d) the α-amino of glutathione is probably required for association of glutathione with the receptor (Lenhoff, 1961a; Lenhoff and Bovaird, 1961).

A preliminary summary of research on the specificity of the glutathione receptor of *H. attenuata* will now be presented. This work involves the use of over 70 analogs, many of which were synthesized specifically for this project.* Full details of these researchers will be available shortly (Cobb *et al.*, 1976).

B. Position 3 (Glycine) Analogs

Analogs of glutathione were synthesized having substituents for glycine in the 3 position. The results (Table III) show that when glycine was replaced by either tyrosine or leucine, the resultant tripeptide activated a perfectly good feeding response. Tripeptides with the other substituents listed also initiated feeding behaviors, but of lesser to barely detectable degrees. We tentatively concluded from these experiments that the nature of the side chain of the 3 position amino acid is not critical for the tripeptide to activate, but that the presence of a charge at that end of the peptide is important. It is of interest that although the hydra GSH receptor responded to tripeptides having either tyrosine or leucine in the 3 position, the γ-glutamyltranspeptidase of hydra did not bind or react with those analogs (Danner *et al.*, 1976).

C. Position 2 (Cysteine) Analogs

Investigations of these analogs showed that many substituents in the 2 position gave peptides with about the same activity as GSH (Table IV). An even greater number of analogs, however, bound sufficiently well to behave as antagonists (competitive inhibitors), but were not able to activate a feeding response.

From these findings, it is apparent that neither the reducing or hydrogen-bonding properties of the sulfhydryl group of cysteine are needed to activate the receptors. The size and hydrophobicity of the 2 position side chain, however,

This research is a collaborative effort of some of my outstanding colleagues—Drs. Garland R. Marshall and Melanie H. Cobb of Washington University, St. Louis, Mo., and Drs. Wyrta Heagy and Jeanne Danner of the University of California at Irvine. Details of this work by these authors will be presented elsewhere.

TABLE III Effect of GSH Analogs with Substitutes for Glycine Moiety[a]

Good activators	Moderate activators	Poor activators
Tyrosine	Proline	Glycinamide
Leucine	Glycylalanine	Aminoisobutyric acid

[a] To avoid complications resulting from the presence of the thiol, these analogs have α-aminobutyric acid substituted for cysteine in position 2.

may play a role in affecting the activating properties of tripeptide analogs. For example, quantitative examination of the data summarized in Table IV shows a corresponding loss in activation and binding potency with norvaline > α-amino-*n*-butyric acid > alanine > glycine. Within this latter series of tripeptides, the potency probably does not depend upon size alone, but on the degree of lipophilicity and conformation flexibility inherent in the substituted groups. It is interesting that γ-glutamylglycylglycine did not activate, but instead was a weak antagonist. Hence, it is not simply the tripeptide structure that is required for activation. Possibly the 2 position glycine affects the conformation of the tripeptide so it will not activate. Addition of a small side chain, like an ethyl or a methyl to that glycine, might lead to a conformation of the tripeptide more compatible with the receptor.

D. Conformation of 1 Position (γ-Glutamyl) Needed for Activation

Because glutamic acid is the only amino acid component of the tripeptide that acts as an antagonist (Lenhoff and Bovaird, 1961) thereby inhibiting feeding behavior in the presence of GSH, we are now obtaining accurate

TABLE IV Effects of Analogs of GSH with Substitutes for the Cysteine Moiety

Strong activators	Moderate activators	Inhibitors
S-Me-cysteine	Alanine	Cysteine-S-S-G
Aminobutyric acid	Serine	Cysteine-S-(*N*-ethylsuccinimido)
Valine	Aminoisobutyric acid	
Norvaline		Tyrosine
Leucine		Phenylalanine
		S-acetylcysteine
		Cysteine sulfinate
		Cysteine sulfonate
		Glycine[a]

[a] γ-Glutamylglycylglycine was only a weak inhibitor.

information of the conformation of this amino acid at the receptor site. Our assay is to determine the concentration of glutamic acid or its analogs that inhibit the response activated by concentrations of GSH close to that of the K_A. With this assay, approximately 2×10^{-5} *M* glutamic acid inhibited the response by 50%. We interpret this inhibition as a reflection of the glutamic acid's ability to bind at the site normally occupied by the γ-glutamyl moiety of GSH.

Examination of the data in Table V shows that: (a) The asymmetric γ carbon was important for binding (inhibition became weaker as the extra methyl was moved from the γ- to the β- to the α-carbon); (b) the α-amino group had to be present and in the α-carbon (inactive as inhibitors were glutamic acid, α-ketoglutaric acid, *N*-methylglutamic acid, and β-aminoglutamic acid); (c) the α-carboxyl must be present (GABA was noninhibitory and isoglutamine was only partially inhibitory); (d) the three charged groups must attach to the receptor (L-glutamate inhibited whereas D-glutamate did not); (e) the three charged groups must be a specified distance from each other (aspartic, proline, and pyroglutamic acid did not inhibit, whereas α-aminoadipic acid was a good inhibitor).

Our investigation of two glutamic acid analogs having cyclopropane rings to restrict rotation around the β-carbon and γ-carbon bond made it possible to determine the approximate conformation of the glutamyl moiety of GSH at the receptor site. The analog *trans-α* (carboxycyclopropyl)glycine, (Tcg), proved to be as effective an inhibitor as is glutamic acid whereas the cis form, (Ccg), did not inhibit.

From the results presented in Table V, four compounds were selected from which we calculated the distances between their respective atoms. The strong inhibitors, glutamic acid and Tcg, were selected to demonstrate the

TABLE V Effect of Glutamic Acid and Its Analogs on Response Activated by GSH

Analog class	Strong inhibitor	Moderate inhibitor	Noninhibitory
A	L-Glutamic acid	—	D-Glutamic acid
B	γ-Me-glutamic acid	β-Me-glutamic acid	α-Me-glutamic acid
C	—	Isoglutamine	γ-Aminobutyric acid
D	—	Glutamine	Norvaline
E	α-Aminoadipic acid	—	Aspartic acid
	—	—	Proline
	—	—	Pyroglutamic acid
F	—	—	Glutamic acid
	—	—	α-Ketoglutaric acid
	—	—	β-Aminoglutamic acid
	—	—	*N*-methyl glutamate
G	*Trans*-(cyclopropyl)glycine	—	*Cis*-(cyclopropyl)glycine

TABLE VI Approximate Distances between Atoms in Glutamic Acid While It Binds to the Receptor

Stand	Distances (Å) of probable active range
α C → δ C[a]	2.89 → 3.87
α C = 0 → δ C	4.38 → 4.97
α N → δ C	4.38 → 4.97

[a] The δ C refers to the γ-carboxyl of glutamic acid.

widest possible ranges of interatomic distances possible. The noninhibitors, aspartic acid and Ccg, which also had the three active ionic groups, were selected to represent interatomic distances that should be excluded from the calculations. From this information, it was possible to approximate the active distances between the atoms while the analogs fit effectively into the receptor site. The results of these calculations are shown in Table VI.

Further calculations were carried out constructing plots similar to the kind developed by Ramachandran. These calculations show the potential energies of glutamic acid and of Tcg as functions of the rotations between α-carbon–β-carbon bonds and the β-carbon–γ-carbon bonds. The area of overlap suggests that most likely torsional angles are $-120 \pm 30°$ for the α-β bond and 180° for the β-γ one. These data are only provisional; we continue to gather information on α-aminoadipic acid and are attempting a synthesis of GSH with Tcg substituted for the γ-glutamyl moiety. From these calculations and experiments, we should be able to prove that the conformation for glutamic acid that acts as an antagonist may also approximate the conformation of the γ-glutamyl moiety of GSH while activating the receptor site.

VII. CONCLUDING REMARKS

In this chapter I summarized five aspects of research on the chemical control of the feeding behavior of hydra and some other cnidarians. The nature of these investigations range from a description of natural history behavioral observations, to a definition of ionic groups in the receptor site, and to a study of the interatomic distances within the activator molecule while it is associated with the receptor. Although the chemical sensory mechanisms involved in feeding may not fit within the accepted definitions of taste and olfaction in higher forms, research on these relatively simple receptor–effector systems may offer some insight into common features of all chemoreceptors—the activation of a receptor by its complimentary activating molecule.

ACKNOWLEDGMENT

I thank my many colleagues and students, all mentioned in the text, who have collaborated with me during the years on various aspects of this project. I give special mention to the special genius of the late W. F. Loomis, who, while discovering the GSH-activated response in hydra, broke down many of the barriers existing between behavioral zoology and biochemistry. The research reported in this paper was supported by various grants from the Nutrition Foundation, National Institutes of Health, and National Science Foundation. Current research is funded by National Science Foundation grant BNS-6810209.

REFERENCES

Asbill, M. (1975). Sodium ion effect on the glutathione induced feeding response of *Hydra attenuata*. *J. Undergrad. Res. Biol. Sci. Univ. Calif., Irvine* **4**, 210–228.

Beidler, L. M. (1954). A theory of taste stimulation. *J. Gen. Physiol.* **38**, 133–139.

Blanquet, R. S., and Lenhoff, H. M. (1968). Tyrosine enteroreceptor of hydra: its function in eliciting a behavior modification. *Science* **159**, 633–634.

Campbell, R. D. (1976). Elimination of *Hydra* interstitial and nerve cells by means of colchicine. *J. Cell Sci.* **21**, 1–13.

Cliffe, E. E., and Waley, S. G. (1958). Effect of analogues of glutathione on the feeding reaction of hydra. *Nature (London)* **183**, 804–805.

Cobb, M. H., Marshall, G. R., Heagy, W., Danner, J., and Lenhoff, H. M. (1976). On the mechanism of the glutathione induced feeding response in hydra: hormone-receptor interaction. *Fed. Proc., Fed. Am. Soc. Exp. Biol.* **35**, 1617. (Abstr.)

Cobb, M. H., Heagy, W., Danner, J., Lenhoff, H. M., and Marshall, G. (1980). Effect of glutathione on cyclic nucleotide levels in *Hydra attenuata*. *Comp. Biochem. Physiol. C.* **65**, 111–115.

Danner, J., Lenhoff, H. M., Cobb, M. H., Heagy, W., and Marshall, G. R. (1976). *Biochem. Biophys. Res. Commun.* **73**, 180–186.

Dixon, M. (1953). The effect of pH on the affinities of enzymes for substrates and inhibitors. *Biochem. J.* **55**, 161–170.

Dixon, M., and Webb, E. C. (1964). "Enzymes," 2nd ed., pp. 154–165. Academic Press, New York.

Døving, K. B., and Holmberg, K. (1974). A note on the function of the olfactory organ of the hagfish *Myxine glutinosa*. *Acta Physiol. Scand.* **91**, 430–432.

Ewer, R. F. (1947). On the functions and mode of action of the nematocyst of *Hydra*. *Proc. Zool. Soc. London* **117**, 365–376.

Fulton, C. (1963). Proline control of the feeding reaction of *Cordylophora*. *J. Gen. Physiol.* **46**, 823–837.

Galun, R., and Kindler, S. N. (1965). Glutathione as an inducer of feeding in ticks. *Science* **147**, 166–176.

Gentleman, S., and Mansour, T. E. (1974). Adenylate cyclase in a sea anemone-implication for chemoreception. *Biochim. Biophys. Acta* **343**, 469–479.

Kater, S., and Rowell, C. (1973). Integration of sensory and centrally programmed components in generation of cyclical feeding activity of *Helisoma trivolvis*. *J. Neurophysiol.* **36**, 142–155.

Lehman, J. T., and Porter, J. W. (1973). Chemical activation of feeding in the Caribbean reef building coral *Montastrea cavernosa*. *Biol. Bull. (Woods Hole, Mass.)* **145**, 140–149.

Lenhoff, H. M. (1961a). Activation of the feeding reflex in *Hydra littoralis*. I. Role played by reduced glutathione and quantitative assay of the feeding reflex. *J. Gen. Physiol.* **45**, 331–344.

Lenhoff, H. M. (1961b). Activation of the feeding reflex in *Hydra littoralis*. *In* "The Biology of Hydra and of Some Other Coelenterates" (H. M. Lenhoff and W. F. Loomis, eds.), pp. 203–232. Univ. of Miami Press, Coral Gables.

Lenhoff, H. M. (1965). Some physicochemical aspects of the microenvironments surrounding hydra during activation of their feeding behavior. *Am. Zool.* **5**, 515–524.

Lenhoff, H. M. (1968a). Behavior, hormone, and hydra. Science **161**, 434–442.

Lenhoff, H. M. (1968b). Chemical perspectives on the feeding response, digestion and nutrition of selected coelenterates. *In* "Chemical Zoology" (M. Florkin and B. Scheer, eds.), Vol. 2, pp. 157–221. Academic Press, New York.

Lenhoff, H. M. (1969). pH profile of a peptide receptor. *Comp. Biochem. Physiol.* **28**, 571–586.

Lenhoff, H. M. (1974). On the mechanism of action and evolution of receptors associated with feeding and digestion. *In* "Coelenterate Biology: Reviews and New Perspectives" (L. Muscatine and H. M. Lenhoff, eds.), pp. 359–389. Academic Press, New York.

Lenhoff, H. M. (1975). On the evolution of receptors associated with feeding. *In* "Sensory Physiology and Behavior" (R. Galun, P. Hillman, I. Parnas, and R. Werman, eds.), pp. 223–236. Plenum, New York.

Lenhoff, H. M., and Bovaird, J. (1959). Requirement of bound calcium for the action of surface chemoreceptors. *Science* **130**, 1474–1476.

Lenhoff, H. M., and Bovaird, J. (1960). Enzymatic activation of a hormone-like response in hydra by proteases. *Nature (Ldonon)* **187**, 671–673.

Lenhoff, H. M., and Bovaird, J. (1961). Action of glutamic acid and glutathione analogues on the hydra glutathione-receptor. *Nature (London)* **189**, 486–487.

Lenhoff, H. M., and Brown, R. D. (1970). Mass culture of Hydra: An improved method and its application to other invertebrates. *Lab. Anim.* **4**, 139–154.

Lenhoff, H. M., and Heagy, W. (1977). Aquatic invertebrates: Model systems for study of receptor activation and evolution of receptor proteins. *Annu. Rev. Pharmacol. Toxicol.* **17**, 243–258.

Lenhoff, H. M., and Schneiderman, H. A. (1959). The chemical control of feeding in the Portuguese man-of-war, *Physalia physalia* L., and its bearing on the evolution of the cnidaria. *Biol. Bull. (Woods Hole, Mass.)* **116**, 452–460.

Lenhoff, H. M., and Zwisler, J. (1963). Zinc activation of a coordinated response in hydra. *Science* **142**, 1666–1668.

Lenhoff, H. M., Heagy, W., and Danner, J. (1976). A view of the evolution of chemoreceptors based on research with *Cnidarians*. *In* "Coelenterate Ecology and Behavior" (G. O. Mackie, ed.), pp. 571–579. Plenum, New York.

Lindstedt, K. J. (1971a). Biphasic feeding response in a sea anemone: control by asparagine and glutathione. *Science* **173**, 333–334.

Lindstedt, K. J. (1971b). Chemical control of feeding behavior. *Comp. Biochem. Physiol.* **39**, 553–581.

Lindstedt, K. J., Muscatine, L., and Lenhoff, H. M. (1968). Valine activation of feeding in the sea anemone *Boloceroides*. *Comp. Biochem. Physiol.* **26**, 567–572.

Loeb, M., and Blanquet, R. S. (1973). Feeding behavior in polyps of the Chesapeake Bay sea nettle, Chrysaora quinquecirrha (Desor, 1848). *Biol. Bull. (Woods Hole, Mass.)* **145**, 150–158.

Loomis, W. F. (1955). Glutathione control of the specific feeding reactions of hydra. *Ann. N.Y. Acad. Sci.* **62**, 209–228.

Loomis, W. F., and Lenhoff, H. M. (1956). Growth and sexual differentiation of *Hydra* in mass culture. *J. Exp. Zool.* **132**, 555–574.

Mackie, G. O., and Boag, D. A. (1963). Fishing, feeding and digestion in siphonophores. *Publ. Stn. Zool. Napoli* **33**, 178–196.

Mariscal, R. N. (1971). The chemical control of the feeding behavior in some Hawaiian corals. *In* "Experimental Coelenterate Biology" (H. M. Lenhoff, L. Muscatine, and L. Davis, eds.), pp. 100–118. Univ. of Hawaii Press, Honolulu.

Mariscal, R. N., and Lenhoff, H. M. (1968). The chemical control of feeding behavior in *Cyphastrea ocella* and some other Hawaiian corals. *J. Exp. Biol.* **49**, 689–699.

Pardy, R. L., and Lenhoff, H. M. (1968). The feeding biology of the gymnoblastic hydroid, *Pennaria tiarella. J. Exp. Zool.* **168**, 197–202.

Reich, E., Rifkin, D. B., and Shaw, E., eds. (1976). "Proteases and Biological Control." Cold Spring Harbor Lab., Cold Spring Harbor, New York.

Reimer, A. A. (1971a). Feeding behavior in the Hawaiian zoanthids *Palythoa* and *Zoanthus. Pac. Sci.* **25**, 512–520.

Reimer, A. A. (1971b). Chemical control of feeding behavior in *Palythoa* (Zoanthidea, Coelenterata). *Comp. Gen. Pharmacol.* **2**, 383–396.

Reimer, A. A. (1973). Feeding behavior in the sea anemone *Callactis polypus* (Forskal, 1775). *Comp. Biochem. Physiol.* **44**, 1289–1301.

Rushforth, N. B. (1965). Inhibition of contraction responses of hydra. *Am. Zool.* **5**, 505–513.

Rushforth, N. B. (1967a). Modification in temporal patterning of pacemaker activity in hydra by sensory input. *J. Gen. Physiol.* **50**, 2484.

Rushforth, N. B. (1967b). Behavioral responses of isolated tentacles of hydra. *Biol. Bull. (Woods Hole, Mass.)* **133**, 449–450.

Rushforth, N. B. (1976). Electrophysiological correlates of fleeding behavior in the Tubularia. *In* "Coelenterate Ecology and Behavior" (G. O. Mackie, ed.), pp. 729–738. Plenum, New York.

Rushforth, N. B., and Hofman, F. (1972). Behavioral and electrophysiological studies of *Hydra*. III. Components of feeding behavior. *Biol. Bull. (Woods Hole, Mass.)* **42**, 110–131.

Rushforth, N. B., Krohn, I. T., and Brown, L. K. (1964). Behavior in hydra: Inhibition of the contraction responses of *Hydra pirardi. Science* **145**, 602–604.

Spencer, A. N. (1974). Behavior and electrical activity in the hydrozoan *Proboscidactyla flavicirrata* (Brandt). *Biol. Bull. (Woods Hole, Mass.)* **146**, 100–115.

Steiner, G. (1957). Über die chemische nahrungswahl von *Actinia equina* (L). *Naturwissenschaften* **44**, 70–71.

Wieland, T. (1954). Chemistry and properties of glutathione. *In* "Glutathione" (S. Colowick, A. Lazarow, E. Racker, and D. R. Schwartz, eds.), pp. 45–59. Academic Press, New York.

23

Sealed Membrane Vesicles from *Torpedo* Electroplax as a Model System for Synaptic Transmission

PAUL R. HARTIG, HSIAO-PING H. MOORE, AND MICHAEL A. RAFTERY

I. INTRODUCTION

Intact membrane vesicles, enriched in acetylcholine receptor, have been isolated from *Torpedo californica* postsynaptic membranes (Hartig and Raftery, 1979). These vesicles are oriented with what was the synaptic face of the electroplaque cells facing out. They therefore provide an ideal system for characterization of synaptic events at the biochemical and biophysical level. *In vitro* studies of ligand interaction and of agonist mediated cation transport can be readily performed using these preparations. Using methods developed for eel electroplax membrane preparations (Kasai and Changeux,

Biochemistry of Taste and Olfaction

ISBN 0-12-154450-8

1971), we have shown that the intact vesicles from *Torpedo californica* electroplax membranes retain ^{22}Na and release it upon addition of agonists (Miller *et al.*, 1978, and unpublished). The ^{22}Na entrapped within these vesicles is also released by osmotic shock (Miller *et al.*, 1978), leading to the conclusion that this occurs by means of rupture of the vesicle membrane, which releases the interior contents.

A variety of techniques have been reported in the literature for the preparation of acetylcholine receptor membrane vesicles that exhibit cation permeability control (Kasai and Changeux, 1971; Miller *et al.*, 1978; Popot *et al.*, 1976; Hess and Andrews, 1977; Schiebler *et al.*, 1977; Moore *et al.*, 1979a). Preparations that have been osmotically shocked (Moore *et al.*, 1979a) under conditions known to rupture vesicles (Hartig and Raftery, 1979) still exhibit such permeability control and retain their vesicular structures (Moore *et al.*, 1979a; Elliott *et al.*, 1980). Crude membrane homogenates, intact vesicle preparation consisting of large, low density inflated vesicles (Hartig and Raftery, 1979), and membranes harvested from sucrose gradients at densities approaching that of pure membranes (possibly collapsed or damaged vesicles), all behave as sealed compartments capable of retaining and releasing ^{22}Na upon interaction with agonist (Miller *et al.*, 1978; Moore *et al.*, 1979a). We find that large, transient pores form in these vesicles during osmotic shock and rapidly and spontaneously reseal; this explains why both intact vesicles harvested at low densities on sucrose gradients and osmotically shocked vesicles harvested at high densities form sealed membrane compartments, which retain ^{22}Na and release it upon addition of agonists.

We will report here some of the ligand binding properties, permeability control properties, and osmotic properties of *Torpedo californica* membrane vesicles.

II. CHOLINERGIC LIGAND-BINDING STUDIES

It has been well-documented (Heidmann and Changeux, 1978) that under equilibrium conditions ligand binding occurs to a ligand-induced high-affinity form of the acetyleholine receptor (AcChR), which seems to be an *in vitro* correlate of the desensitized form of the receptor observed *in vivo*. This conversion occurs on the time-scale of hundreds of milliseconds to a few seconds. Stopped-flow or other rapid-kinetic approaches therefore are required if ligand binding, conformational transitions, and cation transport are to be studied on a physiologically relevant time scale. We have adopted two approaches for the study of ligand binding using fluorescent probes. The first of these (Quast *et al.*, 1978, 1979) utilizes ethidium bromide as an extrinsic probe added to the membranes. The probe responds to cholinergic-ligand binding with an increase in quantum yield and allows study of conformational transitions that result from ligand–receptor complexation. Of the many possible mechanisms, we find that only the one shown in the scheme below fits the data,

$$\begin{array}{ccccc} 2A + R & \rightleftharpoons & AR & \rightleftharpoons & AR_2 \\ & & \updownarrow & & \updownarrow \\ & & C_1 & & C_2 \end{array}$$

where A is an agonist (antagonists yield entirely different kinetics), AR and AR_2 are mono- and diliganded precomplexes that isomerize to C_1 and C_2, respectively, which do not interconvert. The second approach (Dunn *et al.*, 1980) involves covalent attachment of a fluorophore to the AcChR and the study of changes in fluorescence that occur upon binding cholinergic ligands. The mechanism that fits these data is the same as that shown in the scheme below, except that one further monoliganded precomplex is necessary before AR is formed. Thus two independent kinetic approaches yield essentially the same formal mechanism for ligand-induced conformational processes. To date, however, we have not been able to correlate these processes with events involved in cation translocation. This requires direct study of the ion transport process itself.

III. AGONIST-INDUCED ION FLUX: LOADING OF SOLUTES INTO VESICLES BY OSMOTIC SHOCK

The common assay (Kasai and Changeux, 1971) of ^{22}Na efflux used to study transport in AcChR containing vesicles requires equilibration of the radionuclide within the vesicles. Because this process takes several hours (6–8 hr at 4°C), a more rapid method would be desirable. Osmotic shock seemed to be a viable approach to achieve this goal.

If intact *Torpedo* vesicles form transient pores during osmotic shock, which spontaneously reseal, they should entrap radioactive substances from the external shocking solution inside the resealed compartments. Subsequent Millipore filtration of a suspension of vesicles would allow them and their contents to be retained on the filters while the bathing solution is removed. The entrapment of ^{22}Na in *Torpedo* homogenates containing intact vesicles upon osmotic shock was demonstrated by this method (Fig. 1). $^{22}Na^+$ entrapped by the shock and resealing processes was released by a second osmotic shock (Fig. 1, open circle). All experiments were arranged so

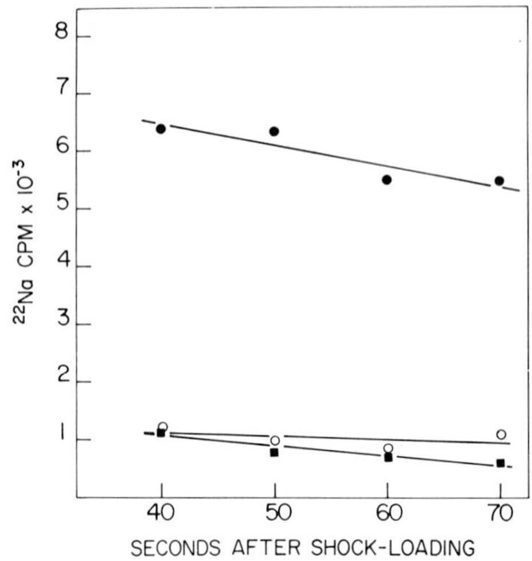

Fig. 1. Entrapment of ^{22}Na in crude *Torpedo* membrane vesicles upon osmotic shock. (●), Membrane vesicles prepared in 400 m*M* NaCl buffer were diluted into a hypotonic medium containing $^{22}Na^+$ (see Table I, exp. I). The entrapped radioactivity was then measured by Millipore filtration. (■), Control experiment without osmotic shock, i.e., isotonic dilution of membranes into a $^{22}Na^+$-containing medium (Table I, exp. II). (○), Release of the loaded $^{22}Na^+$ by a second osmotic shock; membranes were first hypotonically diluted into a $^{22}Na^+$-containing medium, then subjected to a second hypotonic dilution into a nonradioactive medium (Table I, exp. III).

that the compositions of the solutions applied to the Millipore disks were identical; only the order of addition of the components was varied (Table I).

Addition of gramicidin D (McLaughlin and Eisenberg, 1975) to vesicles containing entrapped ^{22}Na rapidly caused its release (Fig. 2) whereas addition of the ionophore to the control (Fig. 2, closed circle) had no effect. ^{22}Na was clearly entrapped in the vesicle interior by a resealing process that followed the formation of transient pores during osmotic shock.

At both 4° and 25°C, the vesicle resealing process was complete within 5 sec, as shown by an experiment in which vesicles were shocked into distilled water and ^{22}Na was added 5 sec later (with the Millipore filtration technique, 5 sec is the minimum time required for solution transfer and mixing). The level of ^{22}Na retained was the same as the control level of nonspecific adsorption (results not shown), demonstrating complete resealing after 5 sec. The size and charge specificities of the transient pores that form during osmotic shock were tested by including a variety of radioactive substances in the shocking solution. Both $^{125}I^-$ and [^{14}C] sucrose were efficiently entrapped during osmotic shock. The transient pores formed during osmotic shock therefore showed no charge specificity since anions, cations, and neutral

TABLE I Osmotic Shock Experiments[a]

Experiment	Time (sec)			
	0	10	20	30
I. Shock entrapment	Add 50μl of membrane homogenate to 90μl of 33 μCi/ml ^{22}NaCl in distilled water	Add 10μl of 4 *M* NaCl	Add 4 ml of 400 m*M* NaCl	—
II. Control	Add 50μl of membrane homogenate to 100 μl of 400 m*M* NaCl containing 29.7 μCi/ml ^{22}NaCl	—	Add 4 ml of 400 m*M* NaCl	—
III. Entrap–release	Add 50μl of membrane homogenate to 90μl of 33 μCi/ml ^{22}NaCl is distilled water	Add 10μl of 4 *M* NaCl	Add 3.6 ml of distilled water	Add 400 μl of 4 *M* NaCl

[a] Membrane homogenate was prepared in 400 m*M* NaCl–1 m*M* EDTA–10 m*M* Na-phosphate, pH 7.4, from *Torpedo* electroplaques by differential centrifugation as described (Elliott *et al.*, 1980). For different experiments described in Fig. 1–3 and 5, the various components were added and mixed with a Vortex mixer at the times indicated. Note that the final solution compositions were identical in all experiments; only the order of addition of the various components was altered. At 40, 50, 60, and 70 sec, 1 ml of the solution was applied to two 0.8 μm Millipore filters in series under vacuum, and the filters were immediately rinsed twice with 7.5 ml of 0.4 *M* NaCl, 10 m*M* Na phosphate, 1 m*M* EDTA, pH 7.5 (unless otherwise specified). For ^{22}Na studies, the filters were counted on a Beckman Gamma 4000 gamma counter; for [^{14}C] and [^{3}H] studies the filters were dried at 40°C, placed in scintillation vials along with 10 ml of toluene-based scintillation fluid containing 25% Triton X-100 and 0.55% Permablend III, and counted on a Packard 3375 liquid scintillation counter.

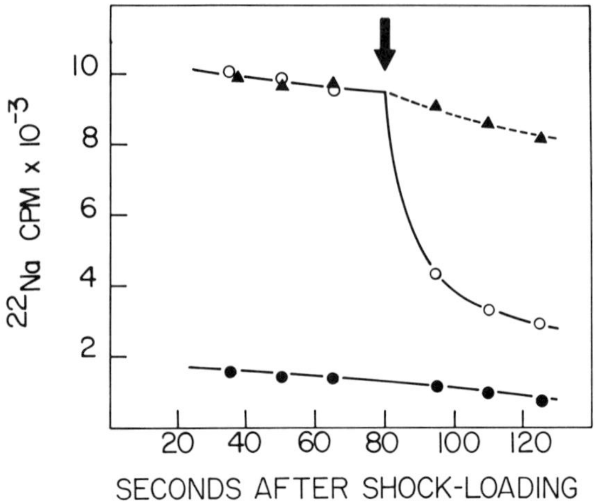

Fig. 2. Release of the $^{22}Na^+$ entrapped by osmotic shock using gramicidin. Vesicles were loaded with $^{22}Na^+$ by osmotic shock as in Fig. 1. (○), At the time indicated by the arrow, gramicidin D (in EtOH) was added to a final concentration of 8 μg/ml. (▲), Same as above, except that ETOH without gramicidin was added. The concentration of ETOH was 0.8% in the final mixture. (●), Control experiment without osmotic shock; addition of gramicidin at the time indicated by the arrow did not affect this baseline.

species were all entrapped. Figure 3 demonstrates the effect of molecular weight on the entrapment process; equal volumes of membranes were shocked into solutions containing the indicated radiolabeled compounds (total cpm were identical). Three curves were obtained as described for Fig. 1, and the average value for each curve was plotted in the form of a bar graph (Fig. 3). The vesicles shock-loaded $^{22}Na^+$ most efficiently. The level of en-

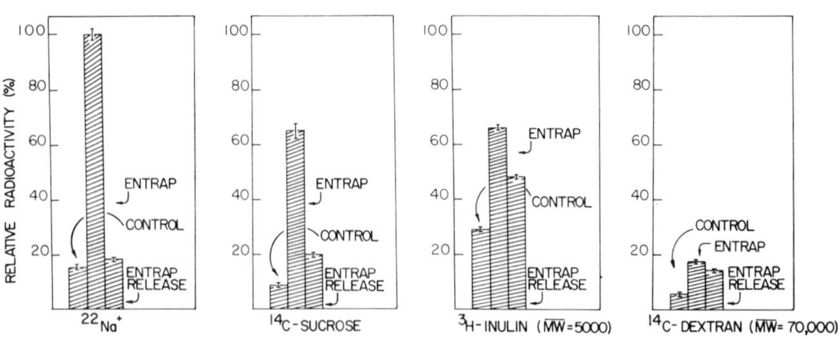

Fig. 3. Effect of molecular weight on the efficiency of loading of vesicles by osmotic shock. The three curves, as in Fig. 1 (also see Table I), were measured for four different radioactive substances: $^{22}Na^+$, [^{14}C]sucrose, [^{3}H]inulin, and [^{14}C]dextran. The bar graph represents the average of the four points as determined in Fig. 1.

trapped counts decreased for [^{14}C] sucrose and [^{3}H]inulin (MW = 5000) and fell off drastically for the [^{14}C]dextran polymer with an average MW of 70,000. It thus appears that the loading efficiency decreases as the molecular weight of the molecules increases.

When the same experiments as those shown in Fig. 3 were carried out for [^{125}I]α-BuTx (α-bungarotoxin) (MW ~8000) on vesicles presaturated with excess unlabeled toxin to block all the specific binding sites, the level of entrapped radioactivity was 10 times higher than that of $^{22}Na^{+}$. The entrapped radioactivity can be released by a second shock and therefore is not due to toxin irreversibly bound to the interior of the vesicle membranes. This preferential accumulation of α-BuTx in the interior of the vesicles may be due to strong protein lipid interactions. Extremely high protein trapping efficiency has also been reported in some liposome systems (Sessa and Weissmann, 1968).

IV. OSMOTIC PROPERTIES OF PURIFIED AcChR-CONTAINING VESICLES

Many cholinergic receptor ion flux studies reported in the literature (Popot *et al.*, 1976; Hess and Andrews, 1977; Schiebler *et al.*, 1977; Moore *et al.*, 1979a) were performed with purified membrane preparations, which had been osmotically shocked and isolated from high-density regions of sucrose gradients. These preparations must contain sealed vesicles formed from AcChR membranes because they retain ^{22}Na and release it upon addition of agonists. These vesicles must contain little interior volume, however, since they equilibrate on sucrose gradients at densities approaching that of pure AcChR membranes (Hartig and Raftery, 1979). From the preceding experiments, we were led to expect that these high-density vesicles had spontaneously resealed following osmotic shock (entrapping a low osmotic strength vesicle interior) and may have collapsed under the high osmotic pressure of the sucrose gradient to a high density form with little interior volume.

To test this hypothesis we examined the properties of osmotically shocked AcChR membranes, which were subsequently purified in reorienting gradients (Elliott *et al.*, 1980) and harvested from high-density regions of the sucrose gradients (density = 1.18). When these membranes were harvested from the sucrose gradient and suspended in a buffer containing 400 m*M* NaCl, they retained very little ^{22}Na, as expected for collapsed vesicles containing little interior volume. When they were resuspended in a small volume (1:4, pellet to buffer) of 10 m*M* Tris, pH 7.4 with the final salt concentration becoming close to 100 m*M* because of the contribution of interstitial fluid in the pellet (see below), the vesicles appeared to expand to a form

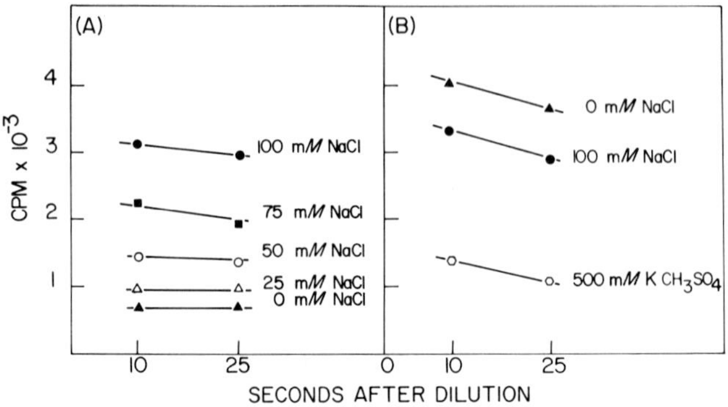

Fig. 4. Osmotic behavior of purified AcChR vesicles harvested from the high density region of the reorienting sucrose gradient. Crude membranes prepared in 400 m*M* NaC1 buffer were osmotically shocked before being layered on a reorienting sucrose gradient in 400 m*M* NaC1 buffer. The membranes that migrated in the high-density region (~1.18) were harvested (Elliott *et al.*, 1980). These vesicles can be inflated to regain considerable interior volume, which contains solutions of medium (~100 m*M*) or low (~10 m*M*) osmotic strength. (A) Vesicles with medium osmolarity. The harvested membranes (in 400 m*M* NaC1-buffer) were resuspended in a small volume of 10 m*M* Tris–C1, pH 7.4, to give a final NaC1 concentration of 100 m*M*. Trace amounts of ^{22}NaC1 was added to the medium, and the mixture was allowed to incubate at 4 °C for 8 hours. The entrapped ^{22}Na was then assayed by the Millipore filter method with assay buffer containing 100, 75, 50, 25, and 0 m*M* NaCl. The decrease in the entrapped radioactivity indicates that osmotic shock occurred during assays in media of osmolarity less than 100 m*M*. (B) Vesicles with low osmolarity: The membranes from (A) were extensively dialyzed against low salt buffer (10 m*M* Tris–C1, pH 7.4). Vesicles were then incubated with ^{22}Na in media of varying osmolarity, and entrapped ^{22}Na was measured as a function of the osmolarity of the incubation and assay buffer. The vesicles shrank in media of high osmolarity as indicated by the decrease in vesicle interior volume.

exhibiting a greater interior volume of approximately 100 m*M* interior osmolarity (Fig. 4A). Such expanded vesicles, once equilibrated with ^{22}Na, displayed increasing degrees of osmotic shock and thus fewer retained counts when they were suspended in solutions containing decreasing salt concentration (use of buffer containing 200 m*M* NaC1 gave essentially the same level of entrapped ^{22}Na as one containing 100 m*M* NaC1). The solution in the vesicle interior can be completely exchanged for a low salt solution by extensive washing or slow dialysis of the membranes against 10 m*M* Tris–C1. In this case, suspension of ^{22}Na equilibrated membranes in solutions of increasing salt concentration caused shrinkage of the vesicles leading to fewer retained counts (Fig. 4B). Thus, these vesicles can be repeatedly osmotically shocked, and each time they are shocked they retain lower salt concentrations in the vesicle interior. The vesicles shrink when suspended in solutions of higher osmotic strength.

V. EFFECTS OF AGONISTS

The results described so far have not differentiated between resealing processes in AcChR containing vesicles and in other membrane vesicles that contaminate the preparations. We tested for resealing of active AcChR vesicles by observing the agonist induced efflux of ^{22}Na that had been entrapped by the osmotic shock and resealing processes. As shown in Fig. 5, carbamylcholine induced the release of ^{22}Na entrapped in resealed vesicles. The level of release correlated well with the Carb induced release of ^{22}Na, loaded by preincubation, from normal sealed vesicles. Therefore membrane vesicles enriched in AcChR form transient pores during osmotic shock, which spontaneously reseal. These AcChR vesicles remain active and respond to addition of agonist by increasing the membrane's cation permeability.

The presence of sealed membrane vesicles was also demonstrated by measuring the percentage of a membrane pellets volume that was accessible to membrane-permeant or to membrane-impermeant radiocompounds. Equal amounts of membranes from reorienting gradients were centrifuged in the presence of ^3H_2O or [^{14}C]sucrose. [We have previously

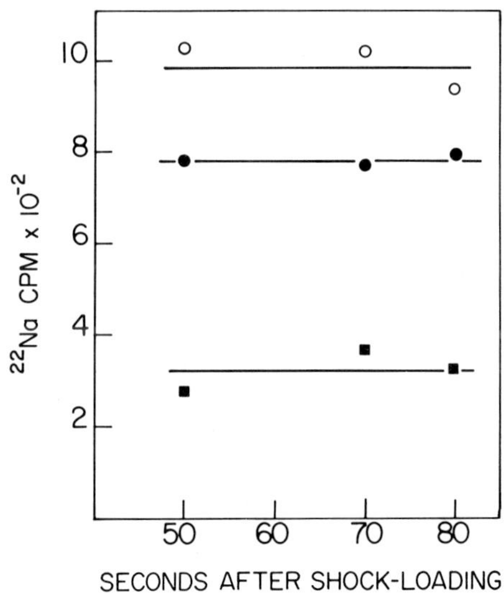

Fig. 5. Carbamylcholine induced efflux of ^{22}Na entrapped in purified AcChR vesicles by osmotic shock. AcChR vesicles purified in reorienting gradients with medium osmolarity (see Fig. 4A) were used for these studies. (○), Entrapment by osmotic shock as described in Table I; (■), control experiments without osmotic shock; (●), entrapment as above, except that 100 μM Carbamylcholine was included in the dilution buffer at t = 20 sec.

shown that intact vesicles are impermeable to sucrose (Hartig and Raftery, 1979).] The supernatants were removed completely, and the pellets were counted. The pellet containing [^{14}C]sucrose entrapped 17% fewer counts than that containing ^3H_2O, suggesting that 17% of the total water space of the pellet was sequestered inside membrane vesicles.

The ability of osmotically shocked *Torpedo* vesicles to spontaneously reseal explains why so many different types of electroplaque membrane preparations are vesicular, as shown by electron microscopy (Strader *et al.*, 1979; Sheridan *et al.*, 1966; Ross *et al.*, 1977) and explains their ability to retain ^{22}Na and exhibit agonist induced cation release.

VI. CORRELATION OF POLYPEPTIDE COMPOSITION AND AcChR CATION TRANSPORT

It has been demonstrated that further purification of AcChR-enriched membrane preparations can be achieved by exposure of the membranes to alkaline pH under conditions of low ionic strength. This results in preparations containing either three (Neubig *et al.*, 1979) or all four (Elliott *et al.*, 1980) of the AcChR polypeptides in intact form. Comparison of the ion-transport properties of the unextracted membranes with those of the extracted membranes provides a method of assigning or eliminating a functional role(s) for polypeptides other than those of the AcChR, such as the

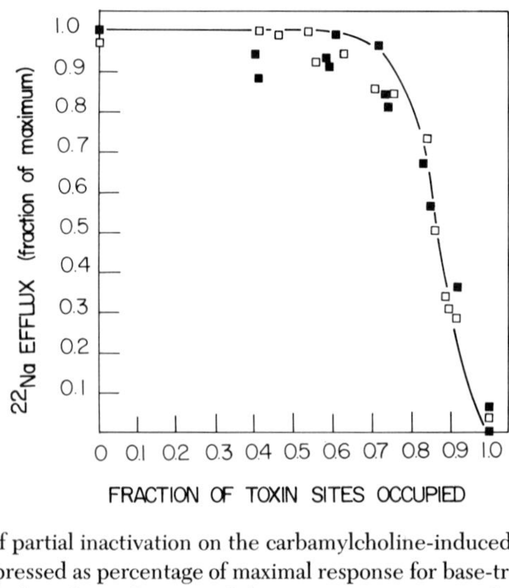

Fig. 6. Effect of partial inactivation on the carbamylcholine-induced $^{22}Na^+$ efflux. The flux amplitudes were expressed as percentage of maximal response for base-treated (■) or untreated (□) membranes and plotted as a function of the α-BuTx sites occupied.

major polypeptide of MW 43,000, which has been proposed as a possible specific ionophore protein (Sobel *et al.*, 1978). Both classes of membrane contain a large excess of AcChR over the necessary level for a full flux response over a 10-sec period (first-time point in the filtration assay). Therefore it was necessary to systematically inactivate the AcChR to reduced levels of flux capability. This was accomplished by use of α-BuTx, which was shown to react randomly under the conditions used. Figure 6 shows the comparative ^{22}Na transport data for the two preparations, demonstrating their essential identity. This result (Moore *et al.*, 1979b) demonstrates that no polypeptides other than those known to be subunits of the AcChR are necessary for ligand binding to and cation transport by the nicotinic acetylcholine receptor.

ACKNOWLEDGMENT

This research was supported by United States Public Health Service Grant NS-10294 and by grants from the Muscular Dystrophy Association of America, the American Heart Foundation, the Pew Charitable Trust, and United States Public Health Service Biomedical Research Grant No. 47123. P. Hartig was a National Institutes of Health Postdoctoral Fellow.

NOTE ADDED IN PROOF: After this work was submitted for publication, we learned that similar observations have been made by others (West and Huang, 1980).

REFERENCES

Dunn, S. M. J., Blanchard, S. G., and Raftery, M. A. (1980). *Biochemistry* **19**, 5645–5652.

Elliott, J., Blanchard, S., Wu, W., Miller, J., Strader, C., Hartig, P., Moore, H. P., Racs, J., and Raftery, M. A. (1980). *Biochem. J.* **185**, 667–677.

Hartig, P. R., and Raftery, M. A. (1979). *Biochemistry* **18**, 1146–1150.

Heidmann, T., and Changeux, J.-P. (1978). *Annu. Rev. Biochem.* **47**, 317–357.

Hess, G. P., and Andrews, J. P. (1977). *Proc. Natl. Acad. Sci. U.S.A.* **74**, 482–486.

Kasai, M., and Changeux, J.-P. (1971). *J. Membr. Biol.* **6**, 1–23.

McLaughlin, S., and Eisenberg, M. (1975). *Annu. Rev. Biophys. Bioeng.* **4**, 335–366.

Miller, D. L., Moore, H.-P. H., Hartig, P. R., and Raftery, N. A. (1978). *Biochem. Biophys. Res. Commun.* **85**, 632–640.

Moore, H.-P., Hartig, P. R., Wu, W.C.-S., and Raftery, M. A. (1979a). *Biochem. Biophys. Res. Commun.* **88**, 735–743.

Moore, H.-P. H., Hartig., P. R., and Raftery, M. A. (1979b). *Proc. Natl. Acad. Sci. U.S.A.* **76**, 6265–6269.

Neubig, R. R., Kdrodel, E. K., Boyd, N. D., and Cohen, J. B. (1979). *Proc. Natl. Acad. Sci. U.S.A.* **76**, 690–694.

Popot, J.-L, Sugiyama, H., and Changeux, J.-P. (1976). *J. Mol. Biol.* **106**, 469–483.

Quast, U., Schimerlik, M. I., and Raftery, M. A. (1978). *Biochem. Biophys. Res. Commun.* **81**, 955–964.

Quast, U., Schimerlik, M. I., and Raftery, M. A. (1979). *Biochemistry* **18**, 1891–1901.

Ross, M. J., Klymkowsky, M. W., Agard, D. A., and Stroud, R. M. (1977). *J. Mol. Biol.* **116**, 635–659.
Schiebler, W., Lauffer, L., and Hucho, F. (1977). *FEBS Lett.* **81**, 39–42.
Sessa, G., and Weissmann, G. (1968). *J. Lipid Res.* **9**, 310–318.
Sheridan, M. N., Whittaker, V. P., and Israel, M. (1966). *Z. Zellforsch. Mikrosk. Anat.* 74, 291–307.
Sobel, A., Heidmann, T., Hofler, J., and Changeux, J.-P. (1978). *Proc. Natl. Acad. Sci. U.S.A.* **75**, 510–514.
Strader, C. D., Revel, J.-P., and Raftery, M. A. (1979). *J. Cell Biol.* **83**, 499–510.
West, L. K., and Huang, L. (1980). *Biochemistry* **19**, 4418–4423.

24

From Receptors to Brain Circuitry

CANDACE B. PERT AND MILES HERKENHAM

I. INTRODUCTION

Basic concepts of pharmacology dictate that all drug actions emanate from the same initial event: the complexing of the drug to a specific receptor site. The jumble of cascading events that follow—the pharmacological response—is the result of how these receptor sites are distributed over various tissue components and the nature of the normal physiological role the receptors play when they are not being perturbed by the drug. Until fairly recently, drug receptors have been merely hypothetical constructs. In recent years, however, we have seen the development of methods for directly measuring cell-surface receptor interactions by direct binding of radiolabeled ligands. The acetylcholine nicotinic receptor, which presented the advantage of having venoms and toxins capable of covalent interactions, was the first to be biochemically labeled (Miledi *et al.*, 1971). Later, iodolabeled peptide hormones possessing high affinities in the nanomolar range were used to directly bind to endocrine receptors (Lefkowitz *et al.*, 1970).

Much has been written about the criteria that must be fulfilled to validate

Biochemistry of Taste and Olfaction
ISBN 0-12-154450-8

a direct binding method for studying receptors. These criteria, which evolved primarily from endocrine receptor identification, such as the insulin receptor, have emphasized kinetic considerations, e.g., binding should be saturable, reversible, and possess high affinity. For the insulin receptor, only a handful of different ligands were available for receptor characterization, these being the insulin molecule for several species (Cuatrecasas, 1971). This deficiency made certain characterizations impossible to perform. When the opiate receptor was characterized (Pert and Snyder, 1973; Simon *et al.*, 1973; Terenhius, 1973), it was possible to utilize many of the literally thousands of synthetic and semisynthetic opiate analogs that had been prepared by medicinal chemists over several decades. From these studies, it became possible to make amazingly excellent correlations between affinities for the opiate receptor determined *in vitro* and relative potencies in eliciting the pharmacological response—analgesia. The construction of such structure–activity correlations became an important criterion for identifying and characterizing a biologically significant receptor.

For example, a homologous series of ketobemidone analogs have been prepared in the laboratory of Dr. Everett May. These opiate analogs had been tested for the ability to cause analgesia in mice on the so-called hot plate test, in which the mouse is placed on a hot plate and the number of seconds are counted until it makes a characteristic "hop, skip, and a jump" indicative of pain. Interestingly, the maximal analgesic potency is achieved when the substituents on the nitrogen atom reach 4 or 5 carbons: less than or greater than that chain length and the analgesic potency rapidly falls off (Wilson *et al.*, 1975). Figure 1 shows a superb correlation between the analgesic potency of these compounds and their ability to displace tritiated naloxone binding in an opiate receptor assay, which provides a parameter directly proportional to the affinity or tightness with which these ligands are capable of binding to the opiate receptor. The possibility of producing such excellent correlations between a biological response and an *in vitro* binding phenomenon leaves little doubt that the binding in the test tube is the same initial binding event that occurs *in vivo* to initiate the biological response. With the more recent characterization of a number of pharmacological receptors for small molecules for which numerous analogs whose known biological potency are available, it has become apparent that such excellent structure–activity correlations are not uncommon, and in fact are *de rigeur*. Thus, while kinetic information in the characterization of these receptors for small ligands can be interesting and useful, in the presence of such compelling correlational data, all other criteria for characterizing receptors seem unconvincing.

Viewed from this perspective, only two things are really necessary to biochemically identify the receptor that is responsible for a given phar-

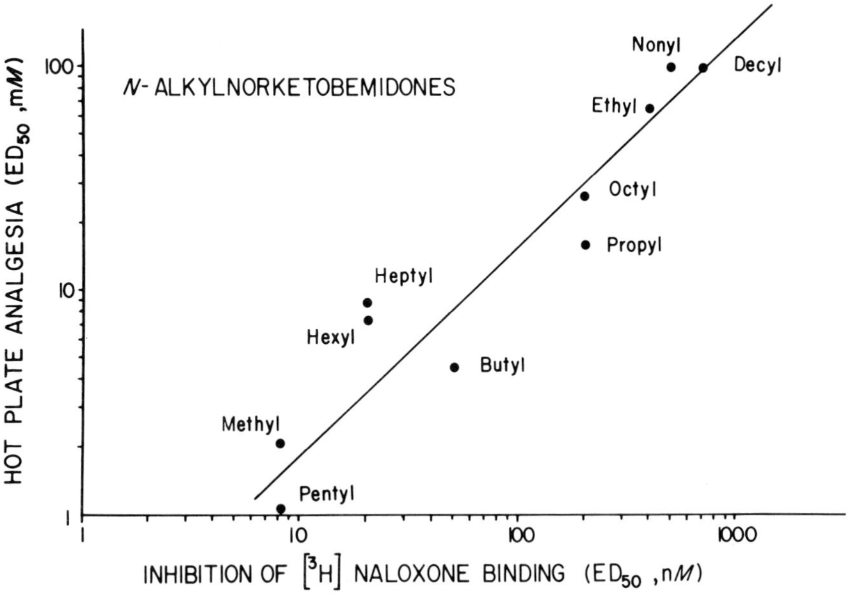

Fig. 1. Correlational analysis of the ability of 10 ketobemidone analogs to displace [^{3}H]naloxone binding to rat brain membranes and to elicit an antinociceptive response in the mouse hot-plate test. Designations of homologs (methyl, etc.) refer to the nature of carbon moieties attached to the nitrogen atom. The correlation ($r = 0.95$) is significant ($p < 0.001$). (Taken from Wilson *et al.*, 1975.)

macological response: first, a highly radiolabeled ligand, which retains its biological activity and high affinity for the receptor in question, and second, a series of structural analogs, which have been previously ranked in some biological test. Figure 2 shows a particularly elegant example of close fit between *in vitro* binding data and a biological response, in this case the ability of rabbit neutrophils to migrate in response to a series of peptide chemoattractants (Aswanikumar *et al.*, 1977). Some white blood cells will migrate to the site of bacterial invasion, drawn by the bacterial degradation products containing the sequence formyl-methionine-, since that sequence is the translational signal in prokaryotes but not eukaryotes and thus constitutes a clear signal. It is apparent that the formyl group is essential for recognition by the neutrophil since the deformylated norLeu-Leu-Phe is about 10,000-fold less potent as a chemoattractant and as an inhibitor of specific chemotactic receptor binding. Thus, the 0.98 correlation illustrated in Figure 2 strongly suggests that neutrophils contain receptors on their surface that can be measured biochemically and are in fact responsible for their ability to migrate toward bacterial invasion.

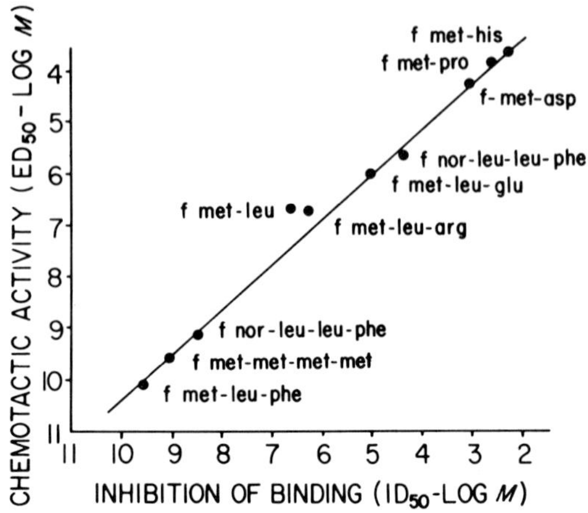

Fig. 2. Comparison of the displacement of ^{3}H-formyl-norLeu-Leu-Phe binding and chemotactic potencies of synthetic peptides ($r = 0.98$, $p < 0.001$). (From Aswanikumar *et al.*, 1977.)

II. BRAIN RECEPTORS

Within the last few years, a major theoretical framework in neuroscience has emerged: The brain contains at least 30 and perhaps even 100 subsets of neurons, each with their own characteristic localization and secreted neurochemical. These neurosecretory chemicals, or neurotransmitters, diffuse across the synaptic cleft after release and combine with their appropriate receptor on the postsynaptic membrane to produce alterations in ionic flux. Apparently, this is an extremely vulnerable process, since many psychoactive drugs appear able to alter this process and produce psychopharmacological effects. Thus, insofar as the mechanism of action of a given psychoactive drug is to mimic or antagonize a given neurotransmitter, characterization of the receptor for that psychoactive drug will reveal the target sites for the affected subset of neurons.

When characterizing structure–activity relationships for brain receptors, most often one must resort to using behavioral phenomena to correlate with binding data. Somewhat surprisingly, correlations can be quite excellent. Figure 3 shows that the receptor for "angel dust," a frequently abused animal tranquilizer, phencyclidine (PCP), correlates reasonably well with two quite different measures of biological response (Zukin and Zukin, 1979). In the mouse rotarod test, shown on the left, mice balance on a slowly turning rod for several minutes. A dose of PCP analog required to make half of the mice fall off within the allotted time is recorded. In a discrimination

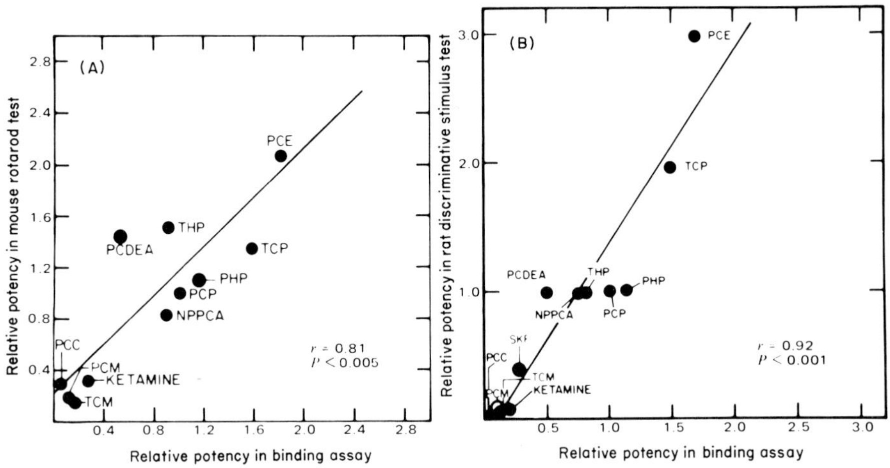

Fig. 3. Relative potencies of a series of PCP-like drugs in displacing bound [^{3}H]PCP versus the relative potencies of these compounds in the mouse rotarod test (A) and the rat discriminative stimulus test (B). The correlations (r) and their significances (p) were determined by the Pearson product-moment correlation analysis. (From Zukin and Zukin, 1979.)

test, shown on the right, rats are required to perform a prescribed response if a drug reminds them of PCP. The threshold dose required to elicit this recognition is recorded. While the correlations of 0.81 and 0.92, respectively, are less impressive than the 0.98 correlation obtained in the above example, it is still fairly amazing that such gross behavioral parameters can be traced to the interaction of one group of small molecules with their specific receptor sites in the brain.

It should be pointed out that for brain receptors, failure to closely correlate receptor binding with biological response data can be attributed to two common sources. First, it must be appreciated that the blood–brain barrier excludes drugs to differential extents depending primarily upon their lipid solubility. Pharmacologically, meperidine (Demerol) is much more potent than it would appear that it should be from its *in vitro* binding properties. This is because, relative to the other opiates, meperidine and its analogs penetrate the blood–brain barrier much more readily due to their relatively greater lipophilicity (Pert *et al.*, 1976b). A second source of error occurs if a drug finally acts in the brain not in its original injected form, but as a metabolite of the original structure. Codeine is only about tenfold weaker than morphine as an analgesic in rats or humans, but is found to be thousands of times weaker in binding to opiate receptors *in vitro*. The explanation for this discrepancy is that in the liver, codeine is first demethylated to form morphine, the active species.

The realization that binding performed *in vitro* can be so well-correlated with potencies in eliciting the pharmacological response has brought the realization that binding data can often be more precise than behavioral data, especially if the behavioral test is not particularly accurate. Figure 4 illustrates the correlation between the binding data and the average therapeutic dose of benzodiazepin analogs required for alleviating anxiety in patients (Möhler and Okada, 1978). The correlation of "only" 0.79 must be considered in light of the crudeness of the response data—it does not take into account errors in prescribing appropriate dosages of Librium, Valium, etc.

Table I shows the affinity of bombesin and its analogs for binding to the receptor as well as the ability of these various analogs to elicit hypothermia or hypoglycemia in the rat (Moody *et al.*, 1978). Bombesin was isolated from the frog skin of *Bombix bombina* over 10 years ago (hence its name) on the basis of its gastrointestinal effects. Only recently it has been shown to be active in brain. Although the affinities of bombesin analogs for the bombesin receptor *in vitro* can be determined quite precisely, up until this time the physiological effects have been quantified rather imprecisely. From the data it is obvious that a rough correlation exists, but since many of the behavioral data are "all or none," the correlation is not so strong.

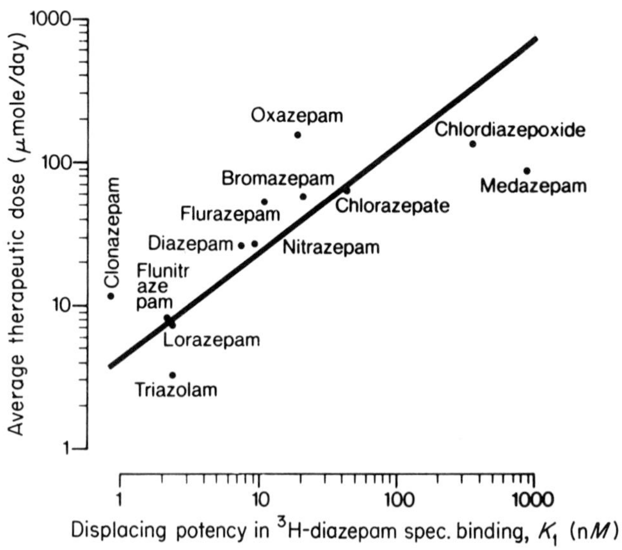

Fig. 4. Correlation between K_i values for the inhibition of [^{3}H]diazepam specific binding by various benzodiazepins and their average therapeutic dose (the dose recommended by the manufacturers for their use as anxiolytics or hypnotics was used). Correlation coefficient $r = 0.79$ ($p < 0.005$). (From Möhler and Okada, 1978.)

TABLE I Comparison of Potencies of Synthetic Bombesin-like Peptides in Inhibiting [^{125}I-Tyr^4]Bombesin Binding to Rat Brain Membrane and Eliciting Hypothermia and Hyperglycemia after iv Injection[a]

Peptide	IC_{50}(nM)	Relative potency to induce	
		Hypothermia	Hyperlycemia
Bombesin (BN)	20	100	100
(Lys^3)BN	5	100	
(Tyr^4)BN	5	100	
(D-Ala^5)BN	15	100	110
(D-Trp^8)BN	>5000	1	<1
(D-Val^{10})BN	>5000	1	
(Pro^{11})BN	>5000	1	<1
(des-His^{12})BN	200	1	
(D-Leu^{13})BN	>5000	<0.1	<0.1
Ranatensin	60	20	
Litorin	40	5	95
(D-Met^{14})BN	600	10	10
BN-OH	600	<1	<1

[a] From Moody *et al.* (1978).

III. IDENTICAL STEREOSELECTIVITY, DIVERSE FUNCTIONAL SIGNIFICANCE DEPENDING ON ANATOMICAL POSITION

The biochemical kinetics of opiate receptor binding in the guinea pig intestine are virtually identical with those of the opiate receptor found in brain. Opiate receptors on the longitudinal muscle of the myenteric plexus of the guinea pig ileum produce dose-dependent suppression of electrically evoked contraction of this muscle when they are occupied by opiate agonists. Kosterlitz *et al.* (1974) have emphasized that the stereoselectivity of opiate receptors in the guinea pig ileum coincides almost exactly with the stereoselectivity of the opiate receptors in brain responsible for producing analgesia in rodent pain tests and in human beings. Opiate receptors are scattered throughout the periphery and the central nervous system. Administration of opiates to the whole animal perturbs receptors in all places, but detection of these perturbations will be restricted to the parameters under scrutiny. While a rodent is being tested on a hot plate to detect the analgesic effects of opiates, the simultaneous opiate effects on locomotor activity elicited by occupation of receptors in the nucleus accumbens (Pert and Sivit, 1977) will not be detected. Thus, a constellation of biological responses may result from administration of a drug, all elicited by the same initial event—

the binding of the drug *in vivo* to chemically similar receptors scattered throughout the brain and body. However, the exact nature of the functional alterations produced by the drug will depend on the normal function of the cells and tissues in which the occupied receptors are located.

IV. FROM RECEPTORS TO BRAIN CIRCUITRY

Opiate-induced analgesia may be mediated by receptors localized to neural circuitry comprising the pain pathways (Pert, 1978). Receptors are found at the sites of termination of high-threshold cutaneous pain afferents in the substantia gelatinosa of the spinal cord dorsal horn. Descending modulation of pain perception may be mediated in part by "opiatergic" enkephalin-containing neurons in the medulla (Bausbaum *et al.*, 1976; Hökfelt *et al.*, 1979). On the other hand, opiate-induced physiological alterations, such as respiratory depression, may be mediated by receptors localized to medullary nuclei receiving visceral afferents (Atweh and Kuhar, 1977a; Pert *et al.*, 1981).

Opiate-receptor participation in other brain functions leading to specifically characterized pharmacological, physiological, and behavioral outcomes may be surmised from analysis of the anatomical localization of receptors in functionally-identified brain circuitry. This approach suggests that anatomical localization studies may provide clues to opiate-sensitive ("opiatergic") functions that have been overlooked or obscured by more obvious and overriding physiological effects. These clues would then lead to hypotheses that are testable by constructing structure–activity correlations.

A good example comes from the results of our newly developed method for the autoradiographic visualization of opiate receptors *in vitro* on slide-mounted tissue slices (Herkenham and Pert, 1980). Briefly, this method utilizes fresh, unfixed, cryostat-cut brain slices that are thaw-mounted and freeze-dried onto subbed slides. Racks of slides are incubated in buffered media, under appropriate ligand binding conditions (ions, temperature, GTP, protease inhibitors, etc.), containing labeled opiate ligands ([^{3}H]naloxone). The slices are then rinsed in phosphate buffer at 0°C to remove nonspecifically bound label, quickly dried, and then fixed in hot paraformaldehyde vapors under a vacuum. The slides can then be dipped in nuclear track emulsion for the visualization of radioactive label by traditional autoradiographic techniques.

The advantages and superb tissue quality and microscopic resolution of receptor localization provided by this *in vitro* technique are evident by the examination of labeling patterns in areas of known discrete and dense opiate receptor distribution, such as the caudate nucleus patches and streaks (cf.

Atweh and Kuhar, 1977b; Herkenham and Pert, 1980; Pert *et al.*, 1976a; Young and Kuhar, 1979). Permitted a global view of opiate receptors distributed throughout the brain and spinal cord, we have noticed and emphasized a general phenomenon: Opiate receptors are found in nearly all primary sensory nuclei, and within these they are localized to superficial or molecular layers in all laminated structures.

This is probably most apparent in the olfactory system (see this volume, Chapters 17 and 19), where the observation also applies to secondary and tertiary nuclei of the basal forebrain olfactory circuitry. Figure 5 shows the high-density receptor distribution indicated by preferential [^{3}H]naloxone binding to the external plexiform layer of the main bulb, and glomeruli of the accessory olfactory bulb, the molecular layers of the anterior olfactory nuclei, the olfactory cortex comprising piriform cortex and olfactory tubercle, and, further caudally, the superficial layers (most densely) of the medial and posterior cortical amygdaloid nuclei. Anatomical tract-tracing studies have revealed a series of successive links originating, as shown by the arrows in Figure 5, in the vomeronasal organ projecting selectively to the accessory olfactory bulb (Shepherd, 1972) and from there to the amygdala and bed nucleus of the stria terminalis (See Fig. 5, nst) (Broadwell, 1975; Scalia and Winans, 1975). Amygdaloid projections to the nst (Krettek and Price, 1978) suggest that the nst is focal center for outflow of this system to other brain areas. One site of termination, along the lateral edge of the medial habenula, is noteworthy for the presence of a dense cluster of opiate receptors in the same location. Such concordances, as illustrated in Fig. 5, offer the possibility that successive links in the olfactory circuit leading to the habenula and ultimately to the interpeduncular nucleus and median raphe are, in fact, opiatergic links.

The vomeronasal organ and its neural connections are crucial for olfaction-induced mating behavior in the hamster (Winans and Powers, 1977; Wysocki, 1979), and the habenula plays a role in the rhythmicity of mating behavior, as judged from data showing disruption of estrous behavior and hormonal function after habenular lesions in rats (Modianos *et al.*, 1974; Zouhar and de Groot, 1963). It would be interesting to speculate that opiates play a role in the mediation of certain, perhaps olfactory, hormonal, or emotional, aspects of mating behavior by their receptor distribution in the olfactory pathways. Structure–activity correlational analysis might bring out the proposed role for opiates in olfaction.

Analysis of distribution of other receptor classes in the olfactory system may provide similar clues. The same *in vitro* autoradiographic technique is not limited to opiate ligands. When incubation conditions are created for visualization of labeled diazepam (Valium) binding in the olfactory bulb, a dense band of labeling is localized to the outer plexiform layer of the main

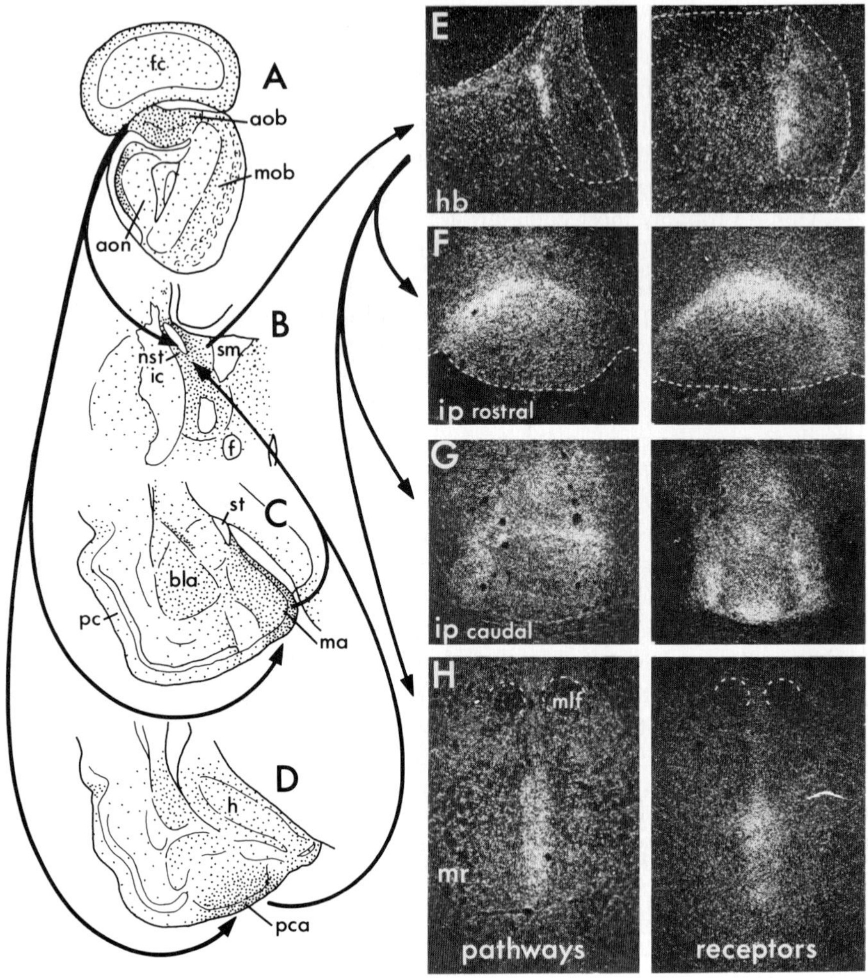

Fig. 5. (A–D) at left are projection drawings of opiate receptor-labeled sections at the levels of (A) the frontal pole showing olfactory structures, (B) the posterior bed nucleus of the stria terminalis, and (C) anterior and (D) posterior levels of the amygdala. Stippling represents opiate receptor distributions in these sections. Arrows indicate neural pathways, derived from anatomical studies, connecting the accessory olfactory bulb (oab) with the bed nucleus of the stria terminals (nst) and the medial (ma) and the posterior cortical (pca) nuclei of the amygdala (see text). These amygdaloid nuclei are known to project to the nst and this structure, in turn, projects to the border zone between the medial and lateral habenular nuclei [M. Herkenham, unpublished data, and (E, left)]. (E–F) are paired photomicrographs of two kinds of autoradiography. In the left pairs, the transport of [^{3}H]amino acids has marked neural pathways from the nst to the habenula (hb), shown in (E), and from this border zone to the interpeduncular nucleus (ip) and the median raphe (mr) as shown in (F–H). The identical corresponding levels shown in the right pairs illustrate opiate receptor distributions marked by [^{3}H]naloxone. The striking concor-

olfactory bulb. The role that the benzodiazepins play in olfaction has yet to be examined, but could be predicted from the anatomical localization of their receptors in olfactory structures. Future studies should be aimed at the combination of anatomical, physiological, and pharmacological data to elucidate brain function mediated at the receptor level.

REFERENCES

Aswanikumar, S., Corcoran, B., Schiffman, E., Day, A. R., Freer, R. J., Showell, H. J., Becker, E. L., and Pert, C. B. (1977). Demonstration of a receptor on rabbit neutrophils for chemotactic peptides. *Biochem. Biophys. Res. Commun.* **74**, 810–817.

Atweh, S. F., and Kuhar, M. J. (1977a). Autoradiographic localization of opiate receptors in rat brain. I. Spinal cord and lower medulla. *Brain Res.* **124**, 53–67.

Atweh, S. F., and Kuhar, M. J. (1977b). Autoradiographic localization of opiate receptors in rat brain. III. The telencephalon. *Brain Res.* **134**, 393–405.

Bausbaum, A. I., Clanton, C. H., and Fields, H. L. (1976). Opiate and stimulus-produced analgesia: functional anatomy of a medullospinal pathway. *Proc. Natl. Acad. Sci. U.S.A.* **73**, 4685–4688.

Broadwell, R. D. (1975). Olfactory relationships of the telencephalon and diencephalon in the rabbit. I. An autoradiographic study of the efferent connections of the main and accessory olfactory bulbs. *J. Comp. Neurol.* **163**, 329–346.

Cuatrecasas, P. (1971). Insulin-receptor interactions in adipose tissue cells: direct measurement and properties. *Proc. Natl. Acad. Sci. U.S.A.* **68**, 1264–1269.

Herkenham, M., and Pert, C. B. (1980). *In vitro* autoradiography of opiate receptors in rat brain suggests loci of "opiatergic" pathways. *Proc. Natl. Acad. Sci. U.S.A.* **77**, 5532–5536.

Hökfelt, T., Terenius, L., Kuypers, H. G. J. M., and Dann, O. (1979). Evidence for enkephalin immunoreactive neurons in the medulla oblongata projecting to the spinal cord. *Neurosci. Lett.* **14**, 55–60.

Kosterlitz, H. W., Waterfield, A. A., and Berthoud, V. (1974). Assessment of the agonist and antagonist properties of narcotic analgesic drugs by their actions on the morphine receptor in the guinea pig ileum. *Adv. Biochem. Psychopharmacol.* **8**, 319–334.

Krettek, J. E., and Price, J. L. (1978). Amygdaloid projections to subcortical structures within the basal forebrain and brainstem in the rat and cat. *J. Comp. Neurol.* **178**, 225–254.

Lefkowitz, R. J., Roth, J., Pricer, W., and Pastan, I. (1970). ACTH receptors in the adrenal: specific binding of ACTH-^{125}I and its relation to adenyl cyclase. *Proc. Natl. Acad. Sci. U.S.A.* **65**, 745–749.

Miledi, R., Molinoff, P., and Potter, L. T. (1971). Isolation of the cholinergic receptor protein of *Torpedo* electric tissue. *Nature (London)* **229**, 554–557.

Modianos, D. T., Hitt, J. C., and Flexman, J. (1974). Habenular lesions produce decrements in feminine but not masculine, sexual behavior in rats. *Behav. Biol.* **10**, 75–87.

---

dances between the terminal distribution of the nst–hb, hb–ip, and hb–mr paths and the receptor distributions suggests that these may be opiatergic tracts. Other abbreviations in this figure are: aon, anterior olfactory nuclei; bla, basolateral nucleus of amygdala; f, fornix; h, hippocampus; ic, internal capsule; mob, main olfactory bulb; mlf, medial longitudinal fasciculus; pc, piriform cortex; sm, stria medullaris; st, stria terminalis. (From Herkenham and Pert, 1980.)

Möhler, H., and Okada, T. (1978). The benzodiazepine receptor in normal and pathological human brain. *Br. J. Psychiatry* **133**, 261–268.

Moody, T. W., Pert, C. B., Rivier, J., and Brown, M. R. (1978). Bombesin: specific binding to rat brain membranes. *Proc. Natl. Acad. Sci. U.S.A.* **75**, 5372–5376.

Pert, A. (1978). Central sites involved in opiate actions. *In* "The Bases of Addiction" (J. Fishman, ed.), pp. 299–332. Dahlem Konf., Berlin.

Pert, A., and Sivit, C. (1977). Neuroanatomical focus for morphine and enkephalin-induced hypermotility. *Nature (London)* **265**, 645–647.

Pert, A., Pert, C. B., Davis, G. C., and Bunney, W. E. (1981). Opiate peptides and brain function. *In* "Handbook of Biological Psychiatry" (H. M. Van Pragg, O. J. Rafaelsen, E. J., Sacher, and M. H. Sabe, eds.), Vol. 2. Dekker, New York. In press.

Pert, C. B., and Snyder, S. H. (1973). Opiate receptor: demonstration in nervous tissue. *Science* **179**, 1011–1014.

Pert, C. B., Kuhar, M. J., and Snyder, S. H. (1976a). Opiate receptor: autoradiographic localization in rat brain. *Proc. Natl. Acad. Sci. U.S.A.* **73**, 3729–3733.

Pert, C. B., Snyder, S. H., and Portoghese, P. (1976b). Correlation of opiate receptor affinity with analgetic effects of meperidine homologues. *J. Med. Chem.* **19**, 1248–1250.

Scalia, F., and Winans, S. S. (1975). The differential projections of the olfactory bulb and accessory olfactory bulb in mammals. *J. Comp. Neurol.* **161**, 31–56.

Shepherd, G. M. (1972). Synaptic organization of the mammalian olfactory bulb. *Physiol. Rev.* **52**, 864–917.

Simon, E. J., Hiller, J. M., and Edelman, I. (1973). Stereospecific binding of the potent narcotic analgesic (3H)etorphine to rat brain homogenate. *Proc. Natl. Acad. Sci. U.S.A.* **70**, 1947–1949.

Terenhius, L. (1973). Characteristics of the "receptor" for narcotic analgesics in synaptic plasma membrane fraction from rat brain. *Acta Pharmacol. Toxicol.* **33**, 377–384.

Wilson, R. S., Rogers, M. E., Pert, C. B., and Snyder, S. H. (1975). Homologous N-alkylnorketobemidones. Correlation of receptor binding with analgesic potency. *J. Med. Chem.* **18**, 240–242.

Winans, S. S., and Powers, J. B. (1977). Olfactory and vomeronasal deafferentation of male hamsters: histological and behavioral analyses. *Brain Res.* **126**, 325–344.

Wysocki, C. (1979). Neurobehavioral evidence for the involvement of the vomeronasal system in mammalian reproduction. *Neurosci. Biobehav. Rev.* **3**, 301–341.

Young, W. S., and Kuhar, M. J. (1979). A new method for receptor autoradiography: [^{3}H]opioid receptors in rat brain. *Brain Res.* **179**, 255–270.

Zouhar, R. L., and de Groot, J. (1963). Effects of limbic brain lesions on aspects of reproduction in female rats. *Anat. Rec.* **145**, 358.

Zukin, S. R., and Zukin, R. S. (1979). Specific [^{3}H]phencyclidine binding in rat central nervous system. *Proc. Natl. Acad. Sci. U.S.A.* **10**, 5372–5376.

PART V

Discussion

Murphy: Why does the pH response profile in hydra necessarily reflect the initial chemoreceptive event rather than a later step in transduction?

Lenhoff: We feel the pH effects are on the surface of the animal because hydra are relatively impermeable to ions, and if the H^+ ions did enter the cells they would most likely be buffered. The observed pH changes cover a wide range, pH 4–8; it seems doubtful that such changes in environmental pH would give comparable pH changes within the cells. One of our goals is to isolate the receptor molecule(s) and see if glutathione binding follows the same pH pattern as with whole animals.

Margolis: You reported that oxidized glutathione inhibits the effect. Glutathione oxidizes more extensively as the pH increases, and the effect in the hydra decreases as the pH increases. How do you know that above neutral pH you are not observing oxidation of the stimulant and subsequent inhibition of the effect?

Lenhoff: There are two reasons. We get the same pH curves using glutathione analogs having *S*-methylcysteine or α-aminobutyric acid substituted for the cysteine. Both analogs have the thiol either blocked or absent; thus, oxidation of the glutathione is impossible. Also, most of the oxidation of reduced glutathione occurs normally above pH 8. It is stable in the pH range we use for our experiments.

Dodd: Dr. Raftery, you must be aware that those of us struggling with taste and smell are envious of the beautiful work on the mechanisms of the acetylcholine receptor. What would you have done if you had not had the electric eels?

Raftery: I would have stayed with enzymology.

Mooser: When you speak of two ligands binding to a receptor, is that per dimer or per monomer?

Raftery: There is controversy about the number of ligands binding, specifically about the number of toxin sites and the number of ligand sites. The answer is that two ligands bind per dimer. In the membranes the numbers of ligand binding sites and toxin binding sites are equal, but in detergent solution another set of toxin binding sites appears to be generated.

Mooser: Is there no interconversion between C_1 and C_2?

Raftery: True. It is the only mechanism that fits, because we have a kinetic phase that increases and then decreases in both rate and amplitude. At high ligand concentration it goes all the way to the pre-equilibrium for C_2, i.e., AR_2, and that then isomerizes to C_2. At intermediate ligand concentrations, the first pre-equilibrium and the second one are both reached, but the rate constant for isomerization to C_2 is faster than to C_1, so there is a transient increased level of

C_2 relative to C_1 compared to the equilibrium level; you then see the slow process as it returns and establishes the correct equilibrium levels of C_1 and C_2 for the intermediate ligand concentration.

Mooser: It might imply that the two ligand binding sites are not identical.

Raftery: They are not identical.

Mooser: In desensitization will the disulfide bond near the binding site ever be broken?

Raftery: We examined that in a preliminary way. By using disulfide and sulfhydryl chemistry, you can stabilize the system either in the low affinity state or the high affinity state so that they do not interconvert. A mechanism can be envisioned in which one has a disulfide with a sulfhydryl nearby, and depending on which disulfide is present you could have one state or the other. But we have no proof.

Paoni: The opiate receptor assay is fascinating, but I have a concern about its specificity. For the assay you paint on the specific ligand, incubate, then wash in several different steps. If a receptor binding event occurs, it seems likely that by diluting out the ligand it would dissociate from the receptor. However, the hydrophobic nature of the opiate compounds makes it less likely that material dissolved into membranes would be washed out. How do you know that what is visualized is not the distribution of membrane which is particularly attractive to the different ligands, and that the opiate receptors represent a very small portion or none of the radiolabel you see?

Herkenham: In our assay, sections are incubated with the tritiated substance, generally naloxone, although we have used other ligands. Low concentrations of ligand are used so that only the high affinity binding sites become occupied. The kinetics of dissociation from these sites, at 4°C, precludes significant loss of specific binding during the wash steps. With excess of unlabeled agonists included in the incubation medium there are little or no counts or radiographic grains. That is a measure of specific binding and suggests that there are a limited number of saturable binding sites.

Paoni: But the agonist may as well inhibit the association into the membranes, because of its hydrophobic nature. Have you solubilized the sections and examined them on SDS gels? You would expect few radioactive bands if a single receptor were labeled.

Herkenham: That has not been done in our tissue slices, but in centrifuged brain homogenates the binding is found in the synaptosomal fraction. In slices there is no binding to white matter, which by its high fat content would be the most hydrophobic tissue component. In gray matter we have no indication that discrete regions, such as those labeled in the striatum, have special hydrophobic qualities. With the proper incubation conditions, different opiate ligands give the same picture, and non-opiate ligands, which are also hydrophobic, give completely different patterns of labeling, consistent with the notion that they mark a different receptor. The autoradiography is exquisite in that sense.

Another point relates to the different types of opiate receptors. Given that there are Type 1 and Type 2 opiate receptors, changing incubation conditions alters the pattern of binding, even with the same ligand. The Type 2 receptor distribution is more homogeneous throughout gray matter, though it is very low in white matter. One of the reassuring things regarding the Type 1 opiate receptor is the high specificity in localization.

Price: Dr. Raftery expressed skepticism as to whether some of the binding is physiological because the association constants were many orders of magnitude lower than the range of the physiological effects. This point is especially pertinent to the chemical senses, where it is frequently assumed that the constituent with the tightest binding must be the physiological receptor.

Lenhoff: Do the inhibitors of methylation also inhibit bacterial swimming?

Paoni: Inhibitors of methylation do not inhibit bacterial swimming, but rather alter its pattern. Normal swimming is characterized by short periods of smooth swimming, or swimming in one direction; then the cell will spontaneously tumble and change direction. If the cells are unable to methylate the MCP proteins, they become incessantly smooth swimming and nonchemotactic. I should point out that the same inhibitors which are so effective in inhibiting methylation in the permeabilized cell assay cannot be used *in vivo*, as they cannot cross the cell membrane. But the effect of abolishing MCP methylation on bacterial swimming behavior has been well documented by using mutants lacking the carboxyl methyltransferase enzyme necessary for MCP methylation, or by starving the cells for *S*-adenosylmethionine, which is the methyl donor for the reaction.

Lenhoff: With ten peptides involved in the effector system, how do you prove that the labeled peptides are actually involved in transduction other than your noting that they become methylated?

Paoni: The conclusion that ten peptides are involved in the signal transduction was drawn from the fact that there are ten different complementation classes in these cells which yield generally nonchemotactic phenotypes. The products of three other genes (tar, tsr, and trg) have been shown to be methylated and demethylated in response to attractants and repellents. The cheR gene product has been identified as a carboxylmethyltransferase enzyme and cheB as a carboxylmethylesterase enzyme; these methylate and demethylate the various MCP proteins. The function of the other chemotaxis gene products is still a mystery.

Raftery: How do you know that the methylation is an increase or is it just different carboxyls on the same polypeptide being methylated? Have you quantitated it? If migration of the polypeptides is influenced by methylation, is your idea that it is different numbers of methylated carboxyls on the same polypeptide? Why couldn't it be different carboxyls methylated to the same level?

Paoni: What we know is that concomitant with an increase in the level of MCP methylation, there is a shift downward in the SDS gel mobility of these proteins. Also, conditions which cause demethylation, such as attractant removal or treatment with mild alkali, result in an upward shift in the protein bands on the gel. Our experiments with the cloned MCP gene products clearly show that changing the level of methylation on the peptide backbone is solely responsible for these mobility differences. It is possible that some of the microheterogeneity we observe in the lower molecular weight bands is due to permutations in the carboxyls modified as you suggest, but the fact that the proteins change their electrophoretic mobility at all is due to increased methylation. I should point out that it is well documented now that charge alterations can cause changes in a protein's electrophoretic profile, especially when the Tris-glycine buffer system is used. In fact, this technique has been suggested as being useful for detecting the covalent modification of proteins.

Raftery: When you used 10–35 μM attractants or 1 mM, it seemed that the kinetics became slower although the level was getting higher.

Paoni: The relationship between the kinetics is interesting. Little, however, can be said about the kinetics from the data I presented, as these experiments were not designed to measure initial velocities. We are presently in the process of studying the kinetics, and hope to determine the relationships between attractant concentration and the rate of MCP methylation.

Raftery: The $t_{1/2}$ at 1 mM was about 2 min. That's very slow.

Paoni: The swimming response correlates well with that. In fact, it appears that both the swimming response and the final level of methylation are proportional to receptor occupancy for a particular attractant. With half of the receptors occupied there is about half the swimming response as with saturation, and there is about half of the final level of methylation.

Dodd: When methylation occurs in the presence of attractant as a function of time, did you show a transient decrease in methylation before the increase started?

Paoni: No, I think that what you are referring to are events prior to attractant stimulation. When the experiment begins we incubate the bacteria with tritiated methionine for 45 min. The methionine is taken into the cells, converted to S-adenosylmethionine, and then the tritiated methyl group from the SAM is transferred to the methylated proteins. Initially what we measure is the approach to the steady state level of MCP methylation. This methylation actually peaks and begins to decline a bit after 45 min; I think that is what you saw. When the attractant is added to the cells, there is a steady increase in methylation until a new equilibrum level is reached.

Koch: Can the small (43,000) subunit of the acetylcholine receptor be considered as a part of the Na^+,K^+-ATPase?

Raftery: It could be.

Koch: You saw no Na^+ action from that portion? There is a report that the 43,000 subunit from Na^+,K^+-ATPase may act as a Na^+ ionophore in the lipoprotein membrane structure.

Raftery: There is no effect. Others also find none.

Karnovsky: In establishing the transmembrane nature of the acetylcholine receptor protein, the outside detection with bungarotoxin appears much more convincing than the picture on the inside. Is there some difficulty in terms of most of the protein being buried in the membrane?

Raftery: Out of every six rabbits only about one will recognize determinants that apparently are associated with the inside. We also used trypsin inside and trypsin outside to degrade the subunits. The kinetics of trypsin degradation outside are very fast, whereas inside they vary for each subunit and in general are slower. Based on that work and on Rosenbluth's electron microscopy or Stroud's X-ray studies, which show the receptor particles protruding far out on the outside, we think there is very little available on the inside, but that it is there.

Gower: If the glutathione receptors for hydra evolved as Dr. Lenhoff suggests, how do you envisage the hydra fed before they had receptors?

Lenhoff: We believe that the nematocyst-glutathione system of coordinating feeding evolved as a secondary adaptation after organisms containing body fluids evolved. Such prey animals include roundworms and up. They have body fluids in a vascular system, coelom, or pseudocoelom, allowing sufficient glutathione to be released after puncture by nematocysts. That activates the feeding response. Before such prey evolved, however, primitive cnidarians might have eaten by filter feeding or another mechanism. Perhaps primitive animals existed with sufficient fluids to serve as prey for cnidarians before higher organisms evolved. Sea anemones, for example, are also filter feeders and some cnidarians are parasites on others.

Wysocki: I noted that the vomeronasal rather than the olfactory epithelium has greater input to the structures labeled with opiates. Are there functional differences with respect to the results with differential labeling? Why would the vomeronasal system require an opiate type receptor and not the olfactory?

Herkenham: The vomeronasal system in rodents has been associated with expression of male sexual behavior induced by the chemosensation of pheromones. Hamsters deprived of this system generally will not show interest in a female in estrus. The role of opiates in olfaction-induced mating behavior might be a testable hypothesis. It would be possible to deposit a series of opiate analogs into the terminal regions of the vomeronasal pathways containing opiate receptors and construct structure–activity relationships in which mating behavior is examined. For example, from examination of precise anatomical distributions of opiate receptors in the amygdala, a limbic structure involved in expression of emotional behavior, we observe that these receptors are predominantly associated with the part receiving the vomeronasal projections via the accessory olfactory bulb. That, I think, is a new insight. The role that opiates play in sexual behavior may involve the amygdala or, more precisely, the medial and posterior cortical nuclei. This is a testable hypothesis that deserves further work.

Kurihara: Do the acetylcholine receptor membrane vesicles discriminate Na^+ and K^+?

Raftery: No. It was shown electrophysiologically with many cations, including Tris and small organic cations, that they are cation selective. There is a definite size with branched compounds where it cuts off. Monovalent and divalent inorganic cations pass through as do small organic cations.

Index

H

I

Aphorismes

I.

L'Univers n'est rien que par la vie; et tout ce qui vit se nourrit.

II.

Les animaux se repaissent; l'homme mange; l'homme d'esprit seul sait manger.

III.

La destinée des Nations dépend de la manière dont elles se nourrissent.

IV.

Dis-moi ce que tu manges: je te dirai ce que tu es.

V.

Le Créateur, en condamnant l'homme à manger pour vivre, l'y invite par l'appétit, et l'en récompense par le plaisir.

VI.

La Gourmandise est un acte de notre jugement, par lequel nous accordons la préférence aux choses qui sont agréables au goût, sur celles qui n'ont pas cette qualité.

VII.

Le Plaisir de la Table est de tous les âges, de toutes les conditions, de tous les pays et de tous les jours; il peut s'associer à tous les autres plaisirs, et reste le dernier, pour nous consoler de leur perte.

VIII.

La Table est le seul endroit où l'on ne s'ennuie jamais pendant la première heure.

IX.

La découverte d'un mêts nouveau fait plus, pour le bonheur du genre humain, que la découverte d'une étoile.

X.

Ceux qui s'indigèrent ou qui s'enivrent, ne savent ni boire ni manger.